POLYOXOMETALATE CHEMISTRY
FROM TOPOLOGY VIA SELF-ASSEMBLY TO APPLICATIONS

Polyoxometalate Chemistry From Topology via Self-Assembly to Applications

Edited by

Michael T. Pope
Georgetown University,
Washington, DC, U.S.A.

and

Achim Müller
University of Bielefeld,
Bielefeld, Germany

SPRINGER-SCIENCE+BUSINESS MEDIA, B.V.

A C.I.P. Catalogue record for this book is available from the Library of Congress.

ISBN 978-90-481-5716-7 ISBN 978-0-306-47625-9 (eBook)
DOI 10.1007/978-0-306-47625-9

Printed on acid-free paper

Coverpictures:
The cover shows the three clusters of the type $\{W_{148}\}$,$\{Mo_{154}\}$ and $\{Mo_{132}\}$. The artwork was prepared by Dr. Hartmut Bögge, University of Bielefeld, Germany.

Originally published by Kluwer Academic Publishers in 2001
Softcover reprint of the hardcover 1st edition 2001

Table of Contents

From Discrete Clusters to Networks and Materials

Applications : Catalysis, Biological Systems, Environmental Studies

Introduction to Polyoxometalate Chemistry : From Topology via Self-Assembly to Applications

M. T. POPE
Department of Chemistry, Georgetown University, Washington DC 20057, USA
A. MÜLLER
Department of Chemistry, University of Bielefeld, D-33501 Bielefeld, Germany

The high abundance of oxygen (55 atom %) in the Earth's Crust can only be partly attributable to the oceans, the silicate-based rocks, and clays. Even when H_2O, SiO_2, and Al_2O_3 are excluded from the accounting, oxygen is still dominant at 47 atom %. Clearly, the chemistry of combined oxygen is an important component of our environment. The bulk of this chemistry is either aqueous solution chemistry of oxoanions of the nonmetals, or the solid-state and surface chemistry of insoluble metal oxides. However, although it is only a very small fraction of the *natural* environment, there exists a third aspect of oxygen chemistry, that of the polyoxometalates, which spans both solution and "metal oxide" realms. As amply demonstrated by the contributions to the present book, this chemistry offers opportunities, insights, properties, and applications that cannot be matched by any other single group of compounds.

Polyoxometalates are the polyoxoanions of the early transition elements, especially vanadium, molybdenum, and tungsten. Although they have been investigated since the last third of the 19th century, it is only within the last four or five decades that modern experimental techniques have begun to reveal the range of structure and reactivity of these substances. Fundamental questions regarding the limits to composition, size and structure, metal incorporation, mechanisms of synthesis and reactivity, remain essentially unanswered at present. In spite of much research activity concerning practical applications of polyoxometalates, especially in heterogeneous and homogeneous catalysis, and in medicine (antiviral and antitumoral agents), it is certainly fair to say, considering the several thousand known polyoxometalates and their derivatives, that their potential in these and other areas remains poorly developed.
In the following chapters current research in several aspects of polyoxometalate chemistry is summarized by some of the leading workers in this field who participated in a workshop held at the Center for Interdisciplinary Research (ZiF) of the University of Bielefeld in October 1999.

Two kinds of polyoxoanions are known, those exemplified by the silicates, and oxoanions of neighboring main-group elements, and those of the early transition elements of groups 5 and 6 (Figure 1). Although both types of polyanions are constructed of linked MO_n polyhedra ($n = 4$-6), polyoxometalates are predominantly characterized by MO_6 octahedra with short "terminal" M=O ($d\pi$-$p\pi$) bonds that tend to result in "closed" discrete structures with such bonds directed outwards. In contrast, the main-group elements, especially

M.T. Pope and A. Müller (eds.), Polyoxometalate Chemistry, 1–6.

phosphates and silicates, exhibit open (cyclic) or polymeric structures based on linked MO_4 tetrahedra.

H																	He
Li	Be											**B**	C	N	O	F	Ne
Na	Mg											Al	**Si**	**P**	**S**	Cl	Ar
K	Ca	Sc	Ti	**V**	**Cr**	Mn	Fe	Co	Ni	Cu	Zn	Ga	**Ge**	**As**	**Se**	Br	Kr
Rb	Sr	Y	Zr	**Nb**	**Mo**	Tc	Ru	Rh	Pd	Ag	Cd	In	Sn	**Sb**	**Te**	**I**	Xe
Cs	Ba	La	Hf	**Ta**	**W**	Re	Os	Ir	Pt	Au	Hg	Tl	Pb	Bi	Po	At	Rn
Fr	Ra	Ac															

Figure 1. Polyoxoanion-forming elements

That polyoxometalates have an extensive solution chemistry in both aqueous and nonaqueous solvents is a consequence of low surface charge densities resulting in weak anion-cation attractions (lattice energies) relative to cation solvation energies. In general, polyoxometalate anion surfaces contain both terminal (M=O) and bridging (M–O–M) oxygen atoms, and although there have been arguments to the contrary,[1] all experimental evidence and recent density functional calculations[2] are in agreement that the bridging oxygens carry a greater negative charge and are protonated in preference to terminal oxygens. The latter atoms may be viewed as part of $O{=}M[O_5]$ or *cis*-$O_2M[O_4]$ groups in the case of polyoxometalates constructed of MO_6 octahedra. The existence of so-called "anti-Lipscomb" polyoxometalate structures in which an MO_6 octahedron has *three* terminal oxygens (always in a *facial* arrangement) has been demonstrated only very rarely.[3] In these cases protonation of one of the oxygens readily occurs, converting $[O_3]MO_3$ to $[O_3]M(OH)O_2$ with two *cis* terminal oxygens.

The formation of polyoxometalates, and especially the rational directed synthesis of specific structures presents a major challenge, but with enormous potential benefits. Some different synthetic strategies in polyoxometalate chemistry are described in the first six chapters of this book. These include processes in both aqueous and nonaqueous solvents, the incorporation of organic and organometallic functionalities, and the synthesis of polyoxo*thio*metalates. The recognition and characterization of extremely large polyoxometalates is a relatively recent development. One of the most challenging problems in contemporary chemistry is the deliberate and especially synthon-based synthesis of multifunctional compounds and materials – including those with network structures – with desirable or predictable properties, such as mesoporosity (well-defined cavities and channels), electronic and ionic transport, ferro- as well as ferrimagnetism, luminescence, and catalytic activity. Transition metal oxide-based compounds are of special interest in that respect. For example, the deeply colored, mixed-valence hydrogen molybdenum bronzes –

with their unusual property of high conductivity and wide range of composition -- play an important role in technology, industrial chemical processes, and materials science. Their fields of applications range from electrochemical elements, hydrogenation and dehydrogenation catalysts, superconductors, passive electrochromic display devices, to "smart" windows. The synthesis of such compounds or solids from preorganized linkable building blocks (synthons) with well-defined geometries and well-defined chemical properties is therefore of special interest to this end. Interestingly, reduced polyoxomolybdates can serve as models for the hydrogen bronzes.

In generating large complex molecular systems we have to realize that natural processes are effected by the linking (directed as well as non-directed) of a huge variety of basic and well-defined fragments. An impressive example of this, discussed in virtually all textbooks on biochemistry, is the self-aggregation process of the tobacco mosaic virus, which is based on preorganized units. This process more or less meets the strategy in controlling the linking of fragments to form larger units and linking the latter again.

In the case of metal-oxide based clusters this means for instance that relatively large molecular fragments can principally be functionalized with groups which allow linking through characteristic reactions: For example, as mentioned above, protonation of highly reactive "anti-Lipscomb" $\{MoO_3\}$ groups positioned on polyoxometalate cluster fragments generates a terminal OH group and results in condensation reactions of the fragment via H_2O formation.[3(b)] The same principle basically applies also to lacunary polyoxotungstates that can be linked by transition metal, lanthanide, and actinide ions to form discrete water-soluble heteropolytungstate anions [4] such as $[(UO_2)_3(H_2O)_6(W_3O_6)(H_2O)_3(\alpha\text{-}AsW_9O_{33})_3]^{15-}$ and $[La_{16}(H_2O)_{36}(W_5O_{18})_4(\alpha\text{-}AsW_9O_{33})_{12}(WO_2)_4(W_2O_6)_8]^{76-}$ ("W_{148}") or recrystallizable linear polymeric arrays (Figure 2).

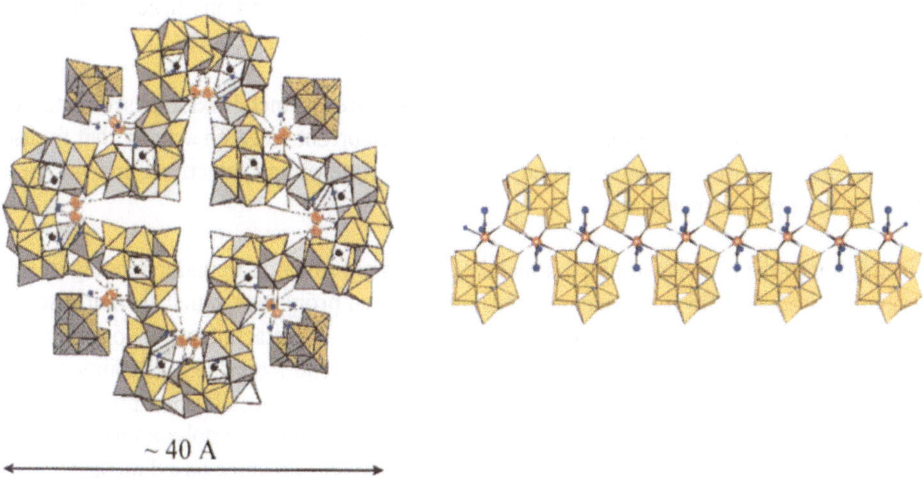

Figure 2. Structures of W_{148} and $\{[Ce(H_2O)_3(\alpha\text{-}SiW_{11}O_{39})]^{5-}\}_n$ (Reference 4)

In the generation of large polyoxometalate clusters, the concept of preorganized units is of particular importance due to the fact that the structural chemistry is often governed by differently transferable building units. For example, the linking of polyoxometalate building blocks containing 17 molybdenum atoms ($\{Mo_{17}\}$ units) results in the formation of cluster anions consisting of two or three of these units. The following basic strategy, which is archetypical for polyoxometalate chemistry, is used for describing or analyzing a solid-state structure. One decomposes, at least mentally, the objects into elementary building blocks (e.g., polygons, polyhedra or aggregates of these) and then tries to identify and explore the local matching rules according to which the building blocks are to be assembled to yield the objects considered. Nanosized polyoxomolybdate clusters now also provide model objects for studies on the initial nucleation steps of crystallization processes, an interesting aspect for solid-state chemists and physicists as the initial steps for crystal growth are not known. This is due to the fact that they represent well-defined molecular systems and have flexible (multi-dimensional) boundary conditions, i.e. clusters with circular and spherical topologies can be considered as potential precursors for such growth. It is envisaged that, with such an approach, it will be possible to unveil some of the mysteries associated with the biomineralization of structures such as the unicellular diatoms. In the context of biomineralization, which takes place at room temperature (whilst chemists need high temperatures), it is remarkable that the linking of 'Giant-Spherical' clusters, described in Chapter 1, to a well-defined solid-state layer structure is also possible at room temperature. Interestingly, even Keggin-type ions can be encapsulated in such cluster shells (Figure 3).

In summary it is important in this context that (1) the above-mentioned nanostructured building blocks can even be isolated (according to their stability) and (2) they have nanostructured cavities and well-defined properties, thus offering the possibility to construct materials with desired emergent properties using characteristic synthons, in accordance with the rule, the whole (due to cooperativity) is more than the sum of the parts. [5]

It is a short conceptual step from large polyoxometalates to metal-oxide-based materials. Eight chapters (13 - 20) demonstrate the intensity of current research activity that focuses on the formation of new materials and on the solid state optical, electrical and magnetic properties of polyoxometalates.

In addition to the promise of polyoxometalate chemistry towards an understanding of self-assembly processes for inorganic materials with desired properties, much current research activity is also directed towards the incorporation or attachment of organic and organometallic groups.[6] Several obvious advantages accrue from the availability of such derivatized polyoxometalates. These include the ability to use established procedures of organic chemistry to assemble large polyanion arrays, to incorporate polyoxometalates into polymer matrices (see for example recent reports of hybrid polymer-based materials [7]), to develop new polyoxometalate catalysts, and to form new, highly specific electron-dense labels, and phasing agents for X-ray crystallographic analysis of large biopolymers. As

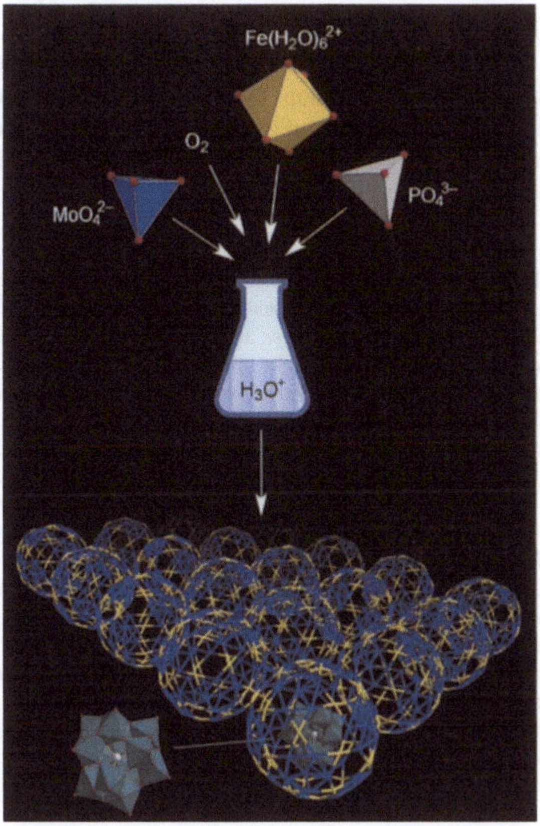

Figure 3. The route to a novel type of supramolecular compound: a layer structure built up by composites containing $\{Fe_{30}Mo_{72}\}$ cluster shells and non-covalently encapsulated Keggin ions. (A. Müller *et al.*, *Angew. Chem.Int.Ed.Engl.* **34**, 3413 (2000))

shown in Chapter 24, even non-functionalized polyoxometalates can provide additional unexpected benefits for analysis of the structure of the ribosome.

Undoubtedly, at present, the most important and promising application of polyoxometalates lies in catalysis, both homogeneous and heterogeneous.[8] Four chapters (21- 23, 25) summarize some recent activity in homogeneous catalysis, and Chapters 7 - 12 describe recent work on the fundamental solution chemistry and spectroscopic properties of polyoxometalates that underlie their catalytic behavior. Driven by environmental concerns, green chemistry becomes a greater imperative for the chemical and pharmaceutical industries, and the demand for more selective and more robust catalysts, especially those that can be employed in aqueous environments is certain to increase. The enormous versatility and variety of polyoxometalates offers considerable opportunities in this and in other areas.[9]

Acknowledgment. We thank the ZiF authorities and the Volkswagen Foundation for generous financial support of the Workshop. Research support from the National Science Foundation and the U.S. Department of Energy (MTP) and from the Deutsche Forschungsgemeinschaft and the Fonds der Chemischen Industrie (AM) is also gratefully acknowledged.

References

1. K.H. Tytko, J. Mehmke, and S. Fischer, *Struct. Bonding (Berlin)* **93**, 129-321 (1999)
2. B.B. Bardin, S.V. Bordawekar, M. Neurock, and R.J. Davis, *J. Phys. Chem. B* **102**, 10817 (1998)
3. (a) L. Ma, S. Liu, and J. Zubieta, *Inorg. Chem.* **28**, 175 (1989) ; (b) A. Müller, E. Krickemeyer, S. Dillinger, J. Meyer, H. Bögge, and A. Stammler, *Angew. Chem. Int. Ed. Engl.* **35**, 171 (1996); (c) R. Klein and B. Krebs, in *Polyoxometalates: from Platonic Solids to Anti-Retroviral Activity*, M.T. Pope and A. Müller, eds.; Kluwer, Dordrecht (1994), p 41
4. (a) K. Wassermann, M.H. Dickman, and M.T. Pope, *Angew. Chem. Int. Ed. Engl.*, **36**, 1445 (1997); (b) M.T. Pope, X. Wei, K. Wassermann, and M.H. Dickman, *C.R.Acad.Sci.Paris, 1, Ser. IIc,* 297 (1998); (c) M. Sadakane, M.H. Dickman, and M.T. Pope, *Angew. Chem. Int. Ed. Engl.* **39**, 2914 (2000)
5. (a) A. Müller, P. Kögerler, and H. Bögge, *Struct. Bonding (Berlin)* **96**, 203 (2000); (b) A. Müller, P. Kögerler, and C. Kuhlmann, *J. Chem. Soc., Chem. Commun.* 1347 (1999); (c) A. Müller and C. Serain, *Acc. Chem. Res.* **33**, 2 (2000)
6. P. Gouzerh and A. Proust , *Chem. Rev.* **98**, 77 (1998)
7. (a) C.R. Mayer, V. Cabuil, T. Lalot, and R. Thouvenot, *Angew. Chem. Int. Ed. Engl.* **38**, 3672 (1999); (b) C.R. Mayer, R. Thouvenot, and T. Lalot, *Chem. Mater.* **12**, 257 (2000)
8. (a) *J. Mol. Catal., A* (special issue, C.L. Hill, ed.) **114**, 1 - 371 (1996); (b) T. Okuhara, N. Mizuno, and M. Misono, *Adv. Catal.* **41**, 113 (1996); (c) R. Neumann, *Prog. Inorg. Chem.* **47**, 317 (1998); (d) I.V. Kozhevnikov, *Chem. Rev.* **98**, 171 (1998); (e) N. Mizuno and M. Misono, *Chem. Rev.* **98**, 199 (1998); (f) M. Sadakane and E. Steckhan, *Chem. Rev.* **98**, 219 (1998)
9. D. Katsoulis, *Chem. Rev.* **98**, 359 (1998)

Rational Approaches to Polyoxometalate Synthesis

R. J. ERRINGTON
Department of Chemistry, The University of Newcastle upon Tyne, NE1 7RU, UK
E-mail: John.Errington@ncl.ac.uk

Abstract

Heteronuclear hexametalates $[XM'M_5O_{18}]^{n-}$ (M = Mo, W), including the first examples of Zr and Hf derivatives, have been prepared by hydrolytic aggregation in non-aqueous media, enabling the reactivity of alkoxide surface groups (X = OMe, OPr^i) to be investigated. Organoimido derivatives result from reactions between $[Mo_6O_{19}]^{2-}$ and organic isocyanates or aromatic amines at elevated temperatures. In studies of vanadate systems we have achieved the quantitative conversion of $[H_3V_{10}O_{28}]^{3-}$ to $[V_{13}O_{34}]^{3-}$ under ambient conditions and the synthesis of a range of new vanadophosphonates. The potential of non-aqueous reductive aggregation for rational polyoxometalate assembly has been demonstrated by the synthesis of $[PV_2Mo_{12}O_{42}]^{3-}$ from $[PMo_{12}O_{40}]^{3-}$ and $[VOCl_3(dme)]$. In the first examples of controlled polyoxometalate halogenation, the hexabromo species $[PW_9O_{28}Br_6]^{3-}$ has been obtained from $[NaPW_{11}O_{39}]^{6-}$ and $[PW_9O_{34}]^{9-}$ by treatment with $C_2O_2Br_2$ or $SOBr_2$. The structure of this anion features a fully brominated face which provides opportunities for further derivatisation.

Keywords: Non-aqueous synthesis, hydrolytic aggregation, alkoxides, tungstates, molybdates, vanadates, vanadophosphonates, reductive aggregation, surface reactivity, organoimido derivatives, bromination.

1. Introduction

The enormous variation in topology, size, electronic properties and elemental composition that is unique to polyoxometalates provides the basis for an expanding research effort into their chemistry and their applications in areas which include catalysis, materials chemistry and biochemistry. However, in order to realise the full potential of these molecular metal oxides, methods must be developed to manipulate their properties in a rational and systematic fashion. This is by no means a trivial challenge, and the fascinating structures of polyoxometalates reflect the complex solution chemistry involved in their aggregation, structural rearrangement and surface reactivity. An understanding of these solution processes is therefore essential if this area is to mature, and several research groups are making progress towards this goal. This article describes recent results from our work on non-aqueous solution aggregation and surface reactivity.

M.T. Pope and A. Müller (eds.), Polyoxometalate Chemistry, 7–22.

2. Hydrolytic Aggregation

Fuchs and coworkers first showed that polyoxometalates could be obtained from metal alkoxides [1], and we have adopted this strategy to develop non-aqueous methods for the rational hydrolytic assembly of polyoxometalates. A feature of this approach is that, provided the extent of hydrolysis can be controlled, alkoxide groups remaining after incomplete hydrolysis are present as reactive sites on the polyoxometalate surface. This was particularly attractive to us because $[W_6O_{19}]^{2-}$ **1** has resisted all of our attempts at surface derivatisation, unlike its molybdenum counterpart $[Mo_6O_{19}]^{2-}$ (see below), and by substituting $[M–OR]^{n+}$ for $[W{=}O]^{4+}$ we hoped to introduce a single reactive heterometal site into an otherwise inert tungsten oxide framework. Another major advantage of this hydrolytic approach is that reactions are conveniently monitored by ^{17}O NMR spectroscopy, provided ^{17}O-enriched water is used for hydrolysis.

2.1 MW$_5$ HEXAMETALATES

Stoichiometric hydrolysis of a 1:5 mixture of $(Bu^n_4N)_2[WO_4]$ and $WO(OMe)_4$ in MeCN gives $(Bu^n_4N)_2$-**1** quantitatively (Equation 1), and the remarkable stability of this hexanuclear structure suggested that the same approach might be used for the preparation of heteronuclear hexametalates from mixtures of their constituent metal alkoxides [2].

$$[WO_4]^{2-} + 5WO(OMe)_4 + 10H_2O \rightarrow [W_6O_{19}]^{2-} + 20MeOH \quad (1)$$

$$3[WO_4]^{2-} + 7WO(OMe)_4 + 2Ti(OMe)_4 + 17H_2O \rightarrow 2[(MeO)TiW_5O_{18}]^{3-} + 34MeOH \quad (2)$$

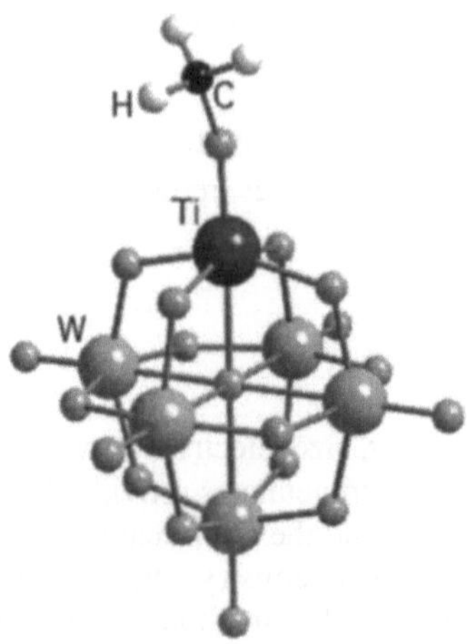

Fig. 1. Structure of $[(MeO)TiW_5O_{18}]^{3-}$ **2**.

Although we had already shown that the dimeric oxoalkoxide $[\{WO(OMe)_4\}_2]$ reacts with $[WO_4]^{2-}$ to give the oxoalkoxoanion $[W_2O_5(OMe)_4]^{2-}$ [3], the complex solution processes occurring during the formation of **1** are not understood, and the complexity was expected to increase upon addition of other metal alkoxides. Nevertheless, the hydrolysis of a mixture of $Ti(OMe)_4$, $WO(OMe)_4$ and $(Bu^n_4N)_2[WO_4]$

(Equation 2) gave good yields of $(Bu^n{}_4N)_3[(MeO)TiW_5O_{18}]$ after recrystallisation to remove small amounts of $(Bu^n{}_4N)_2[W_6O_{19}]$. The structure of $[(MeO)TiW_5O_{18}]^{3-}$ **2** shows a terminal methoxide group bonded to titanium (Figure 1), with an average Ti–O_{μ}W distance of 1.949 Å and a Ti–O(C) bond length of 1.760 Å. The ^{17}O NMR spectrum of **2** (Figure 2) contains two peaks for terminal W=O, a peak for Ti–O–W and two W–O–W peaks in addition to the high field peak due to the central μ_6-O. The small impurity peak indicated by the asterisk is due to **1**. In the ^{183}W NMR spectrum, peaks were observed at δ_W 32.3 and 64.5 in the expected 4:1 ratio. In the IR spectrum of **2**, the strong ν(W=O) band at 947 cm^{-1} is shifted from that of **1** at 978 cm^{-1}.

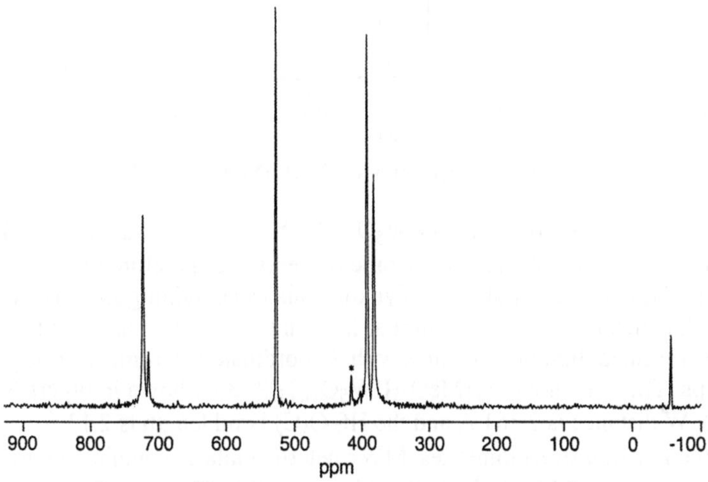

Fig. 2. ^{17}O NMR spectrum of $(MeO)TiW_5O_{18}]^{3-}$ **2**.

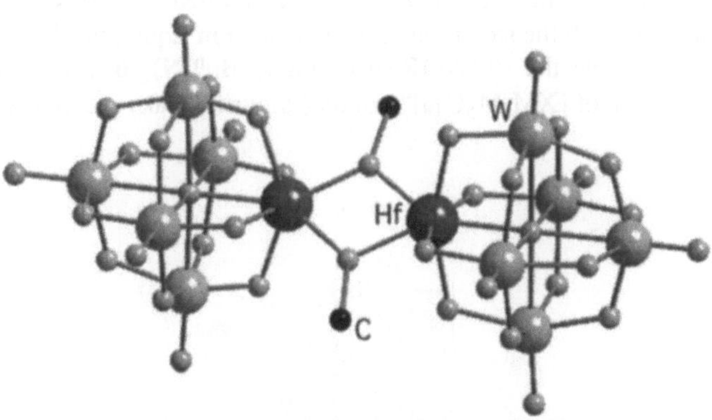

Fig. 3. Structure of $[\{(MeO)HfW_5O_{18}\}_2]^{6-}$ **3**.

$$3[WO_4]^{2-} + 7WO(OMe)_4 + 2VO(OMe)_3 + 17H_2O \rightarrow 2[VW_5O_{19}]^{3-} + 34MeOH \quad (3)$$
$$3[WO_4]^{2-} + 7WO(OMe)_4 + 2Nb(OMe)_5 + 19H_2O \rightarrow 2[NbW_5O_{19}]^{3-} + 38MeOH \quad (4)$$
$$[WO_4]^{2-} + 4WO(OMe)_4 + Nb(OMe)_5 + 10H_2O \rightarrow [(MeO)NbW_5O_{18}]^{2-} + 20MeOH \quad (5)$$

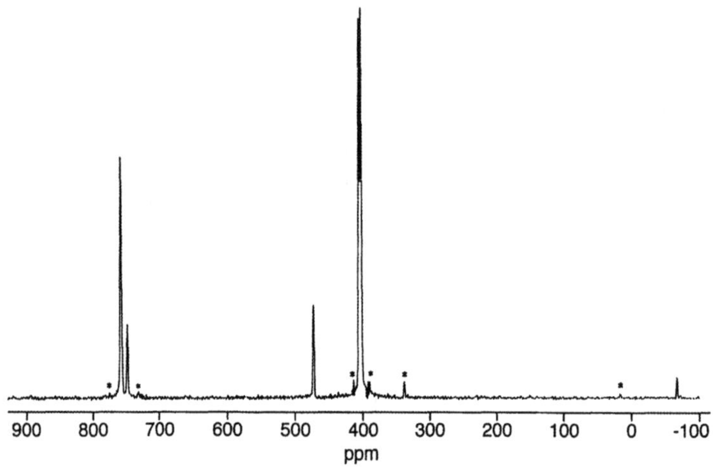

Fig. 4. ^{17}O NMR spectrum of $[(MeO)NbW_5O_{18}]^{2-}$ **6**.

The Zr and Hf analogues of **2** $[(RO)MW_5O_{18}]^{3-}$ (M = Zr, Hf) were expected to provide easier access for incoming nucleophilic reagents and therefore to be more reactive than **2**. These first examples of polyoxometalates containing Zr or Hf were prepared from the metal alkoxides in a similar fashion to **2** and crystal structure determinations revealed dimeric structures with 7-coordinate heterometals bridged by alkoxide groups. The structure of $[\{(MeO)HfW_5O_{18}\}_2]^{6-}$ **3** is shown in Figure 3. The average Hf–O_μW distance is 2.161 Å and the Hf–O(C) bond length is 2.13 Å.
By adjusting the reaction stoichiometries, MW_5 heterometalates containing Group 5 elements were also prepared from their alkoxides using this approach. Equation 3 provides a convenient high yield route to ^{17}O-enriched samples of the known $[VW_5O_{19}]^{3-}$ **4** [4], whilst the niobates $[NbW_5O_{19}]^{3-}$ **5** and $[(MeO)NbW_5O_{18}]^{2-}$ **6** were obtained from reactions with the stoichiometries indicated in Equations 4 and 5 respectively. Figure 4 shows the ^{17}O NMR spectrum of $(Bu^n_4N)_2$-**6** with peaks that are characteristic of this type of $[XM'M_5O_{18}]^{n-}$ anion (impurity peaks are indicated by asterisks).

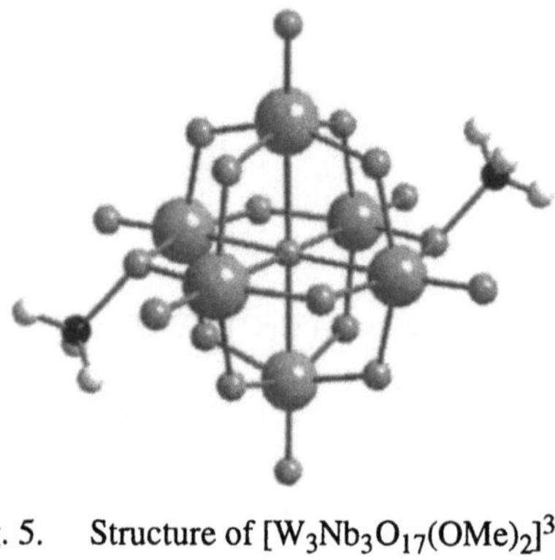

Fig. 5. Structure of $[W_3Nb_3O_{17}(OMe)_2]^{3-}$ **7**.

Our efforts to extend this synthetic approach to hexametalates containing more than one heteroatom have so far produced complex mixtures of products, although an attempt to produce the heteronuclear oxoalkoxoanion $[WNbO_4(OMe)_5]^{2-}$ from the1:1 reaction between WO_4^{2-} and $Nb(OMe)_5$ produced crystals of the tetrabutylammonium salt of $[W_3Nb_3O_{17}(OMe)_2]^{3-}$ **7**. An X-ray crystal structure determination (Figure 5) confirmed the cation:anion ratio of 3:1 and the presence of two methoxide groups, but the metal sites were each occupied approximately equally by W and Nb. We are hoping that ^{17}O NMR studies will reveal whether a single isomer or a mixture of W_3Nb_3 species is present in solution.

2.2 MMo_5 HEXAMETALATES

Given the greater reactivity of $[Mo_6O_{19}]^{2-}$ compared with **1**, we expected that heterometalates $[XMMo_5O_{18}]^{n-}$ would be more reactive than their tungsten analogues. However, the molybdenum oxoalkoxides $[MoO(OR)_4]$ required for reactions analogous to (2)-(5) above are less straightforward to prepare and handle than the corresponding $[WO(OR)_4]$ compounds, so we sought a more convenient route to these hexametalates. The ready availablity of $(Bu^n{}_4N)_2[Mo_2O_7]$ and $(Bu^n{}_4N)_4[Mo_8O_{26}]$ [5] led us to attempt the preparation of $[(Pr^iO)TiMo_5O_{18}]^{3-}$ **8** by a hydrolytic reaction involving $Ti(OPr^i)_4$ as shown in Equation 6. Good yields of $(Bu^n{}_4N)_3$-**8** were obtained after recrystallisation and the structure of the anion is shown in Figure 6. The anion has an average $Ti–O_{\mu}Mo$ distance of 1.936 Å and a Ti–O(C) bond length of 1.785 Å. In the IR spectrum of **8** the main ν(Mo=O) band at 931 cm^{-1} is at a lower wavenumber than the analogous band for the parent $[Mo_6O_{19}]^{2-}$ (951 cm^{-1}), as was also observed for ν(W=O) in **2**.

$$[Mo_2O_7]^{2-} + [Mo_8O_{26}]^{4-} + 2Ti(OPr^i)_4 + 3H_2O \rightarrow 2[(Pr^iO)TiMo_5O_{18}]^{3-} + 6Pr^iOH \qquad (6)$$

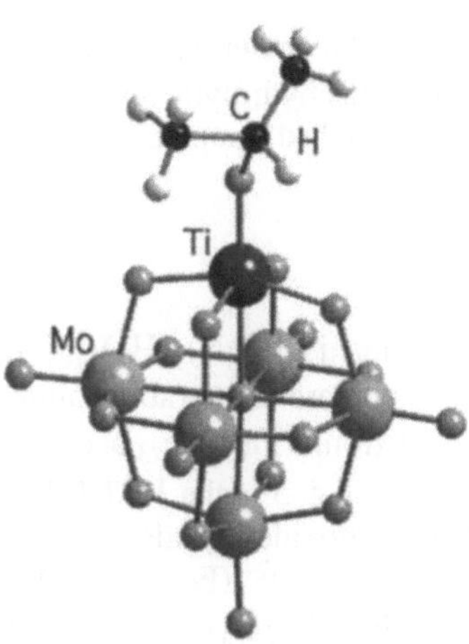

Fig. 6. Structure of $[(Pr^iO)TiMo_5O_{18}]^{3-}$ **8**.

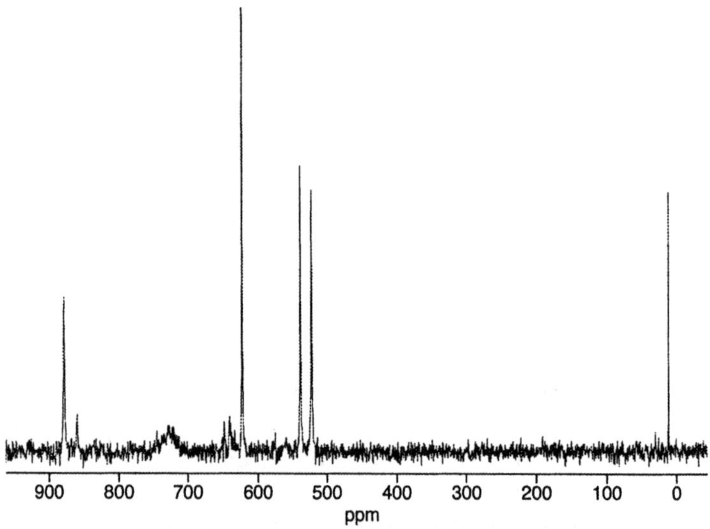

Fig. 7. ^{17}O NMR spectrum of $[(Pr^iO)TiMo_5O_{18}]^{3-}$ **8**.

The ^{17}O NMR spectrum (Figure 7) is characteristic of M′M$_5$ species as discussed above for **2**, although a broad peak at δ_O *ca.* 725 in the region for Ti–O–Ti bonds is possibly due to small amounts of a polynuclear oxoalkoxide such as $[Ti_{11}O_{13}(OPr^i)_{18}]$ [6] produced by hydrolysis of $Ti(OPr^i)_4$. This may explain why, although good yields of **8** are obtained from this reaction, some $(Bu^n{}_4N)_4[Mo_8O_{26}]$ is invariably recovered upon workup.

$$[Mo_2O_7]^{2-} + [Mo_8O_{26}]^{4-} + 2VO(OMe)_3 + 3H_2O \rightarrow 2[VMo_5O_{19}]^{3-} + 6MeOH \qquad (7)$$

As with the tungsten analogue **4**, the known monovanadium species $[VMo_5O_{19}]^{3-}$ **9** [7] can be obtained in high yield by this hydrolytic approach (Equation 7), providing an efficient method of preparing ^{17}O-enriched samples for reactivity studies.

2.3 POLYVANADATES [8]

Although Fuchs has previously obtained $(Bu^n{}_4N)_3[H_3V_{10}O_{28}]$ by basic hydrolysis of $VO(OBu^t)_3$ [1(b)], our attempts to prepare the tetrabutylammonium salts of $[H_3V_{10}O_{28}]^{3-}$ **10** and $[V_5O_{14}]^{3-}$ **11** from $VO(OMe)_3$ according to Equations 8 and 9 produced complex mixtures. Peaks at δ 4–5 in the ^1H NMR spectra of these products indicated the presence of residual methoxide ligands. However, in the attempted preparation of the hexavanadate $[H_6V_6O_{19}]^{2-}$ (Equation 10) hydrolysis proceeded to completion to give the dodecavanadate $[V_{12}O_{32}]^{4-}$ **12** previously characterised by Klemperer [9], indicating that the reaction actually proceeds as in Equation 11. A similar reaction with the stoichiometry shown in Equation 12 aimed at the hexametalate $[V_6O_{13}(OMe)_6]^{2-}$ resulted in the formation of pentavanadate **11** and an insoluble yellow solid.

$$10VO(OMe)_3 + 3OH^- + 15H_2O \rightarrow [H_3V_{10}O_{28}]^{3-} + 30MeOH \quad (8)$$
$$5VO(OMe)_3 + 3OH^- + 6H_2O \rightarrow [V_5O_{14}]^{3-} + 15MeOH \quad (9)$$
$$6VO(OMe)_3 + 2OH^- + 11H_2O \rightarrow [H_6V_6O_{19}]^{2-} + 18MeOH \quad (10)$$
$$12VO(OMe)_3 + 4OH^- + 16H_2O \rightarrow [V_{12}O_{32}]^{4-} + 36MeOH \quad (11)$$
$$6VO(OMe)_3 + 2OH^- + 5H_2O \rightarrow \mathbf{11} \quad (12)$$
$$[H_3V_{10}O_{28}]^{3-} + 3VO(OMe)_3 + 3H_2O \rightarrow [V_{13}O_{34}]^{3-} + 9MeOH \quad (13)$$

In a slightly different approach, we reasoned that the surface OH groups in **10** resulting from protonation of bridging V–O–V sites [10] should react with metal alkoxides and provide a means of expanding the V_{10} structure by hydrolytic aggregation. The reaction between $(Bu^n_4N)_3$-**10** and $VO(OMe)_3$ (Equation 13) gave a 93% isolated yield of $(Bu^n_4N)_3[V_{13}O_{34}]$, a compound previously obtained in only 34% yield by heating $(Bu^n_4N)_3$-**10** in refluxing MeCN [11]. Clearly, controlled hydrolytic assembly under ambient conditions is a much more efficient route to $[V_{13}O_{34}]^{3-}$ **13**. As shown in Figure 8, this aggregation process can be regarded as growth onto one face of a vanadium oxide lattice fragment.

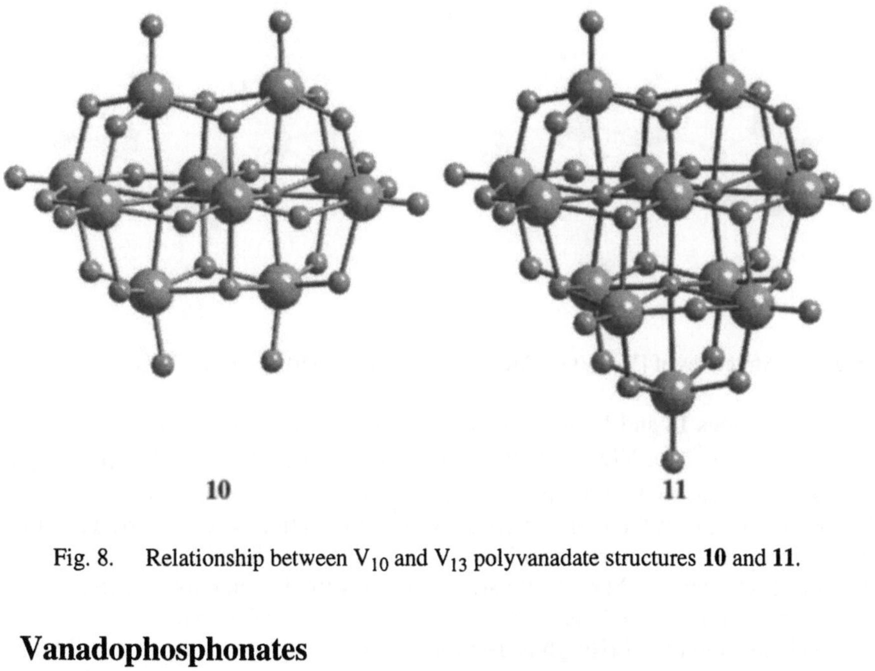

10 **11**

Fig. 8. Relationship between V_{10} and V_{13} polyvanadate structures **10** and **11**.

3. Vanadophosphonates

Zubieta has described a range of vanadium phosphonate complexes prepared by conventional or hydrothermal/solvothermal methods [12]. Results from our efforts to prepare vanadophosphonates by hydrolytic aggregation are described in this section, together with interesting results from reactions which did not involve alkoxide hydrolysis [8].

The 1:1:1 reaction between $VO(OMe)_3$, $PhPO_3H_2$ and OH^-, which was expected to produce oligomeric species $\{[VO_2(O_3PPh)]^-\}_n$, gave the divanadate V_2P_2 species

$[V_2O_4(O_3PPh)_2]^{2-}$ **14** in 82% yield (Equation 14). When the ratio of $VO(OMe)_3$ to $PhPO_3H_2$ in Equation 14 was changed to 3:1, the product was not a V_3P vanado-phosphonate, but instead the pentavanadate **11** was formed in quantitative yield based on vanadium. However, a V_3P species $[V_3O_8(O_3PPh)]^{3-}$ **15** was obtained in 64% yield by treatment of $(Bu^n{}_4N)_3$-**11** with $(Bu^n{}_4N)[PhPO_3H]$ (Equation 15).

$$2VO(OMe)_3 + 2PhPO_3H_2 + 2OH^- \rightarrow [V_2O_4(O_3PPh)_2]^{2-} + 6MeOH \quad (14)$$

$$[V_5O_{14}]^{3-} + 2[PhPO_3H]^- \rightarrow \mathbf{15} \quad (15)$$

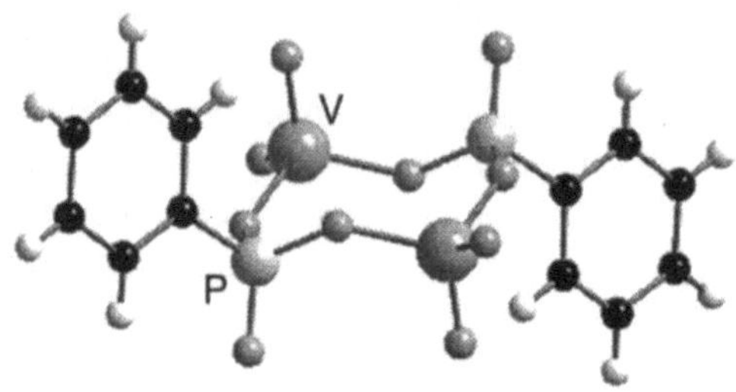

14

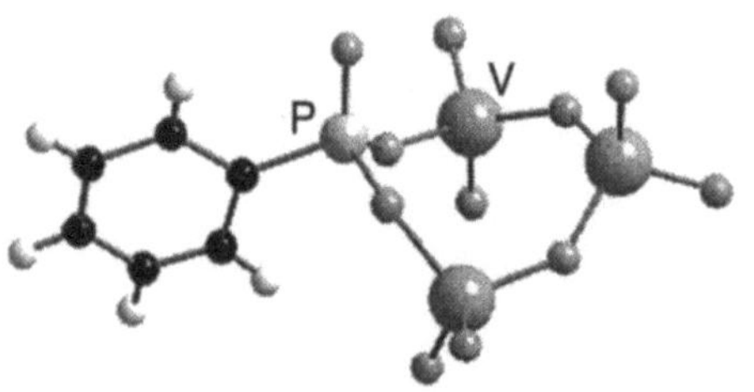

15

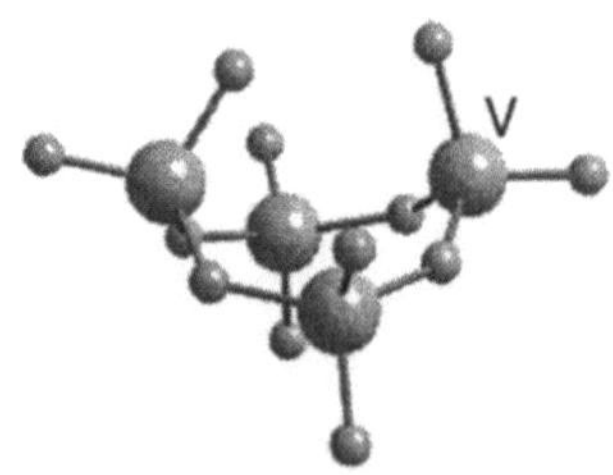

16

Fig. 9. Structures of $[V_2O_4(O_3PPh)_2]^{2-}$ **14**, $[V_3O_8(O_3PPh)]^{3-}$ **15** and $[HV_4O_{12}]^{3-}$ **16**.

The cyclic anions **14** and **15** are related to the parent tetravanadate $[V_4O_{12}]^{4-}$ by substitution of $RPO_3{}^{2-}$ for $VO_4{}^{3-}$ and their structures are shown with that of $[HV_4O_{12}]^{3-}$ **16** in Figure 9. A boat conformation is adopted by **16** with V=O···HO–V hydrogen-bonding across the top of the ring. A twisted boat conformation is adopted by **15** with the phenyl group in an equatorial position, whilst **14** adopts a chair form, again with equatorial phenyl groups. NMR spectra are consistent with the retention of these structures in solution, although there is evidence of fluxional behaviour.

The ready availability of $(Bu^n{}_4N)_2$-**14** prompted us to explore its use as a building block in the preparation of other vanadophosphonates. An attempt to prepare a V_4P_2 species from **14** and $VO(OMe)_3$ (Equation 16) produced the dodecavanadate **12** quantitatively. However, in the absence of water, the same reactants (Equation 17) gave a 76% yield of $(Bu^n{}_4N)_3[V_6O_{11}(O_3PPh)_5]$, which was also obtained from a reaction between $[VOCl_3(dme)]$, $(Bu^n{}_4N)[PhPO_3H]$ and $(Bu^n{}_4N)[OH]$ (Equation 18) in 86% yield. The irregular structure of the green 1-electron reduced $[V_6O_{11}(O_3PPh)_5]^{3-}$ **17** (Figure 10) bears some resemblance to that of red $[V_5O_7(OMe)_2(O_3PPh)_5]^-$ **18** reported by Zubieta [12 (d)]. Both contain an "intrusive" V=O bond and a "dangling" *exo* vanadium site [VO_2 in **17** and VO(OMe) in **18**].

$$[V_2O_4(O_3PPh)_2]^{2-} + 2VO(OMe)_3 + 3H_2O \rightarrow \mathbf{12} \quad (16)$$
$$[V_2O_4(O_3PPh)_2]^{2-} + 2VO(OMe)_3 \rightarrow \mathbf{17} \quad (17)$$
$$[VOCl_3(dme)] + [PhPO_3H]^{-} + 2[OH]^{-} \rightarrow \mathbf{17} \quad (18)$$

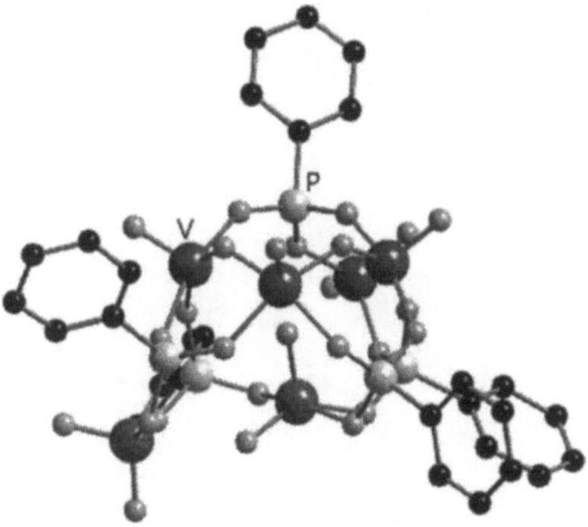

Fig. 10. Structure of $[V_6O_{11}(O_3PPh)_5]$ **17**.

A similar "intrusive" V=O group was also observed in the structure of the trivanadate $[V_3O_3(O_3PPh)_6(thf)]^{3-}$ **19** which we have obtained from a reaction between $(Bu^n{}_4N)[VO_2Cl_2]$ and $(Bu^n{}_4N)[PhPO_3H]$ (Figure 11). The formation of this species is not understood and the crystal structure shows another atom, apparently potassium, interacting with the three P=O groups above the ring (although there was no obvious source of potassium in the reaction).

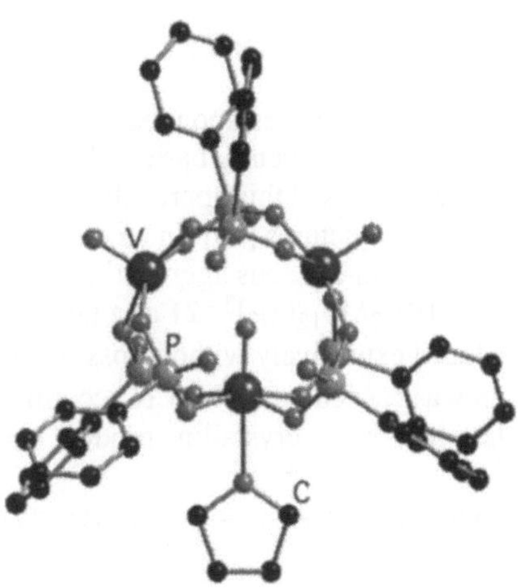

Fig. 11. Structure of $[V_3O_3(O_3PPh)_6(thf)]^{3-}$ **19**.

In another non-alkoxide reaction, a vanadophosphonate cage with an encapsulated chloride $[V_9O_{19}(O_3PPh)_4Cl]^{2-}$ **20** (Figure 12) was obtained in 60% yield by treating a mixture of $[VOCl_3(dme)]$ and Na_3VO_4 with $(Bu^n{}_4N)[PhPO_3H]$ (Equation 19). The ^{51}V NMR of **20** contained peaks at δ_V -583 (4V), -605 (2V), -617 (2V) and -644 (1V), and two peaks (1:1) were observed in the ^{31}P NMR spectrum at δ_P 18.0 and 15.4. It has been proposed that encapsulated molecules or ions within cage-like vanadophosphonates such as **20** act as a templates during aggregation [12 (a)], although the details of such processes are not understood.

$$3VOCl_3(dme) + 3Na_3VO_4 + 2[PhPO_3H]^- \rightarrow \mathbf{20} \qquad (19)$$

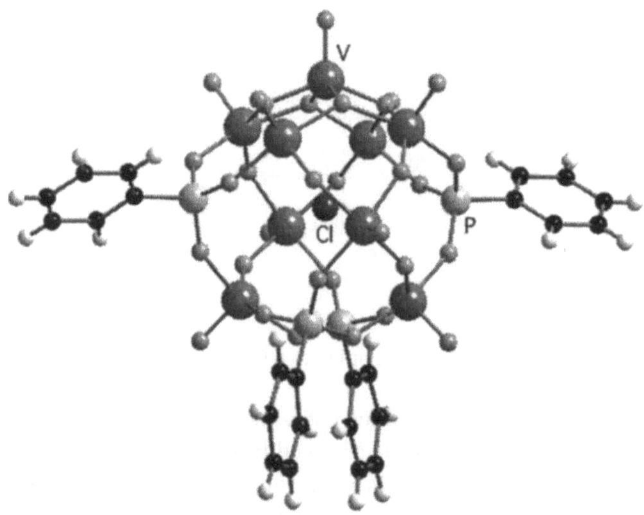

Fig. 12. Structure of $[V_9O_{19}(O_3PPh)_4Cl]^{2-}$ **20**.

4. Reductive aggregation

The aggregation of aqueous oxometalate species upon reduction has been ascribed to the formation of building blocks which are sufficiently basic to bind Lewis acid fragments. Müller and coworkers in particular have used this approach to good effect in the preparation of giant polyoxometalate structures [13]. In an effort to determine whether this strategy is applicable to rational non-aqueous aggregation, we chose the 6-electron reduced bi-capped heterometalate $[PV_2Mo_{12}O_{42}]^{3-}$ **21** as a target because the Keggin anion $[PMo_{12}O_{40}]^{3-}$ can be reduced extensively without loss of structural integrity. The reduction with Na/Hg amalgam was carried out in MeCN according to the stoichiometry shown in Equation 20 and a dark blue-black crystalline product was isolated.

$$[PMo_{12}O_{40}]^{3-} + 2[VOCl_3(dme)] + 6e^- \rightarrow [PV_2Mo_{12}O_{42}]^{3-} + 6Cl^- \qquad (20)$$

Large crystals of $(Bu^n{}_4N)_3$–**21** were obtained on recrystallisation and a crystal structure determination (Figure 13) shows the vanadium atoms to occupy two mutually *trans* positions of the six available square coordination sites on the surface of the Keggin anion. This anion can be regarded as $[PV^{IV}{}_2Mo^{V}{}_4Mo^{VI}{}_8O_{42}]^{3-}$ and has been predicted to be one of the two most stable forms of the free anions $[PV_2Mo_{12}O_{42}]^{n-}$ on the basis of DFT calculations [14]. In the presence of cations that can interact with more highly charged species, extra electrons can be accommodated in this framework, as demonstrated by the 8-electron reduced $(Et_3NH)_5[PV_2Mo_{12}O_{42}]$ which has been obtained from Na_2MoO_4, $VOSO_4$, H_3PO_4 and $[Et_3NH]Cl$ under more vigorous hydrothermal conditions [15]. The synthesis of **21** demonstrates that there is clearly scope for rational reductive aggregation under ambient conditions.

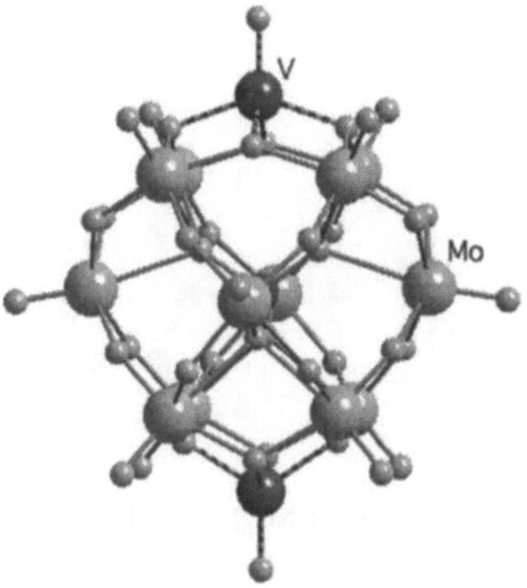

Fig. 13. Structure of $[PV_2Mo_{12}O_{42}]^{3-}$ **18**.

5. Surface Reactivity

5.1 ORGANOIMIDO HEXAMOLYBDATE DERIVATIVES.

We have shown previously that hexamolybdate $[Mo_6O_{19}]^{2-}$ reacts with isocyanates to give aryl- and alkylimido derivatives including $[(AdN)Mo_6O_{18}]^{2-}$ **22** (Ad = adamantyl, Figure 14) and $[(Bu^tN)Mo_6O_{18}]^{2-}$ [16] and Maatta has used similar reactions with bulky isocyanates to obtain multiply substituted anions [17].

We have also demonstrated that aromatic amines react with $[Mo_6O_{19}]^{2-}$ at elevated temperatures[18], providing a route to the amino-derivatised organoimido species $[(H_2NC_6Me_4N)Mo_6O_{18}]^{2-}$ **23** (Figure 15) and $[(H_2NC_6H_4N)_2Mo_6O_{17}]^{2-}$ **24** (Figure 16). We initially hoped that the reactivity of the NH_2 groups in these anions would provide the means to link them into larger assemblies, but results to date suggest that the metal oxide fragments deactivate these amines towards electrophiles. Further studies on these systems are in progress.

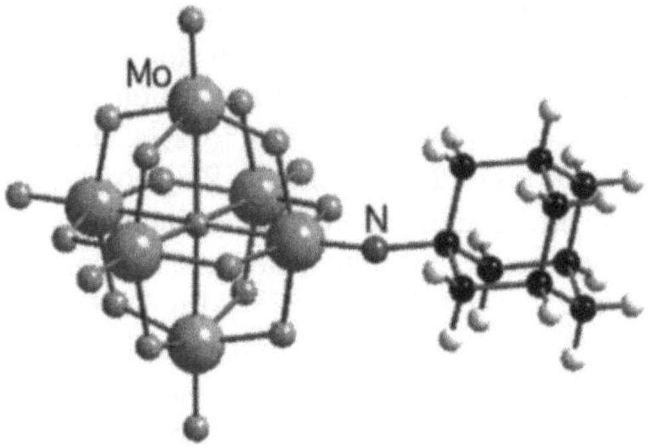

Fig. 14. Structure of $[(AdN)Mo_6O_{18}]^{2-}$ **22**.

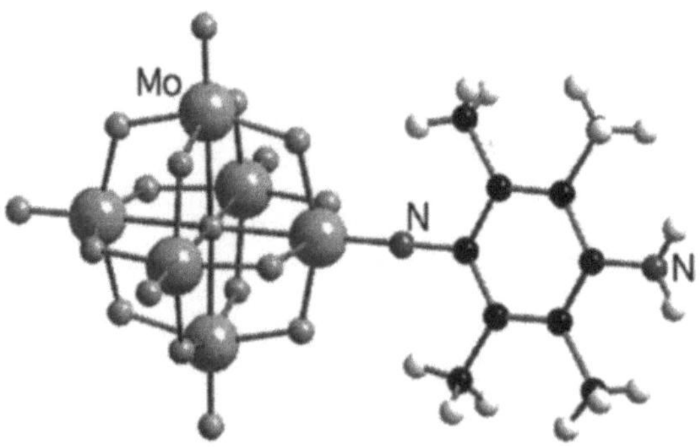

Fig. 15. Structure of $[(H_2NC_6Me_4N)Mo_6O_{18}]^{2-}$ **23**.

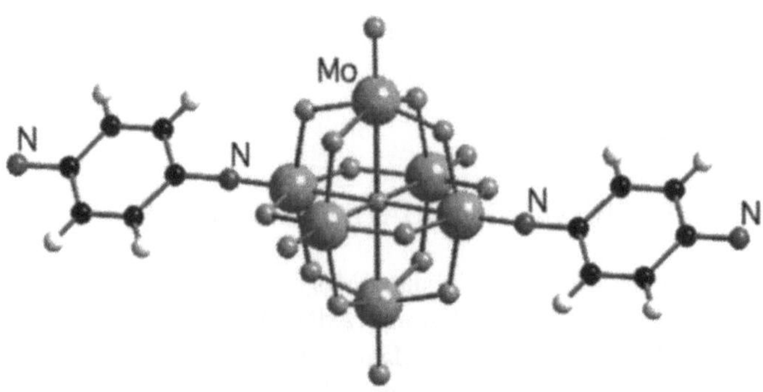

Fig. 16. Structure of $[(H_2NC_6H_4N)_2Mo_6O_{17}]^{2-}$ **24**.

5.2 REACTIVITY OF HEXANUCLEAR HETEROMETALATES.

NMR studies have shown that hydrolysis of the anion $[(MeO)TiW_5O_{18}]^{3-}$ **2** (Figure 1) is slow, requiring an excess of water at room temperature, or overnight reflux if a stoichiometric amount of water is used. In contrast, $(Bu^n{}_4N)_2[(MeO)NbW_5O_{18}]$ was found to be more susceptible to hydrolysis than **2** and attempted recrystallisation by solvent diffusion over several weeks produced $[O\{NbW_5O_{18}\}_2]^{4-}$ **25** (Figure 17).

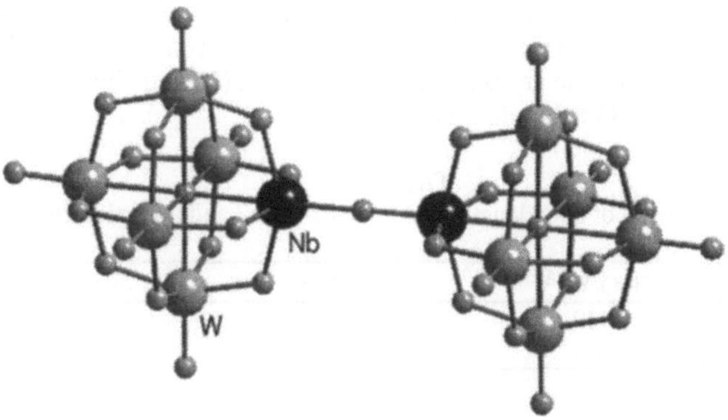

Fig. 17. Structure of $[O\{NbW_5O_{18}\}_2]^{4-}$ **25**.

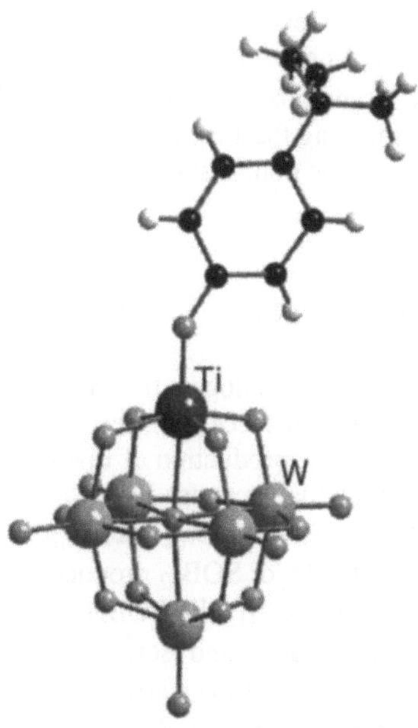

Fig. 18. Structure of $[(4\text{-}Bu^tC_6H_4O)TiW_5O_{18}]^{3-}$ **26**.

It therefore appears that attack at Ti by water in these hydrolysis reactions is inhibited by the higher charge of **2**. The eclipsed orientation of the two oxide cages in **25** indicates significant π-bonding between the bridging oxide and both niobium heteroatoms.

The alkoxohexametalates $[(RO)M'M_5O_{18}]^{n-}$ react with phenols to give aryloxide derivatives, *e.g.* $[(4\text{-}Bu^tC_6H_4O)TiW_5O_{18}]^{3-}$ **26**. Reactions of **8** (M = Mo, R = Pr^i) are faster than those of **2**, which may be due to the greater lability of the secondary alkoxide group or of the Ti–O–Mo bonds in **8** (or both). It is worth noting that the phenoxides $[(PhO)MW_5O_{18}]^{3-}$ (M = Zr, Hf) are monomeric in contrast to the dimeric alkoxide structure shown in Figure 3 (in both cases the phenoxo ligand is disordered over the two axial sites in the crystal structures). This would indicate a reduced availability of the oxygen lone pair for bridging interactions in these aryloxides compared with the corresponding alkoxides, due either to more efficient ligand to metal π-bonding or to C-O π delocalisation in the aryloxide. In this regard, a comparison of the bond lengths in **2** and **26** (Table 1) shows that the aryloxide has longer Ti–O(C) and shorter C–O bonds, indicative of enhanced C–O π-bonding in **26**.

TABLE 1
Comparison of bond lengths (Å) in methoxide **2** and aryloxide **26**.

Anion	Ti–O(W)[a]	Ti–O(C)	C–O
2	1.949	1.760	1.441
26	1.967	1.828	1.322

[a] average value

Treatment of the alkoxohexametalates $[(RO)TiM_5O_{18}]^{3-}$ with arylisocyanates results in the formation of intensely coloured solutions. ^1H NMR and IR spectra of isolated solids are indicative of more than one insertion product, and with an excess of ArNCO the trimers $(ArNCO)_3$ are formed. As expected from the seven-coordinate nature of the reactive site, reactions with the Hf methoxide **3** are faster than those with the Ti methoxide **2**. We are currently studying these and corresponding reactions with alkyl isocyanates in more detail to assess the potential of these polyoxometalates for catalytic isocyanate transformations.

5.3 HALOGENATION REACTIONS.

Previously reported attempts at the direct halogenation of a polyoxometalate surface to produce reactive M–X sites have been unsuccessful, resulting instead in degradation of the polyoxometalate framework and the production of low nuclearity oxohalide complexes [19]. We have now found that lacunary PW_{11} and PW_9 species can be brominated to give the hexabromide $[PW_9O_{28}Br_6]^{3-}$ **27** in good yields [20]. Treatment of $(Bu^n{}_4N)_6[NaPW_{11}O_{39}]$ with $C_2O_2Br_2$ or $SOBr_2$ produced yellow $(Bu^n{}_4N)_3$-**27**, as did the treatment of hydrated $Na_{9-x}H_x[PW_9O_{34}]$ with $SOBr_2$ and $[Bu^n{}_4N]Br$. In the former case, the reaction proceeds with degradation and isomerisation from α-PW_{11} to β, A-PW_9, whereas in the latter the β-structure of the starting material is retained. These reactions probably involve the *in situ* generation of HBr, although this has yet to be established. The structure of **27** (Figure 19) shows a β, A-PW_9 bromooxometalate structure in which one face is fully brominated. We are currently investigating the reactivity of this anion. Initial results from reactions with NaOMe suggest that stepwise substitution gives rise to mixtures of isomers of the type $[PW_9O_{28}Br_{6-x}(OMe)_x]^{3-}$, and a poor quality crystal structure of $(Bu^n{}_4N)_3[PW_9O_{28}(OMe)_6]$ showed the metal oxide framework to have isomerised to the α, A-PW_9 form.

The hexabromide **27** therefore provides an opportunity to study the factors affecting α–β interconversion and to develop the surface reactivity of PW_9 polyoxometalates. We are now extending the methodology employed in the synthesis of **27** to the preparation of bromo derivatives from other highly charged lacunary species.

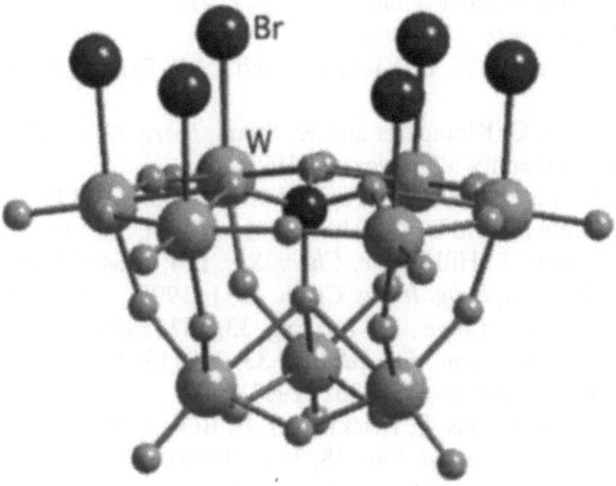

Fig. 19. Structure of $[PW_9O_{28}Br_6]^{3-}$ **27**.

6. Conclusions

The non-aqueous studies described here are beginning to reveal new opportunities for the controlled assembly of polyoxometalates and for systematic studies of their reactivity, although much work remains in order to understand the mechanistic features of aggregation and the factors which determine the underlying stabilities of the various species in solution as well as those isolated in the solid state. An important feature of this work is the ability to introduce specific reactive sites, which has made possible detailed metalorganic studies of the type normally associated with mononuclear organometallic species, thereby providing a better understanding of polyoxometalate surface reactivity. While the full potential of controlled hydrolytic and reductive aggregation has yet to be exploited, the strategies outlined in this article give some indication of the tremendous opportunities for new developments in the synthesis and applications of polyoxometalates.

Acknowledgements

In addition to those postgraduate and postdoctoral researchers whose names appear in the references, undergraduate project students J. L. R. Anderson, T. P. Cranley and S. L. Shaw were involved in the initial work on **8**. Funding was provided by the UK Engineering and Physical Sciences Research Council.

References

[1] (a) K. F. Jahr and J. Fuchs, *Chem. Ber.* **96**, 2457 (1963). (b) K. F. Jahr, J. Fuchs and R. Oberhauser, *Chem. Ber.* **101**, 482 (1968).

[2] W. Clegg, M. R. J. Elsegood, R. J. Errington and J. Havelock, *J. Chem. Soc., Dalton Trans.* 681 (1996).

[3] W. Clegg, R. J. Errington, K. A. Fraser and D. G. Richards, *J. Chem. Soc., Chem. Comm.* 1105 (1993).

[4] C. M. Flynn and M. T. Pope, *Inorg. Chem.* **10**, 2524 (1971).
[5] W. G. Klemperer, *Inorg. Synth.* **27**, 71 (1990).
[6] V. W. Day, T. A. Eberspacher, W. G. Klemperer and C. W. Park, *J. Am. Chem. Soc.* **115**, 8469 (1993).
[7] M. Filowitz, R. K. Ho, W. G. Klemperer and W. Shum, *Inorg. Chem.* **18**, 93 (1979).
[8] R. Bakri, PhD thesis, University of Newcastle (1998).
[9] V. W. Day, W. G. Klemperer and O. M. Yaghi, *J. Am. Chem. Soc.* **111**, 5959 (1989).
[10] V. W. Day, W. G. Klemperer and D. J. Maltbie, *J. Am. Chem. Soc.* 109, 2991 (1987).
[11] D. Hou, K. S. Hagan and C. L. Hill, *J. Am. Chem. Soc.* **114**, 5864 (1992).
[12] (a) M. I. Khan and J. Zubieta, *Prog. Inorg. Chem.* **43**, 1 (1995). (b) J. Salta, Q. Chen, Y.-D. Chang and J. Zubieta, *Angew. Chem., Int. Ed. Eng.* **33**, 757 (1994). (c) Y.-D. Chang, J. Salta and J. Zubieta, *Angew. Chem., Int. Ed. Eng.* **33**, 325 (1994). (d) Q. Chen and J. Zubieta, *Angew. Chem., Int. Ed. Eng.* **32**, 261 (1993).
[13] (a) A. Müller, S. Polarz, S. K. Das, E. Krickmeyer, H. Bögge, M. Schmidtmann and B. Hauptfleisch, *Angew. Chem., Int. Ed. Eng.* **38**, 3241 (1999). (b) A. Müller, E. Krickmeyer, H. Bögge, M. Schmidtmann and F. Peters, *Angew. Chem., Int. Ed. Eng.* **37**, 3360 (1998).
[14] J. M. Maestre, J. M. Poblet, C. Bo, N. Casañ-Pastor and P. Gomez-Romero, *Inorg. Chem.* **37**, 3444 (1998).
[15] Q. Chen and C. L. Hill, *Inorg. Chem.* **35**, 2403 (1996).
[16] R. J. Errington, D. G. Richards, W. Clegg and K. A. Fraser in *Polyoxometalates: From Platonic Solids to Anti-Retroviral Activity* , A. Müller and M. T. Pope Eds., Kluwer, Dordrecht (1994), p 105.
[17] J. B. Strong, B. S. Haggerty, A. L. Rheingold and E. A. Maatta, *J. Chem. Soc., Chem. Comm.* 1137 (1997).
[18] W. Clegg, R. J. Errington, K. A Fraser, S. A. Holmes and A. Schäfer, *J. Chem. Soc., Chem. Commun.* 455 (1995).
[19] Y-J Lu and R. H. Beer, *Polyhedron* **15**, 1667 (1996).
[20] R. L. Wingad, PhD thesis, University of Newcastle (2000).

Functionalization of Polyoxometalates : Achievements and Perspectives

A. PROUST AND R. VILLANNEAU
Laboratoire de Chimie Inorganique et Matériaux Moléculaires, Université Pierre et Marie Curie, 4 Place Jussieu, Case 42, 75252 Paris Cedex 05, France
proust@ccr.jussieu.fr

Abstract. This contribution will focus on the functionalization of polyoxometalates with multiply bonded ligands, notably nitrosyl, imido and cyclopentadienyl ligands.
The first part will define the scope of the different synthetic methodologies, *i.e.* net [2+2] reactions with Mo=O bonds, condensation-type reactions *via* a-hydrogen, and self-assembly reactions *via* the displacement of labile ligands, *e.g.* halide or solvent, from appropriate metal complexes. Selected examples will be presented and the eventual complications, *e.g.* hydrolysis or reduction, will be discussed. Special attention will be paid to the reactivity of phosphonium ylides towards polyoxomolybdates which contrasts that of phosphinimines.
The second part will show that functionalization may provide fine tuning of the electronic properties of the parent anion. Representative examples include the activation of surface oxygen atoms, as demonstrated by methylation of Lindqvist-type anions, and the stabilization of specific compounds, *e.g.* $[M_4O_{13}(OMe)_4\{Mo(NO)\}]^{3-}$ (M = Mo or W), and $[Mo_{10}O_{31-x}(OMe)_x(NO)]^{z-}$, which display their own, interesting, chemistry. Furthermore, NMR and electrochemical data underscore some electronic communication between the attached ligand and the polyoxometalate moiety: a clear example is provided by the series $[Mo_6O_{18}(NC_6H_4\text{-}p\text{-}X)]^{2-}$ where ^{95}Mo and ^{14}N chemical shifts and reduction potentials correlate with the Hammett constant of the substituent.
The last part will deal with the synthesis of cyclopentadienyl derivatives and their potential in the design of strongly interacting bipolar systems for various applications, *e.g.* photochromic or electrochromic material, and sensors,

Key words : functionnalization, nitrosyl derivatives, imido derivatives, cyclopentadienyl derivatives, organometallic oxides, pentamolybdate, EXAFS, metal carbonyl mobility, activation of surface oxygen atoms, methylation, electronic effects.

1. Introduction : functionalization of polyoxometalates, what and why ?

1.1. DEFINITION

In its broadest acceptation, functionalization of polyoxometalates includes :
- formal replacement of some oxo ligands O^{2-}, either terminal or not, by another ligand. The nitrosyl $[Mo_6O_{18}(NO)]^{3-}$[1], imido $[Mo_6O_{18}(NC_6H_4\text{-}pX)]^{2-}$(X = OMe, Me, H, F, Cl, Br, CF_3, NO_2) [2], and cyclopentadienyl $[(C_5Me_5)Mo_6O_{18}]^-$ [3] species are thus related to the Lindqvist anion $[Mo_6O_{19}]^{2-}$.
- formal replacement of some subunits like $\{MoO_2(OR)\}^+$ by another functional group like $\{M(CO)_3\}^+$ (M = Mn, Re) or $\{Ru(p\text{-cymene})\}^{2+}$. The anions $[Mo_4O_{10}(OMe)_6]^{2-}$ [4], on one hand, and $[Mo_2O_6(OMe)_4\{Re(CO)_3\}_2]^{2-}$ and $[Mo_2O_6(OMe)_4\{Ru(p\text{-cymen})\}_2]$ [5], on the other hand, thus display similar molecular structures.

M.T. Pope and A. Müller (eds.), Polyoxometalate Chemistry, 23–38.

- grafting of an organometallic fragment on a polyanion surface, as examplified by the $[\{(C_5Me_5)Rh(H_2O)\}Mo_5O_{13}(OMe)_4(NO)]^-$ [6], $[\{(C_5Me_5)Rh\}_4V_6O_{19}]$ [7] or $[\{(1,5\text{-COD})Ir\}P_2W_{15}Nb_3O_{62}]^{8-}$ [8] species.

The field of functionalization was initiated with the pioneering work of J. Zubieta's group, especially on diazenido, hydrazido and alkoxo derivatives [4], and the groups of W. Klemperer, R. Finke and K. Isobe [9] on organometallic derivatives. More recently, E. Maatta's group has widely explored the chemistry of imido derivatives [10].

1.2. MOTIVATIONS

Functionalization is a matter for polyoxometalate reactivity. First of all, some syntheses relie on the reactivity of the {Mo=O} function in polyoxomolybdates. Besides, the reactivity of functionalized polyoxometalates is then modified when compared to that of their parents and their properties can consequently be adaptated. Functionalization allows to stabilize novel architectures and to activate surface oxygen atom.

Functionalization is also related to surface oxide reactivity modelling since it provides structural and spectroscopic models for substrate or catalyst binding. In some cases, functionalized polyanions even mimic the reactivity of bulk oxides [10e, 11]. Dynamics at an oxide surface can also be reproduced on a polyanion surface.

Finally, functionalization could allow the design of new bipolar systems in strong interaction.

All these points will be developped in the following sections.

2. Synthetic strategies or how to functionalize polyoxometalates

2.1. SOME GUIDES

The simplest idea to achieve functionalization of polyoxometalates is to replace an oxo ligand by an isoelectronic one, like the hydrazido $RR'NN^{2-}$, imido RN^{2-}, cyclopentadienyl Cp^- or alkylidyne CR^{3-}ligands. Such ligands are $\sigma + 2\pi$ donnors. But we have also succeeded in the introduction of the π–acceptor nitrosyl ligand NO^+. In this case, comparison of qualitative energy diagrams showing the interaction between metal d orbitals and σ and π (for the oxo) or π^* (for the nitrosyl) ligand orbitals, leads to the conclusion that d^0-{MO} and d^4-{M(NO)} fragments are isolobal. Such an argument could also account for the formation of d^6-*fac*-{$M'L_3$} (M' = Mn, Re, Ru ; $L_3 = (CO)_3$, ArH) derivatives, d^0-*fac*-{MoO_3} and d^6-*fac*-{$M'L_3$} fragments being probably isolobal [12].

2.2. METHODS

In an effort of rationalization, two main synthetic approaches can be considered : the first one exploits the reactivity of the {Mo=O} function and is thus restricted to isopolymolybdates. It can be divided into metathesis and condensation type reactions and underlines the analogy of reactivity with the carbonyl {>C=O} function. The second approach relies on self-assembly processes and widely applies for the preparation of tungstic derivatives and organometallic oxides.

2.3. METATHESIS REACTIONS

2.3.1 *Imido derivatives*

The mechanism generally invoked in the formation of imido derivatives is a concerted [2+2] pericyclic reaction [13]. Various precursors have been used, phosphinimines, isocyanates, sulfinylamines or even amines and mono- or poly-substituted $[Mo_6O_{19-x}(NR)_x]^{2-}$ Lindqvist derivatives have thus been described [10, 14, 15]. Two examples of two anions linked through a bis-imido ligand have also been published [10d, 15]. We have recently prepared the series of *para*-substituted arylimido anions $[Mo_6O_{18}(NC_6H_4\text{-}pX)]^{2-}$(X = OMe, Me, H, F, Cl, Br, CF_3, NO_2) by reaction of the corresponding arenesulfinylamine $p\text{-}XC_6H_4NSO$ on $[Mo_6O_{19}]^{2-}$, in refluxing acetonitrile for several days. To check the validity of the reaction on another family of polyanions, we then turned to the reaction of tolylisocyanate on $[PMo_{12}O_{40}]^{3-}$, in pyridine at 80°C [16]. After treatment, a mixture of compounds were characterized. It includes the reduced species $[PMo_{12}O_{40}]^{4-}$, the urea by-product $p\text{-}MeC_6H_4NHC(O)NHC_6H_4Me\text{-}p$ resulting from the hydrolysis of the precursor, as well as the azatoluene $p\text{-}MeC_6H_4N{=}NC_6H_4Me\text{-}p$ and the highly substituted species $[Mo_{10}(NC_6H_4Me\text{-}p)_{12}(py)_2O_{18}]$, represented below. This centrosymmetrical complex can be viewed as composed of two '$Mo_3O_7(NC_6H_4Me\text{-}p)_6$' units held together by four extra molybdenum centers. These units are reminiscent of the 'Mo_3O_{13}' building blocks of the starting Keggin anion. Terminal as well as bridging imido ligands are observed at the surface of the compound, which appears as a layer of oxide sandwiched between two organic layers. If not ruled out, the metathesis mechanism fails to explain the formation of the former compound as well as that of azatoluene. This suggests that at least another mechanism is involved, eventually through a competitive pathway.

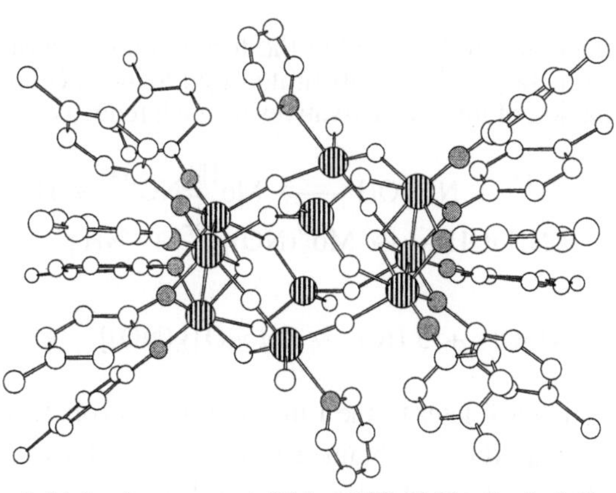

Fig. 1. Molecular structure of $[Mo_{10}(NC_6H_4Me\text{-}p)_{12}(py)_2O_{18}]$.

2.3.2 *Reactions of phosphorus ylides on polyoxomolybdates*

Encouraged by our former results and to asses the analogy of reactivity between the {Mo=O} and {>C=O} functions, we undertook to study the reaction of the phosphorus ylides $R_3P{=}C(R')R''$ (R = Bu, R' = R'' = H ; R = Ph, R' = H and R'' = Ph, C(O)Me, C(O)N(Me)OMe ; R = Ph, R'R'' = $C(CH_2)_4$; R = Ph, R' = Me, R'' = CHO ; R = Ph, R' = R'' = Cl) on various molybdates, including $[MoO_2(acac)_2]$, $[Mo_6O_{19}]^{2-}$ and $[PMo_{12}O_{40}]^{3-}$. Whatever the ylide used, reactions always result in the formation of reduced anions and phosphonium cations. In accordance with ^{31}P NMR data, especially those recorded in the course of the reaction between $[PMo_{12}O_{40}]^{3-}$ and one to two equivalent(s) of Bu_3PCH_2, we propose the following one-electron reduction processes to occur :

$$[PMo_{12}O_{40}]^{3-} + R_3P{=}C(R')R'' \longrightarrow [PMo_{12}O_{40}]^{4-} + [R_3P{=}C(R')R'']^{\cdot +}$$
$$[PMo_{12}O_{40}]^{4-} + R_3P{=}C(R')R'' \longrightarrow [PMo_{12}O_{40}]^{5-} + [R_3P{=}C(R')R'']^{\cdot +}$$
$$[R_3P{=}C(R')R'']^{\cdot +} + H_2O \longrightarrow [R_3P{=}C(R')R''H]^{+} + OH^{\cdot}$$
$$OH^{\cdot} \longrightarrow$$

The radical character of the reaction is further demonstrated by the formation of diphenyldisulfide when reaction proceeds in the presence of thiophenol.

Up to now, we have failed even to suspect the formation of an alkylidene derivative, probably because of lack of adequacy between a reactive but not too reducible polyoxometalate and appropriate R, R' and R'' groups on the ylide. But after all, the reduction observed when reacting $[Mo_6O_{19}]^{2-}$ and Bu_3PN^tBu [17] didn't hinder the development of the imido chemistry of polyoxometalates, as we showed above.

2.4 CONDENSATION REACTIONS

Condensation type reactions are involved in the formation of hydrazido and diazenido derivatives of polyoxometalates from substituted hydrazines [4] and, as far as we are concerned, in the formation of nitrosyl derivatives through reductive nitrosylation :

$$Mo^{VI}{=}O^{4+} + NH_2OH \longrightarrow Mo^{II}{=}NO^{3+} + H_3O^{+}$$
$$Mo^{VI}{=}O^{4+} + 3\,NH_2OH \longrightarrow Mo^{0}(NO)_2{}^{2+} + NH_4{}^{+} + H^{+} + 2\,H_2O$$

$$3\,[Mo_2O_7]^{2-} + NH_2OH + 3\,HCl \rightarrow [Mo_6O_{18}(NO)]^{3-} + 3\,H_2O + 3\,Cl^{-}$$

We will especially come back later to the Lindqvist derivatives $[Mo_6O_{18}(NO)]^{3-}$ and lacunar $[Mo_5O_{13}(OMe)_4(NO)]^{3-}$, obtained from $[\alpha\text{-}Mo_8O_{26}]^{4-}$. As established by ^{95}Mo NMR, they are localized mixed valence species [1].

2.5 SELF-ASSEMBLY OF APPROPRIATE PRECURSORS

2.5.1 *Tungsten derivatives*

The lack of reactivity of the {W=O} function when compared to the {Mo=O} forced us to turn to another strategy for the preparation of functionalized polyoxotungstates : $[W_5O_{18}\{W(NO)\}]^{3-}$ and the bimetallic $[Mo_5O_{18}\{W(NO)\}]^{3-}$ and $[W_5O_{18}\{Mo(NO)\}]^{3-}$ have then be obtained from the reaction of mononuclear nitrosyl complexes $[W(NO)Cl_3(MeCN)_3]$ or $[Mo(NO)(acac)_2\{MeC(NH_2)NO\}]$ and appropriate oxo precursors $[Mo_2O_7]^{2-}$ or $[WO_4]^{2-}$, in acetonitrile. These have been thorougly characterized by multinuclear NMR (^{95}Mo, ^{183}W, ^{17}O et ^{14}N) and electrochemistry [18].

2.5.2 *Organometallic oxides*

Cyclopentadienyl titanium derivatives of polyoxometalates have been described by the groups of W. Klemperer [19] and J. F. W. Keana [20]. On the other hand, the groups of F. Bottomley and A. L. Rheingold have reported on the homonuclear species $[Cp^*Mo_6O_{18}]^-$ [21] and $[Cp^*_2W_6O_{17}]$ [22], respectively. These result from the oxidation of the corresponding carbonyl $[\{Cp^*M(CO)_2\}_2]$ (M = Mo, W) dimer. We have recently proposed an alternative route to these pentamethylcyclopentadienyl compounds starting with the $[Cp^*MO_3]^-$ precursors [23]. The monosubstituted $[Cp^*Mo_6O_{18}]^-$ anion is thus formed through the reaction of $[Cp^*MoO_3]^-$ with $[Mo_4O_{10}(OMe)_4Cl_2]^{2-}$ in refluxing dry methanol. Instable in hot methanol, $[Mo_4O_{10}(OMe)_4Cl_2]^{2-}$ decomposes and liberates acidic "MoO_2^{2+}" units ready to condense with the $[Cp^*MO_3]^-$ organometallic base. Puzzlingly, $[MoO_2(acac)_2]$ or $[MoO_2Cl_2]$ fail to reproduce the same reactivity. We also achieved to prepare in a very similar way the tetramethylcyclopentadienyl derivative $[(C_5Me_4H)Mo_6O_{18}]^-$.

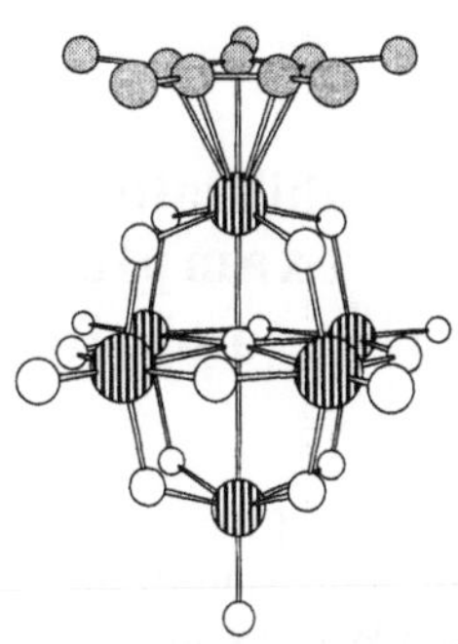

Fig. 2. Molecular structure of $[Cp^*Mo_6O_{18}]^-$.

The use of other organometallic precursors, like $[Cp^*MO_2Cl]$ or $[Cp^*MOCl_3]$, in reactions with $[Mo_2O_7]^{2-}$ or $[WO_4]^{2-}$ is under studies. Triggering of condensation processes in the presence of Brönsted acids will also be investigated. The $[(C_5Me_5)Mo_6O_{18}]^-$ and $[(C_5Me_4H)Mo_6O_{18}]^-$ have been thorougly characterized by single crystal X-ray diffraction, multinuclear NMR and electrochemistry. In particular, their electrochemical behavior, in acetonitrile, is characterized by a reversible reduction process around -0,03 V (referred to ECS). From the comparison of ^{95}Mo and ^{17}O NMR spectra with those of the parent $[Mo_6O_{19}]^{2-}$ and arylimido $[Mo_6O_{18}(NAr)]^{2-}$ species, we could also inferred that the pentamethylcyclopentadienyl ligand is a better $\sigma + 2\pi$ donnor than the arylimido ligand, itself better donnor than the oxo ligand [3]. The ^{95}Mo NMR spectra of $[Cp^*Mo_6O_{18}]^-$ is reproduced below. The most shielded signal at -35 ppm is attributed to the substituted molybdenum.

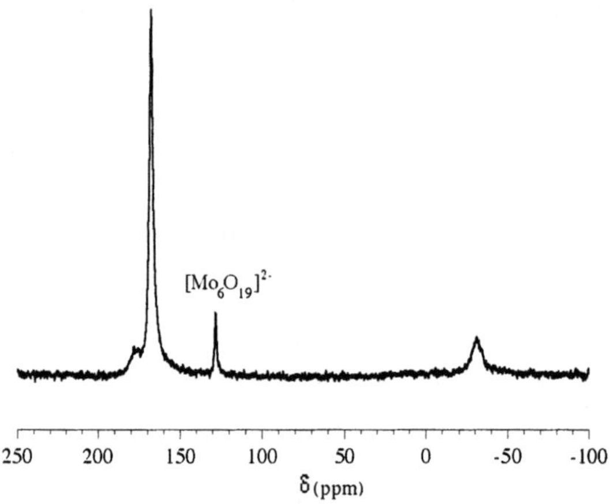

Fig. 3. ^{95}Mo NMR spectrum of $(NBu_4)[Cp^*Mo_6O_{18}]$ in CH_3CN-CD_3CN recorded at 343 K.

Integrated ruthenium *p*-cymen and manganese or rhenium carbonyl derivatives have also been obtained through self-assemble processes. They are described in another contribution of our laboratory. Interplay between cubane-type and rhomb-like structures is especially discussed.

3. Functionalization and stabilization of novel polyoxometallic architectures: the example of the lacunar $[Mo_5O_{13}(OMe)_4(NO)]^{3-}$ and its versatile coordination chemistry

As it modifies the electronic properties of the resulting anions, in particular their whole charge, functionalization may help in the stabilization of otherwise too reactive species. Although the lacunar Lindqvist-like W_5O_{18} structure has been recognized in lanthanide derivatives, the molybdenum analog was unknown. A few years ago, we described the related species $[Mo_5O_{13}(OMe)_4(NO)]^{3-}$, incorporating nitrosyl and methoxo ligands. Similarly, we have reported on a family of nitrosyl decamolybdates, analogs to decatungstate-Y [24].

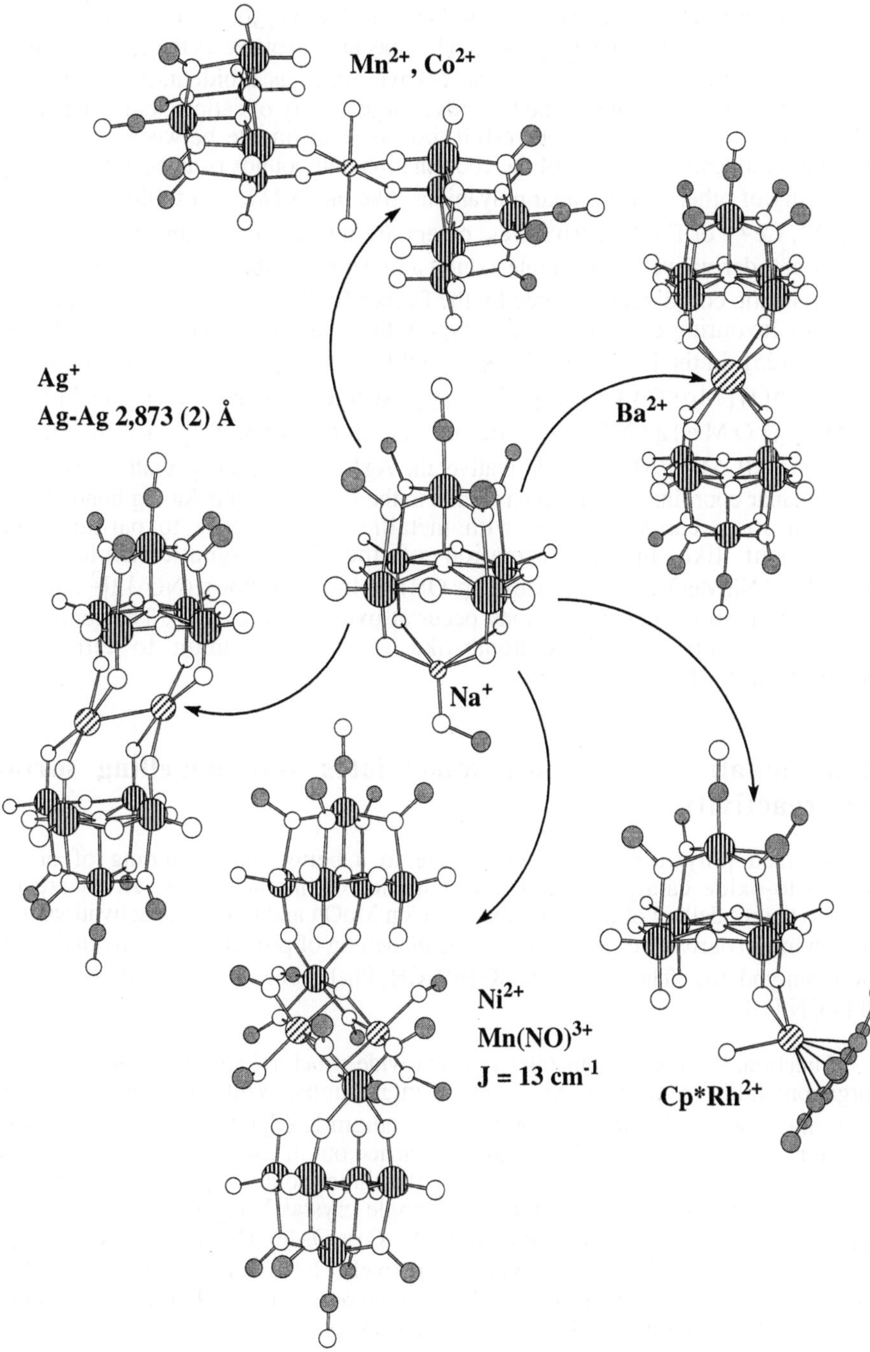

Fig. 4. Coordination chemistry of the lacunar $[Mo_5O_{13}(OMe)_4(NO)]^{3-}$ nitrosyl derivative

The coordination chemistry of the lacunar $[Mo_5O_{13}(OMe)_4(NO)]^{3-}$ anion is remarkable for its diversity and originality. This species is isolated as the sodium complex $[Na\{Mo_5O_{13}(OMe)_4(NO)\}]^{2-}$ and can behave either as a bidentate, bridging bis-bidendate, tri- or tetra-dentate ligand towards a large variety of cations, either metallic or not. IR spectroscopy is then a powerfull tool to discriminate between the different coordination modes.Whilc some of the coordination compounds isolated are common to the chemistry of other monolacunar polyanions, like the $[X\{Mo_5O_{13}(OMe)_4(NO)\}_2]^{3-}$ ($X = Bi^{3+}$, Ce^{3+} or Eu^{3+}) derivatives, others are uncommon or unprecedented. The Ca^{2+} or Ba^{2+} derivatives for example exhibit a rather rare cubic coordination, instead of the antiprismatic coordination shared by the former Bi^{3+}, Ce^{3+} and Eu^{3+} compounds. The factors favouring one coordination type rather than the other are not really well understood [25]. In the $[M(H_2O)_2\{Mo_5O_{13}(OMe)_4(NO)\}_2]^{4-}$ ($M = Mn^{2+}$, Co^{2+}) and $[Cp^*Rh(H_2O)\{Mo_5O_{13}(OMe)_4(NO)\}_2]^-$ species, the lacuna of the bidentate $[Mo_5O_{13}(OMe)_4(NO)]^{3-}$ ligand is not completely filled. In the $[Ag_2\{Mo_5O_{13}(OMe)_4(NO)\}_2]^{4-}$ derivative, the Ag^+ cations display a rather uncommon distorted planar coordination that even results in the formation of a Ag-Ag bond of 2.873 Å [26]. In some cases, reaction with metallic cations leads to partial surface rearrangement like in the formation of the ferromagnetic triple cluster $[\{Mo(NO)\}_2\{Ni(MeOH)_2\}_2(\mu\text{-}OMe)_4(\mu_3\text{-}OH)_2\{Mo_5O_{13}(OMe)_4(NO)\}_2]^{2-}$, and thus illustrates the rearrangements that may occur at oxide surfaces [27]. The central core displays the compact M_4X_{16} rhomb-like structure common to tetranuclear polyoxometalates [4].

4. Functionalization of polyoxometalates and modelling surface oxide reactivity

Functionalized polyoxometalates contribute to a better understanding of organic substrates-to-oxide catalysts interactions. Studies on methoxo derivatives are thus relevant to the modelling of methanol oxidation on MoO_3 and C-H bond activation, while studies on imido derivatives are related to the modelling of propylene ammoxidation over bismuth molybdates. The imido $[Mo_6O_{18}(NCH_2Ph)]^{2-}$ derivative thus decomposes to yield PhCN [10e].

Functionalized polyoxometalates also provide structural and spectroscopic models for organometallic catalyst-to-oxide support interactions. With this in mind, we have characterized by EXAFS spectroscopy at the rhodium K-edge Cp^*Rh^{2+} derivatives of the previous pentamolybdate. This study was carried out in collaboration with F. Villain and M. Verdaguer, from the laboratory. Of the three isolated species, the molecular structures of only two were determined by single crystal X-ray diffraction and were found to obey to the formula $[Cp^*Rh(H_2O)\{Mo_5O_{13}(OMe)_4(NO)\}]$ and $[(Cp^*Rh)_2(\mu\text{-}Cl)\{Mo_5O_{13}(OMe)_4(NO)\}]$, displaying respectively 1 / 1 and 1 / 2 pentamolybdate / rhodium stœchiometry. For this reason, they will be referred to as **1** and **2**. According to elemental analysis, the third, **M**, whose crystals are systematically twinned, contains 2 "Mo_5" / 2 Cp^*Rh^{2+} / 1 Cl^- / 1Na^+. EXAFS signals and Fourier Transforms for the three species are reproduced below.

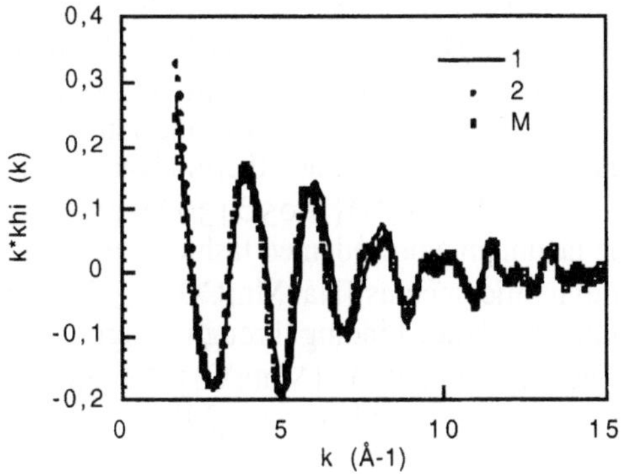

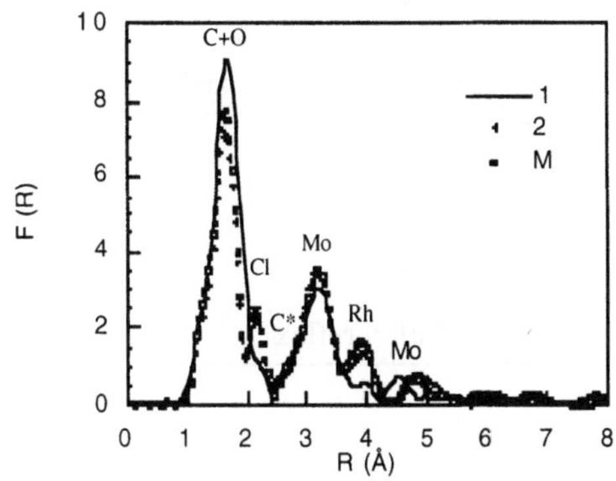

Fig. 5. EXAFS signals (up) and Fourier Transforms (down) for [Cp*Rh(H_2O){$Mo_5O_{13}(OMe)_4(NO)$}] (**1**), [$(Cp^*Rh)_2(\mu\text{-}Cl)${$Mo_5O_{13}(OMe)_4(NO)$}] (**2**) and unknown **M**.
(Collaboration with F. Villain and M. Verdaguer, UPMC)

For the first neighbour sphere, the C + O contribution is higher for **1** than for **2**, due to the presence of the water molecule in the former. Contributions of Cl and Rh peaks are observed only for **2** and **M**. Other peaks are common, with the methyl-carbon and molybdenum contributions. Moreover, the data for **2** and **M** are very similar and suggest that **M** could be a mixture of **2** and the starting pentamolybdate. This was later confirmed by IR and UV-visible spectroscopies. In this study, **1** and **2** were used as models to elucidate the molecular structure of **M**. Beyong, such studies contribute to the building of interrelated structural and spectroscopic data bases of more general use.

Metal carbonyl mobility accross an oxide surface can also be modelled at polyanion level. One example has recently been published in the literature [28]. We present another one encountered in the course of manganese-carbonyl grafting on the pentamolybdate. While at room temperature $[Mn(CO)_3L_3]^+$ reacts with

$[Na\{Mo_5O_{13}(OMe)_4(NO)\}]^{2-}$ in methanol to yield $[Mn(CO)_3(H_2O)\{Mo_5O_{13}(OMe)_4(NO)\}]^{2-}$, the reaction at refluxing methanol results in the formation of $[Na\{Mn(CO)_3\}_2\{Mo_5O_{13}(OMe)_4(NO)\}_2]^{3-}$. Moreover, in the presence of NaBr, the former can be converted to the later by refluxing in methanol, which can be interpreted in terms of kinetic and thermodynamic products, respectively. In the kinetic derivative $[Mn(CO)_3(H_2O)\{Mo_5O_{13}(OMe)_4(NO)\}]^{2-}$, the manganese fragment is linked to the lacuna in a precedented fashion (see paragraph 3. above). On the other hand, the thermodynamic species $[Na\{Mn(CO)_3\}_2\{Mo_5O_{13}(OMe)_4(NO)\}_2]^{3-}$ displays a tri-dentate pentamolybdate binding through bridging oxo and methoxo ligands, in an original fashion. The migration of the $\{Mn(CO)_3\}^+$ fragment thus occurs from hard terminal oxo ligands of the vacancy to softer sites [29].

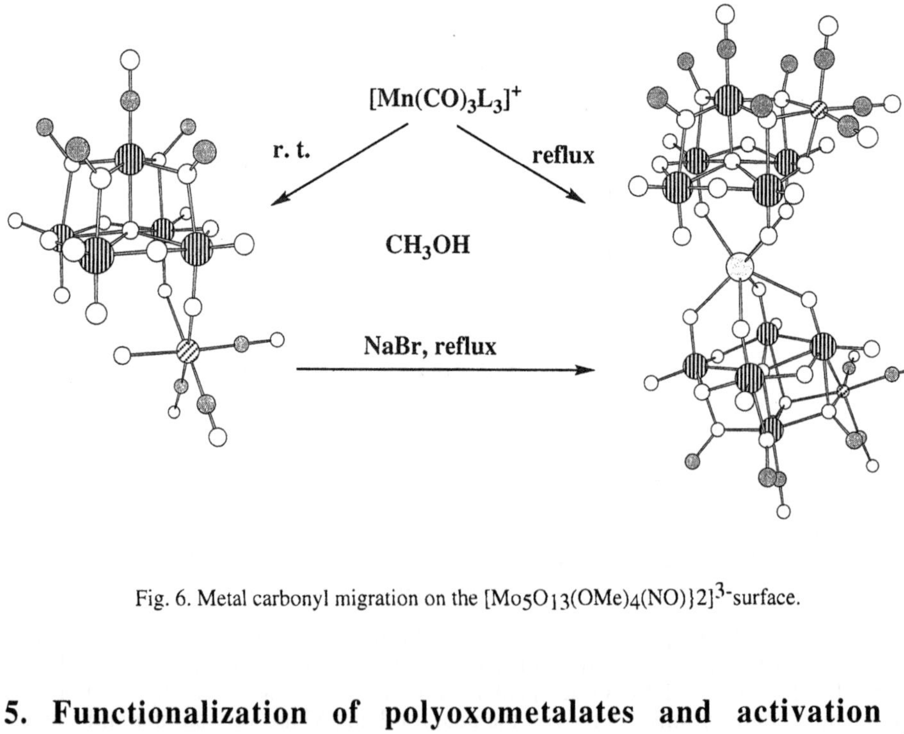

Fig. 6. Metal carbonyl migration on the $[Mo_5O_{13}(OMe)_4(NO)\}2]^{3-}$surface.

5. Functionalization of polyoxometalates and activation of surface oxygen atoms: the example of $[M_5O_{18}\{Mo(NO)\}]^{3-}$ species (M = Mo, W)

The nitrosyl derivatives $[M_5O_{18}\{Mo(NO)\}]^{3-}$ (M = Mo, W) can be methylated by reacting with dimethylsulfate in refluxing acetonitrile, while $[Mo_6O_{19}]^{2-}$ is unreactive upon the same experimental conditions. This is probably a consequence of the whole charge increase induced by the functionalization, which, if it were limited to that effect, would be no more than that observed when replacing Mo or W by V, Nb or Ta. What is more, is the selectivity of the reaction, since only one isomer is formed on the basis of

NMR data. Curiously, the tungstic analogs $[M_5O_{17}(OMe)\{W(NO)\}]^{2-}$ have been obtained in very low yield. According to preliminary characterization, the main products isolated in this case could result from methylation at the nitrosyl site.

Methylation sites within $[M_5O_{17}(OMe)\{Mo(NO)\}]^{2-}$ species could not be established by single crystal X-ray diffraction since the anions are fortuitously located on cristallographic inversion centers. Three sites can be considered: adjacent, or remote, to the nitrosyl ligand, or equatorial. The third one is ruled out by the crystallographic study. To discriminate between the first two ones two parallel studies have been undertaken: multinuclear NMR experiments, carried out in our laboratory in collaboration with R. Thouvenot, and *ab-initio* calculations performed by M.-M. Rohmer and M. Bénard at Louis Pasteur University. Both studies conclude to the methylation at adjacent position.

A projection of the electrostatic potential is presented below for $[Mo_6O_{18}(NO)]^{3-}$. It clearly reveals a deeper potential well for the oxygen atom on the {Mo(NO)} side.

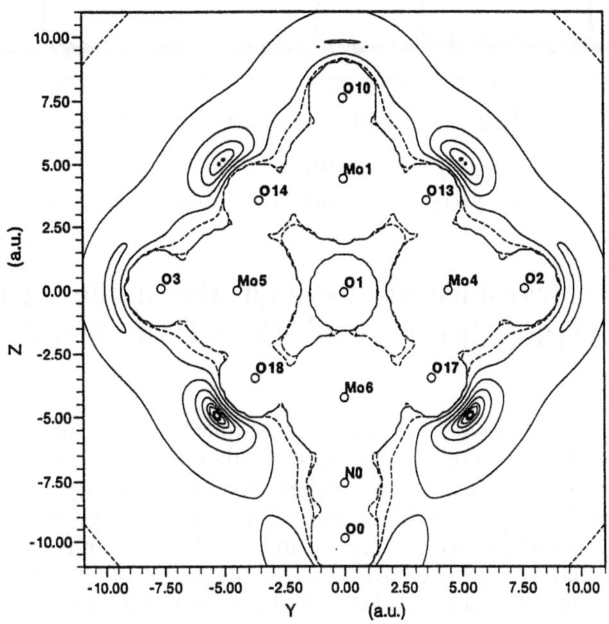

Fig. 7. Electrostatic potential map for $[Mo_6O_{18}(NO)]^{3-}$ (collaboration with M.-M. Rohmer and M. Benard, ULP)

On the other hand, the ^{183}W NMR spectrum of $[W_5O_{17}(OMe)\{Mo(NO)\}]^{2-}$ derivative shows the expected four signals of relative intensities 1/ 2/1/1, one of which appearing as a quartet due to scalar coupling with the three protons of the methyl group ($^3J_{W-H}$ = 1.2 Hz). That only one tungsten was concerned with this coupling was checked by special INEPT and COSY-INEPT polarisation transfer sequences. The observation of tungsten satellites allows a complete assignement of the ^{183}W spectrum.

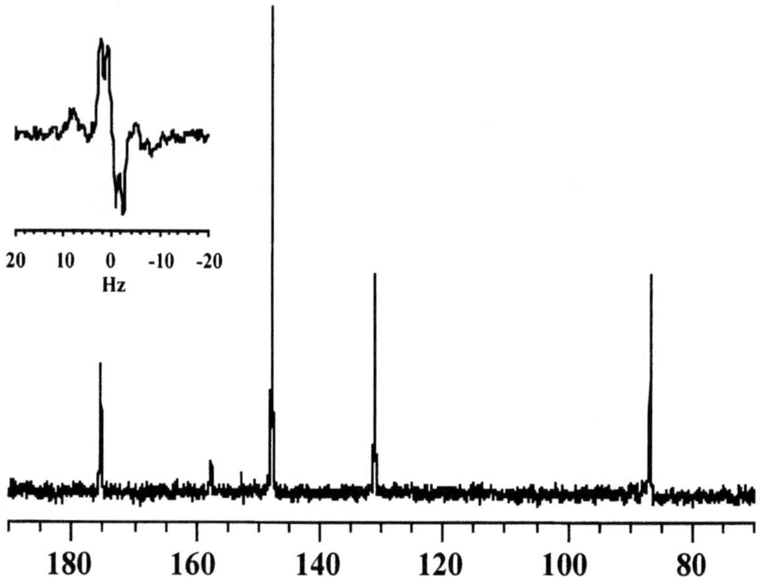

Fig. 8. ^{183}W NMR spectra of $[W_5O_{17}(OMe)\{Mo(NO)\}]^{2-}$ (collaboration with R. Thouvenot, UPMC)

6. Electronic transmission through the imido ligand in the series $[Mo_6O_{18}(NC_6H_4\text{-}pX)]^{2-}$ (X = OMe, Me, H, F, Cl, Br, CF_3, NO_2)

Electronic properties of polyoxometalates can be tuned through functionalization. The reduction potential, for example, strongly depends on the ligand. Imido derivatives $[Mo_6O_{18}(NC_6H_4\text{-}pX)]^{2-}$ are thus more difficult to reduce than the $[Mo_6O_{19}]^{2-}$ parent, consistently with the respective donor abilities of oxo and arylimido ligands. To go further and if we intend to involve functionalized polyanions in the design of bipolar covalently connected systems, we have to assess the degree of communication through the ligand. The observed correlations between the electronic properties of the ligand and those of the substituted anion in the series $[Mo_6O_{18}(NC_6H_4\text{-}pX)]^{2-}$ (X = OMe, Me, H, F, Cl, Br, CF_3, NO_2) indeed demonstrates some transmission of electronic effects.

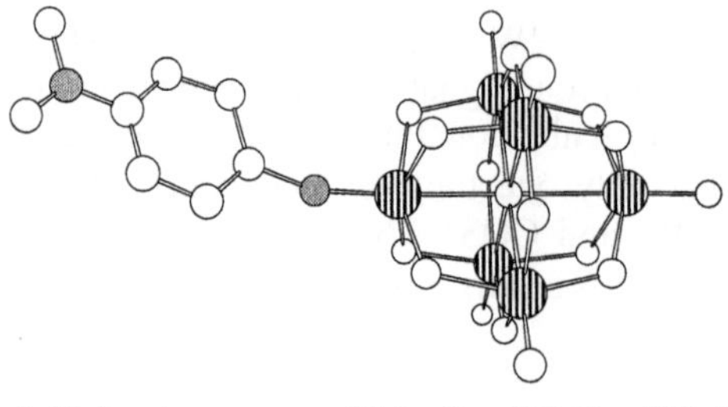

Fig. 9. Molecular structure of $[Mo_6O_{18}(NC_6H_4\text{-}pNO_2)]^{2-}$.

The linear relationship between the reduction potential of the anion $[Mo_6O_{18}(NC_6H_4\text{-}pX)]^{2-}$ and the Hammett constant of the X substituent is presented below. As expected, the more attractive the substituent, the less negative the potential.

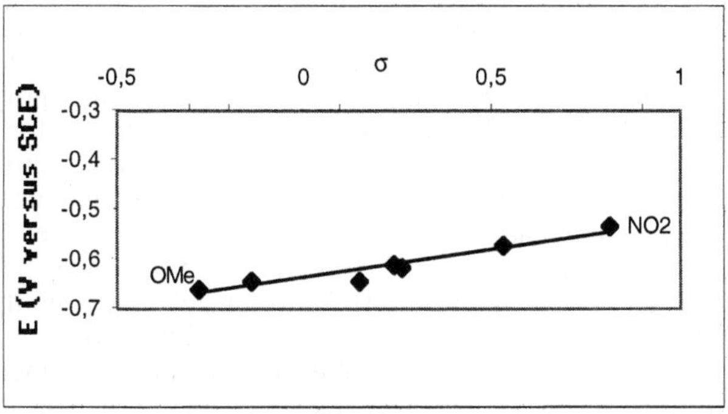

Fig. 10. Correlation between reduction potentials and Hammett constants of the subsituents in the series $[Mo_6O_{18}(NC_6H_4\text{-}pX)]^{2-}$ (X = OMe, Me, H, F, Cl, Br, CF_3, NO_2)

A correlation was also found between the chemical shift of the molybdenum bearing the imido ligand and the Hammett constant of the X substituent.The corresponding signal appears as a triplet due to scalar coupling with nitrogen and is shielded when compared to that of the $[Mo_6O_{19}]^{2-}$ parent. The position of the signal is then modulated by the substituent effect, the more attractive the substituent, the more shielded the signal. This tendency reveals the role of the paramagnetic contribution to the shielding constant and is in agreement with UV-visible spectroscopic data on charge transfer bands. The lower in energy the electronic transitions, the higher the paramagnetic contribution. A similar effect is observed for ^{14}N NMR. Such correlations have also been reported in the literature about ^{51}V NMR study of the $[V(NAr)Cl_3]$ series [30] and ^{14}N NMR study of the $[W(NAr)Cl_4]$ series [31].

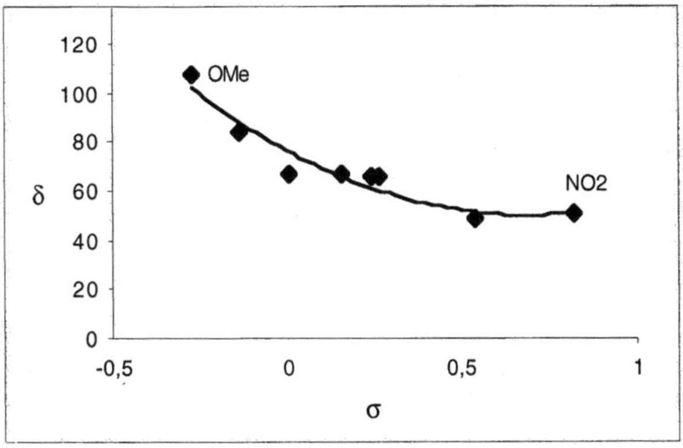

Fig. 11. Correlation between chemical shifts of the functionalized molybdenum and Hammett constants of the subsituents in the series $[Mo_6O_{18}(NC_6H_4\text{-}pX)]^{2-}$ (X = OMe, Me, H, F, Cl, Br, CF_3, NO_2)

7. Perspectives

The class of molecular materials based on polyoxometalates is rapidely expanding [32]. It includes organic or organometallic / inorganic hybrid salts like $\{(BEDTTF)_8[PW_{11}MnO_{39}]\}_n$ [33], $(BEDTTTF)_6H_4[Co_4(H_2O)_2(PW_9O_{34})_2]$ [34] or $[Cp^*Fe_2]_4[XM_{12}O_{40}]$ [35]. In these donnor-acceptor systems a long range magnetic order is expected through indirect exchange between delocalized electrons within the organic sublattice and localized magnetic moments on the polyoxometalates. Polyoxometalates are also implied in the modification of electrodes for redox catalysis, electrocatalysis or sensor applications [32b]. Some selective electrodes have been developed ; those incorporating, for instance, polyoxometalates and macrocycles for the detection of alkylammonium cations [36] could probably be improved in ionic sensors. Polyoxometalates can also be immobilized in hybrid polymers to which they confer their electrochromic and magnetic properties [34]. Unfortunately, the stability of the devices relying on electrostatic interactions can be questionned and the expected synergy between the different components is often weak. We believe that the functionalization of polyoxometalates, because of its covalent character, could improve the design of bipolar systems in strong interaction through the use of bifunctional ligand linked to the polyoxometalate on one side and to another entity, polyanionic or not, on the other. The bisimido $[Mo_5O_{18}\{MoNC_6H_{10}NMo\}Mo_5O_{18}]^{4-}$ complex reported by Maatta's group [10d] can be considered as a prototype of such system. We are now exploring the possibility for substituted-cyclopentadienyl ligands to act as the bridging unit, between polyanions and macrocycles for the design of ionic sensors, between polyanions and ruthenium bipyridine complexes for photochemical applications, between polyanions and conducting polymers ... This could also allow us to explore the interface between organometallic and inorganic chemistry and give us an entry towards supramolecular chemistry.

References

1. A. Proust, R. Thouvenot, F. Robert, P. Gouzerh, *Inorg. Chem.*, 1993, **32**, 5299.

2. S. Taunier, DEA manuscript, Pierre and Marie Curie University, 1996.

3. A. Proust, R. Thouvenot, P. Herson, *J. Chem. Soc., Dalton Trans*, 1999, 51.

4. (a) Q. Chen, J. Zubieta, *Coord. Chem. Rev.*, 1992, **114**, 107 ; (b) M. I. Khan, J. Zubieta, *Progr. Inorg. Chem.*, 1995, **43**, 1.

5. R. Villanneau, R. Delmont, A. Proust, P. Gouzerh, manuscript submitted for publication in *Chem. Eur. J.*

6. A. Proust, F. Robert, P. Gouzerh, *Angew. Chem., Int. Ed. Engl.*, 1993, **32**, 115.

7. (a) H. K. Chae, W. G. Klemperer, V. W. Day, *Inorg. Chem.*, 1989, **28**, 1423 ; (b) Y. Hayashi, Y. Ozawa, K. Isobe, *Chem. Lett.*, 1989, 425.

8. M. Pohl, D. K. Lyon, N. Mizuno, K. Nomiya, R. G. Finke, *Inorg. Chem.*, 1995, **34**, 1413.

9. P. Gouzerh, A. Proust, *Chem. Rev.*, 1998, **98**, 77.

10. (a) Y. Du, A. L. Rheingold, E. A. Maatta, *J. Am. Chem. Soc.*, 1992, **114**, 345 ; (b) J. B. Strong, R. Ostrander, A. L. Rheingold, E. A. Maatta, *J. Am. Chem. Soc.*, 1994, **116**, 3601 ; (c) J. L. Stark, V. G. Young, E. A. Maatta, *Angew. Chem., Int. Ed. Engl.*, 1995, **34**, 2547 ; (d) J. L. Stark, A. L. Rheingold, E. A. Maatta, *J. Chem. Soc., Chem. Commun.*, 1995, 1165 ; (e) J. B. Strong, B. S. Haggerty, A . L. Rheingold, E. A. Maatta, *J. Chem. Soc., Chem. Commun.*, 1997, 1137.

11. E. M. McCarron III, R. L. Harlow I, *J. Am. Chem. Soc.*, 1983, **105**, 6179.

12. T. Szyperski, P. Schwerdtfeger, *Angew. Chem. Int. Ed. Engl.*, 1989, **28**, 1228.

13. (a) P. Jernakoff, G. L. Geoffroy, A. L. Rheingold, S. J. Geib, *J. Chem. Soc., Chem. Commun.*, 1987, 1610 ; (b) R. Pilato, C. E. Housmekenides, P. Jernakoff, D. Rubin, G. L. Geoffroy, A. L. Rheingold, *Organometallics*, 1990, **9**, 2333.

14. A. Proust, R. Thouvenot, M. Chaussade, F. Robert, P. Gouzerh, Inorg. Chim. Acta, 1994, **224**, 81.

15. W. Clegg, R. J. Errington, K. A. Fraser, S. A. Holmes, A. Shäfer, *J. Chem. Soc., Chem. Commun.*, 1995, 455.

16. A. Proust, S. Taunier, V. Artero, F. Robert, R. Thouvenot, P. Gouzerh, *J. Chem. Soc., Chem. Commun.*, 1996, 2437.

17. R. J. Errington, *« Polyoxometalates : From Platonic Solids to Anti-Retroviral Activity »*, M. T. Pope, A. Müller Eds, Kluwer Academic Publishers, Dordrecht, 1994, 112.

18. A. Proust, R. Thouvenot, S.-G. Roh, J.-K. Yoo, P. Gouzerh, *Inorg. Chem.*, 1995, **34**, 4106.

19. V. W. Day, M. F. Fredrich, M. R. Thompson, W. G. Klemperer, R.-S. Liu, W. Shum, *J. Am. Chem. Soc.*, 1981, **103**, 3597 ; (b) T. M. Che, V. W. day, L. C. Francesconi, M. F. Fredrich, W. G. Klemperer, *Inorg. Chem.*, 1985, **24**, 4055 ; (c) V. W. Day, W. G. Klemperer, dans *« Polyoxometalates : From Platonic Solids to Anti-Retroviral Activity »*, M. T. Pope, A. Müller Eds, Kluwer Academic Publishers, Dordrecht, The Netherlands, 1994 , p 87 ; (d) H. Akashi, J . Chen, T. Sakuraba, A. Yagasaki, *Polyhedron*, 1994, **13**, 1091 ; (e) T. M. Che, V. W. Day, L. C. Francesconi, M. F. Fredrich, W. G. Klemperer, D. J. Main, A. Yagasaki, O. M. Yaghi, *Inorg. Chem.*, 1992, **31**, 2920.

20. J. F. W. Keana, M. D. Ogan, *J. Am. Chem. Soc.*, 1986, **108**, 7951 ; (b) J. F. W. Keana, M. D. Ogan, Y.-X. Lü, M. Beer, J. Varkey, *J. Am. Chem. Soc.*, 1986, **108**, 7957.

21. F. Bottomley, J. Chen, *Organometallics*, 1992, **11**, 3404.

22. J. R. Harper, A. L. Rheingold, *J. Am. Chem. Soc.*, 1990, **112**, 4037.

23. M. S. Rau, C. M. Kretz, L. A. Mercando, G. L. Geoffroy, A. L. Rheingold, *Organometallics*, 1993, 12, 3447.

24. A. Proust, F. Robert, P. Gouzerh, Q. Chen, J. Zubieta, *J. Am. Chem. Soc.*, 1997, **119**, 3523.

25. R. Villanneau, A. Proust, F. Robert, P. Gouzerh, *J. Chem. Soc., Dalton Trans.*, 1999, 421.

26. R. Villanneau, A. Proust, F. Robert, P. Gouzerh, *J. Chem. Soc., Chem. Commun.*, 1998, 1491.

27. R. Villanneau, A. Proust, F. Robert, P. Veillet, P. Gouzerh, *Inorg. Chem.*, in press.

28. T. Nagata, M. Pohl, H. weiner, R. G. Finke, *Inorg. Chem.*, 1997, 36, 1366.

29. D. D. Devore, J. D. Lichtenhan, F. Takusagawa, E. A. Maatta, *J. Am. Chem. Soc.*, 1987, **109**, 7408.

30. D. C. Bradley, S. R. Hodge, J. D. Runnacles, M. Hughes, J. Mason, R. L. Richards, *J. Chem. Soc., Dalton Trans.*, 1992, 1663.

31. R. Villanneau, Pierre and Marie Curie University thesis, 1997.

32. (a) Special issue of *Chem. Rev.* devoted to polyoxometalates, 1998, **98** ; (b) D. E. Katsoulis, *Chem. Rev.*, 1998, **98**, 359.

33. E. Coronado, J. R. Galan-Mascaros, C. Gimenez-Saiz, C. J. Gomez-Garcia, S. Triki, *J. Am. Chem. Soc.*, 1998, **120**, 4671.

34. M. Clemente-Leon, E. Coronado, J.-R. Galan-Mascaros, C. Gimenez-Saiz, C. J. Gomez-Garcia, T. Fernandez-Otero, *J. Mater. Chem.*, 1998, **8**(2), 309.

35. P. Le Maguerès, L. Ouahab, S. Golhen, D. Grandjean, O. Pena, J. C. Jegaden, C. J. Gomez-Garcia, P. Delhaes, *Inorg. Chem.*, 1994, **53**, 5180.

36. H.-C. Sheu, J.-S. Shih, *Analytica Chim. Acta*, 1996, **324**, 125.

From the First Sulfurated Keggin Anion to a New Class of Compounds Based on the $[M_2O_2S_2]^{2+}$ Building Block M=Mo,W

E. CADOT, B. SALIGNAC, A. DOLBECQ AND F. SÉCHERESSE
Institut Lavoisier. IREM, UMR 8637. Université de Versailles Saint Quentin, 45 Avenue des Etats-Unis, 78035 Versailles (France)

Abstract. Different strategies of synthesis were developed to introduce sulfur atoms in a polyoxometalate framework. Every synthetic route provides new and specific sulfur-containing compounds, characterized by single crystal X-ray diffraction and multinuclear NMR spectroscopy in solution. The first investigations based on conventional routes of synthesis give predictable oxo-thio Keggin-like clusters while an original strategy, based on the acid-base self-condensation of an oxo-thio building block is the origin of a new generation of polymetalates. Finally, under hydrothermal conditions, successive replacements of sulfur atoms by oxygen atoms take place and unexpected molecular associations between fully oxygenated saturated Keggin anions were obtained.

Keyword: Polyoxothiometalates, cyclic cluster, molybdenum, ^{31}P NMR, ^{183}W NMR

1 Introduction

The synthesis of large and discrete species resulting from transition metal and chalcogenide combinations is still an exciting challenge for the inorganic chemist since this prominent class of compounds is involved in many areas of science and these compounds are often studied for their model character, especially in the field of magnetochemistry, bio-inorganic chemistry, and theoretical problems in materials [1,2]. Such systems display also potential applications in heterogeneous catalysis and in oxidation as in acidic processes [3-5]. Many studies involve the model character of polyoxometalates (POMs) to mimic the reactivity of metal oxide surface as their catalytic properties and their ability to bind covalently functional groups. Conversely, alumina-supported molybdenum and tungsten are industrially used in the hydrotreating of crude oils and the activation of dihydrogen [6,7]. An approach of the catalytic mechanisms consists to regard the molecular Mo-S associations as functional analogues of the active surface of MoS_2 [8,9]. Although a large number of thio- and oxothio-molybdates were reported in the literature [10], most of these compounds are structurally based on archetypal architectures with low nuclearity ranging from the simplest mononuclear anion, $[MoO_{4-n}S_n]^{2-}$ (with n= 1 to 4) to some tetranuclear oxothiomolybdates such as in the dimeric $[Mo_4O_4S_{14}]^{2-}$ anions [11] or in the cubane-like cluster $[Mo_4S_4(CN)_{12}]^{5-}$ [12]. The first step for the preparation of these compounds is generally the direct sulfurization which provides thioanions, precursors for more sophisticated species *via* reactions with electrophiles (H^+ or Cu^+) [13,14], and nucleophiles (cyanide or trialkyl phosphine) [15,16]. Conversely, the functionalization of POMs through the replacement of oxo ligands by functional groups (i.e nitrosyl, organometallic fragment and organic substituant) are still of current interest [17]. On this basis the introduction of sulfur atoms in a polyanionic framework is expected to

M.T. Pope and A. Müller (eds.), Polyoxometalate Chemistry, 39–53.

this basis the introduction of sulfur atoms in a polyanionic framework is expected to modify both the electronic and chemical properties. To complete this stimulating challenge, we have combined polyoxometalate and thiometalate chemistries and developed successively three types of syntheses. The first method we used to introduce sulfur in a polyoxoanionic framework was inspired from the previous work of Klemperer and co-worker [18] and is illustrated here by the synthesis of α-$[PW_{11}NbSO_{40}]^{4-}$. The thio-functionalized anion was obtained by direct reaction of the saturated oxo-parent with an adequate sulfurating agent. Another elegant way consists in the stereo-specific addition of an adapted-preformed thiofragment to polyvacant heteropolyanions. The dinuclear thio-fragment $[M_2S_2O_2]^{2+}$ (M=W, Mo) exhibits interesting properties, especially a good hydrolytic stability, coordination requirements, size and cationic character. Thus, the expected heteropolyoxothiometalates are built on well-defined structural types and exhibit molecular structures imposed by the nature of the polyvacant anion. Depending on the complementarities (geometry, symmetry and coordination requirements) between the POM and the thio-fragment, saturated or sandwich-like compounds have been obtained and characterized. Nevertheless, the compounds designed through this method have a very low sulfur content, always limited by the low nuclearity of the thio-fragment with respect to those of the lacunary oxo-precursors. So, we have now engaged a new strategy, based on the one-step self-condensation of the $[M_2S_2O_2]^{2+}$ sulfur containing building-block. The self-condensation reaction is relevant of acido-basic process and is performed by controlled addition of hydroxide ions to an aqueous solution of the building block. The polycondensation reaction can be monitored in the presence of structurating agent giving the first members of a new class of compounds directly derived from the $\{Mo_2S_2O_2\}$ building block [19].

2 Sulfur in Keggin heteropolyanions

2.1 SULFURIZATION OF PREFORMED KEGGIN UNIT

Introduction of sulfur in polyoxothioanions by direct oxygen-sulfur exchange appeared difficult because the substitution was often accompanied by reduction of the metal centers. The change of both the charge and the coordination of the metal center during the O/S exchange process appeared unfavourable to retain the polyanionic framework and led to the breaking of the POM architecture. The sensible use of the mixed α-$[PW_{11}NbO_{40}]^{4-}$ anion avoids these difficulties because the increase of the negative

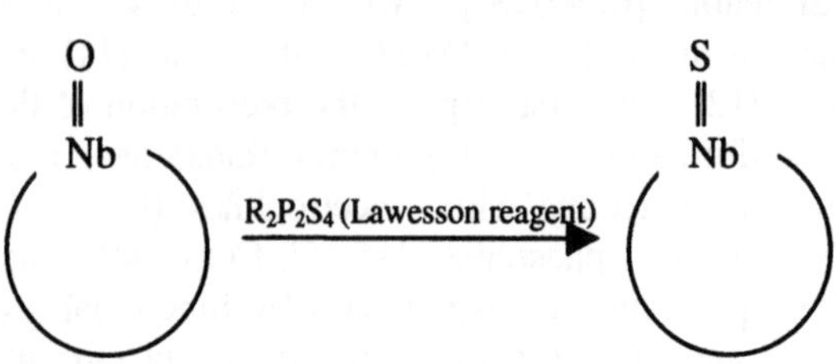

Fig. 1. Schematic representation of the O/S substitution in the oxo α-$[PW_{11}NbO_{40}]^{4-}$ parent

charge, by replacing W^{VI} by Nb^{V}, increases the resistance of the anion toward reduction, and the Nb=O group is renowned to be easy to sulfurate. The mixed α-$[PW_{11}NbO_{40}]^{4-}$ anion was reacted with variable amounts of parametoxyphenylthionophosphine sulfide (noted $R_2P_2S_4$) or hexamethyldisilathiane ($[Me_3Si]_2S$) to give the first Keggin compound with terminal sulfur linked to the Nb atom (Figure 1) [20,21]. The O/S substitution in the Keggin framework was clearly evidenced by ^{31}P NMR, IR and Raman studies.

2.2 ADDITION OF THE $[M_2S_2O_2]^{2+}$ THIOMETALLIC FRAGMENT ON A POLYVACANT POLYANION

The stereospecific addition of the preformed thiometallic cation $[M_2S_2O_2]^{2+}$ on polyvacant heteropolyxotungstates represents an excellent example of matching reactivity and geometry. The thiocation was obtained *via* selective oxidation of terminal $(S_n)^{2-}$ ligand (n=2 or 4) by iodine, in DMF solution [22], or in acidic aqueous solution according to equation (1).

$$[(S_n)M_2S_2O_2(S_n)]^{2-} + 2I_2 \rightarrow [M_2S_2O_2]^{2+} + 2n/8\ S_8 + 4I^- \qquad (1)$$

This redox reaction is remarkable for changing a nucleophile into the strong $[M_2S_2O_2]^{2+}$ electrophile. Then, the dithiocation reacts as a Lewis acid with the basic polyvacant anion to give saturated derived Keggin anions or multi-unit compounds.

Saturated Keggin Oxothio Heteropolyanion

The $[M_2S_2O_2]^{2+}$ thiofragment and the divacant γ-$[XW_{10}O_{36}]^{n-}$ isomers (X=Si or P) represent a quite perfect example of complementary geometries, reacting rapidly to give the γ-dodecametalate isomers (see Figure 2) [23,24]. X-ray diffraction structural analysis showed that the thiofragment in the polyanionic framework has retained the metal-metal bond [d(Mo^V-Mo^V)=2.832Å for γ-$[SiW_{10}Mo_2S_2O_{36}]^{6-}$ and d(W^V-W^V)=2.815Å for γ-$[SiW_{12}S_2O_{36}]^{6-}$]. The γ-oxothio isomers were characterized by ^{183}W

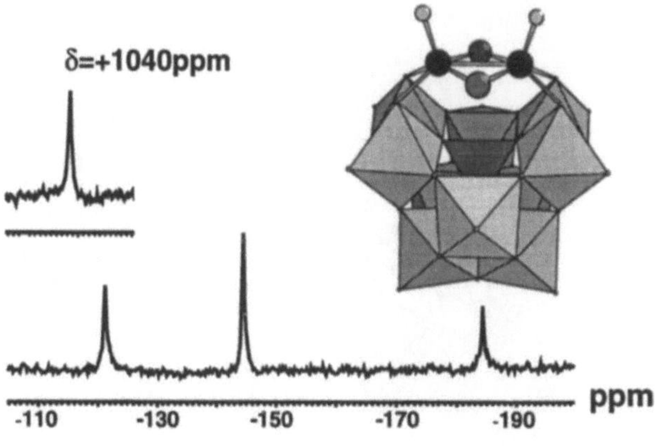

Fig. 2. Sketch of the structure of γ-$[SiW_{12}S_2O_{36}]^{6-}$ with ^{183}W NMR structural analysis: the deshielded resonance (upper part of the spectrum) was assigned to the tungsten nuclei in the $[W_2S_2O_2]$ core and the three shielded lines to the tungsten nuclei in the{SiW_{10}} core.

NMR spectroscopy and exhibit, in the –110 to –200 ppm range, three lines with 2:2:1 intensity ratio corresponding to the resonances expected for the ten W^{VI} atoms belonging to the $\{SiW_{10}\}$ subunit. An additionnal resonance is observed for the homometallic γ-$[XW_{12}S_2O_{38}]^{n-}$ compounds at +1041.2 ppm for X=Si (see Figure 2) and +1078.0 ppm for X=P. Those resonances are characteristic of the two equivalent reduced W^V atoms bridged by sulfur atoms.

The γ-$[SiW_{12}E_2O_{38}]^{6-}$ (E=S or O) is a remarkable system to study the influence of the nature of E upon the electronic delocalization in Keggin heteropolyanion. Indeed, Hervé and Tézé reported that the fully oxygenated compound γ-$[SiW_{12}O_{40}]^{6-}$ exhibits a "heteropoly-blue" behavior highlighted by the presence of the characteristic $W^V \rightarrow W^{VI}$ intervalence charge transfer transition at *ca.* 1100 nm [25]. A careful treatment of the ^{183}W NMR spectrum allowed to calculate the residence times of the two delocalized electrons by the empirical method of Baker [26]. The authors concluded that both electrons are strongly delocalized on the four adjacent tungsten atoms located at the opposite pole of the $\{W_2O_4\}$ fragment. In contrast, the oxo-thio derived γ-$[SiW_{12}S_2O_{38}]^{6-}$ is brown, showing no intervalence transition. In this case, electrons are strongly trapped on the metal centers within the $\{W_2S_2O_2\}$ core. For more informations about the metal-metal bond in those species, see the contribution of M.M. Rohmer and M. Bénard in this issue.

Sandwich-Type Oxothio Heteropolyanion

If the size, symmetry and coordination requirements of the vacancy are not adapted to those of the $[M_2S_2O_2]^{2+}$ thiocation, the direct electrophile-nucleophile addition produces sandwich-like compounds.

With Tricavacant Polyanion. The reaction with α-A-$[PW_9O_{34}]^{9-}$ gives a di-unit anion in which both the $\{PW_9\}$ subunits are bridged by three $[M_2S_2O_2(H_2O)_2]^{2+}$ fragments [27]. From ^{31}P NMR studies, it appears that only one single isomer is obtained, confirming the stereospecificity of the addition. X-ray structural determinations and ^{183}W NMR spectroscopy show that the initial C_{3v} symmetry of the precursor is lowered in the

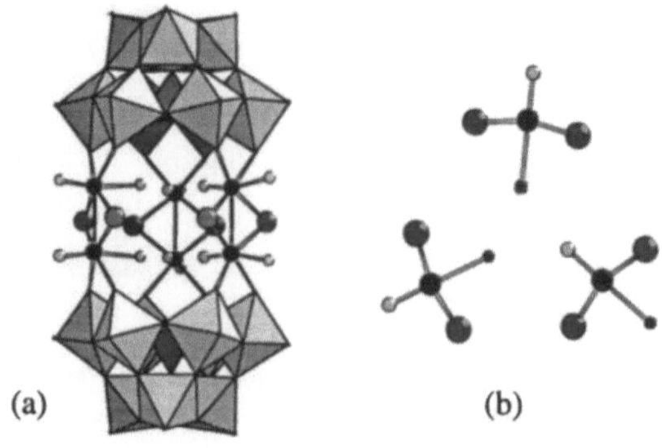

Fig. 3. (a) Polyhedral representation of the sandwich-like $[P_2W_{18}Mo_6S_6O_{74}(H_2O)_6]^{10-}$; (b) Mo plane showing the rotated $\{Mo_2O_2S_2(H_20)_2\}$ core

$[P_2W_{18}M_6S_6O_{74}(H_2O)_6]^{10-}$ adduct (see Figure 3). Because of steric constraints, one fragment $[M_2S_2O_2(H_2O)_2]^{2+}$ is rotated of 180° with respect to the other two, provoking the decrease of the symmetry to C_s. The rotated dinuclear core presents inner-directed Mo=O double bonds characterized by short Mo-O distances [1.65Å] and outer-directed Mo-OH_2 groups with long Mo-O distances [2.28 Å]. The projection of the different atoms attached to the metal centers of the three thio-fragments is shown in Figure 3b.

With a Monovacant Polyanion. The addition of $[Mo_2S_2O_2]^{2+}$ on the monovacant α-$[PW_{11}O_{39}]^{7-}$ leads to two sandwich-like isomers, each isomer being identified by a single ^{31}P resonance, the two resulting lines being separated by less than 0.05 ppm. One isomer (noted Δ_1) can be isolated with 100% purity through selective precipitation or

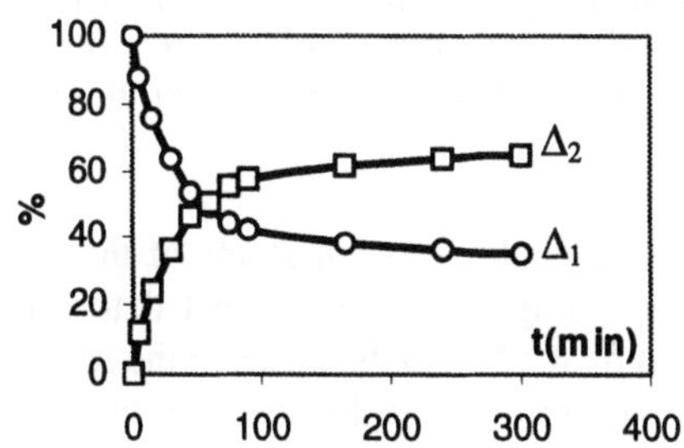

Fig. 4. ^{31}P NMR at 60°C kinetic study of Δ_1 isomerism into Δ_2

crystallization. A kinetic study at 60°C checked by ^{31}P NMR confirms that Δ_1 isomerizes into the second isomer (noted Δ_2) until reaching the thermodynamic equilibrium with molar ratio Δ_1/Δ_2=33/66 (Figure 4). Δ_1 was characterized by single-crystal X-ray diffraction as a potassium salt (see Figure 5). Two $[PW_{11}O_{39}]^{7-}$ subunits are bridged through an unusual tetrameric core resulting from the fusion of two $\{Mo_2S_2O_2\}$ building-blocks. The two Mo-dinuclear units are crystallographically equivalent, related through an inversion center. According to the charge of the cluster, determined by elemental analysis (ten potassium ions per polyanion), four protons must

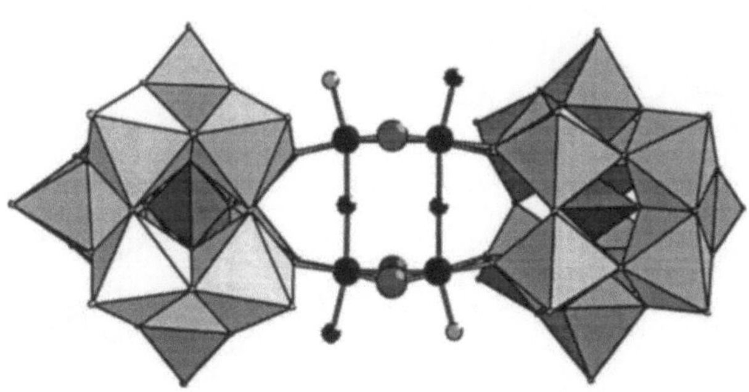

Fig. 5. Polyhedral representation of $[(PW_{11}O_{39})_2(Mo_2S_2(OH)_2O)_2]^{10-}$: the four protons on the tetrameric $\{Mo_4\ S_4O_6\}$ central core are located on oxygen atoms represented as black little spheres.

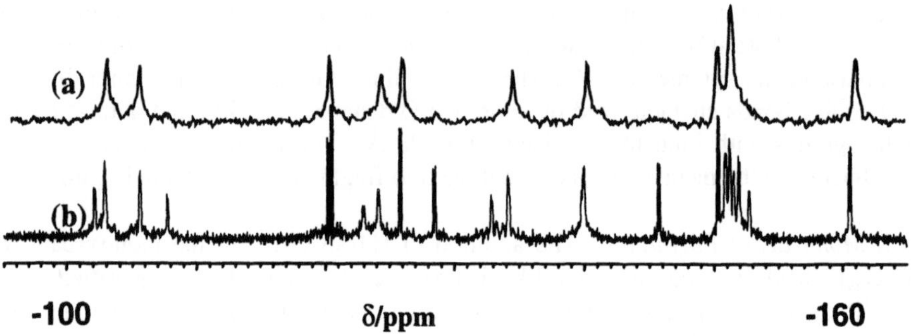

Fig. 6. ^{183}W NMR of $[(PW_{11}O_{39})_2(Mo_2S_2(OH)_2O)_2]^{10-}$: (a) after 1h, the spectrum consists in 11 lines relative to the presence of Δ_1 isomer ; (b) after 25 h, the 22 lines spectrum reveals the presence of the second Δ_2 isomer.

be re-distributed of on the six available oxygen atoms of the central tetrameric cluster $\{Mo_4S_4O_6\}$. The two dinuclear fragments are connected together through two linear hydroxo-bridges as confirmed by the Mo-O distances in the Mo-OH-Mo bridges [1.966-1.938Å]. The two remaining protons are located on two terminal oxygen atoms respectively, [d(Mo-OH)=1.877Å]. In agreement with a strong trans effect, the two remaining terminal Mo-O bonds appear substantially lenghtened [1.808Å]. ^{31}P and ^{183}W NMR studies confirm that both the $[PW_{11}O_{39}]^{7-}$ subunits are equivalent but present a trivial local symmetry because the eleven tungsten atoms in the $\{PW_{11}O_{39}\}$ subunits are unequivalent (Figure 6). The structure of the second isomer can be easily deduced from that of Δ_1 through a 180° rotation of one $[PW_{11}O_{39}]^{7-}$ subunits with respect to the other. Such an assumption is in agreement with the ^{31}P single line and the eleven ^{183}W lines assigned to the Δ_2 isomer. Two enantiomers (noted r-Δ_2 and s-Δ_2) can be deduced from Δ_2 while only a single diastereoisomer is expected for Δ_1. A schematic representation of Δ_1 and Δ_2 isomers is given in Figure 7. The two $\{PW_{11}\}$ subunits can be considered as

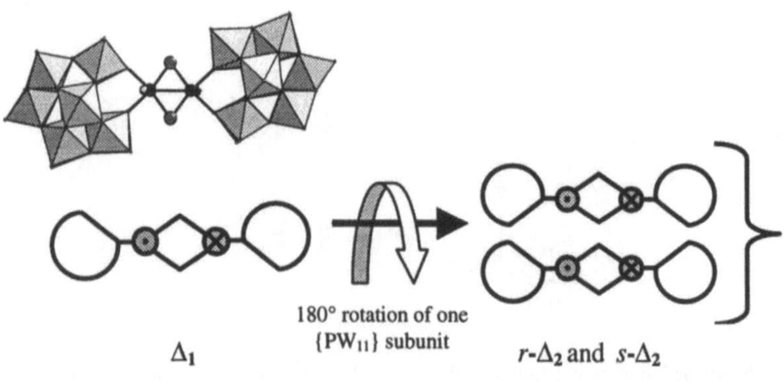

Fig. 7. Schematic top of view of the Δ_1 and Δ_2 isomers : the two possibilities for the rotation of one $\{PW_{11}\}$ in Δ_1 isomer generate the two s-Δ_2 and r-Δ_2 enantiomers.

independent and then, the proportions of the three isomers in equilibria are equal. This assumption agrees with the ^{31}P NMR results since the proportion of Δ_2, about 66%, corresponds to the racemic ratio between the two enantiomeric forms (33% for r-Δ_2 and 33% for s-Δ_2) and 33% for the Δ_1 isomer.

3 Self-condensation of $[Mo_2O_2S_2]^{2+}$

The acidification of basic monomeric oxoanions (WO_4^{2-}, MoO_4^{2-} and VO_3^-) represents a general process for the synthesis of POMs [28]. The polycondensation can be monitored in the presence of a structurating agent or "template" acting as an assembling group, as PO_4^{3-}, SiO_3^{2-} for the commonest. The directed aggregation process is then achieved by the formation of the so-called Keggin or Dawson anions. In fact, we have chosen to adopt a comparable approach usually developed for the POMs synthesis, and demonstrated that the self-condensation of the $[Mo_2S_2O_2]^{2+}$ building block originates a new generation of neutral and anionic molecules. Indeed, the aggregation process appeared to be highly sensitive to the presence of anionic structurating agents.

3.1 THE BEGINING OF THE SERIES

Titration of aqueous mixtures of $[Mo_2S_2O_2]^{2+}$ and KI by solution of potassium hydroxide until *pH* 2.5-3 yielded quantitatively a yellow microcrystalline product which presents a Mo:I ratio = 4:1. After re-crystallization in water, this solid afforded yellow crystals of iodide-free $[Mo_{12}S_{12}O_{12}(OH)_{12}(H_2O)_6]$, namely the dodecameric neutral "wheel"[29]. A striking feature of this structure is the cyclic arrangement of the neutral Mo_{12}-ring shaped cluster with a central cavity of 11Å in diameter (see Figure 8). Six $\{Mo_2S_2O_2\}$ building blocks are connected through hydroxo double-bridges and the

Fig. 8. Polyhedral representation of the neutral ring-shaped $[Mo_{12}S_{12}O_{12}(OH)_{12}(H_2O)_6]$ molecule

coordination of the Mo centers is achieved by six water molecules lining symmetrically the cavity. The lability of the six inner aquo ligands supported by the cationic character of the open cavity due to the twelve Mo^{5+} centers produces a striking host-guest reactivity of the cyclic cluster toward anions. A featuring behavior of the cluster is the reversibility of the self-condensation process : acidification of the dodecameric ring gives back the starting material $[Mo_2S_2O_2]^{2+}$.

3.2 VERSATILITY OF THE RING-SHAPED ARCHITECTURE

The re-crystallization of the former crude yellow microcrystalline solid in DMF containing tetrabutylammonium iodide led to well-shaped crystals. The single-crystal X-ray analysis reveals a decameric ring-shaped architecture $[Mo_{10}S_{10}O_{10}(OH)_{10}(H_2O)_{10}]$ which consists of five $\{Mo_2S_2O_2\}$ units connected to each other by hydroxo double bridges to form a cyclic neutral molecule [30]. Five inner water molecules complete the inter-block connections, lining the open cavity of the ring. The solid state structure exhibits a remarkable supramolecular arrangement involving two iodide ions symmetrically located on both sides of the mean plane defined by the ten Mo atoms (Figure 9). The distance between the two iodide ions and the oxygen atoms of the five inner water molecules are short enough to suggest that the stability of the bis-halide complex is ensured by hydrogen bonds. The five $[I^{-}\cdots H\text{-}O\text{-}H\cdots I^{-}]$ interactions induce a pronounced shortening of the I•••I distances [4.783Å] rather close to the sum of the ionic radii [4.40 Å]. The space filling representation shown in Figure 10 gives a realistic view of the supramolecular close-packing arrangement. Such promising results demonstrate that supramolecular chemistry of anions can be considered in those ring-shaped clusters.

Fig. 9. Molecular structure of $[I_2Mo_{10}S_{10}O_{10}(H_2O)_5]^{2-}$: ball and stick model showing the 10-membered ring with the central iodide anions (black spheres : Mo, light grey spheres : S, dark grey spheres : O, light grey central sphere : I)

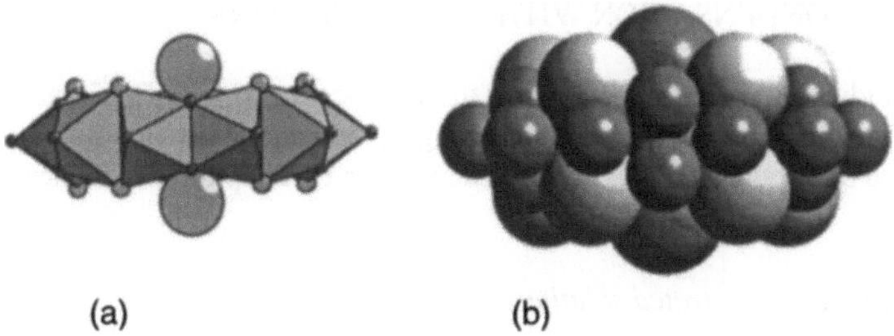

Fig. 10. Side-view of $[I_2Mo_{10}S_{10}O_{10}(OH)_{10}(H_2O)_5]^{2-}$ showing the two interacting iodide anions. (a) polyhedral representation ; (b) space filling sketch

3.3 SELF-CONDENSATION WITH METALATE OR OXALATE

In the presence of stoichiometric amount of $[MoO_4]^{2-}$ or oxalate ions $[C_2O_4]^{2-}$, an octameric ring, $\{Mo_8S_8O_8(OH)_8\}$ encapsulating the structurating anion was isolated [31,32]. The central $C_2O_4^{2-}$ anion is plane while Mo^{VI} is in a distorted octahedron, $\{MoO_3(OH)_2(H_2O)\}^{2-}$. The two anions $[Mo_8S_8O_8(OH)_8(C_2O_4)]^{2-}$ and $[Mo_8S_8O_8(OH)_8(MoO_3(OH)_2(H_2O))]^{2-}$ represented in Figure 11 confirm the nuclearity of the molecular ring is not restricted and can be monitored by templating process. In addition, the possibility to combine high oxidation state oxometalates with the thio-fragment in mixed-valence compounds enlarges the field of the investigations.

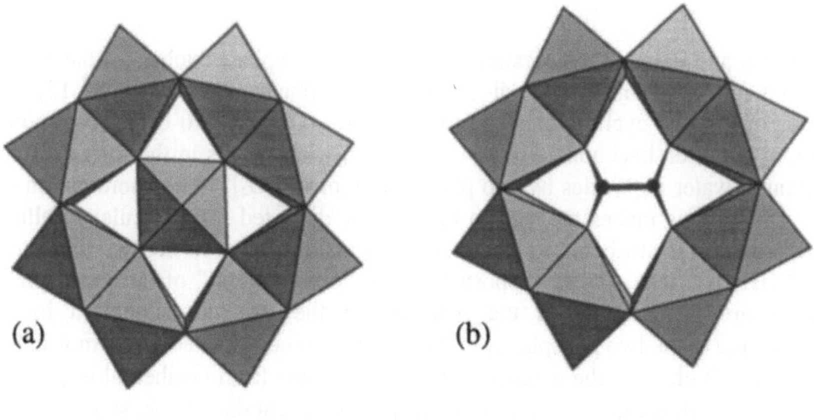

Fig. 11. Polyhedral representations of eight-membered ring (a) with encapsulated metalates $[Mo_8MS_8O_{10}(OH)_{10}(H_2O)]^{2-}$ M=Mo, W ; (b) with encapsulated oxalate $[Mo_8S_8O_8(OH)_8(C_2O_4)]^{2-}$

3.4 SELF-CONDENSATION WITH PHOSPHATE IONS

^{31}P NMR studies of solutions containing the dithiocation $[Mo_2S_2O_2]^{2+}$ and phosphates have revealed that three phosphate-containing compounds exist in solution in the 2-7 *pH* range, the distribution of these species depending on the phosphate concentration.

In low phosphate concentrated solutions

For low concentrations in phosphate (<0.1 mol.L^{-1}), two phosphate-containing species are present in equilibrium in solution. The two compounds were isolated and their structures solved by X-ray diffraction methods [33]. One corresponds to the Mo_{12}-diphosphate ring $[(HPO_4)_2Mo_{12}S_{12}O_{12}(OH)_{12}(H_2O)_2]^{4-}$ (noted P_2Mo_{12}) and the other to the Mo_{10}-monophosphate ring $[(H_2PO_4)Mo_{10}S_{10}O_{10}(OH_2)_2(OH)_{11}]^{2-}$ (noted PMo_{10}), represented in Figure 12. A complete ^{31}P NMR study including variable concentrations

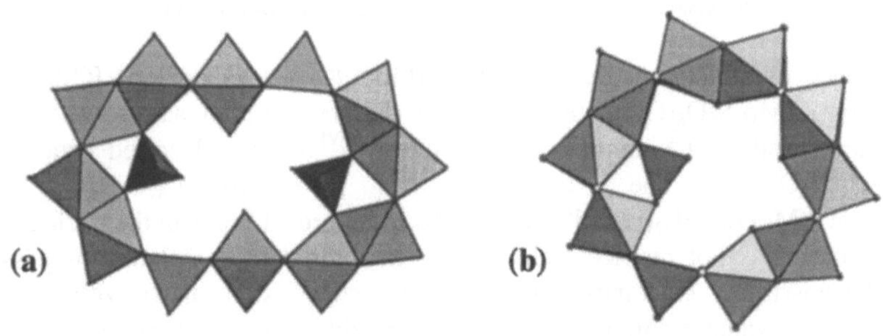

Fig. 12. Polyhedral representations of (a) diphosphato Mo_{12}-ring (P_2Mo_{12}) and (b) monophosphato Mo_{10}-ring (PMo_{10})

in phosphate and variable temperature and arsenate-phosphate exchanges have allowed to assign the ^{31}P resonances to the di- and mono-phosphate rings (see Figure 13). In the two compounds, the cyclic architecture $[Mo_{2n}S_{2n}O_{2n}(OH)_{2n}]$, n=6 or 5, corresponds to the neutral common backbone. For n = 6, the Mo_{12}-ring results in the formal exchange of four inner water molecules by two phosphate ions $[HPO_4]^{2-}$. The coordination of the two phosphate ions makes the ring to be strongly distorted from circular to elliptical. This deformation is attributed to electrostatic repulsions between the two diametrically opposed $[HPO_4]^{2-}$ ions, and is supported by the "pincer effect" of the inner chelating phosphate groups. Because of the flatening of the structure supported by steric constraints due to the two phosphates in the cavity, two coordinated water molecules are displaced which changes the geometry of two Mo atoms from octahedral to pyramidal. Such a behavior illustrates the great flexibility of the cyclic-backbone, supported by the versatility of the Mo-coordination which can adopt octahedral or pyramidal geometries.

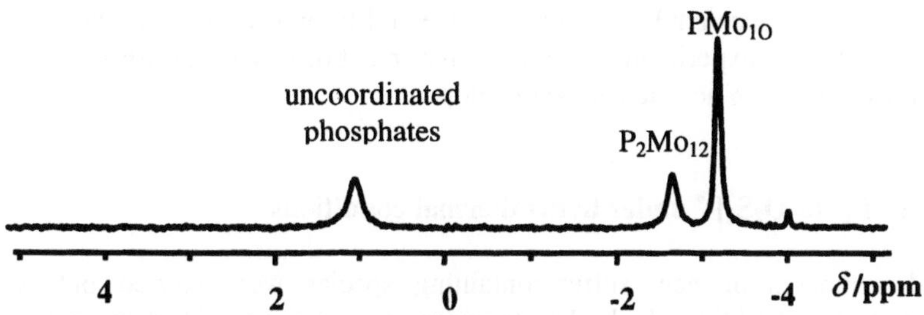

Fig. 13 : ^{31}P NMR characterization of $[(HPO_4)_2Mo_{12}S_{12}O_{12}(OH)_{12}(H_2O)_2]^{4-}$ in solution : the three resonances confirms the existence of an equilibrium between phosphate-containig species

In high phosphate concentrated solutions

For concentrated phosphate solutions (in the 0.5-1 $mol.L^{-1}$ range), the condensation of $[Mo_2O_2S_2]^{2+}$ leads exclusively to a single compound, namely $[(HPO_4)_4Mo_6S_6O_6(OH)_3]^{5-}$ [34]. The molecular structure of this anion, shown in Figure 14 reveals an arrangement similar to that found for the fully oxo analogue $[(HPO_4)_4Mo_6O_{12}(OH)_3]^{5-}$, widely described in the literature [35-37]. Three equivalent $\{Mo_2S_2O_2\}$ units are mutually connected by a peripheral phosphato ligand and by a hydroxo group. The six molybdenum atoms are coplanar and display alternating short Mo-Mo lengths (2.8Å) to

Fig. 14. Polyhedral representation of $[(HPO_4)_4Mo_6S_6O_6(OH)_3]^{5-}$

ensure bonding contact within the building blocks and longer (3.2 Å) to span interblocks connections. Such an arrangement differs from those observed in the previous cyclic wheels since the connections between the building blocks are edge-sharing and not face-sharing. Thus, the $[(HPO_4)_4Mo_6S_6O_6(OH)_3]^{5-}$ anion can be viewed as a hexavacant anion, derived from the ε or δ isomers, according to the Baker-Figgis nomenclature. Study of the $[(HPO_4)_4Mo_6S_6O_6(OH)_3]^{5-}$ anion by ^{31}P NMR in solution reveals that the three peripheral phosphato groups are very labile and can be easily exchanged for acetate or arsenate ions. The controlled substitution of peripheral phosphate by specific groups could be a rational method to prepare new functionalized polyanions. Preliminary experiments confirm that reactions with dicarboxylate or phosphonate ions in place of acetate can be developed.

4 Behavior of $[Mo_2O_2S_2]^{2+}$ under hydrothermal conditions

Hydrothermal syntheses of new sulfur-containing species were carried out with $\{Mo_2O_2S_2\}$ as starting building block. The temperature and initial pH of the medium have a crucial influence on the final product of the reaction. Indeed, above *pH*=5 and *T*> 50°C, a partial replacement of oxygen by sulfur atoms is observed. Such a substitution was evidenced by ^{31}P NMR measurements from solutions containing $[(HPO_4)_4Mo_6S_6O_6(OH)_3]^{5-}$. For *T*> 150°C, a complete degradation of the structure leads to fully oxygenated Keggin derivatives.

4.1 REGIOSELECTIVE O/S SUBSTITUTION IN $[(HPO_4)_4Mo_6S_6O_6(OH)_3]^{5-}$

On heating at T = 110°C and at *pH*=5, a solution of $[(HPO_4)_4Mo_6S_6O_6(OH)_3]^{5-}$ (noted P_4Mo_6) led to the half-substituted $[(HPO_4)_4Mo_6S_3O_9(OH)_3]^{5-}$ anion ($P_4Mo_6S_3O_3$), fully

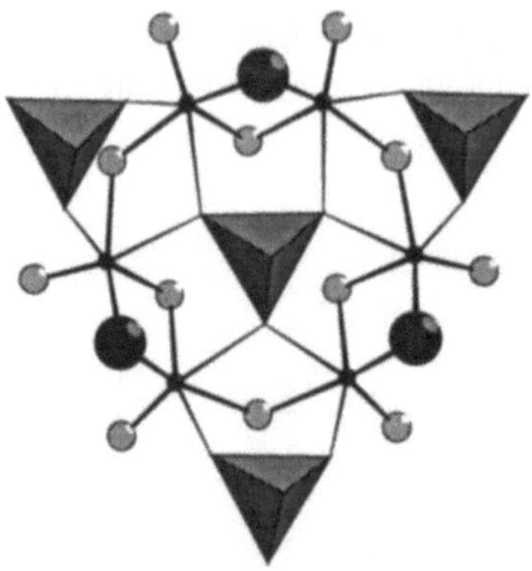

Fig. 15. Representation of the partially desulfurized anion $[(HPO_4)_4Mo_6S_3O_9(OH)_3]^{5-}$: oxygen atom in light grey and sulfur atoms in dark grey,

characterized by X-ray diffraction [38]. The molecular structure shown in Figure 13 closely derives from that of the parent $[(HPO_4)_4Mo_6S_6O_6(OH)_3]^{5-}$ anion. A striking feature of the arrangement is the distribution of the sulfur atoms which lye on the same side of the plane defined by the six Mo atoms, at the opposite side of the four phosphate groups. The O/S substitution takes place at lower temperature (below 100°C) and the reaction was followed by ^{31}P NMR experiments. The O/S substitution is progressive, bridging sulfur atoms being replaced by oxygen atoms for giving $P_4Mo_6S_5O$, $P_4Mo_6S_4O_2$ and $P_4Mo_6S_3O_3$, successively. Furthermore, each substitution step is characterized by only one positional isomer, illustrating the regioselectivity of the reaction which is probably related to the presence of the three peripheral phosphates acting as good protecting groups against the nucleophilic attack of hydroxide ions. The

Compound	$P_4Mo_6S_6$	$P_4Mo_6S_5O$	$P_4Mo_6S_4O_2$	$P_4Mo_6S_3O_3$

Symmetry	C_{3v}	C_s	C_s	C_{3v}
δ/ppm(intensity ratio)				
Central phosphate	5.05(1)	4.15(1)	3.20(1)	1.80(1)
Peripheral phosphate	2.50(3)	2.80(2) 2.75(1)	3.05(2) 3.00(1)	3.40(3)

Fig. 16. ^{31}P NMR data of the O/S substitution in $[(HPO_4)_4Mo_6S_6O_6(OH)_3]^{5-}$

^{31}P NMR characterization of the O/S substituted compounds are summarized in Figure 16.

4.2 UNEXPECTED KEGGIN DERIVATIVES

Under more drastic conditions, the fully reduced oxothio precursor is partially oxidized and decomposes into sulfur-free species. Three fully oxygenated mixed-valence Keggin

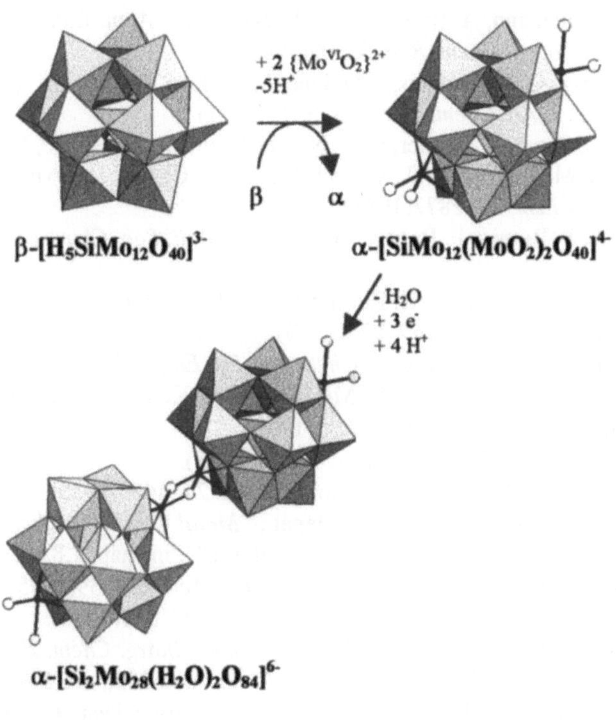

Fig. 17. Schematic stepwise growth process from the reduced β-$[H_5SiMo_{12}O_{40}]^{3-}$

anions (Figure 17) have been synthesized from hydrothermal reactions involving the oxothioprecursor $\{Mo_2S_2O_2\}$, Na_2SiO_3, HCl and $[N(CH_3)_4OH]$. The geometries of these anions derive from the well-known Keggin anion $[SiMo_{12}O_{40}]^{4-}$. In these compounds, the nucleophilic character of the bridging oxygen atoms is enhanced by reduction and the resulting charge increase on the anion is balanced either by protons or by two electrophilic $\{MoO_2\}^{2+}$ groups, depending on the conditions of synthesis (*pH, T°C*). The first case leads to β–$[H_5SiMo^{V}_4Mo^{VI}_8O_{40}]^{3-}$, while the second is encountered in the bicapped α-$[SiMo^{V}_4Mo^{VI}_{10}O_{44}]^{4-}$. At T=180°C and pH=1.5, the bi-capped Keggin unit dimerizes to form the new α-$[Si_2Mo^{V}_{14}Mo^{VI}_{14}O_{84}(H_2O)_2]^{4-}$ [39].

References

1. D. Katsoulis : *Chem. Rev.* **98**, 359 (1998).
2. A. Müller, F. Peters, M. T. Pope and D. Gatteschi : *Chem. Rev.* **98**, 239 (1998).
3. T. Okuhara, N. Mizuno and M. Misono : *Adv. In Catal.* **41**, 113(1996).
4. C. Marchal-Roch, R. Bayer, J.F. Moisan, A. Tézé and G. Hervé : *Topic in Catalysis* **3**, 407(1996).
5. R. Bayer, C. Roch-Marchal, F. X. Liu, A. Tézé and G. Hervé : *J. Mol. Catal.* A **114**, 277(1996).
6. H. Topsøe, B. S. Clausen : *Catal. Rev.-Sci. Eng.* **26**, 395(1984).
7. R. Prins, V. H. J. DeBeer, G. A. Somorjai : *Catal. Rev.-Sci. Eng.* **31**, 1(1989).
8. M. Rakowski Dubois : *Chem. Rev.* **89**, 1(1989).
9. A. Müller, E. Diemann, A. Branding, F. W. Baumann : *Appl. Catal.* **62**, L13(1990).
10. T. Shibahara : *Coord. Chem. Rev.* **5**, 203(1993).
11. D. Coucouvanis, A. Toupadakis, J. D. lane, S. M. Koo, C. G. Kim, A. Hadjikyriakou : *J. Am.Chem. Soc.* **113**, 5271(1991).
12. A. Müller, R. Jostes, W. Elztner, C. S. Nie, E. Diemann, H. Bögge, M. Zimmermann, M. Dartmann, U. Reinsch-Vogell, S. Che, S. J. Cyvin and B. N. Cyvin : *Inorg. Chem.* **24**, 2872(1985).
13. F. Sécheresse, S. Bernès, F. Robert , Y. Jeannin : *Bull. Soc. Chim. Fr.* **132**, 1029(1995).
14. K. Hegetschweiler, T. Keller, H. Zimmermann, W. Schneider, H. Schmalle and E. Dubler : *Inorg. Chim. Acta*, **169**, 235(1990).
15. A. Müller and U. Reinsch : *Angew. Chem. Int. Ed. Engl.* **19**, 72(1980).
16. V. P. Fedin, M. N. Sokolov, Yu. V. Mironov, B. A. Kolesov, S. V. Tkachev and V. Ye. Fedorov : *Inorg. Chim. Acta* **167**,39(1990).
17. P. Gouzerh and A. Proust : *Chem. Rev.* **98**, 77(1998).
18. W. G. Klemperer and C. Schwartz : Inorg. Chem. 24, 4459(1985).
19. F. Sécheresse, E. Cadot, C. Simmonet-Jegat : *"Metal Cluster In Chemistry"* P. Braunstein, L. A. Oro, P. R. Raithby Eds, Wiley-CH, in chap. 1.8, 123 (1999).
20. E. Cadot, V. Béreau, F. Sécheresse : *Inorg. Chim. Acta* **239**, 39(1995).
21. E. Radkov, Y. J. Lu and R. H. Beer : *Inorg. Chem.* **35**, 551(1996).
22. D. Coucouvanis, A. Toupadakis, A. Hadjikyriakou : *Inorg. Chem.* **27**, 3273(1988).
23. E. Cadot, V. Béreau, F. Sécheresse : *Inorg. Chim. Acta* **252**, 101(1996).
24. E. Cadot, V. Béreau, S. Halut and F. Sécheresse : *Inorg. Chem.* **95**, 551(1996).

25. A. Tézé, J. Canny, L. Gurban, R. Thouvenot and G. Hervé : *Inorg. Chem.* **35**, 1001(1996).
26. M. Kozik and L. C. W. Baker: *"Polyoxometalates: From Platonic Solids to Antiretroviral Activity"* Pope, M. T. and Müller, A. Eds, Kluwer Acad. Pub. : Dordrecht, The Netherlands, p.191(1994).
27. V. Béreau, E. Cadot, A. Bögge, A. Müller and F. Sécheresse : *Inorg. Chem* (in press).
28. M. T. Pope : *"Hetero and Isopoly Oxometalates"* Springer-Verlag: New-york (1983).
29. E. Cadot, B. Salignac, S. Halut and F. Sécheresse : *Angew. Chem. Int. Ed.* **37**, 611(1998).
30. E. Cadot, B. Salignac, J. Marrot, A. Dolbecq and F. Sécheresse : *Chem. Commun.* (Submitted).
31. A. Dolbecq, E. Cadot, F. Sécheresse : *Chem. Commun.* 2293(1998).
32. A. Dolbecq, B. Salignac, E. Cadot and F. Sécheresse : *Bull. Pol. Acad. Sci.* **46**, 237(1998).
33. E. Cadot, B. Salignac, T. Loiseau, A. Dolbecq, and F. Sécheresse : *Chem. Eur. J.* (in press).
34. E. Cadot, , B. Salignac, A. Dolbecq and F. Sécheresse : *Chem. Eur. J.* **5**, 2396(1999).
35. R. C. Haushalter, F. W. Lai : *Angew. Chem. Int. Ed.* **28**, 743(1989).
36. A. Guesdon, M. M. Borel, A. Leclaire, B. Raveau : *Chem. Eur. J.* **3**, 1797(1997).
37. M. I. Khan, Q. Chen, J. Zubieta : *Inorg. Chim. Acta* **235**, 135(1995).
38. A. Dolbecq, D. Eisner, E. Cadot and F. Sécheresse : *Inorg. Chim. Acta* (in press).
39. A. Dolbecq, E. Cadot, D. Eisner and F. Sécheresse *: Inorg. Chem.* **38**, 4217(1999).

Organometallic Oxometal Clusters

A. PROUST, R. VILLANNEAU, R. DELMONT, V. ARTERO AND P. GOUZERH
Laboratoire de Chimie Inorganique et Matériaux Moléculaires, Unité CNRS 7071, Case 42, Université Pierre et Marie Curie, 4 Place Jussieu, 75252 Paris Cedex 05, France

Abstract. The presentation focusses on integrated oxometal clusters containing *fac*-$\{M(CO)_3\}^+$ (M = Mn, Re), $\{Rh(\eta^5\text{-}C_5Me_5)\}^{2+}$ or $\{Ru(\eta^6\text{-arene})\}^{2+}$ organometallic subunits and oxo(alkoxo)molybdenum or tungsten subunits. The discussion adresses the following questions: i) structural relationships, ii) structural preferences, and iii) stereochemical non-rigidity. The molecular structures of the organometallic oxometal clusters are discussed in connection with those of previously reported polyoxo(alkoxo)metalates and organometallic clusters. The apparent structural relationships within these clusters underscore the electronic connection between d^6-*fac*-$\{ML_3\}$ (M = Mn, Re, Ru, Rh) and d^0-*fac*-$\{MO_3\}$ or d^0-*fac*-$\{MO_2(OR)\}^+$ (M = Mo, W) units.

Key words: Organometallic oxides, oxometal clusters, polyoxometalates, molybdenum, tungsten, manganese, rhenium, rhodium, ruthenium.

1. Introduction

Since the initial reports of the synthesis of $[PW_{11}O_{39}\{Ti(\eta^5\text{-}C_5H_5)\}]^{4-}$ 20 years ago [1,2], the field of organometallic derivatives of polyoxometalates has expanded significantly, mainly due to incisive contributions from the groups of Klemperer [1,3] and Finke [4], and these derivatives now form a full class of compounds [5]. Significant contributions to this field have also been provided by the groups of Isobe [6] and Siedle [7]. Organometallic derivatives of polyoxometalates are divided into polyoxometalate-incorporated organometallic complexes, *i.e.* integrated clusters, polyoxometalate-supported organometallic complexes [4b], and organometallic cation salts of polyoxometalates [7]. Polyoxometalate-supported organometallic complexes provide discrete analogs of solid-oxide-supported organometallic complexes [3,4]. In this respect, one of the most significant results is the evidence for the relationship of $[(OC)_2M(P_2W_{15}Nb_3O_{62})]^{8-}$ (M = Rh, Ir) to solid oxide-supported $(OC)_2M^+\cdot Al_2O_3$ [8]. In a general way, organometallic derivatives of polyoxometalates are attracting interest as potential polyoxometalate-based catalysts or precatalysts. Representative examples include the one-pot hydroformylation of olefins and subsequent oxidation of aldehydes to carboxylic acids using $[(Ph_3P)_2Rh(CO)]_4[SiW_{12}O_{40}]$ as a bifunctional catalyst [9], the oxydation of cyclohexene with oxygen and $[(1,5\text{-COD})Ir(P_2W_{15}Nb_3O_{62})]^{8-}$ [4b], and the catalytic activity of CO-photoreduced cubane-type molybdenum oxide clusters such as $[(Cp^*Rh)_4Mo_4O_{16}]$ and $[(Cp^*Rh)_2Mo_3O_9(OMe)_4]$ in propene metathesis reaction [10].

M.T. Pope and A. Müller (eds.), Polyoxometalate Chemistry, 55–67.

Integrated complexes have been obtained in several ways including the incorporation of an organometallic moiety into a lacunary polyoxometalate [1], the oxidation of the carbonyl dimers [{Cp*M(CO)$_2$}$_2$] (M = Mo [11a] or W [11b]), and various aggregation processes triggered by protons and/or Lewis acids in aqueous or non-aqueous media [6]. In the course of our study of the coordination chemistry of the defect Lindqvist-type species [Mo_5O_{13}(OMe)$_4$(NO)]$^{3-}$ [12] with organometallic complexes, we have observed that this polyoxoanion eventually liberates oxometal fragments that subsequently combine with organometallic fragments to give novel species. This led us to investigate the reactivity of organometallic complexes, *e.g.* [MBr(CO)$_5$] (M= Mn, Re), [{Cp*RhCl$_2$}$_2$] and [{(η^6-arene)RuCl$_2$}$_2$], towards simple oxometalates, *e.g.* (*n*-Bu$_4$N)$_2$[Mo_2O_7] and (*n*-Bu$_4$N)$_2$[WO_4], in non-aqueous media. Several organometallic oxide clusters containing *fac*-{M(CO)$_3$}$^+$ (M = Mn, Re) or {Ru(η^6-arene)}$^{2+}$ units have been obtained in this way. Their molecular structures will be discussed in connection with those of previously reported polyoxo(alkoxo)metalates and organometallic clusters. Oxomolybdenum clusters containing {RhCp*}$^{2+}$ units will also be considered. The structural relationships within these clusters underscore the electronic connection between d^6-*fac*-{ML$_3$} (M = Mn, Re, Ru, Rh) and d^0-*fac*-{MO$_3$} or d^0-*fac*-{MO$_2$(OR)}$^+$ (M = Mo, W) units.

2. Complexes with *fac*-{M(CO)$_3$}$^+$ units (M = Mn, Re)

Tricarbonylmanganese(I) complexes and the rhenium analogues have played a key role in the development of organometallic chemistry of polyoxometalates. Indeed the complexes [(OC)$_3$M(*cis*-$Nb_2O_4O_{19}$)]$^{3-}$ (M = M, Re) were the very first polyoxometalate-supported organometallic complexes to be reported [13]. The complexes [(OC)$_3$Mn(Mo_5O_{18}TiCp)]$^{2-}$ [14] and [(OC)$_3$Re($P_2W_{15}Nb_3O_{62}$)]$^{8-}$ [8] have been reported subsequently. In all of these complexes the metal tricarbonyl unit is bonded to a triangle of three bridging oxygen atoms. A different binding mode is observed in the complexes [M(CO)$_3$(H_2O){Mo_5O_{13}(OMe)$_4$(NO)}]$^{2-}$ (M = Mn, Re) where the metal tricarbonyl unit is bonded to two terminal oxygen atoms of the lacunary oxo-nitrosyl support and to a molecule of water [15]. Nevertheless these species are properly described as polyoxomolybdate-supported complexes. Thus, the compounds (*n*-Bu$_4$N)[Mo_2O_5(OMe)$_5${M(CO)$_3$}$_2$] (M = Mn, **1a**; M = Re, **1b**), [Mo_2O_4(OMe)$_6${Mn(CO)$_3$}$_2$] (**2**), (*n*-Bu$_4$N)$_2$[Mo_2O_6(OMe)$_4${Re(CO)$_3$}$_2$] (**3**) and the trisalkoxo derivatives [Mo_2O_4{RC(CH_2O)$_3$}$_2${Mn(CO)$_3$}$_2$] (R = Me, **4a**; R = CH_2OH, **4b**), (*n*-Bu$_4$N)[Mo_2O_4{MeC(CH_2O)$_3$}$_2${Mn(CO)$_3$}] (**5**), and (*n*-Bu$_4$N)$_2$[Mo_6O_{16}(OMe)$_2${MeC(CH_2O)$_3$}$_2${Mn(CO)$_3$}$_2$] (**6**), which have been obtained by reaction of [MBr(CO)$_5$] or [M(CO)$_3$(MeCN)$_3$](BF$_4$) (M = Mn, Re) with (*n*-Bu$_4$N)$_2$[Mo_2O_7] in MeOH, appear to provide the first examples of integrated complexes containing {M(CO)$_3$}$^+$ units [16]. In figures 1 to 7, the Mo atoms are shown as hatched spheres while the Re and Mn atoms are shown as dotted spheres.

2.1. TETRANUCLEAR CLUSTERS. RHOMB-LIKE *vs.* CUBANE-TYPE CLUSTERS

Compounds **1a** and **1b** are isostructural. The clusters $[Mo_2O_5(OMe)_5\{M(CO)_3\}_2]^-$ display a tetrahedral arrangement of the metal atoms (Figure 1). They formally derive from the clusters $[\{M(CO)_3(\mu_3\text{-}OH)\}_4]$ (M = Mn [17a] or Re [17b]) by substitution of $\{MoO_2(OMe)\}^+$ units for two $\{M(CO)_3\}^+$ units and one μ_3-oxo ligand and three μ_3-methoxo ligand for the μ_3-hydroxo ligands.

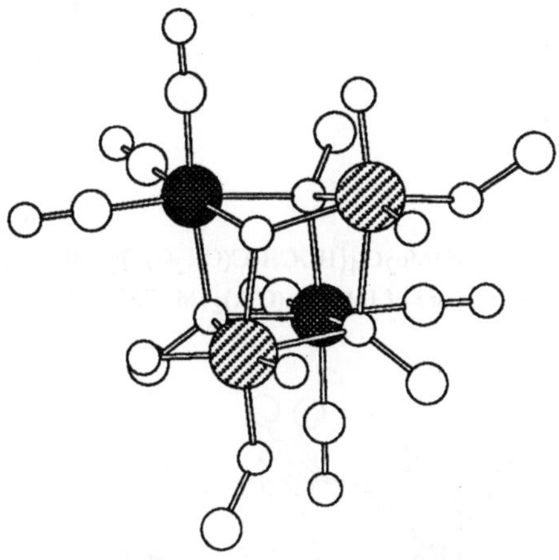

Fig. 1. Molecular structure of $[Mo_2O_5(OMe)_5\{Mn(CO)_3\}_2]^-$ in **1a** [16]

Although their composition is closely related to that of the anion of **1**, clusters **2** and **4b**, and the anion of **3** all adopt a rhomb-like structure based on four edge-sharing octahedra.

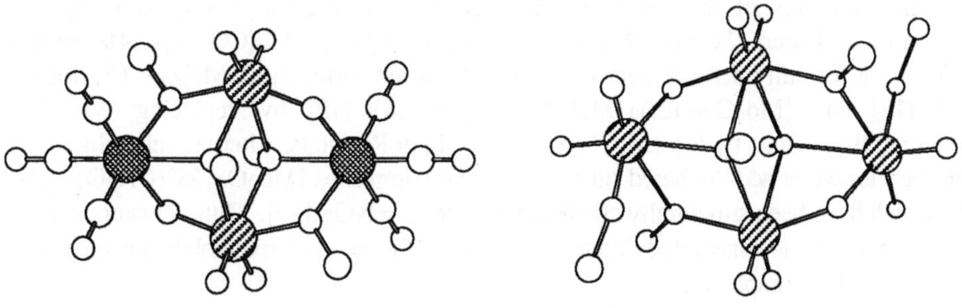

a) b)

Fig. 2. (a) Molecular structure of $[Mo_2O_6(OMe)_4\{Re(CO)_3\}_2]^{2-}$ in **3** [16]
(b) Molecular structure of $[Mo_4O_{10}(OMe)_6]^{2-}$ (**7**) [18]

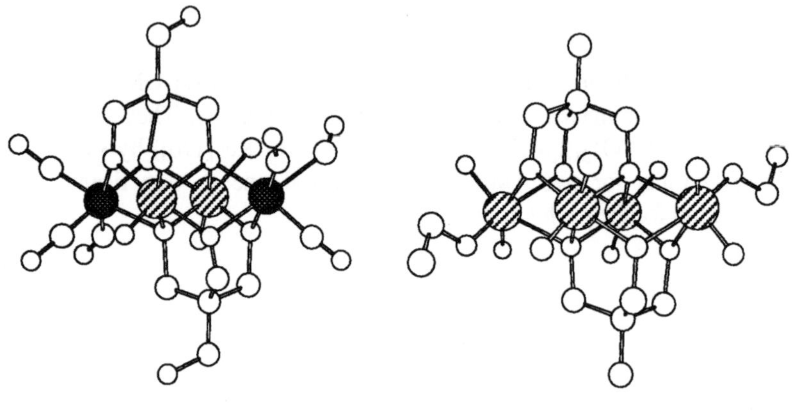

a) b)

Fig. 3. (a) Molecular structure of $[Mo_2O_4\{HOCH_2C(CH_2O)_3\}_2\{Mn(CO)_3\}_2]$ in **4b**·THF [16]
(b) Molecular structure of $[Mo_4O_8(OEt)_2\{MeC(CH_2O)_3\}_2]$ (7) [19]

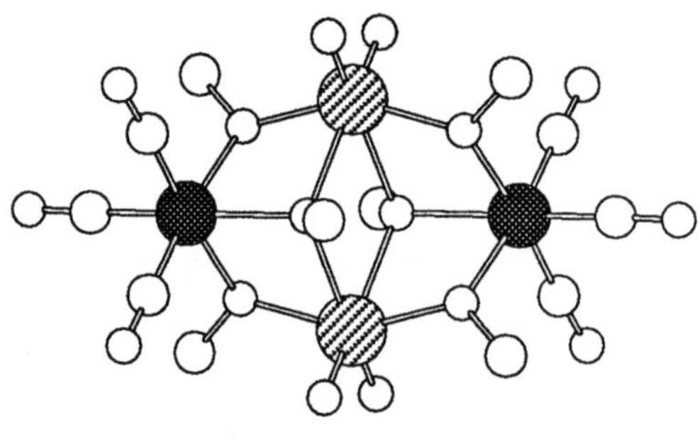

Fig. 4. Molecular structure of $[Mo_2O_4(OMe)_6\{Mn(CO)_3\}_2]$ (**2**) [16]

The structures of the centrosymmetrical clusters $[Mo_2O_6(OMe)_4\{Re(CO)_3\}_2]^{2-}$ (anion of **3**, Figure 2a) and $[Mo_2O_4\{HOCH_2C(CH_2O)_3\}_2\{Mn(CO)_3\}_2]$ (**4b**, Figure 3a) can be formally derived from those of the clusters $[Mo_4O_{10}(OMe)_6]^{2-}$ (**7**, Figure 2b) [18] and $[Mo_4O_8(OEt)_2\{MeC(CH_2O)_3\}_2]$ (**8**) [19] by replacing two *fac*-$\{MoO_2(OR)\}^+$ units by *fac*-$\{M(CO)_3\}^+$ (M = Mn or Re) units. Alternatively, clusters **4** can be viewed as adducts based on the dinuclear complexes $[Mo_2O_4\{RC(CH_2O)_3\}_2]^{2-}$ (**9**) which have been structurally characterized for R = NO_2 [20]. Thus, clusters **4** may be considered as integrated complexes as well as polyoxomolybdate-supported organometallic complexes.

All the tetranuclear oxotrisalkoxo complexes of molybdenum and vanadium that have been reported so far display the rhomb-like planar arrangement although they may differ in the coordination mode of the trisalkoxo ligands. In **4b**, as in **8** [19] and in most other complexes *e.g.* $[V_4O_4(OMe)_6\{MeC(CH_2O)_3\}_2]$ (**10**) [21], the trisalkoxo ligands cap opposite tetrahedral cavities of the oxometal framework and thus display the

μ_4-κ^2:κ^2:κ^3 coordination mode. However, these ligands may also cap metal octahedra, *e.g.* in $[Mo_4O_{10}\{MeC(CH_2O)_3\}_2]^{2-}$ (**11**) [22]. A noteworthy feature of **3**, also observed in **7**, is the marked dissymmetry in the $Mo_2(OMe)$ bridges involving the triply-bridging ligands. The centrosymmetrical complex $[Mo_2O_4(OMe)_6\{Mn(CO)_3\}_2]$ (**2**, Figure 4) only differs from that of $[Mo_2O_6(OMe)_4\{Re(CO)_3\}_2]^{2-}$ in the substitution of Mn for Re and of methoxo ligands for two doubly-bridging oxo ligands. Unlike **3**, the $Mo_2(OMe)$ bridges involving the triply-bridging ligands are nearly symmetrical in **2**.

As far as discrete underivatized polyoxomolybdates are concerned, neither the tetrahedral $[Mo_4O_{12}(\mu_3\text{-}O)_4]^{8-}$ nor the rhomb-like $[Mo_4O_{10}(\mu\text{-}O)_4(\mu_3\text{-}O)_2]^{8-}$ species are known. Derivatization *via* the replacement of either a terminal oxo ligand by an alkoxo ligand or a *fac*-$\{MoO_3\}$ group by an equivalent group, *e.g.* *fac*-$\{M(CO)_3\}^+$ prevents violation of the Lipscomb rule [23] and could lead to the stabilization of both the cubane and the rhombic structures. At present, the reasons for structural preferences are unclear. Indeed the cubane-type arrangement appears to be more easily obtained than the rhomb-like arrangement for manganese (**1a** *vs.* **2**) while the reverse is true for rhenium (**3** *vs.* **1b**). In a general way, rhomb-like structures are much more common than monocubane-type structures within polyoxometalates [5]. To the best of our knowledge, the polyoxoanion-supported organometallic compound composed of a cubane-type $[W_4O_{16}]^{8-}$ unit capped by six $\{Ir(1,5\text{-}COD)\}^+$ groups provides the only unquestionable example of a discrete monocubane-type polyoxometalate derivative [24]. However, cubane-type $\{Mo_4(\mu_3\text{-}O)_4\}^{n+}$ cores may be found in multiple cubane-type organometallic oxide clusters (see 3. below) and in extended solids [25]. The hydrated lithium tungstate $7Li_2WO_4{\cdot}4H_2O$ is presumed to contain tetrahedral $[LiW_4O_{16}]^{7-}$ anions but this is not structurally well characterized [26].

The ^1H and ^{13}C NMR spectra of **1a** and **4b** have been recorded in methanol-d_4 and acetone-d_6, respectively. In both cases, the spectra are consistent with the C_s symmetry of the clusters, which suggests that the solid-state structure is retained in solution. At least, only one form is present in solution and there is no evidence for a fluxional behaviour of these clusters.

2.2 TRINUCLEAR CLUSTERS

The formation of $[Mo_2O_4\{MeC(CH_2O)_3\}_2\{Mn(CO)_3\}]^-$ (anion of **5**) demonstrates the efficiency of trisalkoxo ligands in stabilizing trinuclear clusters [20,27,28]. The structure of this cluster (Figure 5a) is related to that of $[Mo_3O_6(OMe)\{MeC(CH_2O)_3\}_2]^-$ (**11**, Figure 5b) through the formal replacement of a *fac*-$\{MoO_2(OMe)\}^+$ unit by a *fac*-$\{Mn(CO)_3\}^+$ unit. Alternatively, this species can be viewed as deriving from $[Mo_2O_4\{MeC(CH_2O)_3\}_2]^{2-}$ (**9a**) by grafting a $\{Mn(CO)_3\}^+$ unit. The equivalency of the trisalkoxo ligands in the dinuclear precursor, where they display the μ-κ^1O:κ^2O:κ^2O, is lost in $[Mo_2O_4\{MeC(CH_2O)_3\}_2\{Mn(CO)_3\}]^-$ where one ligand displays the μ-κ^2O:κ^2O:κ^2O coordination mode while the second displays the μ-κ^1O:κ^1O:κ^3O coordination mode. The equivalency is restored on coordination of a second $\{Mn(CO)_3\}^+$ unit, which leads to **4a**.

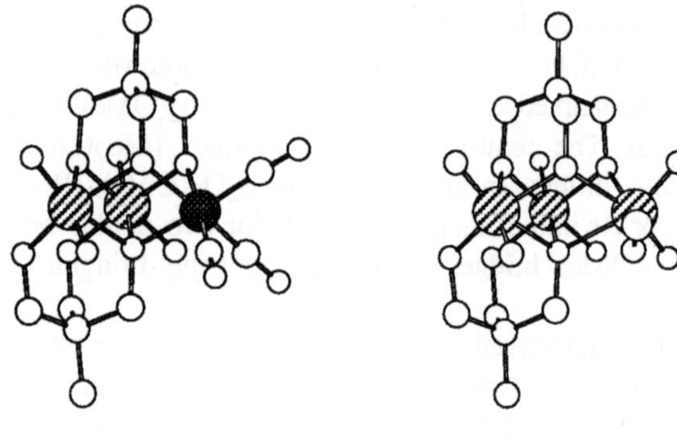

a) b)

Fig. 5. (a) Molecular structure of $[Mo_2O_4\{MeC(CH_2O)_3\}_2\{Mn(CO)_3\}]^-$ in **5** [16]
(b) Molecular structure of $[Mo_3O_6(OMe)\{MeC(CH_2O)_3\}_2]^-$ (**12**) [27]

2.3 OCTANUCLEAR CLUSTERS

In **6**, the anions $[Mo_6O_{16}(OMe)_2\{MeC(CH_2O)_3\}_2\{Mn(CO)_3\}_2]^{2-}$ are located at crystallographic inversion centers (Figure 6a). The two equivalent halves of the anions are connected by two nearly linear Mo-O-Mo bridges with an angle of 162°. The anion may also be viewed as the product of the condensation of two $[Mo_3O_9(OMe)\{MeC(CH_2O)_3\}\{Mn(CO)_3\}]^{3-}$ anions by the sharing of two corners. This tetranuclear subunit displays a rhombic structure. The molecular structure of $[Mo_6O_{16}(OMe)_2\{MeC(CH_2O)_3\}_2\{Mn(CO)_3\}_2]^{2-}$(anion of **6**) is related to that of $[Mo_8O_{20}(OMe)_4\{MeC(CH_2O)_3\}_2]^{2-}$ (**13**, Figure 6b) [29] through the formal replacement of the *fac*-$\{MoO_2(OMe)\}^+$ units by *fac*-$\{Mn(CO)_3\}^+$ units.

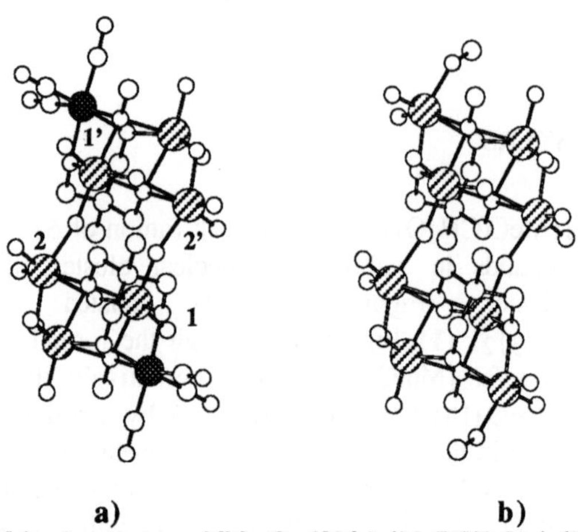

a) b)

Fig. 6. (a) Molecular structure of $[Mo_6O_{16}(OMe)_2\{MeC(CH_2O)_3\}_2\{Mn(CO)_3\}_2]^{2-}$ in **6** [16]
(b) Molecular structure of $[Mo_8O_{20}(OMe)_4\{MeC(CH_2O)_3\}_2]^{2-}$ (**13**) [29]

Both anions are formally related to the hypothetical cluster $[Mo_8O_{30}]^{12-}$ which can be derived from the β-$[Mo_8O_{26}]^{4-}$ anion by sharing of the Mo_4 subunits parallel to one another [5,29]. The tetranuclear $[Mo_3O_9(OMe)\{MeC(CH_2O)_3\}\{Mn(CO)_3\}]^{3-}$ subunit is not expected to be highly stable as a discrete species because the environment of Mo(2) does not fulfill the Lipscomb rule. Dimerization prevents violation of this rule. However it should be noted the Mo(1)-O-Mo(2)' bridge and especially the Mo(1)-O-Mo(2) bridges are clearly dissymmetrical so that the structural parameters for the environment of Mo(1) are reminiscent of those of a unit containing three *terminal* oxo ligands. A similar feature is noticeable in **13**.

2.4 $[Mn(H_2O)_2\{\{Mo_5O_{16}(OMe)_2\}\{Mn(CO)_3\}\}_2]^{2-}$: A UNIQUE CLUSTER

Complex **14** was obtained as green crystals formulated as $(n\text{-}Bu_4N)_2[Mn(H_2O)_2\{\{Mo_5O_{16}(OMe)_2\}\{Mn(CO)_3\}\}_2]\cdot MeOH$ (**14**·MeOH) by refluxing an equimolar mixture of $(n\text{-}Bu_4N)_2[Mo_2O_7]$ and $[MnBr(CO)_5]$ in non-deoxygenated methanol, followed by filtration and cooling to -30 °C. Higher yield in **14** can be achieved by refluxing a mixture of $(n\text{-}Bu_4N)_2[Mo_2O_7]$, $[MnBr(CO)_5]$ and $[Mn(NO_3)_2]\cdot 2H_2O$ in MeOH.

The anion of **14** (Figure 7) can be viewed as formed by binding of $\{Mn(CO)_3\}^+$ and $\{Mn(H_2O)_2\}^{2+}$ cations to $\{Mo_5O_{16}(OMe)_2\}^4$ anions. The latter derive from the hitherto unknown Lindqvist-type anion $\{Mo_5O_{18}\}^{6-}$ by replacing two of the bridging oxo ligands linking the apical Mo atom to the basal Mo atoms by methoxo groups. Each $\{Mn(CO)_3\}^+$ unit is bound to the two methoxo ligands and one bridging oxo ligand of a $\{Mo_5O_{16}(OMe)_2\}^{4-}$ anion. Two such $\{Mo_5O_{16}(OMe)_2\}\{Mn(CO)_3\}^{2+}$ adducts are linked by a Mn(II) ion which achieves six-coordination by binding to two oxo ligands of each adduct and two water molecules [15,30].

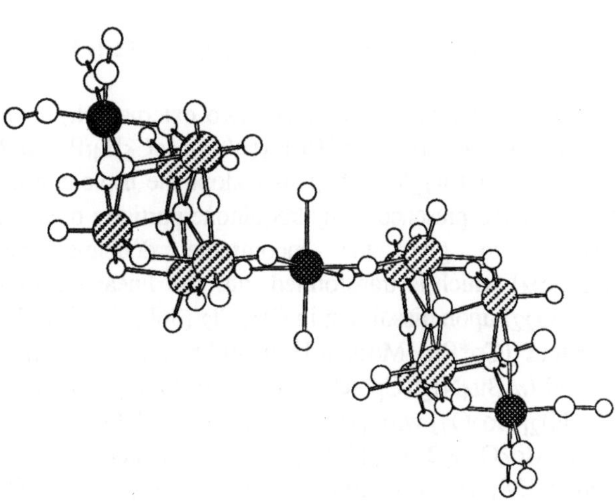

Fig. 7. Molecular structure of $[Mn(H_2O)_2\{\{Mo_5O_{16}(OMe)_2\}\{Mn(CO)_3\}\}_2]^{2-}$ in **14**·MeOH [15,30]

3. Complexes with *fac*-$\{RhCp^*\}^+$ units

Both complete and lacunary polyoxometalate-supported $\{Rh(III)Cp^*\}^{2+}$ complexes have been reported. The former include $[Cp^*Rh(cis\text{-}Nb_2W_4O_{19})]^{2-}$ [31] $[Cp^*Rh(Mo_5O_{18}TiCp)]^-$ [3b], $[(Cp^*Rh)_8(Mo^V{}_{12}O_{36})(Mo^{VI}O_4)]^{2+}$ [32], $[Cp^*Rh(SiNb_3W_9O_{40})]^{5-}$ [4a], $[Cp^*Rh(P_2Nb_3W_{15}O_{62})]^{7-}$ [33], and $[(Cp^*Rh)_4(V_6O_{19})]$ [34]. In these adducts, the $\{RhCp^*\}^{2+}$ unit is bound to three contiguous doubly-bridging oxygen atoms of the support. In $[RhCp^*(H_2O)\{Mo_5O_{13}(OMe)_4(NO)\}]^-$ and $[\{RhCp^*\}_2(\mu\text{-}X)\{Mo_5O_{13}(OMe)_4(NO)\}]$ (X = Cl, Br), the lacunary Lindqvist-type support $[Mo_5O_{13}(OMe)_4(NO)]^{3-}$ acts as a bidentate and a symmetrically bridging tetradentate ligand respectively [35].

On the other hand, integrated $\{RhCp^*\}^{2+}$ oxomolybdenum clusters have been characterized by the groups of Isobe and Süss-Fink. The triple-cubane-type cluster $[(Cp^*Rh)_4Mo_4O_{16}]$ (**15**, Figure 8) has been obtained by reaction of $[\{Cp^*RhCl_2\}_2]$ with $Na_2MoO_4{\cdot}2H_2O$ [36a]. In figures 8 and 9, the Mo atoms are shown as hatched spheres while the Rh atoms are shown as dotted spheres.

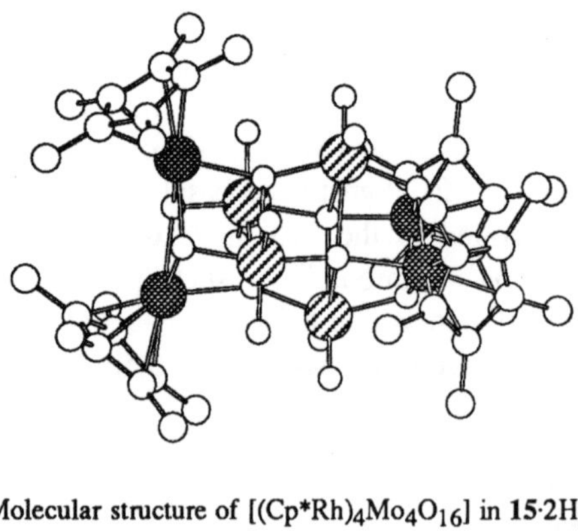

Fig. 8. Molecular structure of $[(Cp^*Rh)_4Mo_4O_{16}]$ in **15**$\cdot 2H_2O$ [36]

This cluster consists of a central Mo_4O_4 core and two external $Rh_2Mo_2O_4$ cores. The mixed rhodium-ruthenium clusters $[(Cp^*Rh)_3\{(\eta^6\text{-}p\text{-}MeC_6H_4iPr)Ru\}Mo_4O_{16}]$ and $[(Cp^*Rh)_2\{(\eta^6\text{-}p\text{-}MeC_6H_4iPr)Ru\}_2Mo_4O_{16}]$ also adopt the triple-cubane-like structure of **15** [37b]. Methanol in the presence of hydroquinone partially breaks the framework of $[(Cp^*Rh)_4Mo_4O_{16}]$ to give the incomplete double-cubane-type cluster $[(Cp^*Rh)_2Mo_3O_9(OMe)_4]$ which is transformed into the linear quadruple-cubane-type cluster $[(Cp^*Rh)_4Mo_6O_{22}]$ upon dissolving in CH_2Cl_2 [36b]. It is hardly expected that mono-cubane-type parent $[(Cp^*Rh)_2Mo_2O_{10}]^{4-}$ would be stable and, in fact, it has not been found. Compound $(n\text{-}Bu_4N)_2[(Cp^*Rh)_2Mo_6O_{20}(OMe)_2]$ (**16**) has been obtained by reaction of $(n\text{-}Bu_4N)_2[Mo_2O_7]$ with $[\{Cp^*RhCl_2\}_2]$ in MeOH [38]. The framework of the anion $[(Cp^*Rh)_2Mo_6O_{20}(OMe)_2]^{2-}$ (Figure 9a) has been connected with that of the γ-$[Mo_8O_{26}]^{4-}$ anion [38]. However comparison with the $[Mo_8O_{24}(OMe)_4]^{4-}$ anion (**17**, Figure 9b) [39] is also appropriate; in this way, the cluster $[(Cp^*Rh)_2Mo_6O_{20}(OMe)_2]^{2-}$ can be formally derived from $[Mo_8O_{24}(OMe)_4]^{4-}$ by replacing two *fac*-$\{MoO_2(OMe)\}^+$ units by $\{RhCp^*\}^{2+}$ units.

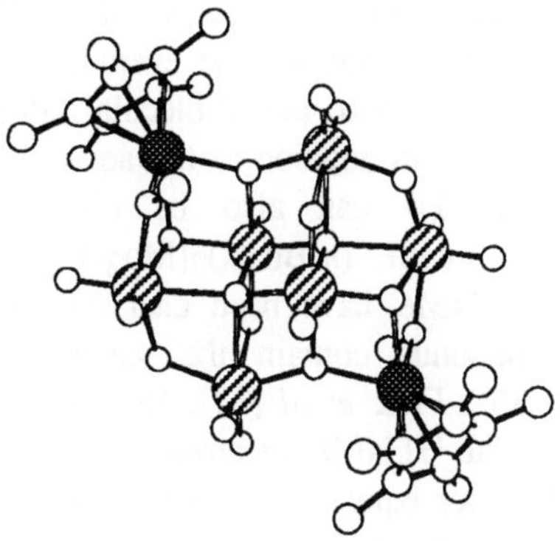

Fig. 9. Molecular structure of $[Mo_6O_{20}(OMe)_2(RhCp^*)_2]^{2-}$ in **16** [38]

We are currently reinvestigating the reaction of $(n\text{-}Bu_4N)_2[Mo_2O_7]$ with $[\{Cp^*RhCl_2\}_2]$ in MeOH. We have found that not only **16** but also cluster **15** can be obtained in this way. However no evidence has been obtained for the formation of the cluster $[(Cp^*Rh)_2Mo_2O_6(OMe)_4]$ which could display either a cubane-like (similar to **1a**) or a rhomb-like structure (similar to **2**).

4. Complexes with $\{Ru(\eta^6\text{-arene})\}^{2+}$ units

A few polyoxometalate-supported $\{Ru(II)(\eta^6\text{-arene})\}^{2+}$ complexes have been reported. These include $[(p\text{-cymene})Ru(Nb_2W_4O_{19})]^{2-}$ [40], $[(\eta^6\text{-}C_6H_6)Ru(W_5O_{18}TiCp^*)]^-$ [40], and $[(\eta^6\text{-}C_6H_6)Ru(P_2Nb_3W_{15}O_{62})]^{7-}$ [33a,33c]. On the other hand, Süss-Fink *et al* have obtained the cluster $[\{(\eta^6\text{-}p\text{-}MeC_6H_4iPr)Ru\}_4Mo_4O_{16}]$ (**18**) by reaction of $[\{(\eta^6\text{-}p\text{-}MeC_6H_4iPr)RuCl_2\}_2]$ with $Na_2MoO_4{\cdot}2H_2O$ in water [37]. Despite its composition is quite similar to that of **15**, **18** adopts a windmill-like structure rather that the fused triple-cubane-like structure of **15** (Figure 10). In figures 10 to 13, the Mo and W atoms are shown as hatched spheres while the Ru atoms are shown as dotted spheres.

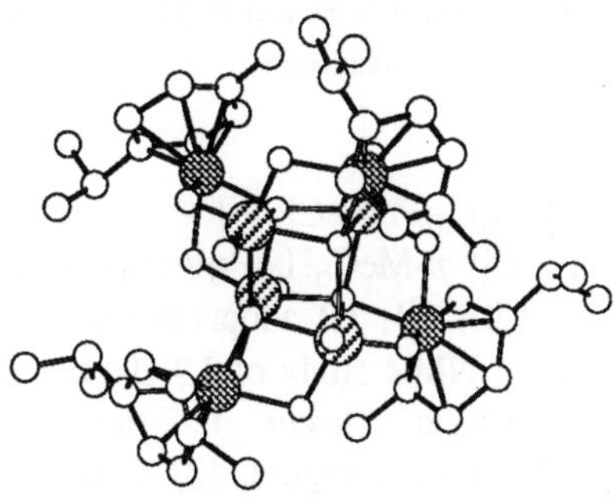

Fig. 10. Molecular structure of $[\{Ru(\eta^6\text{-}p\text{-}MeC_6H_4iPr)\}_4M_4O_{16}]$ in $\mathbf{18}{\cdot}C_6H_5Me$ (M = Mo) [37] and in $\mathbf{19}{\cdot}C_6H_5Me$ (M = W) [41].

We recently undertook a study of the reactions of $[\{(\eta^6\text{-}p\text{-MeC}_6H_4i\text{Pr})RuCl_2\}_2]$ with molybdates and tungstates to examine the eventual influence of the synthesis conditions on the stoichiometry and the molecular structures of the resulting organometallic oxometal clusters. In addition to the aqueous route used by Süss-Fink and coworkers [37], cluster **18** can also be obtained by the reaction of $[\{(\eta^6\text{-}p\text{-MeC}_6H_4i\text{Pr})RuCl_2\}_2]$ with $(n\text{-Bu}_4N)_2[Mo_2O_7]$ in acetonitrile. Products formulated as **18** and **18**·C_6H_5Me have been characterized by single-crystal X-ray diffraction [41]. Both compounds contain the same cluster as the compound **18**·$2C_6H_5Me$ analysed by Süss-Fink *et al* [37]. However, cluster **18** appears to be fluxional in solution. Indeed, the ^1H NMR spectrum of **18** in CD_2Cl_2 reveals, besides the expected signals for the four equivalent *p*-cymene ligands in **18**, another set of signals indicative of another *p*-cymene-containing species. Although this second species is in minority in CD_2Cl_2, the concentration of the two species are nearly equal in $CDCl_3$. A similar conclusion can be drawn from the ^{95}Mo and ^{17}O NMR spectra. The change observed on going from CD_2Cl_2 to $CDCl_3$ is reversible. Such a behaviour is ascribed to a solvent-dependent conformational equilibrium which is slow on the NMR time scale. Moreover, ^{17}O NMR analysis leads to the conclusion that the major species in CD_2Cl_2 is the triple-cubane form, while the minor product is the windmill form. It is noteworthy that the windmill form crystallizes from CD_2Cl_2 although the triple-cubane form is predominant in solution [41].

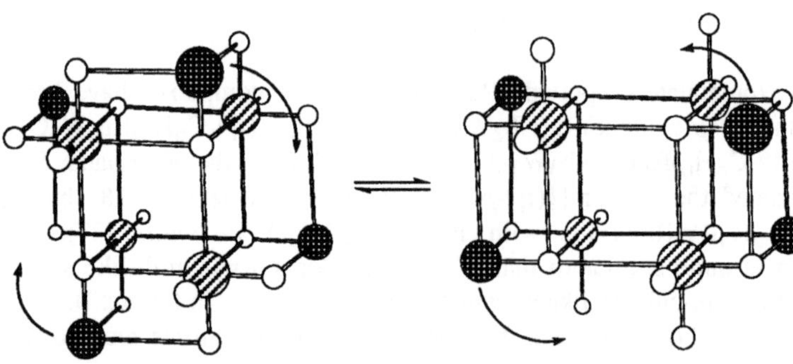

Fig. 11. Postulated mechanism for the fluxional behaviour of $[\{Ru(\eta^6\text{-}p\text{-MeC}_6H_4i\text{Pr})\}_4Mo_4O_{16}]$ (**18**) in solution [41].

The reaction of $[\{(\eta^6\text{-}p\text{-MeC}_6H_4i\text{Pr})RuCl_2\}_2]$ with $(n\text{-Bu}_4N)_2[WO_4]$ in acetonitrile yields two clusters. One is $[\{(\eta^6\text{-}p\text{-MeC}_6H_4i\text{Pr})Ru\}_4W_4O_{16}]$ (**19**) which has been characterized as **19**·C_6H_5Me. Like **18**, **19** adopts the windmill-like structure (Figure 10). However, unlike **18**, the ^1H NMR study of **19** did not reveal any configurational change in solution. The second cluster is the double-cubane-type cluster $[\{Ru(\eta^6\text{-}p\text{-MeC}_6H_4i\text{Pr})\}_4W_2O_{10}]$ (**20**, Figure 12) which has been characterized as **20**·$3H_2O$ [41]. The reaction of $[\{(\eta^6\text{-}p\text{-MeC}_6H_4i\text{Pr})RuCl_2\}_2]$ with $(n\text{-Bu}_4N)_2[Mo_2O_7]$ in MeOH yields the rhomb-like cluster $[\{Ru(\eta^6\text{-}p\text{-MeC}_6H_4i\text{Pr})\}_2Mo_2O_6(OMe)_4]$ (**21**,

Figure 13) [41], which can also be formed by reacting **18** in MeOH in the presence of hydroquinone [42].

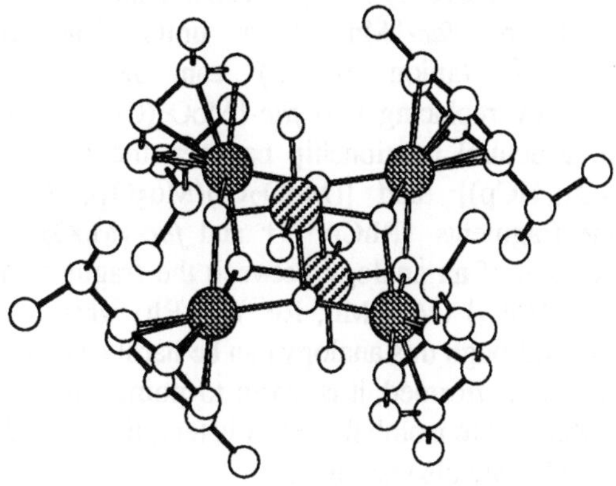

Fig. 12. Molecular structure of $[\{Ru(\eta^6\text{-}p\text{-}MeC_6H_4iPr)\}_4W_2O_{10}]$ in **20**·$3H_2O$ [41]

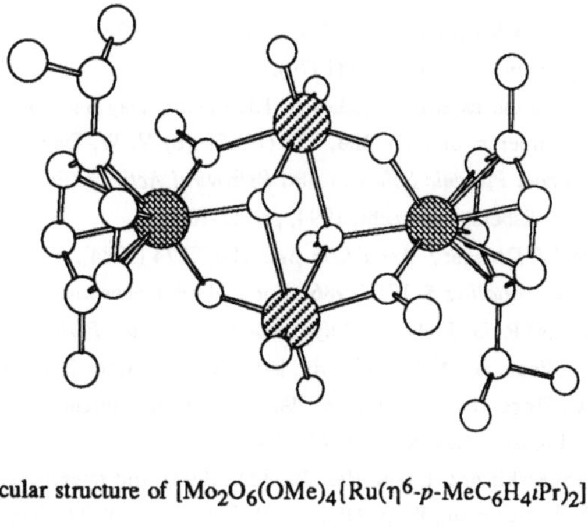

Fig. 13. Molecular structure of $[Mo_2O_6(OMe)_4\{Ru(\eta^6\text{-}p\text{-}MeC_6H_4iPr)_2]$ in **21** [41,42]

5. Concluding remarks

The data that have been discussed show the ubiquity of both the rhomb-like and the cubane-type tetrametallic units in organometallic oxometal clusters. The structural versatility within this field is further illustrated by the fluxionality of the cluster $[\{(\eta^6\text{-}p\text{-}MeC_6H_4iPr)Ru\}_4M_4O_{16}]$ (**18**).

As discussed above, the molecular structures of the clusters $[Mo_2O_6(OMe)_4\{Re(CO)_3\}_2]^{2-}$ (anion of **3**) and $[\{Ru(\eta^6\text{-}p\text{-}MeC_6H_4iPr)\}_2Mo_2O_6(OMe)_4]$ (**21**) are related to that of $[Mo_4O_{10}(OMe)_6]^{2-}$ (**8**) through the formal substitution of *fac*-$\{Re(CO)_3\}^+$ units or $\{Ru(\eta^6\text{-}p\text{-}MeC_6H_4iPr)\}^{2+}$ units for *fac*-$\{MoO_2(OMe)\}^+$ units. Similarly, the clusters $[Mo_2O_4\{MeC(CH_2O)_3\}_2\{Mn(CO)_3\}_2]$ (**4a**), $[Mo_2O_4\{MeC(CH_2O)_3\}_2\{Mn(CO)_3\}]^-$ (anion of **5**) and $[Mo_6O_{16}(OMe)_2\{MeC(CH_2O)_3\}_2\{Mn(CO)_3\}_2]^{2-}$ (anion of **6**) can be derived from $[Mo_4O_8(OMe)_2\{MeC(CH_2O)_3\}_2]$ (**8**), $[Mo_3O_6(OMe)\{MeC(CH_2O)_3\}_2]^-$ (**12**) and $[Mo_8O_{20}(OMe)_4\{MeC(CH_2O)_3\}_2]^{2-}$ (**13**), respectively, by replacement of *fac*-$\{MoO_2(OMe)\}^+$ units by *fac*-$\{Mn(CO)_3\}^+$ units. Furthermore, the cluster $[(RhCp^*)_2Mo_6O_{20}(OMe)_2]^{2-}$ (anion of **16**) can be formally obtained from $[Mo_8O_{24}(OMe)_4]^{4-}$ (**17**) by replacing two *fac*-$\{MoO_2(OMe)\}^+$ units by $\{RhCp^*\}^{2+}$ units. In addition, the structural relationship between the polyoxometalate-supported species $[Cp^*Rh(Mo_5O_{18}TiCp)]^-$ and $[(MoO_2Cl)(Mo_5O_{18}TiCp)]^{2-}$ points out a connection between the fragments $\{RhCp^*\}^{2+}$ and *fac*-$\{MoO_2Cl\}^+$ [14]. Altogether these examples are indicative of an analogy between the fragments d^0-*fac*-$\{MoO_2X\}$ (X = O, OR, Cl) and d^6-*fac*-$\{ML_3\}$ (M = Mn, Re, Ru, Rh) where the charge have been omitted for convenience. Although this analogy can be hardly extended beyond topology before theoretical studies are performed, it is worth to point out that it has been shown that $\{Re(CO)_3\}^+$ and $\{ReO_3\}$ are isolobal [43]. It is tempting to deduce from this result that $\{Re(CO)_3\}^+$ and $\{WO_3\}$ are also isolobal.

References

1. R. K. C. Ho, and W. G. Klemperer: *J. Am. Chem. Soc.* **100**, 6772 (1978).
2. W. H. Knoth: *J. Am. chem. Soc.* **101**, 759 (1979).
3. For lead references to the extensive studies of Klemperer, Day and co-workers, see: (a) V. W. Day, and W. G. Klemperer: *Science* **228**, 533 (1985). (b) V. W. Day, and W. G. Klemperer in *Polyoxometalates: From Platonic Solids to Anti-Retroviral Activity* (Eds.: M. T. Pope, A. Müller), Kluwer Academic Publishers, Dordrecht, 1994, pp. 87-104.
4. R. G. Finke, and M. W. Droege: *J. Am. Chem. Soc.* **106**, 7274 (1984). (b) R. G. Finke, B. Rapko, and P. J. Domaille: *Organometallics* **5**, 175 (1986). For lead references to the extensive studies of Finke and co-workers see: (c) R. G. Finke, in *Polyoxometalates: From Platonic Solids to Anti-Retroviral Activity* (Eds.: M. T. Pope, A. Müller), Kluwer, Dordrecht, 1994, pp. 267-280; (d) H. Weiner, J. D. Aiken III, and R. G. Finke: *Inorg. Chem.* **35**, 7905 (1996) and references 6 and 7 therein.
5. P. Gouzerh, and A. Proust: *Chem. Rev.* **98**, 77 (1998).
6. K. Isobe, and A. Yagasaki: *Acc. Chem. Res.* **26**, 524 (1993), and references therein.
7. (a) A. R. Siedle, R. A. Newmark, W. B. Gleason, R. P. Skarjune, K. O. Hodgson, A. L. Roe, and V. W. Day: *Solid State Ionics* **26**, 109 (1988). (b) A. R. Siedle: *New J. Chem.* **13**, 719 (1989), and references therein.
8. T. Nagata, M. Pohl, H. Weiner, and R. G. Finke: *Inorg. Chem.* **36**, 1366 (1997) and references therein.
9. A. R. Siedle, C.G. Markell, P. A. Lyon, K. O. Hodgson, and A. L. Roe: *Inorg. Chem.* **26**, 219 (1987).
10. Y. Imada, T. Shido, R. Ohnishi, K. Isobe, and M. Ichikawa: *Catal. Lett.* **38**, 101 (1996).
11. (a) F. Bottomley, and J. Chen: *Organometallics* **11**, 3404 (1992). (b) J. R. Harper, and A. L. Rheingold: *J. Am. Chem. Soc.* **112**, 4037 (1990).
12. a) P. Gouzerh, Y. Jeannin, A. Proust, and F. Robert, *Angew. Chem., Int. Ed. Engl.* **28**, 1363 (1989). b) A. Proust, P. Gouzerh, and F. Robert: *Inorg. Chem.* **32**, 5291, (1993).
13. C. J. Besecker, and W. G. Klemperer: *J. Am. Chem. Soc.* **102**, 7598 (1980).
14. V. W. Day, M. F. Fredrich, M. R. Thompson, W. G. Klemperer, R.-S. liu, and W. Shum: *J. Am. Chem. Soc.* **103**, 3597 (1981).
15. R. Villanneau: *Doctoral Dissertation*, Université Pierre et Marie Curie (1997).

16. R. Villanneau, R. Delmont, A. Proust, and P. Gouzerh: submitted for publication.
17. (a) M. D. Clerk and M. J. Zaworotko: *J. Chem. Soc., Chem. Commun* 1607 (1991). (b) M. Herberhold, G. Süβ, J. Ellermnn, and H. Gäbelein: *Chem. Ber.* **111**, 2931 (1978).
18 H. Kang, S. Liu, S. N. Shaikh, T. Nicholson, and J. Zubieta: *Inorg. Chem.* **28**, 920 (1989).
19. A. J. Wilson, W. T. Robinson, and C. J. Wilkins: *Acta Crystallogr. Sect. C.* **39**, 54, (1983).
20. S. Liu, L. Ma, D. McGowty, and J. Zubieta: *Polyhedron* **13**, 1541 (1990).
21. D. Crans, R. W. Marshman, M. S. Gotlieb, O. P. Anderson, and M. M. Miller: *Inorg. Chem.* **31**, 4939 (1992).
22. (a) R. Delmont: *Doctoral Dissertation*, Université Pierre et Marie Curie (1997). (b) R. Delmont, A. Proust, F. Robert, and P. Gouzerh: submitted for publication.
23. W. N. Lipscomb: *Inorg. Chem.* **4**, 132 (1965).
24. Y. Hayashi, F. Müller, Y. Lin, S. M. Miller, O. P. Anderson, and R. G. Finke: *J. Am. Chem. Soc.* **119**, 11401 (1997).
25. (a) E. W. Corcoran: *Inorg. Chem.* **29**, 157 (1990). (b) R. C. Haushalter: *J. Chem. Soc., Chem. Commun.* 1566 (1987). (c) K.-H. Lii, R. C. Haushalter, and C. J. O'Connor: *Angew. Chem. Int. Ed. Engl.* **26**, 549 (1987). (d) L. A. Mundi, and R. C. Haushalter: *J. Am. Chem. Soc.* **113**, 6340 (1991).
26. (a) A. Hüllen: *Angew. Chem.* **76**, 588 (1964). (b) A. Hüllen: *Ber. Bunsengesell.* **70**, 598 (1966).
27. L. Ma, S. Liu, and J. Zubieta: *Inorg. Chem.* **28**, 175 (1989).
28. A. Müller, J. Meyer, H. Bögge, A. Stammler, and A. Botar: *Chem. Eur. J.* **4**, 1388 (1998).
29. L. Ma, S. Liu, and J. Zubieta: *J. Chem. Soc., Chem. Commun* 440 (1989).
30. R. Villanneau, A. Proust, F. Robert, and P. Gouzerh: *manuscript in preparation.*
31. C. J. Besecker, V. W. Day, W. G. Klemperer, and M. R. Thompson: *J. Am. Chem. Soc.* **106**, 4125 (1984).
32. H. K. Chae, W. G. Klemperer, D. E. Paez Loya, V. W. Day, and T. A. Eberspacher: *Inorg. Chem.* **31**, 3187 (1992).
33. (a) D. J. Edlund, R. J. Saxton, D. K. Lyon, and R. G. Finke: *Organometallics* **7**, 1692 (1988). (b) K. Nomiya, C. Nozaki, M. Kaneko, R. G. Finke, and M. Pohl: *J. Organomet. Chem.* **505**, 23 (1995). (c) M. Pohl, Y. Lin, T. J. R. Weakley, K. Nomiya, M. Kaneko, H. Weiner, and R. G. Finke: *Inorg. Chem.* **34**, 767 (1995).
34. (a) H. K. Chae, W. G. Klemperer, and V. W. Day: *Inorg. Chem.* **28**, 1423 (1989). (b) Y. Hayashi, Y. Ozawa, and K. Isobe: *Chem. Lett.* 425 (1989).
35. A. Proust, P. Gouzerh, and F. Robert: *Angew. Chem. Int. Ed. Engl.* **32**, 115 (1993).
36. Y. Hayashi, K. Toriumi, and K. Isobe: *J. Am. Chem. Soc.* **110**, 3666 (1988). (b) Y. Do, X.-Z. You, C. Zhang, Y. Ozawa, and K. Isobe: *J. Am. Chem. Soc.* **113**, 5892 (1991).
37. (a) G. Süss-Fink, L. Plasseraud, V. Ferrand, and H. Stoeckli-Evans: *J. Chem. Soc., Chem. Commun.* 1657 (1997). (b) G. Süss-Fink, L. Plasseraud, V. Ferrand, S. Stanislas, A. Neels, H. Stoeckli-Evans, M. Henry, G. Laurenczy, and R. Roulet: *Polyhedron* **17**, 2817 (1998).
38. S. Tahara, T. Nishioka, I. Kinoshita, and K. Isobe: *J. Chem. Soc., Chem. Commun.* 891 (1997).
39. E. M. McCarron III, and R. L. Harlow, *J. Am. Chem. Soc.* **105**, 6179 (1983).
40. V. W. Day, T. A. Eberspacher, W. G. Klemperer, R. P. Planalp, P. W. Schiller, A. Yagasaki, and B. Zhong: *Inorg. Chem.* **32**, 1629 (1993).
41. (a) V. Artero : *Doctoral Dissertation* (in preparation), Université Pierre et Marie Curie. (b) V. Artero, A. Proust, P. Herson, R. Thouvenot, and P. Gouzerh: *manuscript in preparation.*
42. L. Plasseraud, H. Stoeckli-Evans, and G. Süss-Fink: *Inorg. Chem. Comm.*, **2**, 344 (1999).
43. T. Szyperski, and P. Schwerdfeger: *Angew. Chem. Int. Ed. Engl.* **28**, 1228 (1989).

Spherical (Icosahedral) Objects in Nature and Deliberately Constructable Molecular Keplerates: Structural and Topological Aspects

O. DELGADO, A. DRESS,
Department of Mathematics, University of Bielefeld, D-33501 Bielefeld, Germany

A. MÜLLER
Department of Chemistry, University of Bielefeld, D-33501 Bielefeld, Germany

Abstract:
In polyoxometalate chemistry, a large variety of clusters can be formed by linking together metal-oxide building blocks, including tetrahedra, octahedra, and even pentagonal units with C_5 symmetry. Correspondingly, it is possible to construct spherically shaped polyoxometalates with icosahedral symmetry and predetermined sizes by connecting those pentagonal units using appropriate linkers. Using tools from discrete mathematics, the resulting molecular architectures can be investigated and the basic geometric/topological principles governing their construction can be elucidated.

Key Words: Archimedean solids, clusters, discrete geometry, discrete mathematics, keplerates, magic numbers, molecular architecture, Platonic solids, polyoxometalates, topology, triangulation numbers

1. Introduction

Scientists try to discover common features in the manifold of appearances and search for universal rules that govern their intrinsic architecture. In this context, two observations are of particular interest to the chemist:

(i) When stable structures are created by applying the same matching rule uniformly and repetitively to the same set of basic building units, their shape will often resemble that of regular polyhedra well known from ancient Greek geometry and, in particular, the work of Archimedes of Syracuse [1].

(ii) The resulting basic architectures often scale many sizes – Buckminster Fuller's geodesic structures for instance, like that one housing the US Expo'67 exhibit in Montreal, can be recognized in molecules such as the renowned C_{60}, molecule in enzymes, viruses, organelles, cells (c.f. the title page of a science magazine: "Bucky Balls are Invading our Cells"), and even in small organisms, e.g. in some of Haeckel's celebrated radiolaria [1g]. Regarding the stability of such systems, one should recall Buckminster Fuller's doctrine, that is, his famous *tensegrity* principle: *Don't fight forces, use them*.

Consequently, it is a challenge for the chemist not only to synthesize such regularly shaped objects, but to find routes of synthesis that would allow to scale the size of the resulting objects arbitrarily (within chemically reasonable limits) and to synthesize increasingly larger clusters of a given type. Amazingly enough, the possibility of linking well defined building blocks in predetermined ways allows polyoxometalate chemistry to

M.T. Pope and A. Müller (eds.), Polyoxometalate Chemistry, 69–87.

meet this challenge even with regard to the most complex of these polyhedral shapes, that is, those with icosahedral symmetry, or *keplerates*. It is worth noting in this context that Coxeter doubted in the first edition of his famous book ***Regular Polytopes*** that inorganic molecules with icosahedral symmetry could ever be synthesized; in the preface of the third edition, he writes: *It is, perhaps, worthwhile to mention that the electron microscope has revealed icosahedral symmetry in the shape of many virus macromolecules. For instance, the virus that causes measles looks much like the icosahedron itself. The Preface to the First Edition refers to a passage on Page 13 concerning the impossibility of any inorganic occurrence of this polyhedron. That statement must now be taken with a grain of borax, for the element boron forms a molecule* B_{12} *whose twelve atoms are arranged like the vertices of an icosahedron* [1r].
It is also worth noting in this context that the corresponding mathematical problems of specifying and elaborating the pertinent general building and matching rules have been recognized as a genuine task of topology – already in the very first paper explicitly devoted to this field: In 1847, a young mathematician from Göttingen named Johann Benedict Listing published a paper entitled *'Vorstudien zur Topologie'* [2]; encouraged by his famous teacher Carl Friedrich Gauss, he envisaged a new branch of mathematics that he proposed to call ***Topologie***, its topic being the study and elucidation of *the modal properties of space, that is, the laws of connectivity, mutual position and succession of points, lines, surfaces, solid bodies, and their parts or their aggregates in space, apart from their measure and proportion* ('die modalen Eigenschaften des Raumes, d.h. die Gesetze des Zusammenhangs, der gegenseitigen Lage und der Aufeinanderfolge von Punkten, Linien, Flächen, Körpern und ihren Theilen oder ihren Aggregaten, abgesehen von den Maß- und Größenverhältnissen').
Adopting this point of view, it can for example be understood quite easily that, by referring to Buckminster Fuller's geodesic domes – elaborated later on in a more mathematical way by Coxeter [3] referring to Goldberg [4] – the virologists Caspar and Klug [5] were able to explain the spatial (or, *sensu* Listing and more specifically, the *topological*) structure of certain viruses quite a few years before their first high-resolution electron microscopy images were known. As demonstrated in this article, the same topological approaches can be used also to analyze icosahedral molecules resulting from polyoxometalate chemistry.

2. The Basics of Polyoxometalate Chemistry

Polyoxometalate clusters (that is, inorganic early transition metal oxygen cluster anions) represent a class of inorganic compounds that show an unmatched variety of molecular structure [6]. This is due to the *matchability* of their primary building components, i.e. regular polyhedra of tetrahedral and octahedral shape, allowing the chemist to link the various components in many different ways. Using specific reaction routes, recent insights into molybdenum-oxide based chemistry enable us in particular to construct spherically shaped molecular systems. Based on versatile means of linking cardinal pentagonal units, ball-shaped clusters with icosahedral symmetry of different size are easily accessed, exhibiting extraordinary structural, magnetic, and electronic properties. These novel species are synthesized basically by intertwining (a) pentagonal $\{(Mo)Mo_5\}$-type building blocks P abundant in nearly all larger (nanoscaled) polyoxomolybdate clusters with (b) linker entities L of different size that interconnect and bridge those pentagonal units [7]. The overwhelming structural diversity of polyoxometalates thus provides not only molecular systems of protruded functionality, especially with respect to the demands of current materials science, but also of a truely aesthetic appeal resulting from their amazing spatial structure and symmetry.

3. Structural Principles of Sizeable Spherical Objects Based on Pentagons of the $\{(Mo)Mo_5\}$Type [7-13]

The synthesis of icosahedral molecules of varying size is of particular interest to us, and we will concentrate here on the task of building such species from the pentagonal building blocks mentioned above. The basic $\{(Mo)Mo_5\}$ unit has fivefold rotational symmetry and consists of a central pentagonal bipyramidal MoO_7 unit that shares edges (i.e., pairs of neighbouring oxygen atoms) with five MoO_6 octahedra positioned in the equatorial plane (Fig. 1).

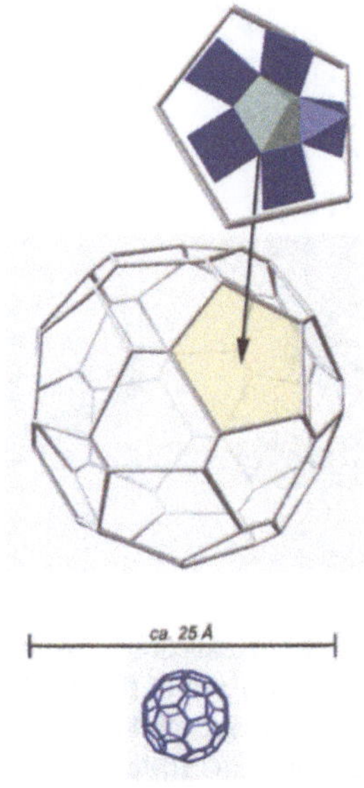

Fig. 1: Structure of the icosahedral $\{Mo^V_2\}_{30}$ fragment of cluster **1** with 12 regular pentagons and 20 trigonal hexagons as well as its coherence to the C_{60} fullerene, which is depicted on the same scale. A single $\{(Mo)Mo_5\}$ pentagon occurring in all clusters discussed here is emphasized in a polyhedral representation.

In an aqueous reaction solution and in the presence of appropriate linkers L each of which interconnects two of the pentagonal units P, the $\{(Mo)Mo_5\}$ units basically self-assemble spontaneously into an icosahedral species of type $P_{12} \cdot L_{30} \equiv \{(Mo)Mo_5\}_{12}L_{30}$, i. e. containing 12 of these pentagons P, and 30 linkers – as well as encapsulated water molecules – in agreement with the pertinent topological requirements. Using different linkers (L = $\{Fe(H_2O)_2\}^{3+}$, $\{OMo(H_2O)\}^{3+}$, and $\{Mo^V{}_2O_4(ac)\}^{2+}$), clusters of different shape and composition are thus formed, later being refered to as the Fe_{30} Mo_{72}, the Mo_{102}, and the Mo_{132} cluster – or just **1**, **2**, and **3**, respectively.
In these systems, the centres of the 12 pentagonal $\{(Mo)Mo_5\}$ units are positioned at the vertices of an icosahedron (Fig. 2a). In **3**, the Mo_{132} cluster, the 30 linker units define a kind of truncated icosahedron with 20 hexagons with C_3-symmetry and 12 pentagons comparable approximately to the C_{60} fullerene.

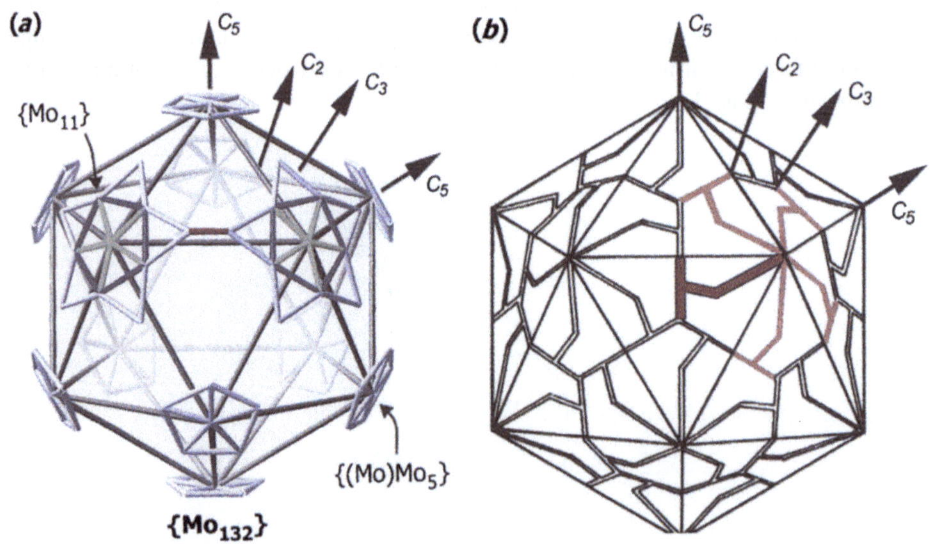

Fig. 2: Schematic representation of the icosahedron spanned by the centres of the $\{(Mo)Mo_5\}$ subfragments of the $\{Mo_{132}\}$-type cluster **3** (a) and of the satellite tobacco necrosis virus (STNV) with triangulation number t=1 (for further literature see ref. 8) highlighting five of the protomers (red) (b). In the case of **3** two $\{Mo_{11}\}$ units each formed by the $\{(Mo)Mo_5\}$ groups and the five related Mo centres of the five neighboring $\{Mo^V{}_2\,O_4\}^{2+}$ linkers are emphasized.

In the two other spherical clusters, **1** and **2**, the mononuclear linkers form an icosidodecahedron – one of the 13 Archimedean solids – having 20 triangles and 12 pentagons (Figs. 3 and 4). In the resulting spherical geometry, the local fivefold symmetry axes of the {(Mo)Mo_5} groups are retained giving rise to global fivefold symmetry axes. In addition, meeting the requirements of icosahedral symmetry, there are also 15 twofold symmetry axes (crossing the centres of the linkers) and 10 threefold axes (crossing the midpoint between three neighbouring units).

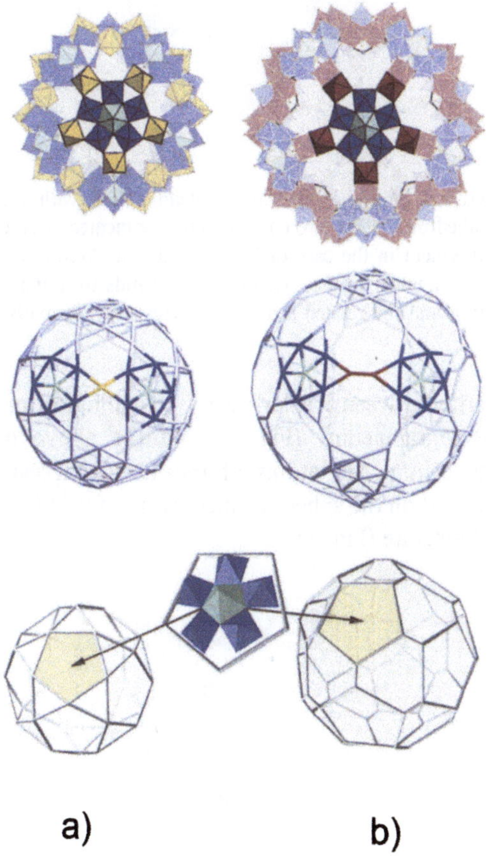

Fig. 3: Top: polyhedral representation of **1** (a) and **3** (b) viewed along a C_5 axis showing five of the openings formed by the (Mo_9O_9) (accessible for very small molecules) and ($Mo_3Fe_3O_6$) rings. ({(Mo)Mo_5} units: blue; central MoO_7 unit: turquoise; linkers {Fe} and {Mo^V_2}: yellow and red, respectively). Centre: comparison of the spherical {$Fe_{30}Mo_{72}$} fragment **1** (a) and the {Mo_{132}} fragment **3** (b) with the corresponding basic fragments {(Mo)Mo_5} and {Fe} in **1** (a) as well as {(Mo)Mo_5} and {Mo^V_2} in **3** (b) (color code as above; viewed along a C_2 axis). Bottom: structures spanned by the fundamental spacer units (Archimedean solids of Fig. 4) of **1** and **3** with one of the 12 capping {(Mo)Mo_5} pentagons: the Fe_{30} icosidodecahedron (12 pentagons and 20 triangular faces) in **1** and a truncated {Mo^V_2}$_{30}$ icosahedron (12 pentagons and 20 hexagons with trigonal symmetry) in **3** (b).

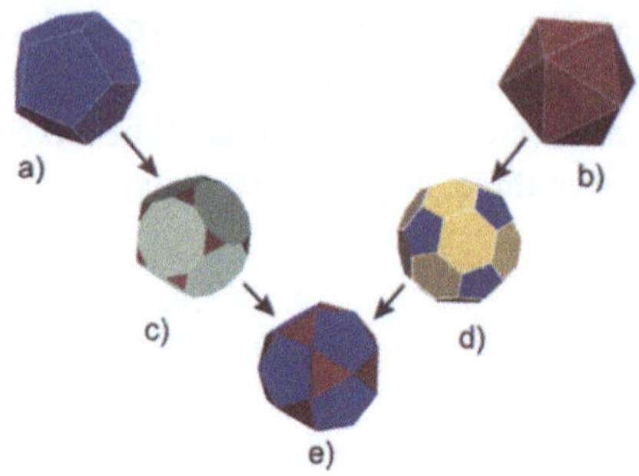

Fig. 4: Relationship between Platonic and Archimedean solids with icosahedral symmetry: the dodecahedron (a) and icosahedron (b), the two corresponding truncated Archimedean solids (in c and d similar to the { $Mo^{V}_{2}\}_{30}$ fragment in the case of **3** as well as the Archimedean icosidodecahedron (e; corresponding to the $\{Fe_{30}\}$ fragment in the case of **1**). The solids in d and e are distinguishable by a different degree of truncation and the related size. This corresponds formally to the relative sizes of **1** and **3**.

Recognizing the similarity between Kepler's early cosmological models discussed in his book "Mysterium cosmographicum" (Fig. 5) and the above mentioned Mo_{132}-type cluster **3** in which the central Mo atoms of the 12 pentagonal bipyramids form an icosahedron and are located on the sphere formed by all of its 132 Mo atoms, the species was the first dubbed a keplerate (Fig. 5).

Fig. 5: Cover picture of the final 1998 issue of Angew. Chem. showing the published keplerate. 3 and the relation to Kepler's early cosm,os model in his Mysterium Cosmographicum.

4. Triangulation and Magic Numbers of Viruses and Keplerates

As mentioned already, the $\{Mo_2\}_{30}$ fragment within the Mo_{132} cluster **3** shows striking similarities with the famous C_{60} fullerene (Figs. 1 and 3). Remarkably, its topology is also reminiscent to that of certain spherical viruses ([8], see also Fig. 2).

These spherical viruses display 60 copies of an identically packed fundamental structural motif that can formally be assigned to pentagonal and hexagonal *capsomers* (morphological units of chemically identifyable oligomers consisting of one or more viral proteins or protein subunits). One of the simplest example is the satellite tobacco necrosis virus (STNV) in which only twelve pentagonal capsomers with a total of 60 identical viral protein subunits – coded by one gene only – are placed at the twelve vertices of an icosahedron (Fig. 2). In the case of the larger spherical viruses – for example those with 3 x 60 = 180 subunits – the motif that occurs 60 times contains three protein monomers, one that is part of a pentagonal capsomer and two that are part of a hexagonal capsomer. In other words, viewing the virus as being organized in capsomers, there are 12 pentagonal capsomers each consisting of five identically connected protein monomers centered at the vertices of an icosahedron, and 20 additional hexagonal capsomers each one built up from six protein monomers.

According to Goldberg [4] and, independently, to Caspar and Klug [5] who were inspired by Fuller's geodesic domes, such structures can be constructed as follows: One takes the icosahedron on the one hand and an Euclidean planar tiling exhibiting sixfold rotational symmetries on the other. Then one replaces each of the 20 triangular faces of the icosahedron by 20 identical copies of an equilateral triangle P, Q, R cut out from the given planar tiling T whose vertices P, Q, R form centres of sixfold rotational symmetries of T (the existence of which we have just presupposed above).

In particular, by applying this simple procedure to the "Platonic" regular tiling T = T(3,6) of the Euclidean plane formed exclusively by equilateral triangles of the same size, one gets polyhedra with icosahedral symmetry whose faces are exclusively triangles, with either exactly five or six of them meeting at each of the polyhedron's vertices. Conversely, as was shown as early as in 1937 by Goldberg, every polyhedron satisfying these conditions can in turn be constructed in this way using the various equilateral triangles that can be found in T(3,6) (cf. [4], see also [14]). The resulting solids are also called icosadeltahedra, their subunits exhibiting, in general, some local sixfold (quasi)-symmetry in addition to the global symmetry inherited from the *parent icosahedron*. Note that the icosahedron itself as well as the capped icosahedron are such solids.

Note also (see Fig. 6) that

(i) the centres of sixfold rotational symmetries of the planar regular tiling T(3,6) are exactly the vertices of that tiling,

(ii) any two of those, say P and Q, determine a unique equilateral triangle PQR in the plane (in, say, counter clockwise orientation) whose third vertex R = R(P,Q) then is also a vertex of T(3,6) and, thus, a centre of a sixfold rotational symmetry of T(3,6),

(iii) for every two vertices P and Q of T(3,6) as above, there exists always a unique *path* p=p(P,Q) from P to Q along the edges of the triangles in T(3,6) that first goes straight for

$$h = h(P,Q) > 0$$

edge lengths, and then either stops or turns left by 60 degrees and then goes on straight for another

$$k = k(P,Q) > 0$$

edge lengths,

(iv) the actual Euclidean distance m between P and Q can be computed from these two numbers h and k (with k:=0 in case no left turn is required) using basic school geometry and turns out to be exactly the square root $\sqrt{t}$ of the so called triangulation number

$$t = t(P,Q) := h^2 + hk + k^2$$

associated with P and Q and used for the classification of viruses [1s, 1t] (times the unit length of the edges of the 'basic' triangles in T(3,6) called 'Δ' in Fig. 6) and that, consequently,

(v) the area of the (large) equilateral triangle T with the three vertices P, Q, and R=R(P,Q) exceeds that of the basic triangles in T(3,6) exactly by the factor m^2=t(P,Q)=hh+hk+kk and, hence, it exceeds the area of the *fundamental domain* of T(3,6) relative to its rotational symmetry group (the dark area within Δ in Fig. 6) by 3 times this factor because each of the basic triangles of T(3,6) (having a threefold centre of symmetry at their centre while no further rotational symmetry of T(3,6) transforms any such triangle into itself) contains exactly three copies of this domain – so altogether, the resulting icosadeltahedron contains 20·3·t=60t copies of this fundamental domain.

It follows that the number of capsomers in an icosahedral virus with triangulation number t is 2+10t, the so called *magic number* associated with t, because – quite generally – there are always 60=12·5 of the altogether 60t protein monomers that form the topologically required 12 pentagonal capsomers while the remaining 60t-60=60(t-1) form 60(t-1)/6=10(t-1) hexagonal capsomers, giving rise to altogether 12+10(t-1)=2+10t capsomers (for further details, see Coxeter [3], Goldberg [4], and Stewart [1q]).

So, following Caspar and Klug and defining the *elementary subunits* in the icosadeltahedron constructed from two points P,Q as above, to be the copies of the – or rather of some, yet fixed – fundamental domain of T(3,6) relative to its rotational symmetry group, we see that the number of these subunits is exactly 60 = 3·20 times the triangulation number t=t(P,Q) of P and Q.

Consequently, the larger – and more abundant – spherical viruses such as TBSV (the tomato bushy stunt virus) can easily consist of more than 60 elementary subunits. A viral

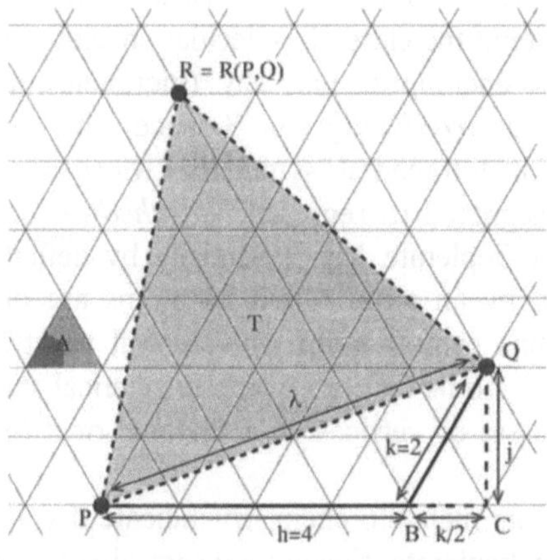

Fig. 6: This figure depicts the tiling T(3,6) together with two arbitrarily chosen vertices to P and Q as well as the third vertex R uniquely determined by the requirement that P, Q and R form a positively oriented equilateral triangle (highlighted by shading), and the two rectangular triangles PCQ and BCQ derived from P and Q, needed to determine the distance m of P and Q by using the theorem of Pythagoras twice: Indeed, the Pythagorean theorem, applied to the triangles PCQ and BCQ, implies

$$m^2 = (h + k/2)^2 + j^2 = h^2 + hk + ((k/2)^2 + j^2) = h^2 + hk + k^2,$$

(see text). In addition, a basic triangle Δ (relative to I) is also highlighted.

structure with t=3 (h=k=1) consists of 180 (= 3 times 60) elementary subunits and 32 capsomers whereas the triangulation number t=1 (h=1,k=0) applies to the small (10 nm) STNV mentioned above, and also for that fragment in each of the three clusters which consists of the altogether 60 MoO_6 octahedral subunits of which five at a time, together with the 12 central pentagonal bipyramids MoO_7, form the 12 {(Mo)Mo_5} pentagonal building blocks P. For Buckminster Fuller's topologically comparable geodesic dome, we have t=36 (h=6,k=0).

Note, by the way, that the parameters h and k are, in general, not determined by their triangulation number $t = t(h, k) = h^2+hk+k^2$, and that t(h,k)=t(h',k') can hold for rather distinct parameter values h,k and h',k': Indeed, we have t(1,11)=133=t(9,4). More generally, it is easy to see that, given some numbers h,k,h',k', one has t(h,k)=t(h',k') whenever some numbers a, b and u, v with

$$h = au - bv, k = av + bu + bv, h' = au + av + bv, \text{ and } k' = bu - av$$

exist because this implies – as one can easily check by direct computation – that t(h,k)=t(a,b)t(u,v) as well as t(h',k')=t(a,b)t(u,v) must hold. In addition, it is a simple consequence of the renowned *Theorem 90* from Hilbert's celebrated ***Zahlbericht*** from 1897 [15] that, conversely, the equation t(h,k)=t(h',k') also implies the existence of such numbers a,b,u,v. In the case above, we have a=2, b=3, u=2, and v=1.

For further extensive literature regarding triangulation numbers and related topics, see [1p, 1q, 3, 5].

In the present context of highly symmetric polyhedral objects of roughly spherical shape, it seems worthwhile to extend the class of molecular species that should be designated to represent a keplerate to encompass just any (inorganic) molecular species with a symmetry group that acts *irreducibly* on 3-space, that is, with (proper or full) icosahedral, octahedral or tetrahedral symmetry. More specifically, such molecular species might also be designated to represent *icosahedral*, *octahedral* and *tetrahedral* keplerates. Clearly, every keplerate has, essentially by definition, one central point – whether or not occupied by an atom – and its atoms are organized in one or more spherical shells around this central point (this central point being also the centre of gravity of the collection of atoms on each of these spherical shells) while each symmetry class of atoms forms the set of vertices of a Platonic or a (generalized) Archimedean solid.[1]

Fascinating new results from magneto chemistry indicate that – not quite unexpectedly – the magnetic behavior of keplerate clusters appears also to exhibit rather specific and highly intriguing properties. At present, this has been established at least for tetrahedral and icosahedral keplerates. Cluster **1**, which contains 150 unpaired electrons (a world record) and which is expected to exhibit the strongest molecular paramagnetic forces yet observed, is the first known molecule that shows properties of bulk materials and that can be treated with classical Heisenberg formalism [16]. The reason for this is probably the particular behavior of a large ensemble of electrons, placed all over the surface of a giant molecular sphere.

5. Topological Considerations

5.1 General Aspects

Based on the definition of keplerates proposed above and motivated by the chemistry of polyoxometalates, we may similarly define – in purely mathematical terms – any convex polyhedron in 3-space with full (or just proper, i.e. rotational) icosahedral, octahedral, or tetrahedral symmetry to be a keplerate or – more precisely – an I_h, O_h, or T_h (or I, O, or T) keplerate, respectively.

[1] Essentially by definition, a Platonic solid is a convex polyhedron whose symmetry group acts transitively on its set vertices, edges, and faces, while an Archimedean solid is a convex polyhedron whose symmetry group acts transitivly just on its set of vertices and whose edges all have the same length. Dropping the latter requirement, we arrive at the class of generalized Archimedean solids. It is easy to see, yet still remarkable, that topologically – that is, apart from measure and proportion – there is no difference between Archimedean solids and generalized Archimedean solids: The various edges in a generalized Archimedean solid can always be rescaled so that a proper Archimedean solid results.

To analyze and to classify such keplerates, we start by observing that every symmetry operation

$$\sigma : P \to P$$

of a convex polyhedron P embedded in $\mathbf{E}^3$ (the 3-dimensional Euclidean space) extends uniquely to an isometry

$$\sigma^* : \mathbf{E}^3 \to \mathbf{E}^3$$

of the full Euclidean space $\mathbf{E}^3$. Hence, assuming that, for any two vertices v_1, v_2 of P, there exists a symmetry operation σ in the symmetry group Symm (P) of P that maps v_1 onto v_2 (or that, in more mathematical terms, the symmetry group of P acts transitively on the set of vertices of P), it is easy to see that we can reconstruct the set of all vertices of P from (i) any single one of its vertices and (ii) the group

$$\Sigma = \mathrm{Symm}^*(P) := \{\sigma^* \mid \sigma \in \mathrm{Symm}(P)\}$$

of extended symmetries of P.
Consequently, when dealing with polyhedra P with a large symmetry group, it makes sense to change perspectives (from a bottom-up to a top-down view) and to start with the group Σ of its extended symmetries considered as just some nice finite subgroup of the group of all isometries of $\mathbf{E}^3$, as the primary object of interest.
One can then choose *any* point v in $\mathbf{E}^3$ to start with, and consider the convex polyhedron $P(\Sigma, v)$ in $\mathbf{E}^3$ that is spanned by the set

$$\Sigma v := \{\sigma(v) \mid \sigma \in \Sigma\}$$

of images of v with respect to the symmetries σ of Σ, the so called Σ-orbit of v.
This allows not only to recover the original polyhedron (provided its symmetry group acts transitively on its set of vertices), but also to view this polyhedron as a particular instance, chosen from a continuous family of such polyhedra, viz. the family of all polyhedra one obtains by varying the point v in $\mathbf{E}^3$.
In addition, this point of view permits us also to deal with polyhedra that exhibit a more complex structure, that is, with polyhedra whose symmetry group – though large – may not act transitively on its set of vertices anymore. In this case, one may just take several points $v_1, v_2, \ldots, v_k$ in $\mathbf{E}^3$ and form the convex polyhedron that is spanned by the union of the Σ-orbits generated by v_1, v_2, ... ,v_k, that is, by all the Σ-images $\sigma(v_i)\ (\sigma \in \Sigma, i = 1, \ldots, k)$ of the all points $v_1, v_2, \ldots, v_k$.
Clearly, given the group Σ to begin with, we may classify the points v in $\mathbf{E}^3$ relative to Σ

according to (the type of) their local, or point symmetry subgroup, i.e. the *stabilizer* subgroup

$$\Sigma_v := \{\sigma \in \Sigma \mid \sigma(v) = v\}$$

of v in Σ. Recall that the *order* (or cardinality) of the Σ-orbit of a point v in $\mathbf{E}^3$ always coincides with the *index* of its stabilizer subgroup in the full symmetry group (i.e. with the integer one gets by dividing the cardinality or order of the full symmetry group by the order of its point symmetry group, which is always a divisor of the former according to Lagrange's Theorem).

Clearly, there is exactly one point $v = v_\Sigma$ – the centre of gravity of P – whose stabilizer group is the full symmetry group Σ. In any coordinate system used to specify Σ (and/or P), this point is generally taken to be the coordinate system's origin, and we will therefore refer to it below as the *coordinate centre*.

In view of the importance of icosahedral structures within the general context referred to above as well as within the context of polyoxometalate chemistry in particular (and also to suppress unnecessary technicalities), we will restrict our attention in the following exclusively to structures with proper icosahedral symmetry, i.e. to I keplerates.

In this case, the stabilizer group Σ_v of any point v in $\mathbf{E}^3$ that is distinct from v_Σ, is either the trivial group C_1, or it is one of the groups C_2, C_3, or C_5, that is, it is a cyclic group consisting of (the identity transformation and) one twofold, two threefold, or four fivefold rotations around a fixed axis connecting v with the coordinate centre.

Moreover, the special structure of the icosahedral symmetry group implies that all points v in $\mathbf{E}^3$ with a given fixed distance to the coordinate centre and local symmetry group of type either C_2, C_3, or C_5 are symmetry equivalent with respect to the full symmetry group I. Consequently, our assumption that keplerates are, by definition, supposed to be *convex* polyhedra (and that they cannot therefore contain two distinct vertices on the same ray originating from the coordinate centre) implies that the set of vertices in an I keplerate contains at most one Σ-orbit of order 30=60/2, at most one of order 20=60/3, and at most one of order 12=60/5 while all other orbits must have order 60. Note however that this does not hold anymore for non-convex structures: In each of the three cluster **1**, **2**, and **3**, the two oxygen atoms at the tips of each of the central pentagonal bipyramids MoO_7, as well as the corresponding central Mo atoms, all lie on the same 5-fold symmetry axis and, together, form three distinct orbits of order 12.

5.2 Keplerates with N=12a+20b+30c+60d Atoms (a, b, c either 0 or 1, d an arbitrary integer)

It follows already from this simple observation that the total number N of atoms in an I keplerate (or the number of equivalent atoms in a fully symmetric fragment) cannot be arbitrary. Instead, it is necessarily one of the following 8 forms:

N=60n: All atoms have trivial local symmetry; examples for n=1 are provided by the Mo_{60} fragments spanned by the 60=12·5 Mo atoms within the MoO_6 octahedra in all three clusters **1**, **2** and **3** discussed above and, for n=2, by the collection of all 120 Mo atoms in the Mo_{132} cluster **3** outside the centres of the 12 pentagonal units.

N=12+60n: Atoms with trivial and fivefold symmetry, only; examples for n=1 are provided by the collection of all 72 Mo atoms in the 12 {$(Mo)Mo_5$} units in each of the three clusters and, for n=2, by all 132 Mo atoms in the Mo_{132} cluster **3**.

N=20+60n: Atoms with trivial and threefold symmetry, only.

N=30+60n: Atoms with trivial and twofold symmetry, only; examples for n=0 are the 30 Fe atoms of **1** and 30 Mo atoms of **2**.

N=32+60n: Atoms with trivial, threefold and fivefold symmetry, only.

N=42+60n: Atoms with trivial, twofold and fivefold symmetry, only; examples for n=1 are provided by all metal atoms in the $Fe_{30}Mo_{72}$ cluster **1** as well as in the Mo_{102} cluster **2**, respectively; an example for n=0 is provided by cluster **1** with the 30 Fe atoms together with the 12 Mo atoms spanning the icosahedron

N=50+60n: Atoms with trivial, twofold and threefold symmetry, only.

N=62+60n: Atoms with all types of local symmetry.

Here, n is any non-negative integer, and one has n=0 if and only if no atoms with non-trivial local symmetry exist. Note that, for a given n, there can be one or more distinct 'combinatorial' types of I keplerates (see Table 1). For n=0, there is – up to equivalence – exactly one I keplerate with 12, 20, or 30 atoms, respectively, the icosahedron, the dodecahedron, and the icosidodecahedron, all actually having full icosahedral symmetry, while there are four distinct types of I keplerates with 60 atoms, all but one having realizations with full icosahedral symmetry. There is also exactly one keplerate with 50 and exactly one keplerate with 42 atoms, while there are three types of keplerates with 32 atoms and three types with exactly 62 atoms, all eight types also exhibiting full icosahedral symmetry.

Table 1 (next page):
For any given number N of atoms up to N=180, this table lists the number of combinatorially (or, equivalently, *topologically*) distinct I and I_h keplerates as well as the (larger) number of combinatorially distinct tilings of the sphere with I und I_h symmetry, respectively, and – finally – the differences between corresponding pairs of numbers, i.e. the number of such tilings that are not combinatorially equivalent to (and, hence, cannot be derived from) a corresponding keplerate.

The algorithms used for computing the numbers in this table actually do not only compute the numbers given, but (the Delaney symbol [17] of) each of these tilings. They are based on the methods developed in [17]. Remarkably, the numbers that result for N = 12 + 30c + 60n and N = 20 + 30c + 60n (c either = 0 or 1, n any natural number) always coincide. For icosahedral tilings, this is a simple consequence of the theory of Delaney symbols [17] or – almost equivalently – the theory of orbifolds [18]. For the keplerates, some additional reasoning is necessary based on Ernst Steinitz' classical theorem (and its more recent elaboration by Peter Mani) that an (equivariant) spherical tiling is (equivariantly) *realizable* as a polyhedron if and only if it is 3-connected.

N	I keplerates	I_h keplerates	I tilings	I_h tilings	non-polyhedral I tilings	non-polyhedral I_h tilings
12	1	1	1	1		
20	1	1	1	1		
30	1	1	1	1		
32	3	3	3	3		
42	1	1	1	1		
50	1	1	1	1		
60	4	3	4	3		
62	3	3	3	3		
72	9	7	11	9	2	2
80	9	7	11	9	2	2
90	7	7	7	7		
92	28	23	32	27	4	4
102	16	15	18	17	2	2
110	16	15	18	17	2	2
120	44	23	50	27	6	4
122	39	31	43	35	4	4
132	123	71	161	101	38	30
140	123	71	161	101	38	30
150	89	67	99	75	10	8
152	368	207	467	285	99	78
162	214	139	263	175	49	36
170	214	139	263	175	49	36
180	656	223	824	311	168	88
Total number	**1970**	**1059**	**2443**	**1385**	**473**	**326**

5.3 A Geometric Construction of Keplerates with Low Atom Numbers

In terms of the procedure for constructing keplerates mentioned above, this can be easily explained:
To construct the four distinct types of I keplerates with 60 atoms, start with some point v somewhere on the surface of an icosahedron Q with $\Sigma = \text{Symm}^*(Q)$ and consider the convex polyhedron P spanned by the Σ-orbit Σv of v. The resulting keplerate will have 60 atoms if and only if v does not lie on any of the rotational axes of Σ, i.e. if and only if v is neither a vertex of Q, nor the midpoint of one of its edges or faces. So, four cases remain: it can either be somewhere properly in between any two adjacent of such points, i.e. on the straight line connecting a vertex w of Q with the midpoint of either an edge or a face of Q adjacent to w, or on the straight line connecting the midpoint of a face of Q with that of one of its edges; or it is somewhere in the interior of the altogether 120 triangles formed by these straight lines. These four cases give rise to the four distinct types of keplerates with exactly 60 atoms.
The keplerates with 50 and 42 atoms result from choosing two points v_1 and v_2 on the surface of Q, one, say v_1, being the midpoint of one of its edges, and the other one, say v_2, being either the midpoint of one of its faces or one of its vertices, respectively. In the first case, one then chooses a point v_2^* slightly above v_2 and, hence, just outside the convex polyhedron spanned by the Σ-orbit of v_1, and considers the convex polyhedron spanned by the union of this Σ-orbit and the Σ-orbit generated by v_2^*. In the second case, one chooses a point v_2^* slightly below v_2 so that the convex polyhedron spanned by the Σ-orbit of v_2^* does not contain v_1, and then considers, as before, the convex polyhedron spanned by the union of that Σ-orbit and the Σ-orbit of v_1.
The three keplerates with 32 atoms result in a similar way from choosing a vertex v_1 of Q and the midpoint v_2 of one of its faces, and then lifting v_2 slightly to a point v_2^* just above it. Considering then the convex polyhedron spanned by the union of the Σ-orbit of v_1 and that of v_2^*, one first gets a *capped* icosahedron, i.e. a polyhedron encompassing 60 triangles and resulting from an icosahedron by capping each of its faces with a trigonal pyramid. Lifting v_2^* a bit further, there is one limiting situation where any two adjacent triangles from two adjacent pyramids become coplanar and form a quadrangle. Lifting v_2^* further, a capped dodecahedron results, also encompassing 60 triangles, now however with six edges emanating from each vertex with local C_3 symmetry and five edges emanating from each vertex with local C_5 symmetry while, in the capped icosahedron, three edges emanate from each vertex with local C_3 symmetry and ten from each vertex with local C_5 symmetry.
In a similar way, one can construct the three types of keplerates with exactly 62 atoms, and one can also construct 9 types of keplerates with 72 atoms and 9 with 80 atoms.

5.4 Icosahedral Tilings of the Sphere

Clearly, every keplerate constructed in this way gives rise to an *icosahedral tiling*, that is a tiling of the sphere with icosahedral symmetry: one can project the straight edges of any such convex keplerate from the coordinate centre onto any sphere with the same centre surrounding the keplerate. It is worth noting that not all tilings of the sphere with I symmetry arise in this way; e.g., there are two tilings with exactly 72 and two with exactly 80 atoms that cannot be derived in this manner. However, combinatorial methods permit actually to enumerate all such spherical tilings even though combinatorial explosion sets in rapidly: there are 7 such tilings with 90 vertices, 50 with 120 vertices, and there are altogether 23691 tilings of the sphere with I symmetry with at most four distinct symmetry classes of vertices of which 16328 are necessarily chiral while the remaining 7363 tilings can be realized so that they exhibit full icosahedral symmetry.
Clearly, this implies that, without further restrictions, classification and enumeration – though mathematically possible – does not give much further insight.
However, there is a way that allows to proceed much beyond icosahedral structures with, say, four or five distinct symmetry classes of vertices/atoms by restricting attention to operations that construct (complex) keplerates from (simple) keplerates which we will discuss now.
In Section 4, we have discussed Goldberg's procedure that allowed to construct keplerates from tilings of the Euclidean plane with sixfold rotational symmetries.
Remarkably, this procedure can be inverted easily:
Given a tiling T of the sphere with icosahedral symmetry (or, for short, an icosahedral tiling), we can always assume that it is realized not on the sphere, but on the icosahedron such that its symmetry group is exactly the symmetry group of that icosahedron.
This is true even for structures like, say, the dodecahedron which can be realized in this way by choosing the midpoints of the various equilateral triangles of the icosahedron as the dodecahedral vertices and by connecting them by broken lines that stretch from those midpoints of the icosahedral triangles first straight to the midpoints of their respective edges and then straight on to the midpoint of the next triangle (see Fig. 7).

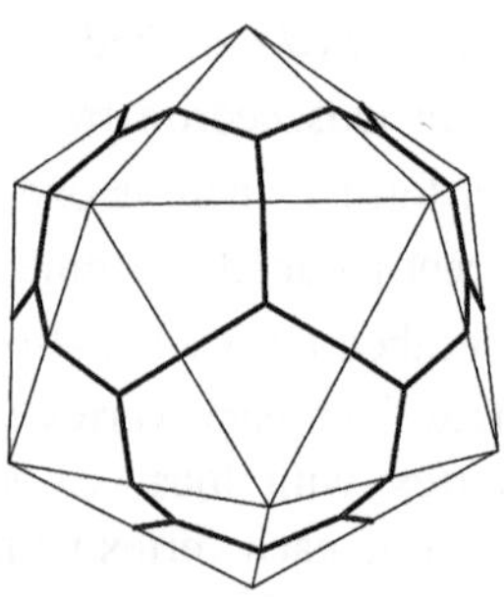

Fig. 7: The dodecahedral tiling realized by implanting 20 triangles into the icosahedron cut out from the planar hexagonal tiling as described in the text

We can now cut out one of the equilateral triangles of the icosahedron, take infinitely many copies of it, and place them in a regular fashion into the Euclidean plane, always six around each vertex – thus using the T(3,6) tiling as a sort of blue print. On applying this for the dodecahedral tiling of the icosahedron described above, the regular hexagonal tiling of the Euclidean plane will thus be created. In any case, this method will always produce a tiling T' of the Euclidean plane with centres of sixfold rotational symmetries at every vertex of the T(3,6) tiling that we used as our blue print.

Thus, we can now apply Goldberg's procedure by choosing two such centres P and Q arbitrarily as described in Section 4 and use them to construct a new tiling T" of the icosahedron (with icosahedral symmetry) by gluing together 20 copies of the equilateral triangle PQR with R=R(P,Q) the third point determined by P and Q, as above. In particular, given two integers $h>0$ and $k \geq 0$, we can choose P and Q so that h=h(P,Q) and k=k(P,Q) holds, which will then give rise to a tiling T" of the icosahedron that only depends on the input tiling T and the two integers h and k. Denoting this tiling by, say, G(h,k;T) (with 'G' for Goldberg), we see then that this construction yields, for given h and k, an operator G(h,k) that creates more complex icosahedral tilings from simpler ones.

If we apply these operators to the icosahedron itself, we will get exactly the family of icosadeltahedra; if we apply them to the dodecahedron, we get exactly the family of fullerene structures with icosahedral symmetry (see [14]).

It can also be shown that by applying the operator G(h',k') to G(h,k;T), we get the tiling G(h",k";T) with h"=hh'-kk' and k"=hk'+kh'+kk' in case hh'>kk', and with h"=hk'+kh'+hh' and k"=kk'-hh' otherwise.

Clearly, there are many further options for defining such operators all of which can most easily be defined by (a) referring to the theory of *Delaney symbols* (cf. [17]) and *orbifolds* (cf. [18]) and (b) combining this theory with methods for determining specific subgroups of crystallographic groups as studied in *symmetry breaking*, see for instance [19].

Acknowledgements:

We thank Dr. H. Bögge for his help in preparing the artwork and the Deutsche Forschungsgemeinschaft and the Fonds der Chemischen Industrie for highly appreciated financial support.

References:

1. a. I. Hargittai and M. Hargittai: *In Our Own Image: Personal Symmetry in Discovery*, Kluwer, Dordrecht (2000);
   b. H. Weyl: *Symmetry*, Princeton University Press, Princeton (1952);
   c. A. Müller: *ZiF-Mitteilungen* (Zentrum für interdisziplinäre Forschung der Universität Bielefeld) issue 4 (1999);
   d. H. Hubert et al.: *Nature* **391**, 378 (1998);
   e. D'Arcy W. Thompson: *On Growth and Form*, Cambridge University Press, Cambridge (1961);
   f. K. Mainzer: *Symmetries of Nature*, de Gruyter, Berlin (1996);
   g. E. Haeckel: *Kunstformen der Natur* (1904) Prestel, München (1998);
   h. M. Kemp: *Nature* **396**, 123 (1998);
   i. P. Sitte: *Biol. i. u. Zeit* **14**, 161 (1984);
   j. J. Glanz: *Science* **276**, 678 (1997);
   k. D. E. Ingber: *Scient. Amer.*, p. 30, (January 1998) (*Spektrum d. Wiss.*, p. 32 (March 1998));
   l. L. R. MacGillivray, J. L. Atwood: *Angew. Chem. Int. Ed.* **38**, 1018 (1999);
   m. A. J. Maniotis, C. S. Chen, and D. E. Ingber: *Chem. Proc. Nat. Akad. Sci.* (USA) **14**, 849 (1997);
   n. M. Brookes: *New Scientist* **2206**, 42 (1990);
   o. A. L. Mackay: *Symmetry* **1**, 3 (1990);
   p. W. Krätschmar and H. Schuster (eds.): *Von Fuller bis zu Fullerenen, Beispiele einer interdisziplinären Forschung*, Vieweg, Braunschweig (1996);
   q. I. Stewart: *Spiel, Satz und Sieg für die Mathematik*, Birkhäuser, Basel (1999);
   r. H. S. M. Coxeter: *Regular Polytopes*, Dover, 3rd Ed., New York (1973);
   s. D. Voet and J. G. Voet: *Biochemie*, VCH Weinheim (1992);
   t. P. Ball: *The Self-Made Tapestry, Pattern Formation in Nature*, Oxford University Press, Oxford (1999);
   u. D. E. Ingber: *Annual Review of Physiology*, **59**, 575 (1997).
2. B. Listing: *Gött. Studien* **1**, 811 (1847).
3. H. S. M. Coxeter: in J. C. Butcher (ed.): *A Spectrum of Mathematics*, Oxford University Press / Auckland University Press, Oxford/Auckland (1971).
4. M. Goldberg: *Tôhoku Math. J.* **43**, 104 (1937).
5. D. Casper and A. Klug: . *Quant. Biol.* **27**, 1 (1962).
6. M. T. Pope and A. Müller: *Angew. Chem. Int. Ed. Engl.* **30**, 34 (1991).
   A. Müller, F. Peters, M. T. Pope, and D. Gatteschi: *Chem. Rev.* **98**, 239 (1998).
7. A. Müller, P. Kögerler, and H. Bögge: *Structure and Bonding* **96**, 203 (2000).
8. A. Müller, E. Krickemeyer, H. Bögge, M. Schmidtmann, and F. Peters: *Angew. Chem. Int. Ed. Engl.* **37**, 3360 (1998).
9. A. Müller, P. Kögerler, and C. Kuhlmann: *J. Chem. Soc., Chem. Commun.* 1347 (1999).
10. A. Müller, S. Sarkar, S. Q. Nazir Shah, H. Bögge, M. Schmidtmann, Sh. Sarkar, P. Kögerler, B. Hauptfleisch, A. X. Trautwein, and V. Schünemann: *Angew. Chem. Int. Ed. Engl.* **38**, 3238 (1999).
11. A. Müller, S. Polarz, S. K. Das, E. Krickemeyer, H. Bögge, M. Schmidtmann, and B. Hauptfleisch: *Angew. Chem. Int. Ed. Engl.* **38**, 3241 (1999).
12. A. Müller, S.-Q. Nazir-Shah, H. Bögge, M. Schmidtmann, P. Kögerler, B. Hauptfleisch, S. Leiding, and K. Wittler: *Angew. Chem. Int. Ed.* **39**, 1614 (2000).
13. A. Müller, E. Krickemeyer, S. K. Das, S. Sarkar, H. Bögge, M. Schmidtmann, P. Kögerler, and Sh. Sarker: *Angew. Chem. Int. Ed.* **39**, 1612 (2000).

14. P. W. Fowler and D. E. Manolopoulos: *An Atlas of Fullerenes*, Oxford University Press, Oxford (1995).
15. D. Hilbert: *Die Theorie der Algebraischen Zahlkörper*, Jahresbericht der DMV **4**, 175 (1897).
16. A. Müller, P. Kögerler, and M. Luban: *Inorg. Chem.* (in press);
A. Müller, M. Luban et al. (in preparation).
17. O. Delgado Friedrichs, A. W. M. Dress and D. H. Huson in: R. Corriu and P. Jutzi (eds.): *Tailor-Made Silicon-Oxygen Compounds: From Molecules to Materials*, Vieweg, Braunschweig (1996);
O. Delgado Friedrichs, A. W. M. Dress, D. H. Huson, J. Klinowski, and A. L. Mackay: *Nature* **400**, 644 (1999);
O. Delgado Friedrichs, A. W. M. Dress, A. Müller, and M. T. Pope: *Molecular Engineering* **3**, 9 (1993);
A. W. M. Dress in L. Smith (ed): *Algebraic Topology, Lecture Notes in Mathematics* **1172**, Springer, Heidelberg, **56** (1984);
A. W. M. Dress: *Advances in Mathematics* **63**, 196 (1987).
18. P. Scott: *Bull. London Math. Soc.* **15**, 401 (1983).
19. A. W. M. Dress and D. H. Huson: *Structural Topology* **17**, 5 (1991).

Syntheses and Crystal Structure Studies of Novel Selenium- and Tellurium-Substituted Lacunary Polyoxometalates

B. KREBS, E. DROSTE AND M. PIEPENBRINK
Westfälische Wilhelms-Universität, Anorganisch-Chemisches Institut, Wilhelm-Klemm-Straße 8, D-48149 Münster, Germany

(Received: 20 October 1999)

Abstract. The enormous potential of polyoxometalates in fields of catalytic, technical or medical applications is dependent on the synthesis and structural characterization of new heteropolytungstates and -molybdates. As we are aware of the necessity to suggest new models for catalytic processes we put emphasis on the detailed characterization of so far unknown heteropolyoxoanions. Emphasis is given to a comprehensive investigation of the structures of polytungstates formed by Keggin derived fragments.
The following chapter deals with syntheses and structures of novel selenium- and tellurium-substituted heteropolyoxometalates. Hitherto, only a few polyoxometalates containing Se or Te are reported; this field is hardly investigated.
Giving results of the structural characterization of $[(Mn(H_2O))_2OSe_6W_{24}O_{94}]^{18-}$ (**6**), $[Fe_4Se_8W_{36}O_{136}(OH)_6]^{24-}$ (**7**) and the $[Na_9Se_{13}(VO_4)_3V_{23}Mo_{19}O_{151}]^{21-}$ -anion (**8**), three new structural arrangements of macro-heteropolyanions are described. Furthermore, we focus on sandwich-like polyoxotungstates consisting of defect Keggin fragments like B-XW_9O_{33}. In the anion $[Se_2V_2W_{16}O_{60}]^{8-}$ lacunary subunits are directly linked together whereas in the other ten compounds (**1-3, 5, 9-14**) connection is reached by oxygen atoms coordinated to transition metal atoms. All crystal structures were determined by single crystal X-ray methods.

Key words: polyoxometalates, selenium, tellurium, structural characterization.

1. Introduction

In recent years there has been a growing interest in the remarkable properties and structures of polyoxoanions of tungsten, molybdenum and vanadium [1]. Polyoxometalates can be described as molecular blocks of metal oxide formed by MO_6 octahedra sharing corners, edges and faces. There are two generic families, the isopolyoxometalates, which contain Mo, W or V in their highest oxidation states, and the heteropolyoxometalates, which contain at least one p- or d-block element as heteroatom. The chemistry of heteropolyanions and their related isopoly compounds mostly concerns with their synthesis, structural characterization, properties and applications [2].
Research efforts on isopolymetalates range from mono- and polynuclear compounds to giant species such as $[(MoO_3)_{176}(H_2O)_{80}H_{32}]$ have been reviewed during the last decade [2-5]. Another important field are the studies on polyoxoanions serving as ligands in organometallic complexes [6,7].

M.T. Pope and A. Müller (eds.), Polyoxometalate Chemistry, 89–99.

Regarding the class of heteropolyanions, more than 70 different elements have been reported as constituents. The structural type most investigated is the Keggin anion, typically represented by the formula $[XM_{12}O_{40}]^{x-8}$, where X is the central atom (Si^{IV}, P^{V}, etc.), x is its oxidation state, and M is the metal ion (W^{VI} or Mo^{VI}). It is composed of a XO_4 tetrahedron surrounded by twelve edge- and corner-sharing metal-oxygen octahedra [8]. A family of closely related species has been generated by modification of this fundamental framework [9].
One of them is the lacunary or "defect" Keggin structure which can be obtained by removal of MO_6 octahedra. Dependent on which addenda atoms are removed, two isomers termed A- and B-XM_9O_{33} are formed. These subunits can be linked by transition metal ions leading to a sandwich-like heteropolyanion. We compiled our latest works on the systematic syntheses of Sb^{III} and Bi^{III} containing heteropolyanions such as $[SbW_9O_{33}]^{9-}$, $[Na_2Sb_8W_{36}O_{132}(H_2O)_4]^{22-}$, $[X_2W_{22}O_{74}(OH)_2]^{12-}$ (X = Sb^{III}, Bi^{III}) and $[X_2W_{20}M_2O_{70}(H_2O)_6]^{(14-2n)-}$ (X = Sb^{III}, Bi^{III}; M^{n+} = Fe^{III}, Co^{II}, Zn^{II}, Ni^{II}, Mn^{II}) [10-13]. A number of literature reports discuss the successful use of these transition metal substituted polyoxometalates in oxidation chemistry [11,14-18].
Another important structural type of heteropolyanions is the Dawson structure. It consists of two A-type α-XW_9 units with six "polar" and twelve "equatorial" tungsten atoms. The syntheses of $[P_2W_{18}O_{62}]^{6-}$ and $[As_2W_{18}O_{62}]^{6-}$ afford chiefly the α-isomer and variable amounts of β-isomer in which one group of three polar MO_6 octahedra has been rotated by $\pi/3$ [19-21].
Only a few examples of polyoxometalates with selenium as heteroatom have been identified. Selenotungstates with a ratio of Se:W = 1:6 and 1:12 have been reported [22,23] although structural characterizations by X-rays were not included. Various parts of other research efforts based on organometallic complexes such as $[PhSeMo_4O_{15}H]^{3-}$ [6,7]. Sasaki and co-workers described the anion $[Se_2Mo_5O_{21}]^{4-}$ [24] consisting of two SeO_3 units and five MoO_6 octahedra. During the last decade the structure of $[Se_2Mo_2V_6O_{28}]^{6-}$ [25] has been successfully described. Robl *et al.* compiled their work on the preparation of selenium substituted heteropolyanions with the formula $[XY_nMo_6O_{21+4n}(OH)_{6-2n}]^{8-}$ (n = 0-3; X = Se^{IV}, Y = Se^{VI}, S^{VI}) [26].
Polyoxometalates containing tellurium are even less investigated. TeW_6 and TeW_{18} have been mentioned by Ganelina *et al.* [27] and R. Ripan and N. Calu [28].
During the late 1980s Sasaki and co-workers carried out potentiometric studies on telluric and selenic acids plus molybdates [29,30].

In this chapter, we give a comprehensive outlook on new aspects in the field of selenium- and tellurium-substituted lacunary polyoxometalates which involve incorporation of low-valent transition metals.

2. Selenium Substituted Heteropolytungstates

2.1 $Na_{10}[Cu_3(H_2O)_3(SeW_9O_{33})_2] \cdot 15H_2O$ (**1**), $(Me_4N)_7[Cu_2(H_2O)_2W(OH)(SeW_9O_{33})_2] \cdot 12H_2O$ (**2**) AND $Na_2(Me_4N)_3[CuW_2(OH)_3(SeW_9O_{33})_2] \cdot 18H_2O$ (**3**)

Compounds **1-3** can be prepared by reacting $Na_2WO_4 \cdot 2H_2O$ and H_2SeO_3 in an aqueous, acidic solution with stoichiometric amounts of $Cu(NO_3)_2 \cdot 3H_2O$.
The X-ray structure analyses confirm sandwich-like double Keggin structures. The fundamental common feature is represented by the trivacant α-B-SeW_9O_{33} unit which can be derived from the Keggin-structure by removing one triad of edge-sharing WO_6 octahedra. The centre of each lacunary subunit is occupied by a selenium atom with a pseudotetrahedral environment. Linking of the two Keggin-type halves occurs through Cu-O as well as W-O bridges, depending on the stoichiometric ratio of tungstate and copper in the solution. The bridging copper ion reveals a square-planar coordination sphere and causes the observed green colours.

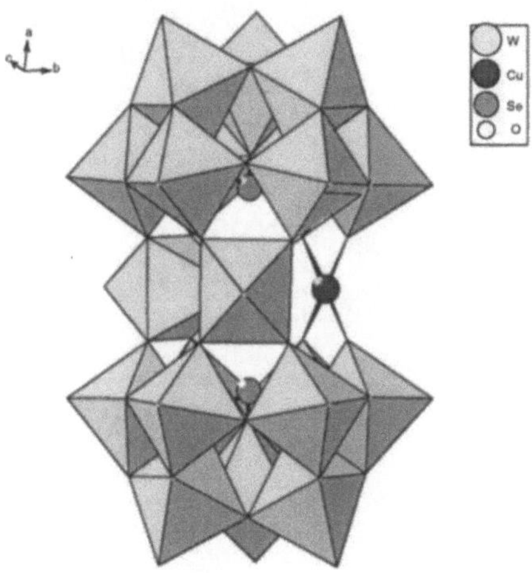

Figure 1: Molecular structure of $[CuW_2(OH)_3(SeW_9O_{33})_2]^{5-}$

A similar structure $[Cu_3(H_2O)_2(AsW_9O_{33})_2]^{12-}$ consisting of α-B-type AsW_9 units is reported by G. Hervé *et al.* [31].

2.2 $(Me_4N)_7Na[Se_2V_2W_{16}O_{60}] \cdot 21H_2O$ (**4**)

An aqueous solution of Me_4NCl is added to a mixture of $Na_2WO_4 \cdot 2H_2O$, H_2SeO_3 and $VOSO_4 \cdot 5H_2O$ in the same solvent. After a few days red crystals of **4** can be obtained.
Each $[Se_2V_2W_{16}O_{60}]^{8-}$ -anion consists of two SeW_9 units linked by six oxygen atoms whereby one half reveals a 60° rotation. The SeW_9 group can be identified as α-B-Keggin fragment with a tri-coordinated Se^{IV} atom in the centre. As seen in other

reported vanadium substituted polyoxotungstates, the vanadium atoms occupy statistically the positions of the tungsten atoms. Elemental analyses confirm the composition obtained by X-rays.

2.3 $Na_{12}[WCo_3(H_2O)_2(CoW_9O_{34})_2] \cdot 36H_2O$ (**5**)

Dark green crystals of **5** can be obtained by heating acidified sodium tungstate solution with Na_2SeO_3 and $CoCl_2 \cdot 6H_2O$ in stoichiometric ratio.
The structure of $[WCo_3(H_2O)_2(CoW_9O_{34})_2]^{12-}$ is closely related to that of earlier reported tetranuclear sandwich complexes [32-35]. The anion consists of two analogous α-B-$[CoW_9O_{34}]^{12-}$ fragments which are connected by a system of four coplanar metal atoms. These four metal atoms reveal a fairly regular octahedral coordination site in a close-packed arrangement. Remarkably, two metal atom sites are related by an inversion symmetry and are surrounded by six oxygens belonging to the lacunar units.

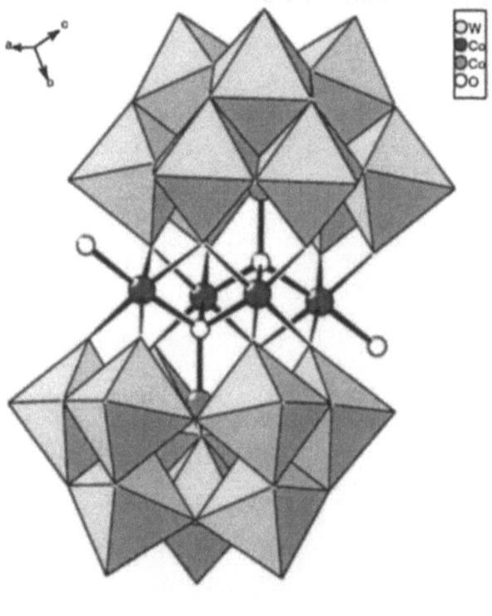

Figure 2: Molecular structure of $[WCo_3(H_2O)_2(CoW_9O_{34})_2]^{12-}$

The crystallographic investigation reveals a random occupation either by tungsten or cobalt atom. Refinement of a disorder model resulted in occupancies of 50 % Co and 50 % W. The remaining two cobalt atoms are surrounded octahedrally by five oxygens belonging to the CoW_9 units and one terminal oxygen atom from a coordinated water molecule.

2.4 $Na_{18}[(Mn(H_2O))_2OSe_6W_{24}O_{94}] \cdot 53H_2O$ (**6**)

The yellow sodium salt of the $[(Mn(H_2O))_2OSe_6W_{24}O_{94}]^{18-}$-anion was obtained by reaction of $Na_2WO_4 \cdot 2H_2O$, H_2SeO_3 and $MnCl_2 \cdot 6H_2O$ in stoichiometric amounts.
The crystallographic investigation of **6** reveals two identical $Se_3W_{12}O_{47}$ fragments which are joined by two manganese atoms and two Se-O-Na bridges. Figure 3 shows the

structure of the $[(Mn(H_2O))_2OSe_6W_{24}O_{94}]^{18-}$-anion. In detail, the fragments derived from the well-known γ-Dawson complex by removing six WO_6 octahedra in parallel with the threefold axis. The γ-Dawson complex consists of two XW_9 subunits in which the heteroatom is linked to a W_3O_{13} group and three W_2O_{10} groups. All W_3O_{13} groups within the γ-isomer have been rotated by π/3. Corresponding to this, the tungsten atoms can be divided into sixteen "equatorial" and eight "polar" tungsten atoms.

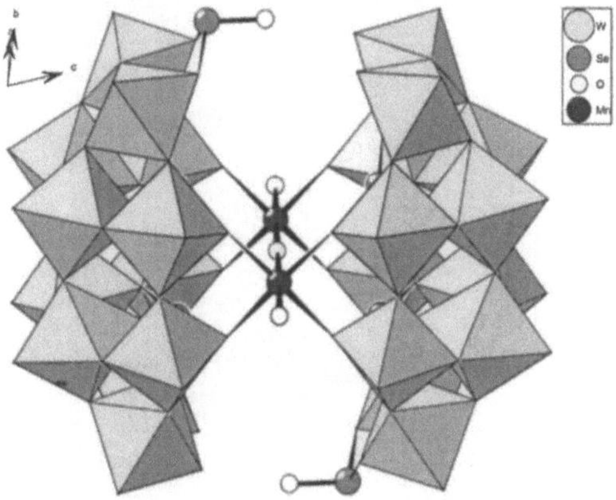

Figure 3: Molecular structure of $[(Mn(H_2O))_2OSe_6W_{24}O_{94}]^{18-}$

The Se^{IV} atoms in the centre of each XW_9 unit are surrounded pyramidally by three oxygen atoms. Respectively, the unshared pair of electrons of each selenium atom is pointing towards the open side and thus prevents the formation of the complete spherical structure. The octahedral coordination sphere of the manganese atoms is formed by two oxygen atoms of each $Se_3W_{12}O_{47}$ unit and two aquo ligands. The two manganese atoms are μ-oxo bridged.

2.5 $Na_4(H_3O)_{14}[Fe_4Se_8W_{36}O_{136}(OH)_6] \cdot 18H_2O$ (**7**)

The remarkable polyoxoanion **7** was prepared by mixing stoichiometric amounts of H_2SeO_3 and $Fe(NO_3)_3 \cdot 9H_2O$ in an aqueous solution and adding a solution of $Na_2WO_4 \cdot 2H_2O$ in destilled water. After several days yellow crystals of **7** were obtained.
The anion $[Fe_4Se_8W_{36}O_{136}(OH)_6]^{18-}$ (**7**) can be described as a dimer of two Dawson-like-units which are connected by additional four WO_6 octahedra. Formally, each unit is built up by sixteen tungsten and two iron atoms. During the refinement we found a disorder concerning the four metal atoms located at the bridging site of the subunit. Occupancies of 50% tungsten and 50% iron were found. The dimerization is due to a rotation of two equatorial WO_6 octahedra by 60°. This leads to a considerable expansion and a formation of additional terminal oxygen atoms. The latter atoms are saturated via the detected dimerization.
The fragments are interlinked by two W_2O_{11} groups to form a 40-cored anion. Within each W_2O_{11} group the WO_6 octahedra are bridged directly by two SeO_3 groups.

Subsequently four selenium atoms are located in the periphery of **4**. Within the $Fe_2Se_2W_{16}$ fragments Se is surrounded pyramidally by three oxygen atoms and occupies typical positions of the Dawson structure.
The $[Fe_4Se_8W_{36}O_{136}(OH)_6]^{18-}$-anion is given in Figure 4.

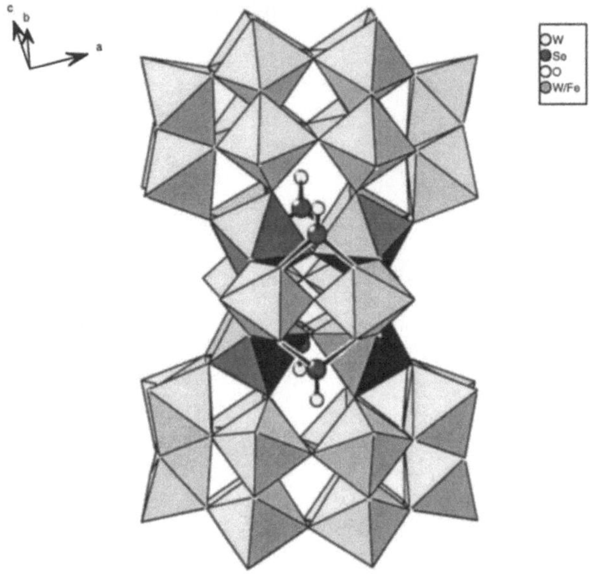

Figure 4: Molecular structure of $[Fe_4Se_8W_{36}O_{136}(OH)_6]^{18-}$

2.6 $(Me_2NH_2)_{13}Na_4H_4[Na_9Se_{13}(VO_4)_3V_{23}Mo_{19}O_{151}] \cdot 58H_2O$ (**8**)

The mixed sodium salt of $[Na_9Se_{13}(VO_4)_3V_{23}Mo_{19}O_{151}]^{21-}$ was crystallized by adding an aqueous solution containing $Na_2MoO_4 \cdot 2H_2O$ and H_2SeO_3 to an acidic solution of $NaVO_3 \cdot H_2O$. After several days red crystals of **8** for X-ray structure analysis could be obtained.
Microanalysis established a vanadium(V)/molybdenum(VI) ratio of 26:19. Corresponding to this, the occupation factors of the octahedrally coordinated molybdenum were fixed at 0.75 for common and 0.375 for special atomic positions.
Compound **8** consists of three identical $Se_4M_9O_{36}$ units with the known α-B-SeM_9O_{33} Keggin lacunar structure which are located around a central $Se(VO_4)_3M_{15}O_{43}$ unit.
The $Se_4M_9O_{36}$ subunits contain three corner-sharing M_3O_{13} triads, formed by three edge-sharing MO_6 octahedra. In the centre of each trivacant unit one Se^{IV} atom is surrounded pseudotetrahedrally by three oxygen atoms. The remaining three selenium atoms are coordinated via one oxygen atom to each of two M_3O_{13} subunits. One more terminal oxygen atom completes the pseudotetrahedral coordination sphere of these subvalent selenium atoms. Their unshared pair of electrons is oriented in the direction of the open side of each $Se_4M_9O_{36}$ unit. The vacancies between the selenium atoms are filled with sodium ions achieving an additional connection to the central $Se(VO_4)_3M_{15}O_{43}$ unit.

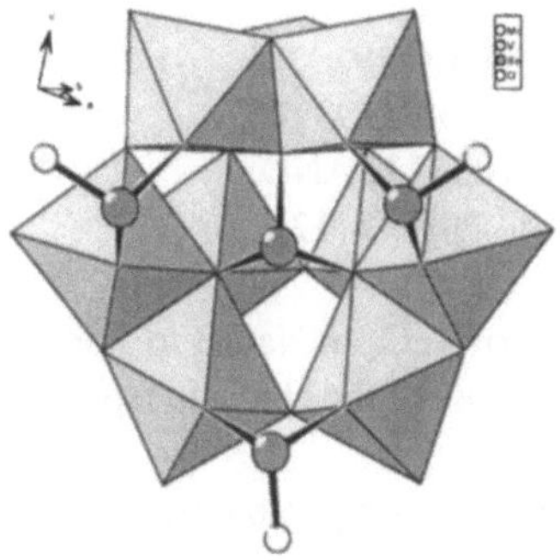

Figure 5: Molecular structure of $Se_4M_9O_{36}$ unit

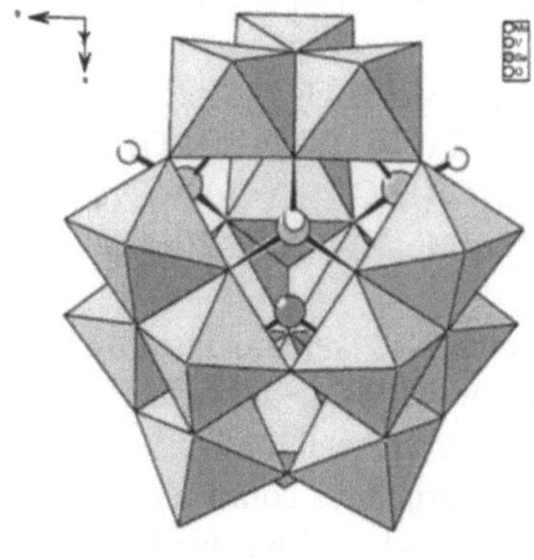

Figure 6: Molecular structure of $Se(VO_4)_3M_{15}O_{43}$ unit

The central fragment contains fifteen MO_6 octahedra, three VO_4 tetrahedra and one SeO_3 group. In particular, the MO_6 octahedra can be divided into three tetranuclear and one trinuclear group. The tetranuclear group consists of four edge-sharing MO_6 octahedra forming a distorted rectangular arrangement of the molybdenum atoms. The trinuclear subunit is comprised of three edge-sharing MO_6 octahedra. In this M-O framework three free coordination sites arise where the VO_4 tetrahedra are located. In detail, each vanadium atom is coordinated tetrahedrally via one oxygen to the trinuclear, two oxygens to the tetranuclear subunits and one terminal oxygen atom. Additionally, a selenium atom with a pseudotetrahedral coordination sphere consisting of three oxygen atoms of the tetranuclear subunits is located in the centre of the $Se(VO_4)_3M_{15}O_{43}$ unit.
Polyhedral plots of each subunit are given in Figures 5 and 6; a plot of the $[Na_9Se_{13}(VO_4)_3V_{23}Mo_{19}O_{151}]^{21-}$-anion in Figure 7.

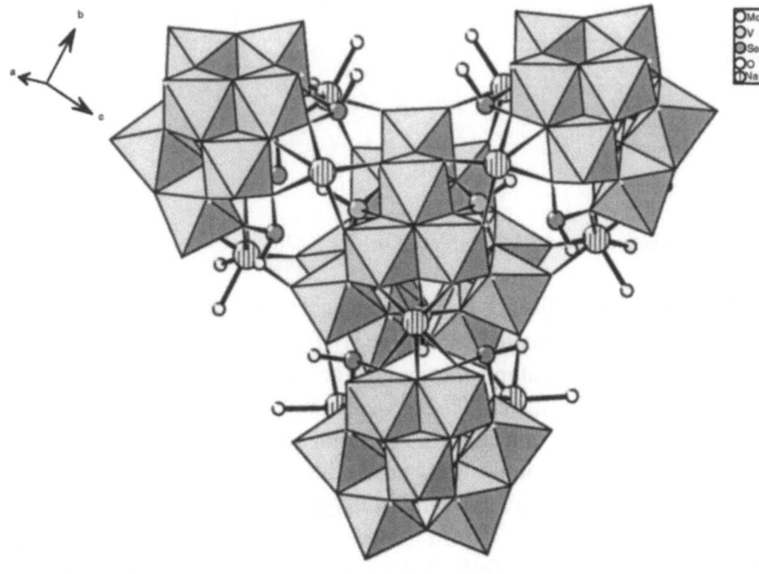

Figure 7: Polyhedral plot of $[Na_9Se_{13}(VO_4)_3V_{23}Mo_{19}O_{151}]^{21-}$

3. Tellurium Substituted Heteropolytungstates

3.1 $Na_8[(Mn(H_2O)_3)_2(Mn(H_2O)_2)_2(TeW_9O_{33})_2] \cdot 34H_2O$ (**9**), $K_8[(Co(H_2O)_3)_2(Co(H_2O)_2)(WO_2)(TeW_9O_{33})_2] \cdot 30H_2O$ (**10**), $K_6[(V(H_2O)_3)_2(VO_2)(WO_2)(TeW_9O_{33})_2] \cdot 22H_2O$ (**11**) AND $Na_6(Me_4N)_2[(Zn(H_2O)_3)_2(WO_2)_2(TeW_9O_{33})_2] \cdot 29H_2O$ (**12**)

Compounds **9-12** were obtained by adding solutions of TeO_2 in NaOH to acidic aqueous solutions of $Na_2WO_4 \cdot 2H_2O$ and stoichiometric amounts of the chloride rsp. sulfate salts of the transition metal ions.
The X-ray structures confirm a double Keggin structure comparable to earlier reported structures [10-13]. The fundamental common feature of the tellurium-substituted heteropolytungstates **9-12** is the trivacant β-B-TeW_9O_{33} Keggin fragment as the structural building unit. In detail these units can be derived from the α-Keggin structure by a 60° rotation of one of the three edge-sharing fragments. The Te^{IV} is surrounded pyramidally by three oxygen atoms.

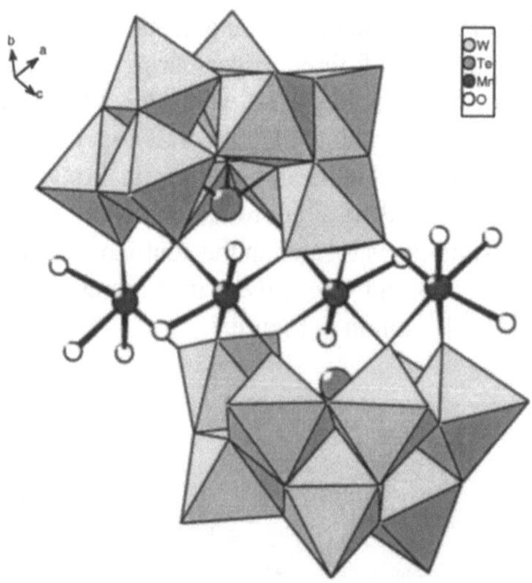

Figure 8: Molecular structure of $[(Mn(H_2O)_3)_2(Mn(H_2O)_2)_2(TeW_9O_{33})_2]^{8-}$

The two β-B-TeW_9O_{33} subunits are formally connected by a belt of four MO_6 octahedra. Two outside positions are occupied with $MO_3(H_2O)_3$ groups (M = Mn^{II}, Co^{II}, V^{IV}, Zn^{II}). Formally, the transition metal atoms form a linkage with two oxygens of one subunit and one of the subtended fragment. Three water molecules complete the octahedral coordination sphere of each transition metal atom. In contrast, compounds **9-12** show different compositions of the inner MO_6 octahedra: Within the manganese-substituted sodium bis-decatungsto tellurate $Na_8[(Mn(H_2O)_3)_2(Mn(H_2O)_2)_2(TeW_9O_{33})_2] \cdot 34H_2O$ (**9**) two $MnO_4(H_2O)_2$ groups form the octahedra; in zinc-substituted $Na_6(Me_4N)_2[(Zn(H_2O)_3)_2(WO_2)_2(TeW_9O_{33})_2] \cdot 29H_2O$ (**12**) two WO_2 groups link the TeW_9O_{33} subunits.

During the refinement of the cobalt- and vanadium-substituted tungstates (**10, 11**) disorders became evident. Using a disorder model, occupancies of 50% W and 50% Co/V were detected. From a formal point of view, both structures reflect a composition of disordered $[(M(H_2O)_3)_2(WO_2)_2(TeW_9O_{33})_2]^{n-}$ and $[(M(H_2O)_3)_2(MO_2)_2(TeW_9O_{33})_2]^{m-}$ ($M = Co^{II}$, V^{IV}). Cobalt in heteropolytungstate **10** completes its octahedral coordination sphere with two (rsp. three) water molecules.

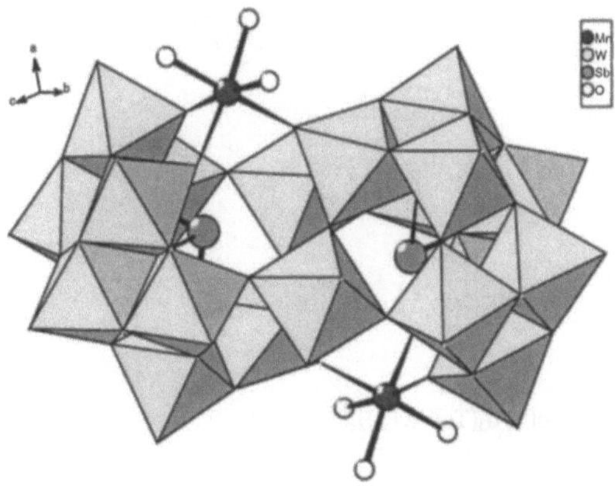

Figure 9: Molecular structure of $[(Zn(H_2O)_3)_2(WO_2)_2(TeW_9O_{33})_2]^{8-}$

3.2 $K_9Na[Cu_3(H_2O)(TeW_9O_{33})_2] \cdot 14H_2O$ (13) AND $K_4Na_3(Me_4N)_3[Pd_3(TeW_9O_{33})_2] \cdot 11H_2O$ (14)

Copper- and palladium- containing bis-nonatungsto tellurates $K_9Na[Cu_3(H_2O)(TeW_9O_{33})_2] \cdot 14H_2O$ (**13**) and $K_4Na_3(Me_4N)_3[Pd_3(TeW_9O_{33})_2] \cdot 11H_2O$ (**14**) were prepared by reacting TeO_2 in NaOH with $Na_2WO_4 \cdot 2H_2O$ in an acidic aqueous solution. After dropwise addition of transition metal ions (Cu^{II} respectively Pd^{II}) containing solutions green (rsp. brown) coloured single crystals were obtained.
X-ray structure analyses reveal the sandwich-like double Keggin structure shown in Figure 10. The anions consist of two α-B-TeW_9O_{33} fragments bonded by three MO_4 groups. Copper and palladium atoms are in a square-planar coordination by two oxygen atoms of each TeW_9 subunit. One copper atom shows pyramidal coordination; an oxygen of one water molecule forms the top of the pyramid.

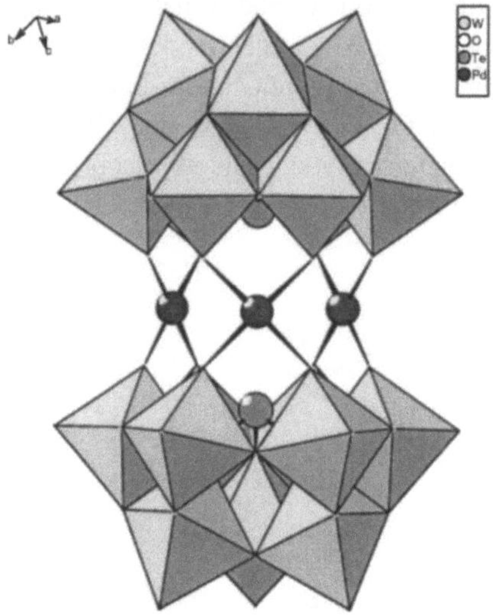

Figure 10: Molecular structure of $[Pd_3(TeW_9O_{33})_2]^{10-}$

4. Conclusion

Systematic variations of the reaction conditions yielded novel polyoxometalates containing selenium or tellurium as subvalent main group elements and various transition metals ($V^{IV/V}$, Mn^{II}, Co^{II}, Fe^{III}, Cu^{II}, Zn^{II} and Pd^{II}). We established the structures of 14 compounds composed of lacunary Keggin and Dawson ions as fundamental structural building units. Detailed structural characterization was accomplished via X-ray structure analyses. All novel polyoxometalates were investigated by IR-, Raman-, UV-spectroscopy and even by cyclovoltammetric studies.
The presented synthetic work opens a wide field in the synthesis of new polyoxometalates with interesting properties and have already been investigated successfully to catalytic applications [11].

Acknowledgements

This work was supported by the Deutsche Forschungsgemeinschaft and the Fonds der Chemischen Industrie. We like to thank R. Böhner, M. Bösing, R. Klein and C. Thülig who contributed essentials to this presented research.

References

1. M.T. Pope, A. Müller: *Angew. Chem.* **103**, 56 (1991); *Angew. Chem. Int. Eng.* **30**, 34 (1991).
2. *Chemical Reviews* **98**, No. 1 (1998).
3. A. Müller, E. Krickemeyer, H. Bögge, M. Schmidtmann, C. Beugholt, P. Krögerler, C. Lu: *Angew. Chem.* **110**, 1278 (1998); *Angew. Chem. Int. Eng.* **37**, 1220 (1998).
4. (a) B. Krebs, I. Paulat-Böschen: *J. Chem. Soc., Chem. Comm.*, 780 (1979); (b) B. Krebs, I. Paulat-Böschen: *Acta Cryst.* **B38**, 1710 (1982); (c) K. H. Tytko, B. Schönfeld, B. Buss, O. Glemser: *Angew. Chem.* **85**, 305 (1973); *Angew. Chem. Int. Ed.* **12**, 330 (1973); (d) B. Krebs, S. Stiller, K. H. Tytko, J. Mehmke: *Eur. J. Solid State Inorg. Chem.* **28**, 883 (1991).
5. A. Müller, E. Krickemeyer, J. Meyer, H. Bögge, F. Peters, W. Plass, E. Diemann, S. Dillinger, F. Nonnenbruch, M. Randerath, C. Menke: *Angew. Chem.* **107**, 2293 (1995); *Angew. Chem. Int. Ed. Engl.* **34**, 2122 (1995).
6. B. Krebs, B. Lettmann, H. Pohlmann and R. Fröhlich: *Z. Kristallogr.* **196**, 231 (1991).
7. B. Krebs, B. Lettmann and H. Pohlmann: *Z. Kristallogr.* **186**, 233 (1989).
8. J. F. Keggin: *Proc. R.*, Soc. London, **A144**, 75 (1934).
9. B. Krebs, I. Loose, M. Bösing, A. Nöh, E. Droste: *C. R. Acad. Sci. Ser. IIc*, 351 (1998).
10. I. Loose, E. Droste, M. Bösing, H. Pohlmann, M. H. Dickmann, C. Rosu, M. T. Pope, B. Krebs: *Inorg. Chem.* **38**, 2688 (1999).
11. M. Bösing, A. Nöh, I. Loose, B. Krebs: *J. Am. Chem. Soc.* **120**, 7252 (1998).
12. M. Bösing, I. Loose, H. Pohlmann, B. Krebs: *Chem. Eur. J.* **3**, 1232 (1997).
13. B. Krebs, R. Klein, in: Pope M. T., Müller A. (Eds.) *Polyoxometalates: From Platonic Solids to Anti-Retroviral Activity*, Kluwer Academic Publishers, Dordrecht, The Netherlands (1994) pp.41; B. Krebs, R. Klein: *Mol. Eng.* **3**, 43 (1993).
14. R. Neumann, M. Gara: *J. Am. Chem. Soc.* **116**, 5509 (1994).
15. R. Neumann, A. M. Khenkin: *J. Mol. Catal.* **114**, 169 (1996).
16. R. Neumann, D. Juwiler: *Tetrahedron* **47**, 8781 (1996).
17. R. Neumann, A. M. Khenkin, D. Juwiler, H. Miller, M. Gara: *J. Mol. Catl.* **117**, 169 (1997).
18. A. M. Khenkin, C. L. Hill: *Mendeleev Commun.* **140** (1993).
19. H. D`Amour: *Acta Cryst.* **B32**, 729 (1976).
20. B. Dawson: *Acta Cryst.* **6**, 113 (1953).
21. E. Van Dalen, M. G. Mellon: *Anal. Chem.* **36**, 1068 (1963).
22. G. Petrini, O. Pilanti, N. Giordano: *Chim. Ind. (Milan)* **50**, 1002 (1968).
23. Z. F. Shakhova, S. A. Morosanova, V. F. Zakharova: *Russ. J. Inorg. Chem.* **14**, 1609 (1969).
24. H. Ichida, H. Fukushima, Y. Sasaki: *Nippon Kagalu Kaishi*, 1521 (1986).
25. L. V. Derkach, A. Marques Rios, R. I. Maksimovskaya, A. V. Muzychenko: *Zh. Neorg. Khim.* **34**, 3094 (1989).
26. C. Robl., K. Haake: *J. Chem. Soc., Chem. Commun.* 397 (1993).
27. E. S. Ganelina, N. I. Nerevyatkina: *Russ. J. Inorg. Chem.* **10**, 483 (1965).
28. R. Ripan, N. Calu: *Stud. Univ. Babes-Bolyai Chem.* **10**, 135 (1965).
29. A. Yagasaki, Y. Sasaki: *Bull. Chem. Soc. Jpn.* **60**, 763 (1987).
30. T. Ozeki, A. Yagasaki, H. Ichida, Y. Sasaki: *Polyhedron* **7**, 1131 (1988).
31. F. Robert, M. Leyrie, G. Hervé: *Acta Cryst.* **B38**, 358 (1982).
32. C. M. Tourné, G. F. Tourné, F. Zonnevijlle: *J. Chem. Soc. Dalton Trans.* 143 (1991).
33. R. Neumann, A. M. Khenkin: *Inorg. Chem.* **34**, 5753 (1995).
34. T. J. R. Weakley, H. T. Evans, jun., J. S. Showell, G. F. Tourné, C. M. Tourné: *J. Chem. Soc., Chem. Commun.* 139 (1973).
35. H. T. Evans, C. M. Tourné, G. F. Tourné, T. J. R. Weakley: *J. Chem. Soc., Dalton Trans.* 2699 (1986).

Vibrational Spectroscopy of Heteropoly Acids

H. RATAJCZAK,[a,b] A.J. BARNES,[c] A. BIELAŃSKI,[d] H.D. LUTZ[e], A. MÜLLER[f] and M.T. POPE[g]

[a] *Institute of Low Temperature and Structure Research, Polish Academy of Sciences, PO Box 1410, 50-950 Wrocław, Poland*
[b] *Faculty of Chemistry, University of Wrocław, F. Joliot-Curie 14, 50-383 Wrocław, Poland*
[c] *School of Sciences – Chemistry, University of Salford, Salford M5 4WT, Great Britain*
[d] *Institute of Catalysis and Surface Chemistry, Polish Academy of Sciences, Niezapominajek 3, 30-329 Kraków, Poland*
[e] *Inorganic Chemistry I, University of Siegen, D-57068 Siegen, Germany*
[f] *Inorganic Chemistry I, Faculty of Chemistry, University of Bielefeld, Pf. 100131, D-33501 Bielefeld, Germany*
[g] *Department of Chemistry, Georgetown University, Washington DC 20057-1227, U.S.A.*

Abstract The vibrational spectra of the heteropoly acids can be conveniently regarded as composed of contributions from the polyoxometalate anion (the primary structure) and from the water of crystallisation and hydrated protons (the secondary structure). Following a brief general survey of vibrational spectra of hydrogen-bonded systems, the spectra of water and hydrated protons in crystalline solids are reviewed. The vibrational spectra of the primary structure of heteropoly acids (the Keggin anion) are described and the observed spectra of the secondary structures of highly hydrated, hexahydrated and dehydrated heteropoly acids are discussed in relation to the spectra expected for protons in different environments.

Keywords: heteropoly acids, vibrational spectroscopy, hydrogen bonding, hydrated proton, oxonium ion, water of crystallisation.

1. Introduction

Heteropoly acids and their salts, containing the corresponding heteropoly anions, constitute a large class of compounds, which have important applications in catalysis [1]. Heteropoly acids are also of special interest as new materials due to their high proton conductivity [2]. Heteropoly compounds are very strong Brönsted acids and also efficient oxidising agents. Since they are very soluble in polar solvents and have relatively good thermal stability in the solid state they can act as either homogeneous or heterogeneous acid or redox catalysts. Commercial applications include hydration of alkenes such as propene, polymerisation of tetrahydrofuran and oxidation of methacrolein to methacrylic acid.

A general heteropoly anion $[X_xM_yO_z]^{n-}$ comprises a central (hetero) atom or atoms X, typically a P or Si atom tetrahedrally coordinated by oxygens, surrounded by between 2 and 18 oxygen-linked hexavalent metal atoms M, usually Mo or W although some of these metal atoms may be replaced by other transition metals such as V, Nb or

M.T. Pope and A. Müller (eds.), Polyoxometalate Chemistry, 101–116.

Ta [3]. The structure of such a unit is known as the primary structure. The best known heteropoly molybdates and tungstates are those containing an $[X^{(n+)}M_{12}O_{40}]^{(8-n)-}$ anion with the Keggin structure [4]. In this structure, the central XO_4 tetrahedron is surrounded by 12 MO_6 octahedra arranged in 4 groups of 3 edge-shared octahedra; these groups are linked to each other and to the central XO_4 tetrahedron by shared corners leaving 12 terminal (unshared) O atoms. The solid heteropoly acids are ionic crystals consisting of these primary units together with cations (H^+, H_3O^+, $H_5O_2^+$), water of crystallisation and, in some cases, other molecules linked to form the secondary structure. In the heteropoly acids, the protons are directly balancing the negative charges on the anions; in the corresponding salts, H_3O^+ or $H_5O_2^+$ ions are replaced by metal ions. The crystal structure of heteropoly compounds varies with the countercation and, especially, with the extent of hydration, for example 12-tungstophosphoric acid exhibits a variety of different packing arrangements as hydration water is lost (table 1).

TABLE 1
Crystal structures of 12-tungstophosphoric acid hydrates.

hydrate	space group	ref.
$H_3PW_{12}O_{40}\cdot 29H_2O$	cubic, $Fd\bar{3}m$	5
$H_3PW_{12}O_{40}\cdot 21H_2O$	orthorhombic, $Pcca$	6
$H_3PW_{12}O_{40}\cdot 14H_2O$	triclinic, $P\bar{1}$	7
$H_3PW_{12}O_{40}\cdot 6H_2O$	cubic, $Pn3m$	8

The primary and secondary structures of the heteropoly acids, and their salts, in the solid phase and in solution have been investigated by a variety of techniques: X-ray and neutron diffraction, infrared and Raman spectroscopy, inelastic and quasielastic neutron scattering, NMR spectroscopy and UV/visible absorption spectroscopy. It is well known that vibrational spectroscopy provides valuable information for hydrogen-bonded systems such as these [9]. However the interpretation can be complicated by the unusual type of spectra typically generated by strongly-hydrogen-bonded species such as the $H_5O_2^+$ ion; also vibrational modes of water and protonated water species may overlap with the modes of the primary polyoxometalate unit. In this article we shall focus particularly on the vibrational spectra of O–H···O hydrogen bonds found in water and hydrated proton - H_3O^+, $H_5O_2^+$, $H^+(H_2O)_n$ ($n > 2$) - species.

2. Vibrational spectra of hydrogen-bonded systems

The main intramolecular vibrational modes of a hydrogen-bonded system of the type X–H···Y may be approximately represented as follows:

$$\overset{\rightarrow}{\text{X—H}}\cdots\text{Y} \qquad \text{XH stretching, } \nu(\text{XH}) \text{ or } \nu_a(\text{XHY})$$

$$\overset{\uparrow}{\text{X—H}}\cdots\text{Y} \qquad \text{XH in-plane bending, } \delta(\text{XH})$$

$$\overset{\oplus}{\text{X—H}}\cdots\text{Y} \qquad \text{XH out-of-plane bending, } \gamma(\text{XH})$$

The formation of a hydrogen bond also leads to the appearance of 6 low frequency intermolecular modes relating to vibrations of the X–H···Y bridge, of which the most important are:

$$\overset{\leftarrow}{\text{X}}\text{—}\overset{\leftarrow}{\text{H}}\cdots\overset{\rightarrow}{\text{Y}} \qquad \text{hydrogen bond stretching, } \nu_s(\text{XHY})$$

$$\overset{\uparrow}{\text{X}}\text{—}\overset{\uparrow}{\text{H}}\cdots\underset{\downarrow}{\text{Y}} \qquad \text{hydrogen bond in-plane bending, } \delta(\text{XHY})$$

$$\overset{\oplus}{\text{X}}\text{—}\overset{\oplus}{\text{H}}\cdots\overset{\ominus}{\text{Y}} \qquad \text{hydrogen bond out-of-plane bending, } \gamma(\text{XHY})$$

There are many experimental data relating to the spectroscopic behaviour of the ν(XH) stretching vibration. Its characteristic changes are:

- the position of the band shifts to lower frequency;
- the infrared intensity increases strongly, the Raman intensity decreases;
- the band becomes broader and sub-maxima may appear.

These spectral effects are strongly dependent upon the X–H···Y bond distance (i.e. the strength of the hydrogen bond) and are illustrated in fig. 1. The XH in-plane and out-of-plane bending modes generally shift to higher frequency, but comparatively little systematic work has been reported on these vibrations [10].

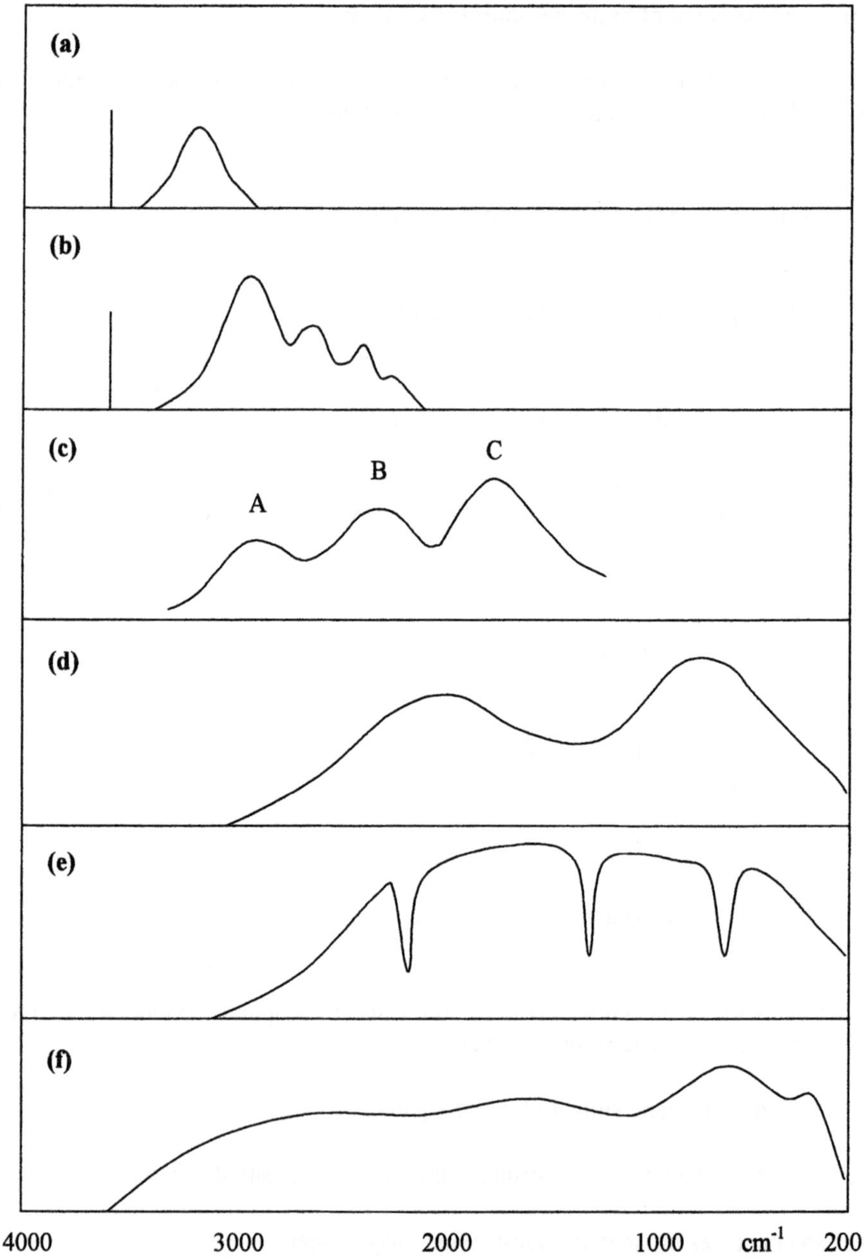

Fig. 1. Infrared spectra of the XH stretching vibration for different hydrogen bond strengths: (a) weak; (b) medium, Millen type; (c) strong, Hadži type; (d) very strong; (e) very strong, with Evans holes; (f) very strong, Zundel type.

The shift of the XH stretch to lower frequency and the increase of its infrared intensity are of electronic origin and are related to the change of electronic structure of the hydrogen-bonded system. However, the band shape represents a particularly complex problem and should be related to the dynamics of the X–H···Y system. A great deal of effort has been made over the past 60 years to understand the complex infrared absorption profiles of hydrogen-bonded stretching bands in gas, liquid and solid phases with many theories being proposed [11]. Finally in recent years a consensus has been reached concerning the origin of this complex spectral behaviour. It has been recognised, both theoretically and experimentally, that the shape of the XH stretching band is generated mainly by mechanical anharmonic coupling between the high frequency, ν(XH), and low frequency, ν_s(XHY) and δ(XHY) / γ(XHY), vibrational modes of the hydrogen bond X–H···Y. In condensed phases, the system is further coupled to the thermal bath.

However, for medium-strong and strong hydrogen bonds, one must take into account an additional mechanism in the generation of the complex shape of the XH stretching band profile, namely Fermi resonances between the ν(XH) stretching mode and some other internal modes [11]. Recently it has been shown that in strongly coupled systems Fermi resonances generate complicated shapes with sub-structure [12]. In principle, on the basis of these approaches, it is possible to understand the origin of the complex shapes of the XH stretching bands and to reproduce them semi-quantitatively. This is true only for the spectra of the types shown in fig. 1, (a) to (e).

It should perhaps be mentioned that all the working theories are developed on the basis of the double Born-Oppenheimer approximations [11]. This means on the assumption of separation of electron and nuclei movements in the system as well as the separation of the high and low frequency vibrations of the hydrogen bond X–H···Y.

In the case of Zundel-type spectra, fig. 1(f), probably additional mechanism(s) should be introduced in order to understand the origin of this extremely broad quasi-continuous intense absorption which can appear in the range *ca.* 3600 cm^{-1} to 200-150 cm^{-1} [13]. This absorption occurs in aqueous solutions of acids and bases, in $(X\cdots H\cdots X)^+$ or $(X\cdots H\cdots X)^-$ hydrogen-bonded ions, and in systems where tautomeric equilibria X–H···Y ⇔ $X^-\cdots H{-}Y^+$ occur in non-aqueous solutions. This type of absorption has also been found in strongly hydrogen-bonded solids. On the basis of a very large body of experimental data one can reach the conclusion that such very broad quasi-continuous infrared absorption occurs in hydrogen-bonded systems where the proton is delocalised along the X–H···Y bond. The question arises whether the proton fluctuation influences the generation of breadth of the XH stretching band. Zundel *et al.* [13] were the first to show that in the hydrogen-bonded systems described by double minimum energy surfaces, or energy surfaces with flat broad wells, so-called "protonic polarisabilities" appear which are about 2 orders of magnitude larger than the electronic polarisabilities. Of course, such strongly polarised systems easily interact with their environment leading to a change in the distribution of energy levels. This mechanism can generate the breadth and also contribute to the shape of the absorption band [14]; however see also [15].

3. Vibrational spectra of water, H_3O^+, $H_5O_2^+$ and $H^+(H_2O)_n$ (n > 2) ions

3.1 *Water*

In a salt hydrate the water of crystallisation may be coordinated to metal ions (e.g. as ligands in aqua complexes) or to anions and other proton accepting groups (by hydrogen bonds) or it can be present as weakly bound solvate molecules ("lattice water") [16]. A free (gas phase) water molecule has C_{2v} symmetry and three modes of vibration: antisymmetric stretching (ν_3) at 3756 cm^{-1}, symmetric stretching (ν_1) at 3657 cm^{-1} and bending (ν_2) at 1595 cm^{-1}. In a molecular complex with a base, B···H–O–H, the water stretching modes become uncoupled to give a "free" OH stretch and a hydrogen-bonded OH stretch. However, in solid hydrates typically both hydrogen atoms are involved in hydrogen bonds. The effects of hydrogen bonding are to shift and broaden the vibrations in a manner dependent on the strength of the hydrogen bond (as discussed earlier), thus the bands are observed over fairly wide wavenumber ranges (table 2). As well as these internal vibrational modes of the water molecule, relatively low frequency bands are observed in solid hydrates due to librational and translational modes of the molecule.

TABLE 2
Ranges of observed band positions (cm^{-1}) for H_2O, HDO and D_2O in solid hydrates [16].

mode	H_2O	HDO	D_2O
stretches ν_3, ν_1	3600-3000	3600-3000, 2650-2300	2650-2300
bend ν_2	1660-1590	1460-1400	1225-1175
librations R	900-350	900-260	680-260
translations T	350-100	350-95	330-95

3.2 *H_3O^+, the oxonium ion*

In the majority of solids containing H_3O^+ ions, the 3 hydrogen atoms all form hydrogen bonds to neighbouring anions giving a pyramidal configuration which may or may not retain the C_{3v} symmetry of the isolated ion [17]. A lower site symmetry will lift the degeneracy of the antisymmetric stretching (ν_3) and bending (ν_4) vibrations (table 3); also the factor group may increase the number of vibrational modes according to the number of ions in the primitive cell. As for water, the effects of hydrogen bonding are to shift and broaden the vibrations in a manner dependent on the strength of the hydrogen bond. The OH stretching region may also be complicated by Fermi resonance with overtones or combination bands. Since the oxonium ion often participates in rather shorter and stronger hydrogen bonds than those involving water molecules, the bands are observed over rather extensive wavenumber ranges (table 3).

TABLE 3
Modes, symmetries under different site point groups, and ranges of observed band positions and band half-widths for H_3O^+ and D_3O^+ ions in solids [17].

mode	symmetry			H_3O^+		D_3O^+	
	C_{3v}	C_s	C_1	ν / cm^{-1}	$\Delta\nu_{½}$ / cm^{-1}	ν / cm^{-1}	$\Delta\nu_{½}$ / cm^{-1}
ν_3	E	A'+A''	2A	3510-2450	80-1200	2600-1945	350-1000
ν_1	A_1	A'	A	*	*	*	*
$2\nu_2$ †	A_1	A'	A	2250-2050	100	1700-1430	100
ν_4	E	A'+A''	2A	1730-1575	100-250	1350-1160	30-80
ν_2	A_1	A'	A	1180-900	50-200	885-710	50-200

* included within the ranges quoted for ν_3.
† or $\nu_4 + \nu_R$ where R is a librational mode.

As well as these internal vibrational modes of the H_3O^+ ion, various low frequency bands observed in oxonium salts have been identified as librational (820-420 cm^{-1}) or translational (< 400 cm^{-1}) modes of the ion.

3.3 $H_5O_2^+$ *and* $H^+(H_2O)_n$ $(n > 2)$ *ions*

Probably the first evidence that the $H_5O_2^+$ species can exist in a crystalline solid came from an X-ray study of $[Co(en)_2Cl_2]Cl.HCl.2H_2O$ by Nakahara *et al.* [18] in 1952. Now there are many experimental data (X-ray and neutron diffraction, inelastic neutron scattering, NMR spectroscopy) on the existence of the $H_5O_2^+$ cation in a variety of different crystalline systems. Assignment of the vibrational spectra has been the subject of much debate; the ranges in which the various modes are typically assigned are shown in table 4.

TABLE 4

Ranges of band positions assigned to the $H_5O_2^+$ ion [17].

assignment	ν / cm^{-1}
ν_a(OHO)	2000–700
δ(OHO), γ(OHO)	1400–1100
ν_s(OHO)	≈450

For $H_5O_2^+$, mainly two different possibilities are conceivable as the equilibrium structure: either a structure with a symmetrical hydrogen bond, $(H_2O\cdots H\cdots OH_2)^+$, or a structure with an asymmetric hydrogen bond, $H_2O{-}H^+\cdots OH_2$. It is known that in protonated hydrate crystals the preferred structure is of the latter type [17]. As far as the geometrical structure of such a species is concerned one could expect possibilities such as those shown below.

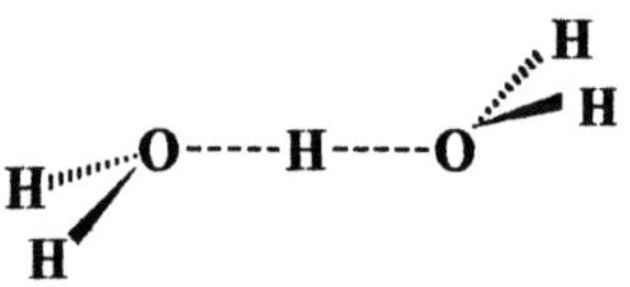

C_2 (trans)

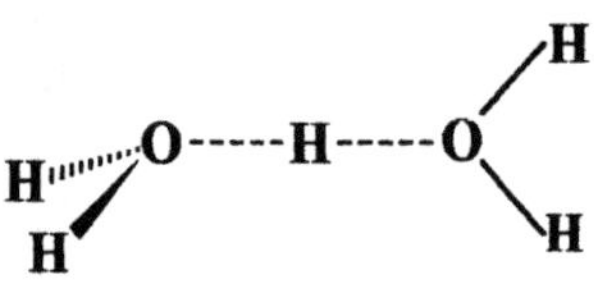

C_s (gauche)

Highly advanced quantum chemical calculations [19, 20], carried out for the isolated $H_5O_2^+$ ion, favour the form with C_2 symmetry, however the global minimum lies only *ca.* 2 kJ mol^{-1} lower in energy than the transition state with C_s symmetry. The potential energy surface of the $H_5O_2^+$ ion is highly anharmonic with a single, symmetric, flat minimum at an O···H···O distance of 2.40Å. The calculated hydration enthalpy is *ca.* -141.5 kJ mol^{-1}, which is in excellent agreement with the experimental values: -132.2 [21], -133.1 [22] and -138 [23] kJ mol^{-1}. The calculated vibrational wavenumbers are listed in table 5. As can be seen, the ν_a(OHO) vibrational mode is strongly shifted to low frequency, *ca.* 1275 cm^{-1}. This vibration is highly anharmonic and coupled with ν_s(OHO) as well as with many external modes [24].

TABLE 5

Calculated harmonic and anharmonic vibrational wavenumbers for the OHO antisymmetric and symmetric stretching modes of $H_5O_2^+$ [20].

approximation	ν_a(OHO)	ν_s(OHO)
harmonic	983	619
anharmonic	1623	615
coupled	1275	572

Zundel *et al.* [25] were the first to postulate the appearance of $H_5O_2^+$ species in aqueous solutions of acids. Since the seventies, the $H_5O_2^+$ ion has been studied by many quantum chemists [26], using different methods, since it is treated as a simple solvated proton system which lies at the heart of acid-base chemistry. In recent years, the hydrated proton has been the subject of renewed interest, while a variety of theoretical studies have been reported ranging from quantum chemical calculations of gas phase $H^+(H_2O)_n$ clusters, through Car-Parrinello simulations with as many as 32 water molecules solvating an excess proton, to molecular dynamics simulations with semi-empirical force fields comprising hundreds of water molecules [24, 27-29].

It has been shown that the $H_5O_2^+$ ion is responsible for the large absorption between 2000 and 1000 cm^{-1}. This absorption appears for $H_5O_2^+$ dissolved in liquid water as well as for $H_5O_2^+$ in vacuo. The absorption band is made up of 3 peaks at *ca.* 1750, 1400 and 1200 cm^{-1} [24]. The 1200 and 1750 cm^{-1} features are due to ν_a(OHO) and δ(OH) of the terminal water molecules, which are strongly coupled to each other, whereas the 1400 cm^{-1} feature is attributed to δ(OH) of the $(O\cdots H\cdots O)^+$ bond. These three modes were located at 1849, 1539 and 947 cm^{-1} in the MCSCF normal mode calculation carried out by Muguet [26]. However, so far the calculations do not show any continuous absorption between 3000 and 2000 cm^{-1} or below 1000 cm^{-1}, which are characteristic features of the hydrated proton [24].

The species $H_5O_2^+$ in crystalline solids exhibits an absorption extending over the region *ca.* 2000-700 cm^{-1}, which is attributed to the ν_a(OHO) vibrational mode [13, 17, 30]. However, on the basis of infrared spectra, it is difficult to distinguish between the symmetric and pseudo-symmetric structures of this ion. A good example of this situation is provided by $HClO_4{\cdot}2H_2O$ and $HCl{\cdot}2H_2O$ [30]. For both compounds the $H_5O_2^+$ ion appears in the crystalline solids with a very short $(O_1\cdots H_1\cdots O_2)^+$ bond (*ca.* 2.41-2.42Å), but in the former case the cation configuration is trans (symmetric) whereas in the latter case it is gauche (pseudo-symmetric), with O_1–H_1 and O_2–H_1 distances of *ca.* 1.19 and 1.21Å respectively, i.e. nearly symmetrical. Nevertheless the infrared spectra of the two crystals are similar as far as their hydrogen bond absorptions are concerned.

It is already well known that a symmetrical structure in which the excess proton is equally shared between two water molecules has roughly the same total energy as the asymmetric H_2O–$H^+\cdots OH_2$ structure in which it is closer to one of the oxygen atoms. Probably such a situation occurs quite often in crystalline solids, where the symmetry of the $H_5O_2^+$ ion depends on the symmetry of the crystal and its geometrical parameters depend also on the interactions of the cation via hydrogen bonds with the surrounding anions. Such changes should be observable by very careful analysis of the infrared spectra.

4. Vibrational spectra of the primary structure of heteropoly acids: the Keggin ion

The vibrational spectra of the heteropoly acids can be conveniently regarded as composed of contributions from the polyoxometalate anion (the primary structure) and from the water of crystallisation and hydrated protons (the secondary structure).

The isolated Keggin anion (symmetry T_d) is constructed from a tetrahedral XO_4 group surrounded by 12 MO_6 groups (symmetry C_s) which form 4 linked M_3O_{13} units (symmetry C_{3v}). The oxygen atoms fall into one of four types [31]: 4 O_a atoms (C_{3v}) linking the XO_4 tetrahedron to the 3 octahedra of an M_3O_{13} group, 12 O_b atoms (C_s) in M-O-M bridges between 2 different M_3O_{13} groups, 12 O_c atoms (C_s) in M-O-M bridges within the same M_3O_{13} group, and 12 O_d atoms (C_s) in terminal (unshared) positions. The normal modes of vibration of the anion may be represented as:

$$\Gamma_{vib} = 9A_1 + 4A_2 + 13E + 16F_1 + 22F_2$$

of which the F_2 modes are active in both infrared and Raman while the A_1 and E modes are Raman active only. Although there must inevitably be some mixing of the vibrations of the XO_4 tetrahedra and the MO_6 octahedra, the observed bands are usually assigned as though they are group vibrations. Strictly, however, this should only be applicable to the terminal $M{=}O_d$ stretching vibrations. The contributions of the various types of stretching mode are:

$$\Gamma_{X\text{-}O} = A_1 + F_2$$

$$\Gamma_{M\text{-}Oa} = A_1 + E + F_1 + 2F_2$$

$$\Gamma_{M\text{-}Ob} = A_1 + A_2 + 2E + 3F_1 + 3F_2$$

$$\Gamma_{M\text{-}Oc} = A_1 + A_2 + 2E + 3F_1 + 3F_2$$

$$\Gamma_{M\text{-}Od} = A_1 + E + F_1 + 2F_2$$

The principal bands observed [31] for the 12-molybdophosphate and 12-tungstophosphate Keggin ions in the infrared and Raman spectra of the corresponding heteropoly acids are listed in table 6. The precise wavenumbers of the various vibrations vary with the solvent, with the degree of hydration in the solid heteropoly acids and with the size of the cation in the corresponding salts. Distortion of the Keggin anion from its tetrahedral symmetry leads to the triple degeneracy of the F_2 modes being lifted and consequently additional bands being observed in the spectra.

TABLE 6
The principal bands observed (cm^{-1}) in the infrared and Raman spectra of the 12-molybdophosphate and 12-tungstophosphate Keggin ions [31].

		$[PMo_{12}O_{40}]^{3-}$		$[PW_{12}O_{40}]^{3-}$	
assignment	sym.	IR	Raman	IR	Raman
P-O_a asym. stretch	F_2	1070(s)		1080(s)	
M=O_d sym. stretch	A_1		997(vs)		1011(vs)
M=O_d asym. stretch	F_2	965(s)	981(m)	985(s)	996(m)
P-O_a sym. stretch	A_1		978(m)		993(m)
M-O_b-M asym. stretch	F_2	870(s)	900(w)	887(s)	925(w)
			870(w)		900(w)
M-O_c-M asym. stretch	F_2	790(vs)		807(vs)	
M-O-M sym. stretch	A_1		605(m)		536(w)
O_aPO_a asym. deformation	F_2	598(m)		598(m)	
		500(m)		525(s)	522(w)
		380(s)	374(w)	382(s)	372(w)
		340(s)	350(vw)	338(s)	337(vw)
		282(m)		278(sh)	
		265(m)		263(m)	
M-O_a sym. stretch	A_1		248(s)		217(s)

(vs = very strong, s = strong, m = medium, w = weak, vw = very weak, sh = shoulder)

Rocchiccioli-Deltcheff and Fournier [32] studied the effect of dehydration on the Keggin ion vibrations of 12-molybdo- and 12-tungstophosphoric acids and related vanadium substituted compounds. They found that the M-O_b-M asymmetric stretching mode is particularly sensitive to the extent of hydration, decreasing in frequency by *ca.* 20 cm^{-1} on dehydration.

5. Vibrational spectra of the secondary structure of heteropoly acids

The secondary structure of the heteropoly acids is largely determined by hydrogen bonding of the polyoxometalate anions via water and hydrated proton species. It is therefore self-evident that the secondary structure will be strongly influenced by the degree of hydration of the heteropoly acid. This is clearly manifested in the variation of crystal structure with extent of hydration for 12-tungstophosphoric acid (table 1) and related compounds. Kanda *et al.* [33] differentiated three different states of the protons in $H_3PW_{12}O_{40}{\cdot}nH_2O$ from solid-state ^1H and ^{31}P NMR spectroscopy: (i) protons present in highly hydrated samples ($n > 6$), (ii) protonated water which is hydrogen-bonded to terminal oxygen $W{=}O{\cdots}H^+(H_2O)_2$ ($n \approx 6$), and (iii) protons which are directly bonded to bridging oxygen W–O(H)–W ($n \approx 0$). It therefore seems appropriate to discuss the vibrational spectra of the heteropoly acids following this classification.

5.1 *Highly hydrated heteropoly acids, $H_wX_xM_yO_z{\cdot}nH_2O$ ($n > 6$)*

Most of the heteropolyacids containing Keggin anions form isomorphous 29 or 30-hydrates which melt in their own water of crystallisation at 40-100°C [3]. In for example $H_3PW_{12}O_{40}{\cdot}nH_2O$, with $n > 6$, the water molecules are generally loosely bound lattice water. Three of the water molecules should in principle be identifiable as oxonium ions; however the NMR spectra show that the protons are in a uniform state and highly mobile, consistent with the water being pseudo-liquid [33].

Bielański *et al.* [34-36] have used infrared spectroscopy to examine the dehydration of several highly hydrated heteropoly acids ($H_3PMo_{12}O_{40}{\cdot}nH_2O$, $H_4PMo_{11}VO_{40}{\cdot}nH_2O$, $H_4PMo_{10}V_2O_{40}{\cdot}nH_2O$ and $H_4SiW_{12}O_{40}{\cdot}nH_2O$) under room temperature evacuation and observed the loss of bands at ca. 3550 and 1615 cm^{-1}, characteristic of relatively weakly hydrogen-bonded water. At room temperature, the infrared spectra show the Zundel type absorption observed in aqueous acid solutions [13]. A similar type of phenomenon has been found by Highfield and Moffat [37] in $H_3PW_{12}O_{40}{\cdot}nH_2O$, and by us [38] in $H_3PMo_{12}O_{40}{\cdot}nH_2O$, $H_3PW_{12}O_{40}{\cdot}nH_2O$ and $H_4SiW_{12}O_{40}{\cdot}nH_2O$ using photoacoustic spectroscopy in the infrared region. These observations suggest that the state of the protons in highly hydrated heteropoly acids is similar to that in liquid water.

5.2 *Heteropoly acid hexahydrates, $H_wX_xM_yO_z{\cdot}nH_2O$ ($n \approx 6$)*

The hexahydrate of 12-tungstophosphoric acid was shown by Brown *et al.* [8], using single crystal X-ray and neutron diffraction, to have a secondary structure in which four Keggin anions are linked by hydrogen bonds between terminal W=O oxygens and the four free hydrogens of an $H_5O_2^+$ cation:

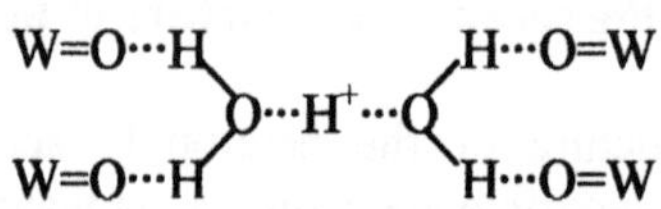

The $H_5O_2^+$ ions are almost planar with a linear symmetric $O{\cdots}H^+{\cdots}O$ bond having an O···O distance of 2.37Å. Kearley *et al.* [39] assigned the vibrational modes of this species from infrared, Raman and inelastic neutron scattering (INS) spectra (table 7).

In a later study of $H_3PMo_{12}O_{40}{\cdot}6H_2O$ by Mioc *et al.* [40], using incoherent inelastic neutron scattering (IINS) and infrared spectroscopy, sharp features in the IINS spectra were assigned to the $H_5O_2^+$ species at 25K but to H_3O^+ species at 120K. They suggested that above *ca.* 100K the $H_5O_2^+$ species dissociate:

$$H_5O_2^+ \Leftrightarrow H_3O^+ + H_2O \text{ (or } H_2O\text{–}H^+\cdots OH_2)$$

A broad background above 300 cm^{-1} in the IINS spectra was taken as evidence of mobile protons (proton gas).

TABLE 7
Bands observed (cm^{-1}) for the $H_5O_2^+$ cation in $H_3PW_{12}O_{40}{\cdot}6H_2O$ [39].

infrared	Raman	INS	INS [40]	assignment
3400	3266		3585	ν(OH)
3200	3165		3260	ν(OH)
	1900	1760	1730	$\delta(H_2O)$
1705	1650	1669		$\delta(H_2O)$
1620	1583	1566		$\delta(H_2O)$
	1450	1389	1360	δ(OHO)
	1258	1250		
		1142		
1080		1050	1070	γ(OHO)
	1011	1029		$\rho(H_2O)$
982	990			
888	905	927		$\rho(H_2O)$
799		794	830	$\omega(H_2O)$?
		590	620	$\omega(H_2O)$?
	550		560	$\tau(H_2O)$?
	522	534	530	$\tau(H_2O)$?
		454	485	ν_s(OHO)

Bielański *et al.* [35, 36] have recorded infrared spectra of $H_4SiW_{12}O_{40}{\cdot}{\approx}6H_2O$, in the temperature range 25-80°C, which show bands at 3150, 1702, 1107 and 1017 cm^{-1} that may be assigned to vibrations of the $H_5O_2^+$ or $H_2O\text{–}H^+\cdots OH_2$ species. The bands at 3150 and 1702 cm^{-1} may be readily assigned to ν(OH) and $\delta(H_2O)$ vibrations of terminal water molecules in the cation (*cf.* table 7), but the assignment of the 1107 and 1017 cm^{-1} bands is less clear.

5.3 *Dehydrated heteropoly acids,* $H_wX_xM_yO_z \cdot nH_2O$ *($n \approx 0$)*

The thermal dehydration of heteropoly acids has been studied by several vibrational spectroscopic methods: infrared spectroscopy [32, 34-36, 40, 41], photoacoustic spectroscopy in the infrared region [37, 38] and incoherent inelastic neutron scattering [40]. Mioc *et al.* [40] found that after dehydration at 200°C, leading to the loss of about 5 water molecules, the IINS spectrum of $H_3PMo_{12}O_{40}{\cdot}6H_2O$ was drastically modified. The bands observed are consistent with the presence of H_3O^+ ions in the dehydrated sample. A broad background above 50 cm^{-1} in the IINS spectrum was taken as evidence of mobile protons. After further dehydration at 300°C, an entirely different IINS spectrum was obtained showing a strong peak at 1150 cm^{-1} with overtones at 2290 and 3330 cm^{-1}, assigned to an isotropic oscillator. This spectrum would be consistent with the following structure, postulated by Kozhevnikov [1a] on the basis of ^{17}O NMR spectroscopy:

```
W=O ······· O=W
 :           :
 :    H+     :
 :           :
W=O ······· O=W
```

in which the proton is shared by 4 equivalent terminal oxygens from 4 different heteropoly anions. On the other hand, on the basis of significant shifts in the bands in the infrared spectrum due to the M–O_b–M stretching vibrations, Mizuno and Misono [1b] suggested that on dehydration above 100°C the protons migrate from the terminal oxygens to bridging oxygens (in the free heteropoly anions in solution the preferred protonation sites are the bridging oxygens, which have a higher electron density than the terminal atoms).

The infrared spectroscopic studies carried out on $H_4SiW_{12}O_{40}{\cdot}nH_2O$ by Bielański *et al.* [35, 36] showed that complete dehydration of the heteropoly acid is already apparent at *ca.* 90°C. At that temperature the 1702 cm^{-1} band, due to $\delta(H_2O)$ vibrations of terminal water molecules in $H_5O_2^+$ hydrogen-bonded to the Keggin anion, disappears. However the corresponding 3150 cm^{-1} band is still present, but shifted to slightly lower frequency (*ca.* 3100 cm^{-1}). Thus the four remaining protons must be engaged in hydrogen bonding between O atoms of the Keggin anions. In such a case, one should expect to observe also two bending vibrations, δ(OH) and γ(OH), for this protonated Keggin unit. Using the correlations between the band positions of these three modes and the O–H···O distance [42], one can estimate the positions of the bands due to the bending vibrations as $\delta(OH) \approx 1220$–1150 cm^{-1} and $\gamma(OH) \approx 950$–920 cm^{-1}. A relatively weak band appears in the infrared spectra at *ca.* 1107 cm^{-1}, which is also apparent at lower temperature (although with a different band shape) and whose origin is not clear. However the 950-920 cm^{-1} range is obscured by an intense absorption due to the Si–O stretching vibration. A similar situation applies in the case of $H_4PMo_{11}VO_{40}$ [34].

6. Conclusions

From a combination of vibrational spectroscopic investigations with structure determination using X-ray and neutron diffraction and the use of other techniques, such as NMR spectroscopy, the overall picture relating to structural variations in the heteropoly acids is reasonably well understood. There are, however, a number of areas which we have discussed where there is still controversy over details of the structure, particularly in relation to the location of the protons and a better understanding of the broad, quasi-continuous intense absorption which appears in many systems with very strong hydrogen bonds in liquids and solids (there is as yet almost no experimental data for the gas phase, which would be very helpful in clarifying the origin of the absorption). Further careful studies by infrared and other vibrational spectroscopic techniques should help to elucidate these problems.

7. References

1. (a) I.V. Kozhevnikov: *Chem. Rev.* **98**, 171 (1998); N. Mizuno and M. Misono: *Chem. Rev.* **98,** 199 (1998).
2. N. Tjapkin, M. Davidovics, Ph. Colomban and U. Mioc: *Solid State Ionics* **61**, 179 (1993)
3. T.J.R. Weakley: *Structure and Bonding* (Berlin) **18**, 131 (1974); G.A. Tsigdinos: *Topics in Current Chemistry* **76**, 1 (1978)
4. J.F. Keggin: *Proc. R. Soc. London* **A144**, 75 (1934)
5. M.-R. Noe-Spirlet, G.M. Brown, W.R. Busing and H.A. Levy: *Acta Cryst.* **A31**, S80 (1975)
6. M.-R. Spirlet and W.R. Busing: *Acta Cryst.* **B34**, 907 (1978)
7. M. Fournier, C. Feumi-Jantou, C. Rabia, G. Hervé and S. Launay: *J. Mater. Chem.* **2**, 971 (1992)
8. G.M. Brown, M.-R. Noe-Spirlet, W.R. Busing and H.A. Levy: *Acta Cryst.* **B33**, 1038 (1977)
9. P. Schuster, G. Zundel and C. Sandorfy (Eds): *The Hydrogen Bond* **2**, North-Holland, Amsterdam, 1976; H. Ratajczak and W.J. Orville-Thomas (Eds.): *Molecular Interactions* **1**, Wiley, Chichester, 1980
10. A. Novak: *Struct. Bonding (Berlin)* **18**, 177 (1974)
11. S. Bratos, H. Ratajczak and P. Viot: in *Hydrogen-bonded Liquids*, J.C. Dore and J. Teixeira (Eds.), Kluwer, 1991, p.221, and references cited therein
12. H. Ratajczak and A.M. Yaremko: *Chem. Phys. Lett.* **314**, 122 (1999) ;*J. Mol. Struct. (Theochem)* **500**, 413 (2000); A.M. Yaremko, D.I. Ostrovskii, H. Ratajczak, and B.Silvi: *J. Mol.Struct.* **482-483**, 665 (1999); H. Ratajczak, W.J.Orville-Thomas, A.M. Yaremko and D.I. Ostrovskii: *Bull. Pol. Acad. Sci.,Chem.* **47**, 193 (1999)
13. G. Zundel: *Adv. Chem. Phys.* **111**, 1 (2000), and references cited therein
14. N.B. Librovich, V.P. Sakun and N.D. Sokolov: *Chem. Phys.* **39**, 351 (1979).
15. H. Abramczyk: *Chem. Phys.* **144**, 305 and 319 (1990)
16. H.D. Lutz: *Struct. Bonding (Berlin)* **69**, 97 (1988)
17. C.I. Ratcliffe and D.E. Irish: in *Water Science Reviews* **2**, F. Franks (Ed.), Cambridge University Press, Cambridge, 1986, p.149.
18. A. Nakahara, Y. Saito and H. Kuroya: *Bull. Chem. Soc. Japan* **25**, 331 (1952).
19. E.F. Valeev and H.F. Schaefer III: J. Chem. Phys. **108**, 7197 (1998)
20. L. Ojamäe, I. Shavitt and S.J. Singer: *Int. J. Quantum Chem.: Quantum Chem.Symp.* **29**, 657 (1995) and *J. Chem. Phys.* **109**, 5547 (1998).

20. A.J. Cunningham, J.D. Payzant and P. Kerbale: *J. Am. Chem. Soc.* **94**, 7627 (1972)
21. M. Meot-Ner (Mautner) and C.V. Speller: *J. Phys. Chem.* **90**, 6616 (1986).
22. M. Meot-Ner (Mautner) and F.H. Field, *J. Am. Chem. Soc.* **99**, 998 (1977).
24. R. Vuilleumier and D. Borgis: *J. Chem. Phys.* **111**, 4251 (1999) and *J. Mol Struct.* in press (2000), and references cited therein
25. G. Zundel and H. Metzger: Z. Physik. Chem. (Frankfurt) **58**, 225 (1968)
26. F.F. Muguet: *J. Mol. Struct. (Theochem)* **368**, 173 (1996), and references cited therein
27. H.-P. Cheng and J.L. Krause: *J. Chem. Phys.* **107**, 8461 (1997).
28. H.-P. Cheng: *J. Phys. Chem.* **A102**, 6201 (1998).
29. D.E. Sagnella and M.E. Tuckerman: *J. Chem. Phys.* **108**, 2073 (1998).
30. J.M. Williams: in *The Hydrogen Bond* **2**, P. Schuster, G. Zundel and C. Sandorfy (Eds.), North-Holland, Amsterdam, 1976, p.655.
31. C. Rocchiccioli-Deltcheff, R. Thouvenot and R. Franck: *Spectrochim. Acta* **32A**, 587 (1967); C. Rocchiccioli-Deltcheff, M. Fournier, R. Franck and R. Thouvenot: *Inorg. Chem.* **22**, 207 (1983)
32. C. Rocchiccioli-Deltcheff and M. Fournier: *J. Chem. Soc. Faraday Trans.* **87**, 3913 (1991).
33. Y. Kanda, K.Y. Lee, S.-I. Nakata, S. Asaoka and M. Misono: *Chem. Lett.*, 139 (1988).
34 A. Bielański, A. Małecka and L. Kubelkova: *J. Chem. Soc., Faraday Trans. I* **85**, 2847 (1989).
35. A. Bielański, J. Datka, B. Gil, A. Małecka-Lubańska and A. Micek-Linicka: *Catal. Lett.* **57**, 61 (1999).
36. J. Baran, A.J. Barnes, A. Bielański, A. Müller and H. Ratajczak: unpublished results.
37. J.G. Highfield and J.B. Moffat: *J. Catal.* **88**, 177 (1984).
38. A.J. Barnes, A. Bielański, H. Ratajczak and M. Wiewiórowski: unpublished results.
39. G.J. Kearley, R.P. White, C. Forano and R.C.T. Slade: *Spectrochim. Acta* **46A**, 419 (1990).
40. U.B. Mioc, Ph. Colomban, M. Davidovic and J. Tomkinson: *J. Mol. Struct.* **326**, 99 (1994).
41. U.B. Mioc, R. Dimitrijevic, M. Davidovic, Z. Nedic, M. Mitrovic and Ph. Colomban: *J. Mater. Sci.* **29**, 3705 (1994).
42. H. Ratajczak and W.J. Orville-Thomas: *J. Mol. Struct.* **1**, 449 (1967-68).

Bond-Stretch Isomerism in Polyoxometalates?

M.-M. ROHMER AND M. BENARD*
Laboratoire de Chimie Quantique, UMR 7551, CNRS and Université Louis Pasteur, Strasbourg, France.

E. CADOT AND F. SECHERESSE
I.R.E.M., UMR 173, CNRS and Université Versailles Saint-Quentin, Versailles, France.

Abstract. In spite of thorough investigations, the most debated issue of bond-stretch isomerism has remained elusive up to now in transition metal chemistry. DFT calculations are reported on the reduced Keggin oxothio heteropolyanions γ-$[SiW_{10}Mo_2S_2O_{38}]^{6-}$ (**1**), γ-$[SiW_{12}S_2O_{38}]^{6-}$ (**2**), γ-$[SiW_{10}Mo_2O_{40}]^{6-}$ (**3**) and γ-$[SiW_{12}O_{40}]^{6-}$ (**4**), obtained from the stereospecific reaction between a preformed $[M^V_2X_2O_2]^{2+}$ cation (M=Mo,W; X=S,O) and a polyvacant γ-Keggin anion. The calculations show that those four clusters display the distinctive signature of bond-stretch isomerism, namely the presence of a double minimum on their potential energy surface depending on a single metal-metal distance. The energy minima are assigned to the localisation of the metal electron pair into the cationic moiety giving rise to a metal-metal bond, and to its transfer to the γ-Keggin core, respectively. The energy barriers separating the two minima do not exceed 6kcal.mol^{-1}, which precludes a physical separation of the isomers. At variance with small clusters containing a limited number of metal atoms, supramolecules made of the assembly of several organometallic/inorganic fragments could be well suited to bond-stretch isomerism due to the possibility of *intramolecular* electron transfers with structural consequences similar to those of standard oxido-reduction.

Key words: Bond-stretch isomerism, distortional isomerism, Keggin oxothio polyanions, electronic structure, DFT calculations, potential energy surfaces.

1. Bond-stretch isomerism: the historical background.

Bond-stretch, or distortional isomerism, as defined by Parkin,[1] is "the unusual phenomenon whereby molecules differ only in the length of one or more bonds". Since this concept was introduced in the early seventies[2,3] it has been the subject of intense controversies at the interface of various areas of chemistry: theoretical modelling,[2,4-6] molecular topology,[1-6] preparative chemistry, mainly inorganic/organometallic,[3,7] X-ray cristallography.[1,8] The original idea of Stohrer and Hoffmann stemmed from the possibility of obtaining an *avoided crossing* on the lowest potential energy curve of tricyclo(2,2,2,0) octane C_8H_{12} and other tricyclic molecules by just varying a critical C-C distance.[2] The energy surface corresponding to the ground state of such molecules should then display a *double minimum* as a function of the bond distance (Figure 1). This double minimum is the signature of bond-stretch isomerism. The phenomenon could not be observed on the tricyclic systems because of concurrent electronic processes, but theoretical investigation on small organic or inorganic systems is still going on.[4,9]

M.T. Pope and A. Müller (eds.), Polyoxometalate Chemistry, 117–133.

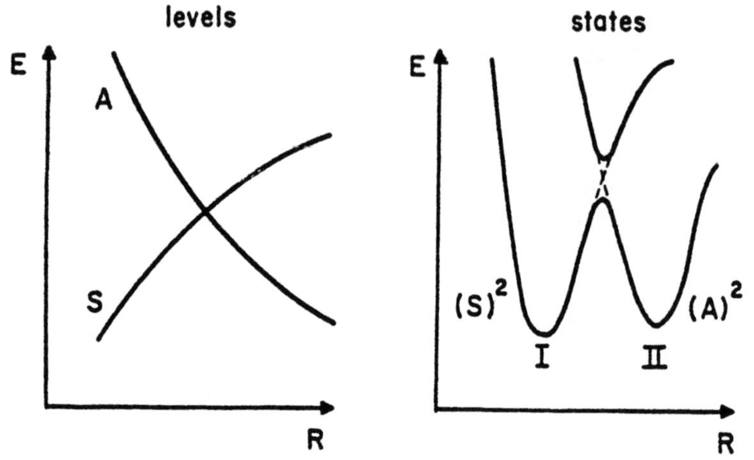

Fig. 1. Orbital diagram (left) and state diagram (right) characteristic of bond-stretch isomerism, according to Stohrer and Hoffmann. S and A represent two molecular orbitals belonging to different irreducible representations and accommodating a total of two electrons. Reproduced from reference 2b, with permission.

At the time of the first paper by Stohrer and Hoffmann, experimental evidence in favor of bond-stretch isomerism had already been given by Chatt[3a] in the field of transition metal chemistry. Molybdenum complex *mer*-$MoOCl_2(PEt_2Ph)_3$ and related systems differing in the phosphine substituents could be isolated in two isomeric forms, one blue, and one green. Surprisingly, a characterization by X-ray diffraction showed that both isomers displayed the same *cis* conformation of the chloride ligands. The only dissemblance in their structures was a significant difference of the Mo=O bond length: 1.803(11)Å for the green isomer of *cis-mer*$MoOCl_2(PEt_2Ph)_3$; and 1.676(7)Å for the blue form of *cis-mer*-$MoOCl_2(PMe_2Ph)_3$. Subsequent studies reported exemples of "distortional isomerism" involving different types of metal-oxo, metal-nitrido and metal-sulfido complexes.[7,10]

The emergence of what appeared as a new class of isomers triggered an interesting discussion on the concept of isomerism and its evolution, summarized in the paper by Jean *et al.*[5] The characterization of *isomers* requires the distinct conformations to persist as separate entities during a few minutes at room temperature. This puts a constraint on the energy barrier separating the two equilibrium geometries, whose height should reach at least 30kcal.mol^{-1}. A lower energy barrier entails a rapid interconversion between the two forms, which are then considered as *conformers*. More than two conformations are often involved and the compound undergoes a dynamic behavior called *fluxionality*. The nature of the geometry difference observed between distinct molecular frameworks made of the same atoms was also a major criterion for defining isomerism. A change in the three-dimensional arrangement of atoms implying at least a rotation, was requested. Consequently, neither the complete rupture of a bond, nor the small structural differences observed between molecules sharing the same asymmetric unit in a crystal were considered as isomers. Then, the new class of bond-stretch isomers was viewed as the "missing link" between those various forms of interconvertible systems, abolishing the topological change as a criterion for isomerism.[5]

Although two possible electronic mechanisms had been proposed by Jean *et al* to explain the distortional isomerism of $MoOCl_2(PR_3)_3$,[5] careful ab initio calculations by Song and Hall failed to produce the expected two-minima potential curve, and the title of their report: "Bond-Stretch Isomers of Transition-Metal Complexes. Do They Exist?"[6] sounded as a first crack in the success story of distortional isomerism. Soon after came the collapse: after a meticulous investigation, G. Parkin was able to demonstrate that the "long Mo=O distance" characterized for the "green" form of *cis-mer*-$MoOCl_2(PMe_3)_3$ was an artifact due to the presence of a small and variable amount of yellow *mer*-$MoCl_3(PMe_3)_3$ cocrystallized with blue *cis-mer*-$MoOCl_2(PMe_3)_3$.[1,8] The presence of similar contaminants was then detected, or suspected in all crystal samples for which long M=O, M=S, or M=N bonds had been characterized.[1]

Little was left of bond-stretch isomerism after this final stroke, even though Parkin made a point of noting that the concept was still alive and could be illustrated further.[8a] A controversy developed concerning a number of well-established cases in which changes in bond lengths are associated with changes in the spin state[11] but Parkin and Hoffmann preferred to define those compounds as *spin-state isomers* and to stress that the concept of bond-stretch isomerism was "introduced for isomers of the same spin state, on the same potential energy curve, for which there was no obvious explanation".[12]

Since then, bond-stretch isomerism has made an unexpected come-back, particularly in the last two years, although without clear evidence up to now, either from experience or from theory. In the field of organic chemistry, two energy minima corresponding to strikingly different distributions of the C-C bond lengths were characterized from DFT and ab initio MP2 calculations on benzodicyclobutadiene.[9] However, only one of those geometries, displaying an abnormally long C-C aromatic bond, was characterized for the existing derivative. In transition metal chemistry, dynamic processes involving mobile metal-metal bonds were evidenced by Rauchfuss in cubane-like M_4S_4 clusters.[13,14] In most cases, this mobility resulted in a simple fluxional behavior,[13] but the low-temperature ^1H NMR spectrum of a mixed valence dication with the Ru_4S_4 core was assigned to a pair of distinct isomers differing in the arrangement of their Ru-Ru bonds.[14] The term "geometric isomerism" was coined by Rauchfuss to label such distinct molecules exclusively differing in the relative positions of their M-M bonds.[14]

Finally, the linear trimetallic complexes of Co(II) first synthesized by Yang *et al* [15] and then characterized and investigated by Cotton *et al*[16,18-20] appear extremely promising. Those linear chains are supported by four dipyridylamine (dpa) anions which adopt a spiral conformation because of internal steric strain.[16,19] The neutral tricobalt complex, axially coordinated to halogen atoms or BF_4^- ions, was structurally characterized in two forms differing essentially in the Co-Co distances.[16] The first type of crystal (Type I) shows a symmetrical arrangement of the linear chain, with two short Co-Co distances (2.25 to 2.32Å) indicative of metal-

metal bonds. In the second type of crystal (Type II) the Co-Co-Co chain is highly unsymmetrical, with a short (2.28Å) and a long (2.46-2.50Å) Co-Co separations. Both crystal structures were characterized for the same complex, $Co_3(dpa)_4Cl_2$ which was tentatively assigned to be the first case of bond stretch isomerism.[16] However, DFT calculations failed again to characterize the expected double minimum on the potential energy curve corresponding to a low-spin (doublet) state: all geometry optimization processes converged toward a symmetric form very close to the experimental type I.[17] Very recently, an investigation of the magnetic susceptibility in solution showed that a spin transition from a doublet to a sextet state was taking place in the temperature range 193 to 308K.[18] The reported NMR spectrum however indicates that the molecule seems to remain symmetric in this temperature range.

The case of the $(Cr_3(dpa)_4X_2)^+$ complexes still more dramatically illustrates the great variability of the linear trimetallic unit. According to the nature of the counter-ion, four isomers have been characterized, revealing a smooth transition from a symmetric Cr-Cr-Cr arrangement (Cr1-Cr2 = Cr2-Cr3 = 2.30Å) to an extremely unsymmetrical one (Cr1-Cr2=1.90Å; Cr2-Cr3 = 2.60Å).[19] The magnetic and spectroscopic behavior of those complexes are presently being investigated, and theoretical studies are being carried out, that will possibly allow, for the first time after Parkin's demonstration, to remove the question mark after the words "bond-stretch isomers".

2. The Case of Heteropolyanions: γ-$[SiW_{10}M_2X_2O_{38}]^{6-}$ ($M=Mo^V,W^V$; X=O,S).

This series of four γ-Keggin heteropolyanions was synthesized at the University of Versailles through stereospecific addition of the dication $[M^V{}_2X_2O_2]^{2+}$ to the divacant $[SiW^{VI}{}_{10}O_{38}]^{8-}$ anion in dimethylformamide. The oxothio anions (X = S; M = Mo, W) could be isolated in crystal form with various counterions and structurally characterized from X-ray diffraction.[20] A typical structure is displayed in Figure 2 and a selection of geometrical parameters is displayed in Table 1. The coordination geometry about the M^V atoms is that of two square pyramids sharing a common basal edge formed by the two sulfur atoms. A short M^V-M^V separation is observed for γ-$[SiW_{10}Mo_2S_2O_{38}]^{6-}$(1; Mo-Mo = 2.832Å) as for : γ-$[SiW_{12}S_2O_{38}]^{6-}$ (2; W-W = 2.815Å). Those values are comparable with the M-M distances observed in M^V-M^V dinuclear compounds showing similar $[M_2X_2\,\mu\text{-}S_2]^{2+}$ cores,[21] including the parent precursor[22] and are consistent with the presence of a metal-metal single bond.

The oxo anions γ-$[SiW_{10}Mo_2O_{40}]^{6-}$ (3) and γ-$[SiW_{12}O_{40}]^{6-}$ (4) were prepared according to similar procedures, but could not be characterized crystallographically. It appeared however that the dodecatungsten oxoanion (4) could behave differently from the three other Keggin clusters. More specifically, the blue color of the solutions and solid residues containing (4) contrasts with the red or red-brown aspect of the equivalent material obtained with (1), (2) and (3). Since the blue color in reduced polyoxoanions ("heteropolyblues") is generally characteristic of an important delocalization of the metal electrons,[23] the hypothesis of a structure without a direct metal-metal bond connecting the W^V centers was considered and investigated by means of quantum theoretical calculations within the Density Functional Theory (DFT) framework. This contribution represents a preliminary account of the calculations carried out on complexes (**1**), (**2**) (**3**) and (**4**).

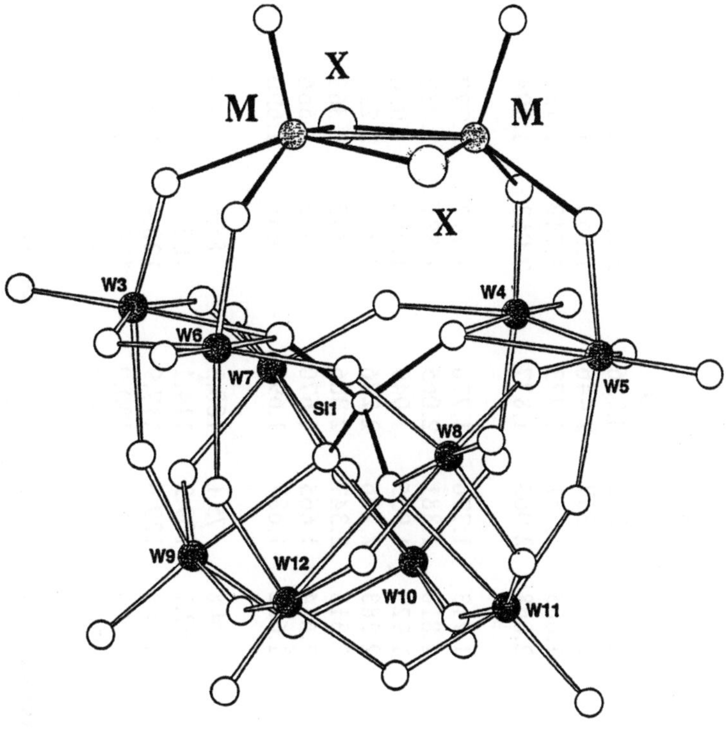

Fig. 2. CAMERON view of γ-$[SiW_{10}M_2X_2O_{38}]^{6-}$ (M=Mo,W; X=S,O) with a M-M bond from the X-ray structure characterized for γ-$[SiW_{10}Mo_2S_2O_{38}]^{6-}$ (ref. 20).

3. Computational Details

All geometry optimizations on the various electronic configurations considered for (**1**), (**2**) (**3**) and (**4**) have been carried out with the ADF program.[24] The formalism is based upon the local spin density approximation characterized by the electron gas exchange ($X\alpha$ with α=2/3) together with Vosko-Wilk-Nüsair[25] parametrization for correlation. Nonlocal corrections due to Becke for the exchange energy[26] and to Perdew for the correlation energy[27] have been added. For nonmetal atoms, a frozen core composed of the 1s shell for oxygen; of the 1s, 2s and 2p shells for sulfur, was described by means of single Slater functions. The Slater basis set used for the valence shell of S and O was of triple-ζ quality and completed by a d-type

TABLE 1

Selected geometrical parameters for γ-$[SiW_{10}Mo_2S_2O_{38}]^{6-}$ (**1**), γ-$[SiW_{12}S_2O_{38}]^{6-}$ (**2**), γ-$[SiW_{10}Mo_2O_{40}]^{6-}$ (**3**) and γ-$[SiW_{12}O_{40}]^{6-}$ (**4**) in the Localized (L) and Delocalized (D) electronic configurations. X-ray values are reported for (**1**) and (**2**).

	1			**2**			**3**		**4**	
	Exp.	L	D	Exp.	L	D	L	D	L	D
M-M	2.832	2.998	3.865	2.815	2.872	3.756	2.653	3.103	2.569	3.088
M-X	2.302	2.359	2.464	2.318	2.357	2.446	1.952	1.965	1.957	1.971
M-O(1)[a]	1.680	1.740	1.759	1.685	1.761	1.773	1.749	1.751	1.764	1.781
M-O(2)	2.068	2.060	1.864	2.068	2.065	1.877	2.063	1.894	2.068	1.901
W-O(1)[a]	1.716	1.777	1.774	1.713	1.776	1.776	1.777	1.777	1.779	1.775
W-O(M)[b]	1.825	1.881	2.077	1.818	1.881	2.065	1.881	2.033	1.876	2.028
W-O(2)	1.930	1.955	1.948	1.923	1.955	1.949	1.956	1.950	1.956	1.955
W-O(3)	2.283	2.270	2.270	2.283	2.296	2.286	2.269	2.299	2.296	2.309
W-O(4)	2.332	2.343	2.362	2.318	2.347	2.366	2.349	2.363	2.353	2.367
Si-O(3)	1.633	1.662	1.660	1.614	1.665	1.670	1.662	1.666	1.665	1.661
Si-O(4)	1.634	1.688	1.690	1.662	1.687	1.693	1.688	1.690	1.686	1.687
M-X-M	75.9	78.9	103.3	74.8	75.1	100.3	85.6	104.3	82.0	103.1
M-M-O(1)	107.0	112.3	117.1	107.1	109.2	114.1	107.7	107.9	105.1	107.8
M-M-O(2)	129.2	126.0	114.5	129.8	129.6	117.9	128.4	123.5	131.3	124.0
M-O(2)-W	128.4	132.9	132.8	128.0	128.2	130.9	129.1	127.7	125.2	127.2
O(2)-M-O(2)	78.9	78.8	95.0	78.2	78.9	93.1	81.9	92.6	81.8	92.5

[a] M-O(n) and W-O(n) represent the average distances between the metal and oxygen atoms bonded to (n-1) other atoms, including Si, for n=3, 4.
[b] O(M) represents the bridging oxygen connecting the $[M_2X_2O_2]$ fragment to the γ-Keggin moiety.

polarization function.[28] The frozen core of the metal atoms composed of the 1s to 3sp shells for molybdenum, of the 1s to 4sp shells for tungsten, was also modelled by a minimal Slater basis. The ns and np shells of metal were described by a double–ζ Slater basis; the nd and (n+1)s shells by a triple–ζ basis and the (n+1)p shell by a single orbital. No f-type polarization function is added. First order, scalar relativistic corrections were included in all calculations by diagonalizing the Pauli Hamiltonian in the space of the non-relativistic solutions. The geometry optimization processes have been carried out by minimizing the energy gradient by the BFGS formalism[29] combined with a DIIS-type convergence acceleration method.[30] The optimization cycles were continued until all of the three following convergence criteria were fulfilled: (i) the difference in the *total energy* between two successive cycles is less than 0.001 hartree; (ii) the difference in the *norm of the gradient* between two successive cycles is less than 0.01 hartree.$Å^{-1}$; (iii) the maximal difference in the *Cartesian coordinates* between two successive cycles is less than 0.01Å.

4. The Two Energy Minima.

Geometry optimization processes carried out on clusters (**1**) to (**4**) show that double minima can be characterized on the potential energy surfaces of all four γ-Keggin anions. The convergence toward one or the other minimum depends on the starting geometries.

A first series of geometry optimizations was carried out on the oxothio clusters (**1**) and (**2**), starting from the structure determined from X rays, slightly modelled in order to take advantage of the C_{2v} symmetry. The X-ray structures were then adapted to the case of the oxo clusters (**3**) and (**4**) in order to provide reasonable starting geometries with short M^V-M^V distances. After convergence of the optimization processes, it was noted that all four structures calculated from those starting geometries are associated with similar electronic configurations, referred to as L, for Localized. The wavefunctions obtained for (**1**), (**2**), (**3**) and (**4**) are characterized by a highest occupied molecular orbital (HOMO) with a_1 symmetry and localized on the $(\mu\text{-}X_2M_2O_2)^{2+}$ fragment. This orbital, schematized in Figure 3, displays a strong σ-bonding character between the M^V centers and some repulsive interaction with the bridging oxo, or thio ligands. In agreement with the presence of this metal-metal single bond, short distances were obtained at equilibrium between the metals of the $(\mu\text{-}X_2M_2O_2)^{2+}$ fragment (Figure 4 and Table 1). For γ-$[SiW_{12}S_2O_{38}]^{6-}$, the optimized W-W distance, 2.872Å, agrees well with the experimental value of 2.815Å. For the oxothio complex with a dimolybdenum capping fragment, the optimized Mo-Mo distance (2.998Å) is appreciably longer than the observed bond length (2.832Å), but remains quite compatible with a metal-metal bond. The other geometrical parameters are in good agreement with the X-ray structures (Table 1). The replacement of the bridging thio by bridging oxo ligands results in an important decrease of the metal-metal bond length, of the order of 0.3Å for tungsten, and still larger for molybdenum. The calculated Mo-Mo distance in (**3**) (2.653Å) however remains substantially longer than the W-W bond length in (4) (2.569Å). Geometrical modifications in the Keggin core are limited (Table 1).

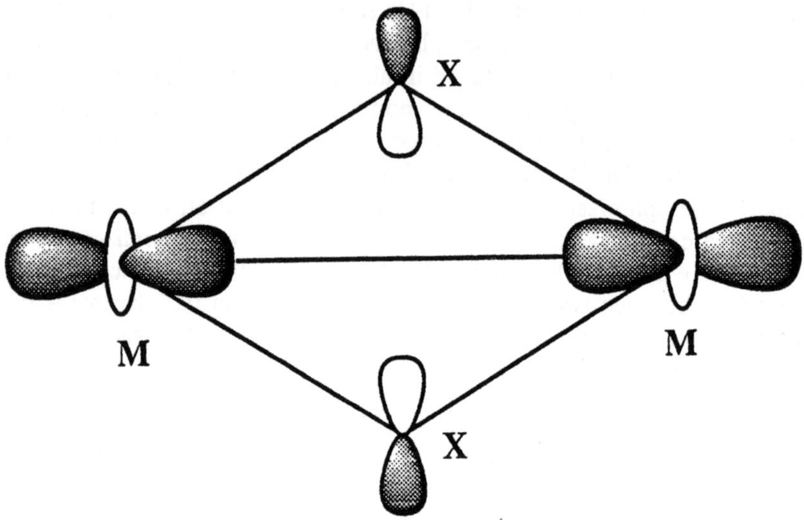

Fig. 3. HOMO, localized on the $(\mu\text{-}X_2M_2O_2)^{2+}$ fragment, in the metal-metal bonded electronic configuration.

The electronic structure associated with the energy minimum characterized above and common to the four clusters refers to the standard orbital sequence of saturated polyoxometalates. The nonreduced species display an extremely large energy gap between the highest occupied levels, delocalized over the formally complete valence shells of the oxo - or thio - ligands, and the unoccupied d shells of the metal atoms, destabilized by the donation interactions. In clusters (1) to (4), this gap can be easily recognized between the HOMO-1 and the LUMO (Figure 4). The energy values associated with those orbitals are practically constant for the whole series (between +8.48eV and +8.87eV for the HOMO-1; between +10.96eV and +11.10eV for the LUMO). The average energy gap of 2.4eV can be considered as very high by DFT standards.[31] Since the four species undergo a two-electron reduction, those additional electrons have to be accommodated somewhere. In the electronic configuration presently considered, the additional level is provided by the metal-metal bonding orbital represented in Figure 3. The energy of that orbital falls in the large gap separating the occupied band of the oxo ligands from the empty metal d band (Figure 4). This orbital energy is not constant along the series of clusters; the LUMO is increasingly destabilized as the metal-metal bond length becomes shorter (Figure 4). This can be explained quite easily: the HOMO is the metal-ligand antibonding counterpart of a low-energy, all-bonding MO with major weight on the ligand p orbitals pointing toward the center of the M_2X_2 core. A contraction of the metal-metal distance increases the metal-ligand overlap and the strength of the metal-ligand interaction. The antibonding term of this interaction is therefore rejected to higher energies (Figure 4). Such a four-electron interaction is known to be globally unfavorable. When the contraction of the M_2X_2 core pushes the HOMO close enough to the empty d band, it could become advantageous to delocalize the electrons over the γ-Keggin tungsten framework, even at the expense of the M^V-M^V bond.

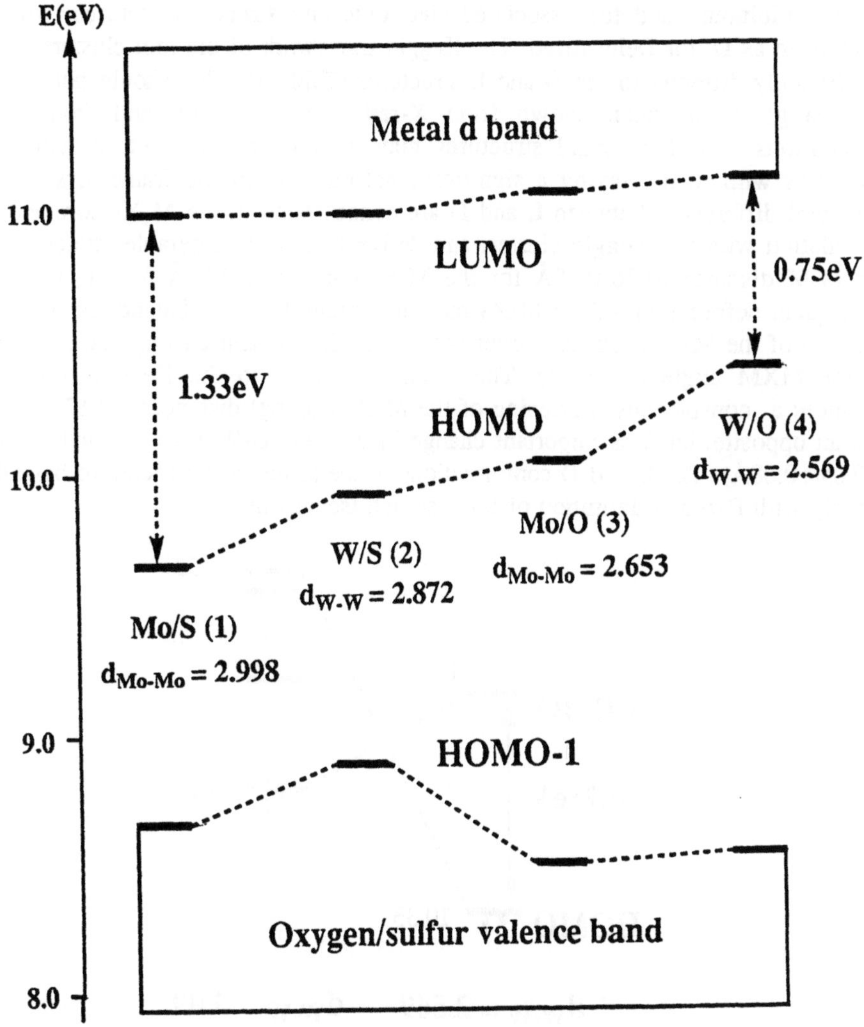

Fig. 4. Energies of the Kohn-Sham frontier orbitals calculated for γ-$[SiW_{10}M_2X_2O_{38}]^{6-}$ (M=Mo,W; X=S,O) at the equilibrium geometry associated with a single M-M bond.

To test this hypothesis, another series of geometry optimization processes has been carried out on (1)-(4) now assuming in the starting geometries an elongated M_2X_2 core with M^V-M^V distances stretched by ~0.4Å. Rather surprisingly, the geometry of the bridged dimetal fragments evolved toward still longer M^V-M^V distances, not only for (4), but for all the clusters. Since a closed-shell electronic configuration was imposed to the clusters in that series of calculations as in the former one, this clearly means that a double energy minimum does exist on the potential energy hypersurface of the diamagnetic spin state. The second minimum, and the associated electronic and structural features, will be referred to as D, for Delocalized. The Keggin framework of the four clusters is not significantly different in the D and L structures (Table 1). This should not appear surprising; it has been shown from X-ray crystallography and from DFT calculations that those rigid structures may undergo many electron reduction processes without displaying a significant deformation of the framework.[32] The structural differences between L and D are concentrated in the M_2X_2 core and its articulation with the Keggin cluster, and derive from a considerable stretching of the M-M distance (0.86-0.88Å for the M_2S_2 core; 0.45-0.52Å for M_2O_2). The subsequent deformation of the M_2X_2 rhombus results from the balance between an increase of the M-X distances, larger with X=S (0.09Å) and an important opening of the MXM angle (Table 1). The electron transfer to the Keggin core also produces a conspicuous shortening of the M-O(bridging) distances (-0.17/-0.19Å) and an opposite, but less important change in the next O-W(3,4,5,6) bond lengths (+0.05/+0.08Å). The L and D conformations of the same cluster therefore basically comply with Parkin's definition of bond-stretch isomerism.[1]

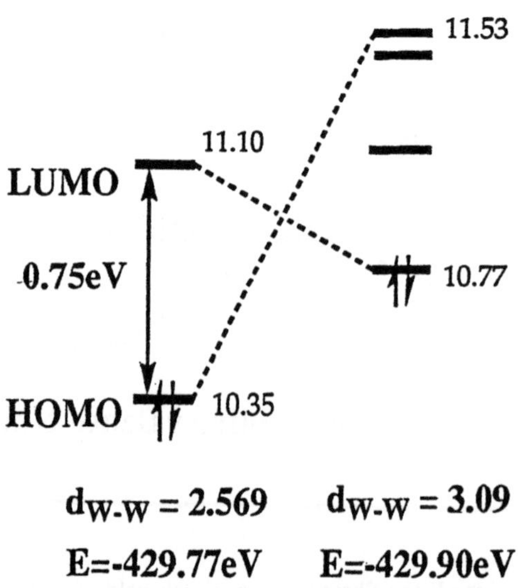

Fig. 5. $[SiW_{12}O_{40}]^{6-}$: Schematic representation of the energy crossing occurring between the HOMO (a_1 symmetry) and the LUMO (a_2 symmetry) along the displacement coordinate connecting the two minima of the potential energy surface.

The change in the electronic configuration which goes along with the stretching of the M-M distance is illustrated in Figure 5 for the case of (**4**). As the M^V-M^V distance increases, the energy of the HOMO rises due to the disappearance of the metal-metal bond. It eventually reaches the level of the metal d band in which it is embedded at the D equilibrium conformation. In this conformation, the metal electron pair is accommodated in the lowest orbital of the d band, a MO with a_2 symmetry delocalized over the tungsten atoms W9, W10, W11, W12, and also, to some extent, on W3, W4, W5 and W6 (see Figure 2 for the atom numbering).

5. Isomers or Conformers? An Approach to the Potential Energy Surfaces.

5.1. COMPUTATIONAL STRATEGY

In order to discuss the relative stabilities of the L and D conformations, and to clarify the issue of bond-stretch isomerism, it is necessary to calculate the critical points of the potential energy surfaces (PES) of (**1**), (**2**), (**3**) and (**4**). The two *minima* characterized on the ground state PES are the most important of those critical points. Their relative energies are reported in Table 2. Since those minima have been obtained with different closed-shell electronic configurations, the *saddle point* separating those minima corresponds to an *avoided crossing* between the ground state and an excited state, the electronic nature of which is interchanged in the region of the saddle point. The energy of the saddle point is a crucial parameter, since it determines the height of the barrier separating the two minima, and therefore conditions the possibility to physically separate the bond-stretch isomers. Since the ground state wavefunction in the vicinity of the saddle point correponds to a mixture of the D and L electronic configurations, a convenient and well-balanced description of the critical points requires an ab initio multiconfigurational treatment, i.e. a CASSCF calculation with an appropriate active space populated with two electrons.[33] This treatment must be completed with a multiconfigurational MP2 calculation accounting for dynamic correlation. Those calculations are presently in progress. The DFT formalism, which is basically monodeterminantal, is not supposed to provide an adequate description of the closed-shell singlet state of lowest energy in the region of the saddle point. One must however consider that *in the present case*, the electron densities associated with the two crossing states are confined in separate parts of the cluster, namely the $M_2X_2O_2$ moiety and the silicotungstate Keggin cluster, where they are described by *very weakly overlapping fragment orbitals*. The energy separation between the *saddle-point* of the ground state and the *minimum* of the closed-shell excited state (interstate gap) is expected to vanish at the limit of zero overlap. When this condition holds, the energy of the saddle point is also degenerate with that of two monodeterminantal open-shell states, either singlet or triplet obtained by populating the *fragment orbitals* with one electron each. Those monodeterminantal states are accessible to DFT calculations and the point of degeneracy with the - elusive -

closed-shell states along the displacement coordinate (the M^V-M^V distance) will be detected at the crossing point of the fragment orbital energies.

In order to complete this preliminary description of the potential energy surfaces within the DFT formalism, two points were characterized on the surface of the *closed shell excited state* by just imposing the orbital occupancy of the D form *at the equilibrium M-M distance calculated for the L form* and vice-versa. All other geometrical parameters were reoptimized. This dual exploration of the two potential surfaces as a function of the M-M distance provides another way of estimating the energy barrier separating the two minima. Still assuming a negligible overlap between the wavefunctions which describe the two states, the crossing point of the potential energy curves will provide a basically correct representation of the saddle point within the DFT formalism, with a slightly underestimated value of the energy due to the separate reoptimization of the cluster geometries for both states, at the considered M-M distance.

TABLE 2

γ-$[SiW_{10}Mo_2S_2O_{38}]^{6-}$ (1), γ-$[SiW_{12}S_2O_{38}]^{6-}$ (2), γ-$[SiW_{10}Mo_2O_{40}]^{6-}$ (3) and γ-$[SiW_{12}O_{40}]^{6-}$ (4): M-M distances (Å) and relative energies (kcal.mol^{-1}) calculated for some points of the potential energy curves. Values in bold represent the critical points : minima (singlet) and saddle points, estimated either from the triplet state or from the crossing point between the singlet potential curves (see text). The estimates for the *barrier energies* are given in parentheses.

	Localized Singlet		Delocalized Singlet		Triplet		Curve Crossing	
	d_{M-M}	E	d_{M-M}	E	d_{M-M}	E	d_{M-M}	E
(1)								
	2.998	**0.0**	2.998	+58.5	**3.80**	**24.4**	**3.77**	**+24.2**
	3.865	+30.4	**3.865**	**+23.8**		*(0.6)*		*(0.4)*
(2)								
	2.872	**0.0**	2.872	+52.7	**3.42**	**+14.4**	**3.50**	**+13.8**
	3.756	+27.5	**3.756**	**+10.2**		*(4.2)*		*(3.6)*
(3)								
	2.653	**0.0**	2.653	+47.2	**3.03**	**+13.5**	**3.02**	**+13.6**
	3.103	+20.3	**3.103**	**+12.5**		*(1.0)*		*(1.1)*
(4)								
	2.569	**+2.9**	2.569	+41.9	**2.86**	**+8.7**	**2.85**	**+8.5**
	3.088	+21.2	**3.088**	**0.0**		*(5.8)*		*(5.6)*

5.2 RESULTS

The relative energies of the five points characterized for each cluster are displayed in Table 2. The potential energy curves that can be deduced from those points, assuming the harmonic approximation, are represented in Figure 6. An uniform interstate gap of 2kcal.mol-1 has been assumed to visualize the avoided crossing.

A first important result concerns the relative energies of the two minima: for (1), (2) and (3), the lowest minimum corresponds to the L form, characterized by a short M^V-M^V bond. The energy difference between the two minima is large: more than 10kcal.mol-1 for (2) and (3), and close to 24kcal.mol-1 for (1) (Table 2). Without considering yet the energy barrier, the L isomer can be considered as the normal form of those three clusters, in agreement with the X-ray characterization of (1) and (2) and with the red-brown color observed for the three complexes in solution. In contrast with this high relative stability of the L conformation, the two minima characterized for cluster (4) are close in energy, with the D conformation favored by 2.9kcal.mol-1 (Table 2). The delocalized form is therefore expected to dominate, in agreement with the blue color of the solution, but a thermodynamic mixture of the two conformations cannot be excluded.

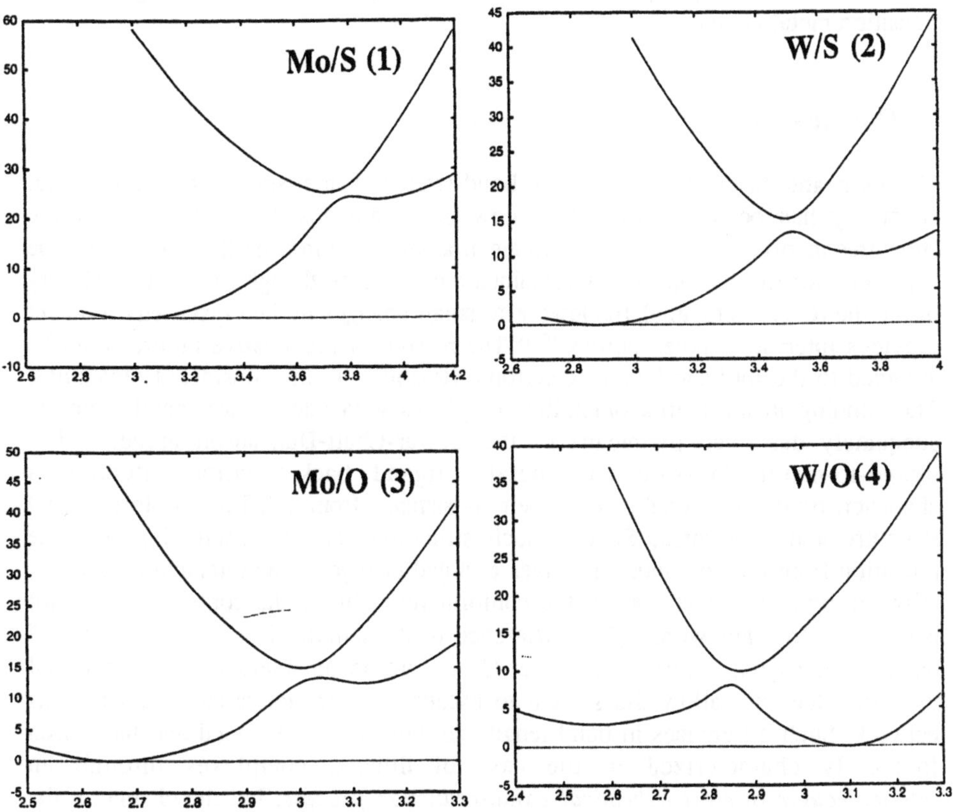

Fig. 6. Aspect of the potential energy curves of (1), (2), (3) and (4) along the M^V-M^V distance, from the critical points calculated using the DFT formalism. An interstate gap of 2 kcal. has been assumed at the avoided crossing. Energies in kcal.mol-1, distances in Å.

An examination of the estimates obtained for the energy barriers indeed suggests that all four clusters will rather behave as *conformers* than as *isomers*. Both methods used to approximate the position and the height of the energy barrier within the DFT formalism agree to predict a barrier of ~1kcal.mol^{-1} or less, occurring at very long M-M distances for clusters (**1**) and (**3**), both synthesized from a cationic moiety containing *molybdenum* (Table 2). The barriers calculated for the all-tungsten clusters are more important: about 4kcal.mol^{-1} for (**2**), and 5.6-5.8kcal.mol^{-1} for (**4**). The position of the barriers is found at large M-M distances and relatively low energies even for cluster (**4**), due to the relatively shallow character of the potential energy curves with metal-metal bonding character. Those barriers are clearly not sufficient to make a physical separation of the isomers possible. Their values are not negligible however and could possibly be enhanced by an appropriate tuning of the dimetal cationic moiety. A possible direction should be to make steeper the potential energy curve associated with the L conformation by designing an oxo- or oxothiometallic cation involving first-row transition metal atoms.

6. Conclusion

The long and fascinating quest for bond-stretch isomerism in transition metal chemistry has been focused up to now on complexes with relatively simple structures involving one, two or three metal atoms. This may be the reason why the expected property has up to now remained elusive, even though the high sensitivity of the the $(Cr_3)^{6+}$ structural framework to small changes in the crystal environment provides interesting potentialities.[15-19] The reason for this relative failure could be assigned to the intrinsic lack of electronic flexibility in the considered molecules. The bonding in most organometallic complexes with one or two metal atoms is adequately described by means of the Dewar-Chatt-Duncanson model.[34] The ligand-to-metal donation and metal-to-ligand back-donation interactions characteristic of this model receive their consistency from a definite conformational structure and vice-versa. Several such structures may be competing, but the transition from one to another involves a global change of the interaction network, mirrored by a global change in the conformation: this is the context of standard isomerism. In other words, the "difference in the length of one or more bonds" basically does not modify the Dewar-Chatt-Duncanson interaction network and therefore does not allow the system to escape the attractor associated with this network. Such differences in bond length can however be obtained and have been frequently characterized in the case of dimetal complexes through an *intermolecular* or *electrochemical oxido-reduction process*, but the change in the total electron count obviously generates a distinct system.

Larger systems, i.e. nanostructures or "supramolecules" made from the assembly of several organometallic/inorganic fragments offer more potentialities, since *intramolecular* electron transfers may occur with structural consequences similar to those of standard oxido-reduction. Moreover, the extension in space of the heterogeneous nanostructures and the possiblity to gather in the same system loosely connected metallic cores makes the competing electronic configurations very weakly overlapping, thus favoring the preservation of energy barriers. Those conditions are clearly fulfilled in the four clusters investigated in the present work: $[SiW_{10}Mo_2S_2O_{38}]^{6-}$, $[SiW_{12}S_2O_{38}]^{6-}$, $[SiW_{10}Mo_2O_{40}]^{6-}$ and $[SiW_{12}O_{40}]^{6-}$. Those clusters are made of two quite different moieties: i) the Keggin core, γ-$[SiW_{10}O_{36}]^{n-}$ which, as most polyoxoanions behaves as a rigid electron reservoir, and ii) the dimetallic fragment, $M_2X_2O_2$, a flexible rhombus that can be either stretched or squeezed depending on the oxidation state of the metal atoms. The balanced - or unbalanced - electrophilic character of the two moieties will therefore condition the localisation of the metal electron pair in the dimetallic fragment or its transfer to the Keggin core. If the transfer eventually occurs, as in γ-$[SiW_{12}O_{40}]^{6-}$, it has little structural impact on the Keggin framework, but the oxidation of the dimetallic fragment and the subsequent vanishing of the metal-metal bond have dramatic consequences on the geometry of the M_2X_2 rhombus. In the present case, the potential energy curves associated with the stretching of the Mo^V-Mo^V or W^V-W^V bonds are relatively shallow and are unable to generate high enough energy barriers permitting a physical separation of the bond-stretch isomers. A tuning of the cationic species involving the replacement of Mo or W by metals with short-range d-d overlap could possibly open the way to the first real case of bond-stretch isomerism in inorganic chemistry.

References

1. G. Parkin: *Chem. Rev.* **93**, 887 (1993).
2. (a) W.-D. Stohrer and R. Hoffmann: *J. Am. Chem. Soc.* **94**, 779 (1972); (b) W.-D. Stohrer and R. Hoffmann: *J. Am. Chem. Soc.* **94**, 1661 (1972).
3. J. Chatt, L. Manojlovic-Muir, and K. W. Muir: *J. Chem. Soc. (D)* 655 (1971); L. Manojlovic-Muir and K. W. Muir: *J. Chem. Soc., Dalton Trans.* 686 (1972).
4. M. N. Paddon-Row, L. Radom, and A. R. Gregory: *J. Chem. Soc., Chem. Commun.* 427 (1976); P. v. R. Schleyer, A. F. Sax, J. Kalcher, and R. Janoschek: *Angew. Chem., Int. Ed. Engl.* **26**, 364 (1987); J. A. Boatz and M. S. Gordon: *J. Phys. Chem.* **93**, 2888 (1989); S. Nagase and T. Kudo: *J. Chem. Soc., Chem. Commun.* 54 (1988); S. Collins, R. Dutler, and A. Rauk: *J. Am. Chem. Soc.* **109**, 2564 (1987); W. W. Schoeller, T. Dabisch, and T. Busch: *Inorg. Chem.* **26**, 4383 (1987); E. Kaufmann and P. v. R. Schleyer: *Inorg. Chem.* **27**, 3987 (1988); P. V. Sudhakar, O. F. Güner, and K. Lammertsma: *J. Phys. Chem.* **93**, 7289 (1989); K. Lammertsma and O. F. Güner: *J. Am. Chem. Soc.* **110**, 5239 (1988); P. V. Sudhakar and K. Lammertsma: *J. Phys. Chem.* **96**, 4830 (1992).
5. Y. Jean, A. Lledos, J. K. Burdett, and R. Hoffmann: *J. Am. Chem. Soc.* **110**, 4506 (1988).
6. J. Song and M. B. Hall: *Inorg. Chem.* **30**, 4433 (1991).
7. K. Wieghardt, G. Backes-Dahmann, B. Nuber, and J. Weiss: *Angew. Chem., Int. Ed. Engl.* **24**, 777 (1985); A. Bashall, B. C. Gibson, T. P. Kee, M. McPartlin, O. B. Robinson, and A. Shaw: *Angew. Chem., Int. Ed. Engl.* **30**, 980 (1991); S. Lincoln and S. A. Koch: *Inorg. Chem.* **25**, 1594 (1986); I. A. Degnan, J. Behm, M. R. Cook, and

W. A. Herrmann: *Inorg. Chem.* **30**, 2165 (1991); G. Backes-Dahmann and K. Wieghardt: *Inorg. Chem.* **24**, 4044 (1985).
8. G. Parkin: *Acc. Chem. Res.* **25**, 455 (1992); K. Yoon, G. Parkin, and A. L. Rheingold: *J. Am. Chem. Soc.* **113**, 1437 (1991); K. Yoon, G. Parkin, and A. L. Rheingold: *J. Am. Chem. Soc.* **114**, 2210 (1992).
9. R. Boese, J. Benet-Buchholz, A. Stanger, K. Tanaka, and F. Toda: *Chem. Commun.* 319 (1999).
10. W. A. Nugent and J. M. Mayer: *Metal-Ligand Multiple Bonds*, Wiley, New York (1988); J. M. Mayer: *Inorg. Chem.* **27**, 3899 (1988).
11. P. Gütlich, H. A. Goodwin, and D. N. Hendrickson: *Angew. Chem. Int. Ed. Engl.* **33**, 425 (1994).
12. G. Parkin and R. Hoffmann: *Angew. Chem. Int. Ed. Engl.* **33**, 1462 (1994).
13. E. J. Houser, T. B. Rauchfuss, and S. R. Wilson: *Inorg. Chem.* **32**, 4069 (1993); E. J. Houser, J. Amarasekera, T. B. Rauchfuss, and S. R. Wilson: *J. Am. Chem. Soc.* **113**, 7440 (1991); A. Venturelli and T. B. Rauchfuss: *J. Am. Chem. Soc.* **116**, 4824 (1994); E. J. Houser, A. Venturelli, T. B. Rauchfuss, and S. R. Wilson: *Inorg. Chem.* **34**, 6402 (1995).
14. Q. Feng, T. B. Rauchfuss, and S. R. Wilson: *J. Am. Chem. Soc.* **117**, 4702 (1995).
15. E.-C. Yang, M.-C. Cheng, M.-S. Tsai, and S.-M. Peng: *J. Chem. Soc., Chem. Commun.* 2377 (1994).
16. F. A. Cotton, L. M. Daniels, and G. T. Jordan IV: *Chem. Commun.* 421 (1997); F. A. Cotton, L. M. Daniels, G. T. Jordan IV, and C. A. Murillo: *J. Am. Chem. Soc.* **117**, 10377 (1997).
17. M.-M. Rohmer and M. Bénard: *J. Am. Chem. Soc.* **120**, 9372 (1998).
18. F. A. Cotton, C. A. Murillo, and X. Wang: *Inorg. Chem.*, in press.
19. F. A. Cotton, L. M. Daniels, C. A. Murillo, and I. Pascual: *J. Am. Chem. Soc.* **119**, 10223 (1997); F. A. Cotton, L. M. Daniels, C. A. Murillo, and I. Pascual: *Inorg. Chem. Commun.* **1**, 1 (1998).
20. E. Cadot, V. Béreau, B. Marg, S. Halut, and F. Sécheresse: *Inorg. Chem.* **35**, 3099 (1996).
21. W. Clegg, N. Mohan, A. Müller, A. Neumann, W. Rittner, and G. M. Sheldrick: *Inorg. Chem.* **19**, 2066 (1980); W. H. Pan, M. A. Harmer, T. R. Halbert, and E. I. Stiefel: *J. Am. Chem. Soc.* **106**, 459 (1984); W. Clegg, G. Christou, C. D. Garner, and G. M. Sheldrick: *Inorg. Chem.* **20**, 1562 (1981); F. Sécheresse, J. Lefebvre, J. C. Daran and Y. Jeannin: *Inorg. Chem.* **21**, 1311 (1982); C. Simonnet-Jegat, R. A. Toscano, F. Robert, J. C. Daran, and F. Sécheresse: *J. Chem. Soc., Dalton Trans.* 1311 (1994); T. Shibahara, Y. Izomuri, R. Kubota, and H. Kuroya: *Chem. Lett.* 2327 (1987); R. Bhattacharyya, P. K. Chakrabarty, P. N. Ghosh, A. K. Mukherjee, D. Podder, and M. Mukherjee: *Inorg. Chem.* **30**, 3948 (1991).
22. S. A. Cohen and E. I. Stiefel: *Inorg. Chem.* **24**, 4657 (1985); A. Müller: *Polyhedron* **5**, 323 (1986).
23. C. L. Hill, ed. *Chem. Rev.* **98** (special issue), 1-389 (1998).
24. *ADF 2.3 User's Guide;* Chemistry Department, Vrije Universiteit, Amsterdam, The Netherlands, 1997: E. J. Baerends, D. E. Ellis, and P. Ros: *Chem. Phys.* **2**, 41 (1973); G. te Velde and E. J. Baerends: *J. Comput. Phys.* **99**, 84 (1992); C. Fonseca-Guerra, O. Visser, J. G. Snijders, G. te Velde, and E. J. Baerends: *Methods and Techniques in Computational Chemistry: METECC-95*, eds. E. Clementi and G. Corongiu, STEF, Cagliari, Italy, pp. 305-395 (1995).
25. S. H. Vosko, L. Wilk, and M. Nusair: *Can. J. Phys.* **58**, 1200 (1980).

26. A. D. Becke: *J. Chem. Phys.* **84**, 4524 (1986); A. D. Becke: *Phys. Rev.* **A38**, 3098 (1988).
27. J. P. Perdew: *Phys. Rev.* **B33**, 8882 (1986); J. P. Perdew: **B34**, 7406 (1986).
28. J. G. Snijders, E. J. Baerends, and P. Vernooijs: *At. Nucl. Tables* **26**, 483 (1982); P. Vernooijs, J. G. Snijders, and E. J. Baerends: *Slater type basis functions for the whole periodic system*, Internal Report, Free University of Amsterdam, The Netherlands (1981).
29. T. H. Fisher and J. Almlöf: *J. Phys. Chem.* **96**, 9768 (1992).
30. L. Versluis: *Ph. D. Thesis*, University of Calgary, Calgary, Alberta, Canada (1989).
31. R. Stowasser and R. Hoffmann: *J. Am. Chem. Soc.* **121**, 3414 (1999).
32. J. M. Maestre, J. M. Poblet, C. Bo, N. Casañ-Pastor, and P. Gomez-Romero: *Inorg. Chem.* **37**, 3444 (1998).
33. The orbitals to be selected in the CASSCF active space are i) the σ-bonding and σ-antibonding MOs localized on the M^V centers and ii) the four phase combinations of the tungsten d orbitals involved in a delocalized description of the cluster at the ab initio level.
34. M. J. S. Dewar: *Bull. Soc. Chim. Fr.* **18**, 679 (1951); J. Chatt and L. A. Duncanson: *J. Chem. Soc.* 2939 (1953).

Quantum-chemical studies of electron transfer in transition-metal substituted polyoxometalates

SERGUEI A. BORSHCH[1], HÉLÈNE DUCLUSAUD
Institut de Recherches sur la Catalyse, UPR 5401 CNRS, 2, avenue Albert Einstein, 69626 Villeurbanne Cedex, and Laboratoire de Chimie Théorique, Ecole normale supérieure de Lyon, 46, allée d'Italie, 69364 Lyon Cedex 07, FRANCE

Abstract. The results of the quantum-chemical DFT (density functional theory) studies of the electron transfer between the substituted transition metal and the polyoxoanion addenda atoms are presented. This work is motivated by the experimental research on the catalytic activity of the iron-substituted 12-molybdophosphoric acid and its salts in the oxidation of alcanes. Two cases have been considered corresponding to two experimental situations extensively studied by Mössbauer spectroscopy:
1) An iron(II) ion substitutes for molybdenum within a Keggin unit. The calculations of a cluster model show that the potential surface has two minima close in energy and corresponding to valence configurations Fe^{II}+Mo^{VI} and Fe^{III}+Mo^{V}. It gives an interesting example of participation of heteroatoms in polyoxoanion addenda in the electron delocalization processes.
2) Iron (III) ions play the role of counter-ions in secondary structure of acid. The experimental studies have shown that electron transfer from the reduced Keggin unit to iron becomes possible only after hydration. On the base of our calculations we explain the role of water. The hydration pushes the iron ion toward a terminal oxygen of the Keggin unit in position more suitable for the electron transfer.
Key words: molybdophosphoric acid, DFT, electronic structure, electron transfer.

I. INTRODUCTION

An extremely rich redox chemistry of polyoxometalates justifies numerous experimental and theoretical studies of the electron transfer with participation of these metal-oxygen clusters. Two different types of the electron transfer should be distinguished. One of them concerns reduction/oxidation of polyoxoanions [1, 2]. The theoretical description of these processes is usually performed on the basis of the Marcus theory of the electron transfer in solution [1,3]. Another phenomenon widely discussed in the polyoxoanion literature is the electron transfer between addenda atoms of reduced clusters known as the "heteropoly blues". The main question under study for such systems is whether reducing electron(s) is (are) localized or delocalized over several atoms or the whole structure. The main tool in the theoretical treatment of reduced systems is the model Hamiltonian method. The model Hamiltonians used include basic electronic interactions (the resonance intercenter interaction, the Coulomb repulsion) as well as the vibronic interactions [4-8]. In this approach the topology of the polyoxoanion is directly taken into account, and the chemical individuality of the system is hidden in the model Hamiltonian parameters. The type of "blue" electron behavior is defined by the nature of ground electronic state.

M.T. Pope and A. Müller (eds.), Polyoxometalate Chemistry, 135–144.

This paper is aimed at analysis of the electron transfer in polyoxometalates on the basis of quantum-chemical DFT calculations. Numerous calculations of the electronic structure of polyoxoanions are known in the literature. Although the earliest theoretical work on polyoxoanions was performed with the simplest semi-empirical Hückel method [9-14], recently the more rigorous *ab initio* or DFT calculations of rather large polyoxoanions have begun to appear [15-22]. Our work is motivated by the experimental studies of the molybdophosphoric acid (MPA) and its derivatives as active catalysts in the oxidation of alkanes [23-28]. It was shown that a total or partial substitution of the protons of the acid by alkaline and transition metals drastically affects catalytic activity and selectivity. The substituting transition metal cations play the role of a reservoir exchanging by electrons with a polyoxoanion. This electron transfer promotes reducibility of the solid and, as a consequence, its catalytic activity. We will consider in details the electron transfer between an Fe counter-ion and the Keggin polyoxoanion $PMo_{12}O_{40}^{3-}$.

The addition of a large enough amount of iron into the MPA cesium salt leads to the partial substitution of both the Cs counter-ion and molybdenum atoms in the addenda of the polyoxoanion. The substitution of molybdenum by Fe^{2+} creates the possibility of intramolecular electron transfer within the Keggin unit. As noted by Goodenough [29] several years ago, the closeness of redox potentials Mo^{6+} / Mo^{5+} and Fe^{3+} / Fe^{2+} in octahedral environment allows one to anticipate an equilibrium $Fe^{2+}+Mo^{6+} \leftrightarrow Fe^{3+}+Mo^{5+}$ in oxides containing both ions in octahedral sites. This electron transfer is analogous to the electron delocalization in the "heteropoly blues". We will also study this case which is not so important for catalytic activity but gives an interesting example of the participation of heteroatoms in the intramolecular electron delocalization.

The DFT calculations presented here are carried out by two different quantum-chemical programs: Gaussian 94 [30] (intramolecular electron transfer) and ADF [31] (intermolecular electron transfer). The technical details are given in our previous publications [32, 33].

II. IRON-MOLYBDENUM ELECTRON DELOCALIZATION IN THE KEGGIN POLYOXOANION.

The fully optimized calculations of such large molecules as the Keggin unit by first-principles methods present rather difficult issues and can be achieved only by assuming some symmetry. However, substitution at an addenda position, as well as a close-lying counter-ion, and structural relaxation accompanying electronic redistribution lead to loss of symmetry. As a result, the calculations become more complex and cost much more CPU time. Therefore, we model the whole Keggin unit by sub-units, representing a quarter or a half of the whole cluster (Fig. 1). In order to represent the rest of the Keggin unit we also add the phosphorus atom with its coordination sphere completed to tetrahedral by three OH groups. Hanging bonds were completed by the placement of hydrogen atoms. These models reproduce in a rather satisfactory way the structural characteristics of the full molecule [32]. The composition of the blocks of highest occupied and lowest free orbitals of our model clusters is the same as for the complete Keggin unit [15]. The former are composed of oxygen orbitals and the latter contain mainly metal orbitals.

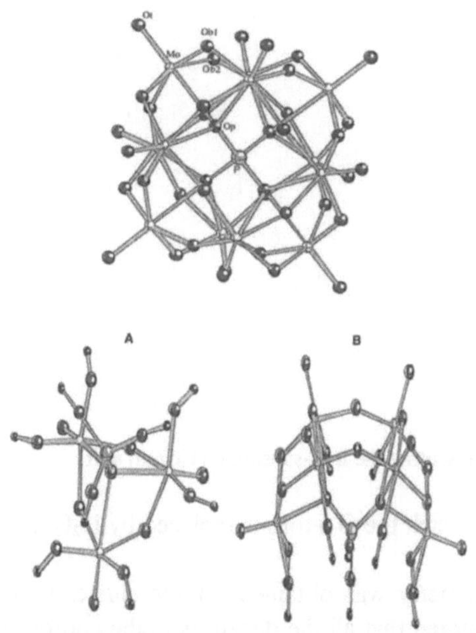

Fig.1. Ball-and-stick representation of the Keggin structure and of the model clusters used in the present work. The different types of oxygen atoms are indicated (O_t is a terminal oxygen, O_{b1} and O_{b2} are bridging oxygens in one trimer and between two trimers, respectively, O_P stands for the oxygen in phosphorus tetrahedral environment).

If the concentration of the substituting iron atoms is not too high, the probability of finding two neighboring Fe atoms within a polyoxoanion is negligibly small. Therefore we considered the substitution of only one molybdenum atom in our model clusters.

Two pathways for the electron transfer between nearest-neighbor metal sites can be suggested. They correspond to the two types of contacts of the adjacent MoO_6 octahedra: through an edge and through a corner. The average Mo-Mo distance within a trimer is 3.41Å, whereas that between corner-sharing Mo atoms is 3.71Å. One can suppose that the electron transfer is more probable in the former case. As both clusters A and B give a satisfactory description of the total Keggin unit we considered only the simplest cluster A, having the formula $PMo_3O_{16}H_9$ and belonging to the symmetry group C_{3v}. In the non-substituted cluster the Mulliken charge of molybdenum atoms is equal to 1.51, giving the reference for +6 oxidation state.

Although there are no X-ray data for the iron-substituted cluster, it is commonly admitted that the general structure of POA is preserved. We supposed that iron ion simply substitutes for molybdenum, and to compensate the negative extra charge corresponding to the valence configuration Fe^{2+}+Mo^{6+} four protons were added (Fig.

2). In solution, depending on the pH, hydroxo or water ligands may replace terminal oxygen in the coordination sphere of substituted atom [34-36]. However, our calculations for the

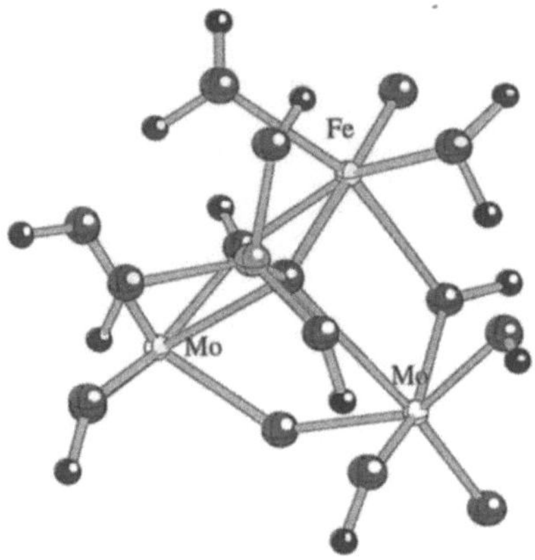

Fig. 2. Model cluster representing the iron-substituted Keggin unit.

substituted cluster in which the O_t atom is replaced by H_2O do not lead to qualitatively different results.

The equilibrium geometry was obtained for the cluster (Fig.2) by the optimization procedure. It can be noted that all the distances in the coordination spheres around the Mo atoms besides Mo-O_{b2} change slightly. The Mulliken charges for Mo atoms are equal to 1.52, i. e. very close to the value found for cluster A. One can conclude that the found energy minimum describes Mo atoms in the oxidation state Mo^{6+}, and the calculated iron charge 0.57 corresponds to Fe^{2+}.

The iron d orbitals fall into the energy gap between nonbonding oxygen and antibonding Mo-O_b orbitals. The spin β HOMO is localized on iron and oxygens and is followed by the LUMO containing Mo and bridging oxygen orbitals (Fig. 3 a).

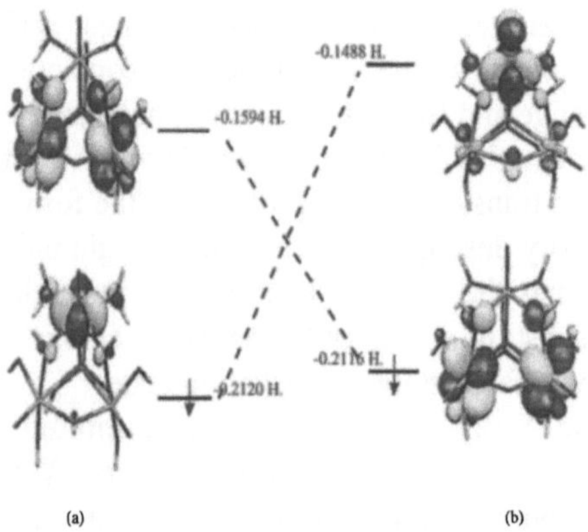

Fig. 3. Composition of HOMO and LUMO of the iron-substituted cluster in two energy minima.

The most straightforward way to produce the electron transfer between Fe and Mo is to interchange these two orbitals. It is difficult to specify *a priori* a distortion which can result in this inversion. Recent studies of heterobimetallic oxygen-bridged complexes suggested that some kind of resonance can occur in these species between the structures $M^{4+}=O\cdots M'^{2+}$ and $M^{2+}\cdots O=M'^{4+}$ [37]. The bridging oxygen atoms play a dominating role in such dynamics. Following these ideas we forced the bridging oxygens to shift toward iron and used this new departure point in the geometry optimization procedure. A new minimum was found in this way with a total energy (-1476.8187 Hartree) very close to that of the previous one (-1476.8177 Hartree). However, the structures and charge distributions corresponding to the two minima are quite different. The distances for the second minimum clearly indicate a contraction of the coordination sphere around the iron atom. The average Fe-O distance goes from 2.06 Å to 1.97 Å. This is in agreement with the increase of the iron Mulliken population which changes from 0.57 to 0.73. Correspondingly, the populations of the two equivalent Mo atoms decrease to 1.40. The average Mo-O distance in the first coordination sphere passes from 2.03 Å to 2.06 Å. The composition of the spin β HOMO and LUMO clearly indicates their inversion comparatively to the first minimum (Fig. 3 b). A certain transfer of electron density toward Mo is also found in α orbitals. One can conclude that the second minimum describes a state where an electron has been transferred from Fe to two Mo atoms. The less pronounced change of the average Mo-O distance occurs due to the delocalization of the transferred charge between two metal atoms. The presence of two potential surface minima with different localization of "extra" electron is reminiscent of multi-minima surfaces of class II mixed-valence compounds, such as the "heteropoly blues".

Since the energies of the two minima are very close fast electron transfer is likely to occur at room temperature. We tried to localize a transition state for this transfer to estimate the barrier energy. However, we did not succeed in this research, probably due to the complexity of our model system. We may assume that the potential surface is flat enough, corresponding to a flexible structure.

The results of our calculations agree with the Mössbauer data [38] for the iron-substituted cesium salt of molybdophosphoric acid containg about 0.6 Fe atom as a counter-ion and 0.9 Fe atom substituting for molybdenum per Keggin unit. The ^{57}Fe Mössbauer spectrum at room temperature presents two quadrupole doublets attributed to iron(III) either in counter-ion position or incorporated into the Keggin unit. However, at 4.2 K some of the intensity of the second doublet is transferred to a new doublet characteristic for iron(II). It suggests that one can have a coexistence of valence configurations $Fe^{2+}+Mo^{6+}$ and $Fe^{3+}+Mo^{5+}$.

III. ELECTRON TRANSFER BETWEEN IRON COUNTER-ION AND THE KEGGIN POLYOXOANION. THE ROLE OF WATER.

The experimental results [28] unambiguously show that the catalytic activity of iron-doped Keggin type molybdophosphoric heteropoly compounds for the oxidation of isobutane depends on their reducibility. The iron counter-ions participate in electron transfer processes to and from polyoxoanions. At the same time, the experimental data indicate that electron transfers depend on the hydration state of the catalyst. In order to explain this effect we undertook quantum-chemical calculations of the interaction of the Fe counterion with the Keggin polyoxoanion in hydrated and non-hydrated solid.

There is no direct structural data either for the non-hydrated acid, or for its iron-substituted derivative. At the same time, the secondary structure of the cesium salt of MPA is isomorphous with the structure of the tungstophosphoric acid with six water molecules [39]. It is a cubic structure with each cesium atom surrounded by four Keggin units. The shortest distances from cesium atoms are to bridging oxygen atoms in the Keggin structure. Different species were proposed to describe the state of iron as a counter-ion in substituted salts. The interanionic cavities are too large to contain simple Fe^{3+} ions. Also the charge equilibrium does not allow any simple substitution of an alkali metal in salts or a proton in the acid by Fe^{3+} ion. We choose for the iron entity the $Fe(OH)_2^+$ cation which has a +1 charge and is approximately the same size as a Cs cation. The same species was proposed by Trifiro and coworkers [40]. The polyoxoanion in our model was represented by cluster B (Fig. 1). In such sub-unit all types (terminal and two bridging) of oxygen atoms existing in the full anion are represented. The complete system $Fe(OH)_2^+ + (PMo_6O_{26}H_{10})^{-1}$ was calculated. Different positions of iron relative to anion oxygen atoms were checked. The energy minimum was found only for the position when iron more strongly interacts with bridging oxygens similarly to cesium in the corresponding salt (Fig.4). The optimized main distances are listed in Table I.

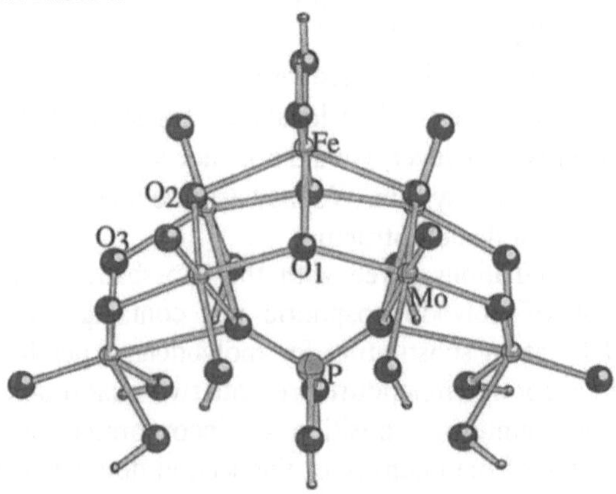

Fig.4. Model cluster representing the interaction between the iron counter-ion and the Keggin unit in non-hydrated solid. O_1, O_2, and O_3 indicate possible positions for a vacancy.

Table I. Optimized distances (in Å) for the coordination sphere of iron in counter-ion in non-hydrated solid.

$Fe-O_{b1}$	$Fe-O_{b2}$	Fe-OH
2.05	2.22	1.82

The Mulliken charges of Mo atoms vary from +2.22 and +2.38 and correspond to formal oxidation state +6. The same value for Fe^{3+} was found equal to 1.14. We must note that these values of the Mulliken charges are different from those presented in Section II. This results from the use of different basis sets in the quantum-chemical programs : Slater functions for ADF (Section III) and Gaussian functions for GAUSSIAN 94 (Section II).

The reduction of catalyst under the hydrogen flow with a loss of water is accompanied by formation of oxygen vacancies in the anion [28]. The calculations show that the energetically favorable localization of the vacancy is in the position of bridging oxygen between two trimers (O_1 on Fig. 4). The most important change in the lacunary structure concerns the distance between molybdenum atoms neighboring the vacancy. This value increases from 3.76Å to 3.86Å, leaving the Keggin structure slightly "open". The negative charge initially localized at lost oxygen atom is mainly redistributed between other atoms of Keggin unit.

Experimental data indicate that reduction of an iron counter-ion is conditioned by hydration of the catalyst. One way to induce electron transfer between the Keggin unit and the counter-ion can be looked for in filling the lacuna by a water molecule. One can hope in this way to free one of two "excess" electrons toward the iron ion. However, our calculations did not show any appreciable electron transfer between the Keggin unit with the oxygen vacancy filled by water molecule and the counter-ion. Only molybdenum atoms undergo reduction in comparison with the non-reduced cluster. Thus, this mechanism cannot explain the iron reduction in the presence of water.

Another possibility consists in the direct hydration of the iron coordination sphere. Similar models were proposed by Hervé and coworkers for systems containing vanadyl as a counter-ion [41]. The experimental data show the stabilization of the bulk acid with about five water molecules. One can suppose that these five water molecules participate in the formation of the iron coordination sphere. So, we performed calculations of the system $Fe(H_2O)_5^{3+}+(PMo_6O_{26}H_{10})^{-1}$. The optimized structure corresponds to the preferential interaction of iron with the terminal oxygen rather than with bridging ones (Fig.5, Table II). Next we studied the interaction of hydrated iron with the reduced Keggin unit, keeping a vacancy in the same position as before. In the optimized structure the position of iron near the terminal oxygen is preserved. However, the coordination sphere of iron is expanded, changing the average iron-oxygen distance to 2.13 Å, as compared to 2.05 Å in the non-reduced system. Meanwhile, the iron Mulliken charge decreases from 1.18 to 0.97. These changes suggest that the oxidation state of the hydrated iron counter-ion goes from +3 to +2. A similar conclusion can be drawn from calculation of the electronic density on the Mössbauer nucleus Fe^{57} [33].

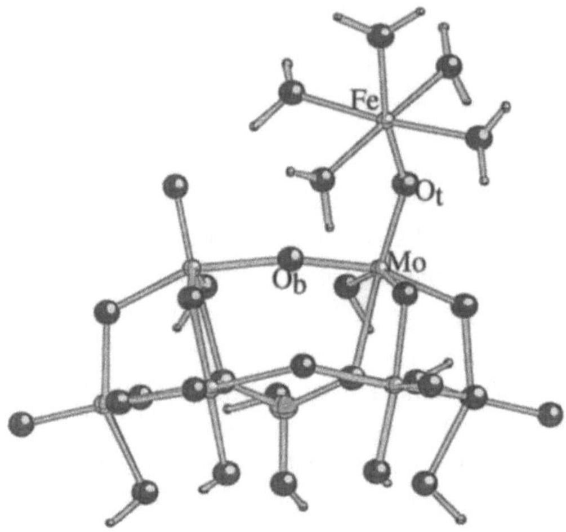

Fig.5. Model cluster representing the interaction between the iron counter-ion and the Keggin unit in hydrated solid.

Table II. Optimized distances (in Å) for the coordination sphere of iron in counter-ion in hydrated solid.

$Fe-O_t$	$Fe-OH_2$
1.81	2.07-2.13

Our study show, that the role of water in the modulation of the reducibility of the iron substituted MPA results from the modification of the position of the counter-ion relatively to the Keggin unit. In this new iron coordination, the electron transfer $Mo^{5+}+Fe^{3+} \longrightarrow Mo^{6+}+Fe^{2+}$ becomes possible. So, the presence of water promotes the reducibility of the solid.

IV. CONCLUSIONS.

We hope that the present work provides a clear example of the efficacy of quantum-chemical methods in studies of electron transfer events involving participation of polyoxometalates. The quantum-chemical analysis of the intramolecular electron transfer can be further expanded to "heteropoly blues". We also give, for the first time, the quantum-chemical description of the interaction of polyoxoanions with counter-ions. It shows that the electron-reservoir function of the transition metal counter-ions can be modulated by some other factors, such as hydration in our case. Of course, our models are limited by a number of restrictions. Now, when tackling of full polyoxoanion structures becomes more and more feasible with the latest computer hardware, quantum chemistry should become a credible tool in analysis of physical and chemical properties of polyoxometalates.

Acknowledgment

The authors wish to thank J. M. M. Millet for helpful discussions and for providing us with the experimental data of the iron-substituted MPA.

References

1. I. Weinstock: *Chem. Rev.* **98**, 113 (1998).
2. S. K. Saha, M. Ali, P. Banerjee: *Coord. Chem. Rev.* **122**, 41 (1993).
3. M. Kozik, L. C. W. Baker: *J. Am. Chem. Soc.* **112**, 7604 (1990).
4. J. J. Girerd, J. P. Launay: *Chem. Phys.* **74**, 217 (1983).
5. S. A. Borshch, B. Bigot: *Chem. Phys. Lett.* **212**, 398 (1993).
6. J. J. Borras-Allmenar, J. M. Clemente, E. Coronado, B. S. Tsukerblat: *Chem. Phys.* **195**, 1, 17, 29 (1995).
7. S. A. Borshch: *Inorg. Chem.* **37**, 3116 (1998).
8. H. Duclusaud, S. A. Borshch: *Chem. Phys. Lett.* **290**, 526 (1998).
9. J. B. Moffat: *J. Mol. Catal.* **26**, 385 (1984).
10. D. Masure, P. Chaquin, C. Louis, M. Che, M. Fournier: *J. Catal.* **119**, 415 (1989).
11. E. N. Yurchenko, H. Missner, A. Trunschke: *Zh. Struct. Khimii* **30**, 29 (1989). (in Russian)
12. R. S. Weber: *J. Phys. Chem.* **98**, 2999 (1994).
13. S. H. Wang, S. A. Jansen: *Chem. Mater.* **6**, 2130 (1994).
14. M. K. Awad, A. B. Anderson: *J. Am. Chem. Soc.* **112**, 1603 (1990).
15. H. Taketa, S. Katsuki, K. Eguchi, T. Seiyama, N. Yamazoe: *J. Phys. Chem.* **90**, 2959 (1986).
16. K. Eguchi, T. Seiyama, N. Yamazoe, S. Katsuki, H. Taketa: *J. Catal.* **111**, 336 (1988).
17. T. L. Chen, J. Ji, S. X. Xiao, T. X. Cai, G. S. Yan: *Int. J. Quant. Chem.* **44**, 1015 (1992).
18. M. M. Rohmer, J. Devémy, R. Wiest, M. Bénard: *J. Am. Chem. Soc.* **118**, 13007 (1996).
19. J. M. Maestre, J. P. Sarasa, C. Bo, J. M. Poblet: *Inorg. Chem.* **37**, 3071 (1998).
20. J. M. Maestre, J. M. Poblet, C. Bo, N. Casañ-Pastor, P. Gomez-Romero: *Inorg.Chem.* **37**, 3444 (1998).
21. B. B. Bardin, S. W. Bordawekar, M. Neurock, R. J. Davis: *J. Phys. Chem. B* **102**, 10817 (1998).
22. M. M. Rohmer, M. Bénard, J. P. Blaudeau, J. M. Maestre, J. M. Poblet: *Coord. Chem. Rev.* **178-180**, 1019 (1998).

23. M. Akimoto, K. Shima, H. Ikeda, E. Echigoya: *J. Catal.* **86**, 173 (1984).
24. N. Mizuno, M. Misono: *Current Opinion in Solid State & Materials Sci.* **2**, 84 (1997).
25. J. B. Moffat: *Appl. Catal.* **146**, 65 (1996).
26. M. Langpape, J. M. M. Millet, U. S. Ozkan, M. Boudeulle: *J. Catal.* **181**, 80 (1999).
27. M. Langpape, J. M. M. Millet, U. S. Ozkan, P. Delichère: *J. Catal.* **182**, 148 (1999).
28. M. Langpape, J. M. M. Millet: *Appl. Catal. A: General* (in press).
29. J. B. Goodenough: in *Chemical Uses of Molybdenum, Proc. 4th Intern. Conf. on Molybdenum*; eds. H. F. Barry and P. C. H. Mitchell, CLIMAX Molybdenum Co, Michigan (1982), p. 1.
30. M. J. Frisch, G. W. Trucks, H. B. Schlegel, P. M. W. Gill, B. G. Johnson, M. A. Robb, J. R. Cheeseman, T. Keith, G. A. Petersson, J. A. Montgomery, K. Raghavachari, M. A. Al-Laham, V. G. Zakrzewski, J. V. Ortiz, J. B. Foresman, C. Y. Peng, P. Y. Ayala, W. Chen, M. W. Wong, J. L. Andres, E. S. Replogle, R. Gomperts, R. L. Martin, D. J. Fox, J. S. Binkley, D. J. DeFrees, J. Baker, J. J. P. Stewart, M. Head-Gordon, C. Gonzales, J. A. Pople: *Gaussian 94* Gaussian, Inc., Pittsburgh PA (1995).
31. E. J. Baerends, D. E. Ellis, P. Ros: *Chem. Phys.* **2**, 41 (1973); P. M. Boerrigter, G. te Velde, E. J. Baerends: *Int. J. Quant. Chem.* **33**, 87 (1988); G. te Velde, E. J. Baerends: *J. Comp. Phys.* **99**, 84 (1992).
32. H. Duclusaud, S. A. Borshch: *Inorg. Chem.* **38**, 3489 (1999).
33. S. A. Borshch, H. Duclusaud, J. M. M. Millet: Appl. Catal. A: General (in press).
34. C. L. Hill, C. M. Prosser-McCartha: *Coord. Chem. Rev.* **143**, 407 (1995).
35. F. Zonnevijlle, C. M. Tourné, G. F. Tourné: *Inorg.Chem.* **21**, 2742 (1982).
36. J. E. Toth, F. C.Anson: *J.Electroanal.Chem.* **256**, 361 (1988).
37. S. N. Dean, J. K. Cooper, R. S. Czernuszewicz, D. Ji, C. J. Carrano: *Inorg.Chem* **36**, 2760 (1997).
38. M. Langpape: Ph. D. Thesis, Lyon (1997).
39. G. M. Brown, M. R. Noe-Spirlet, W. R. Busing, H. A. Levy: *Acta Cryst. B* **33** 1038 (1977).
40. F. Cavani, E. Etienne, M. Favaro, A. Galli, F. Trifiro, G. Hecquet: *Catal. Lett.* **32**, 215 (1995).
41. R. Bayer, M. Marchal, F. X. Liu, A. Tézé, G. Hervé: *J. Mol. Catal.* **110**, 65 (1996).

Aqueous Peroxoisopolyoxometalates

OLIVER W. HOWARTH,
Department of Chemistry, University of Warwick, Coventry CV4 7AL, U.K.

LAGE PETTERSSON AND INGEGÄRD ANDERSSON
Department of Inorganic Chemistry, Umeå University, S-90187, Sweden

Abstract. New ^{17}O NMR, potentiometry and ESI-MS measurements on aqueous peroxomolybdates and peroxotungstates reveal many new species, including diperoxo hepta-anions and also confirm other proposals based on potentiometry alone, such as a monoperoxo monomer. They also show the presence of many anions previously identified only in the solid state, including both forms of the $Mo_2(peroxide)_4$ anion. Comparison with recent work on peroxovanadates and peroxoniobates shows a marked preference in all cases for each metal atom to be coordinated to two peroxo ligands. Peroxotungstates, like tungstates themselves, are generally more complex than peroxovanadates and -molybdates.

Key words: ^{17}O NMR, peroxomolybdates, peroxotungstates, peroxovanadates, anion ESI-MS

1. Introduction

Peroxopolyoxometalates have long been recognised as sources of active dioxygen ($^1O_2$) in reactions such as epoxidation [1]. However, the precise nature of the species involved has been less clear. An important, recent application of peroxopolyoxometalates has been in the delignification of wood pulp, for paper manufacture. Acidified, aqueous hydrogen peroxide in the presence of Mo^{VI} constitutes an environmentally friendly replacement for chlorine: the Mo can be recycled and the only other waste product is water. Also, a new area of application has developed in biochemistry and medicine, where peroxovanadates mimic bromoperoxidase enzymes, not only as simple oxidising agents, but also as brominating agents in the presence of bromide [2]. In a further development of this, Sels *et al.* [3] have shown that tungstate, attached within a layered double hydroxide matrix, also functions analogously to a bromoperoxidase, giving usefully selective brominations. Dinuclear peroxovanadates have also been shown to possess antitumour activity [4]. This chapter therefore describes the known aqueous chemistry of V^V, Mo^{VI} and W^{VI}, with particular reference to current studies using ^{17}O NMR, potentiometry and anion ESI-MS.

2. Vanadium^V

The aqueous chemistry of peroxovanadates has recently been described in detail elsewhere [5], and will therefore only be summarised here. This quantitative study, carried out in a 0.15 M NaCl medium to reflect physiological conditions, permits the peroxovanadate speciation to be calculated over a wide range of possible V

M.T. Pope and A. Müller (eds.), Polyoxometalate Chemistry, 145–159.

concentrations and pH values. It confirms and extends earlier, more qualitative studies [6, 7] and its general conclusions also broadly hold good for other peroxometalate systems. Very broadly speaking, $[O_2]^{2-}$ binds more readily than $[O]^{2-}$, although the resulting pattern of metal NMR chemical shifts shows that the metal to oxo bond is more covalent than metal to peroxo [8].

In the absence of other ligands, the peroxovanadates are dominated by the yellow diperoxo anions $[VO(OH)(O_2)_2(H_2O)_2]^{2-}$ and, from pH 7.5 to 2, $[V(OH)_2(O_2)_2(H_2O)_2]^-$. The site of protonation implied in these formulae has been tentatively deduced using ^{17}O NMR [7] as has the inner-sphere coordination of water. Other ligands, such as NH_3 and imidazole [5] can readily displace this water, which is probably significant for the physiological activity of peroxovanadates [4]. Below pH 2, the red cation $[VO(O_2)(H_2O)_n]^+$ predominates, analogously to $[VO_2(H_2O)_n]^+$ in the absence of peroxide. However, decavanadate has no known peroxo analogues. Thus peroxide tends to break up larger polyoxoanions. This is partly confirmed below for peroxomolybdates and peroxotungstates. The only observed peroxovanadate oligomers are a range of symmetrical and unsymmetrical dimers, such as $[\{VO(O_2)_2(H_2O)_n\}_2OH]^{3-}$ whose ^{51}V chemical shifts have been assigned. Here, ^{51}V-^{51}V COSY allows all the unsymmetrical dimer resonances to be assigned unambiguously. The peaks from the symmetrical dimers are then also assignable. All the dimers are relatively minor species at normal concentrations, although a crystal structure has been obtained for the bis(diperoxo) dimer [9].

It remains unclear why vanadium shows this marked preference for two peroxo ligands, even though the same preference is also seen to a lesser extent with Mo^{VI} and W^{VI}. There is some evidence that the minor monoperoxo anion $[HVO_3(O_2)]^{2-}$ has tetrahedral coordination, presumably with monodentate peroxide, because its ^{51}V linewidth is far closer to that of $[HVO_4]^{2-}$, a species known to be tetrahedral, than to any diperoxovanadate. Furthermore, the corresponding Mo species $[HMoO_3(O_2)]^-$ (see below) is also unique in showing relatively rapid exchange with $[MoO_4]^{2-}$, and also in having a markedly narrow ^{17}O resonance for bound peroxide. Thus the preference for two peroxide ligands may arise from the creation of a sterically convenient coordination sphere.

Notwithstanding this preference, it is possible to persuade V^V, Mo^{VI} and W^{VI} to accept three or even four peroxo ligands. The resulting anions are difficult to study because they decompose in a matter of seconds or minutes, and also some only exist in the narrow pH range around 9 where $[O_2H]^-$ is significant but $[OH]^-$ less so. However, the reversible replacement of the final oxo ligand by peroxo leads to marked electronic changes, from pale yellow to deep orange in the case of Mo and from yellow to blue-purple for V. The crystal structure of $[V(O_2)_4]^{3-}$ has been published [10] and shows coordination based on an icosahedral frame.

3. Molybdenum^{VI}

Peroxomolybdates are been harder to study in solution than peroxovanadates, because ^{95}Mo NMR is a relatively insensitive technique, and gives broad resonances in all but the most favourable cases. A potentiometric study has unambiguously identified the anions MoX_2^-, MoX_2 (pK_a 2.15), MoX_2S^{2-}, MoX_2S^- (pK_a 1.00), $Mo_2X_4^{2-}$ and Mo_7X^{6-} through to Mo_7X^{3-} (pK_a 5.26, 3.79, 2.48). Here X = H_2O_2 and S = SO_4^{2-}. It also identified more minor MoX^- and MoX species (pK_a 2.70) and left open the possibility of a parallel $Mo_7X_2^{n-}$ series of other anions being present, especially at higher concentrations of Mo and peroxide, and of transient tri- and tetraperoxo anions[11]. ^{17}O NMR, aided by modest (typically 3-5%) isotopic enrichment, has already been applied successfully to aqueous molybdates [12] and molybdovanadates [13]. Also, almost all aqueous molybdate and peroxomolybdate species equilibrate within seconds or at most minutes at ambient temperatures. Therefore they can also be studied by potentiometry, for this operates by applying the laws of chemical equilibrium. Raman spectroscopy has also been of some use in concentrated solutions [14, 15], although it relies heavily on comparisons with solid samples of known structure.

The known crystal structures, notably the work of Stomberg's group, include the following: $[MoO(O_2)_2(H_2O)_2]^0$ [16]; $[Mo_2O_3(O_2)_4(H_2O)_2]^{2-}$ [17]; $[Mo_3O_7(O_2)_4]^{6-}$ [18]; $[Mo_4O_{12}(O_2)_2]^{4-}$ and $[Mo_7O_{22}(O_2)_2]^{6-}$ [19, 20]; $[HMo_7O_{22}(O_2)_2]^{5-}$ and $[Mo_8O_{24}(O_2)_2(H_2O)_2]^{6-}$ [17]; $[Mo_5O_{10}(O_2)_8]^{6-}$ [21]; $[Mo_{10}O_{22}(O_2)_{12}]^{8-}$ [22]; $[Mo(O_2)_4]^{2-}$ [23]. Many of these species have also now been identified in solution. These anions all have chelating peroxo ligands, as with the related peroxovanadates. Griffith identifies $[Mo_2O_3(O_2)_4(H_2O)_2]^{2-}$ as the main agent responsible for organic oxidation processes [14]. The expanded coordination sphere of the diperoxo monomer, as seen in ref. [16] may be unusual, because it is a doubly protonated species with zero charge. In general, one would expect the coordination number to decrease with increasing negative charge, either because the ligands become larger on average, or because the M-O bonds shorten.

Most of the above anions have now been identified in aqueous solution, although some others may be hard to see because of oxygen exchange broadening, or simply because the crystallised species are minor in solution, but also less soluble than the competing species. Fig. 1 shows a typical ^{17}O NMR spectrum. When the overall ratio [Mo]/[peroxide] exceeds 2, then only the monomer, $[MoO(OH)(O_2)_2(H_2O)_n]^-$ or (1,2), and the dimer, $[Mo_2O_3(O_2)_4(H_2O)_2]^{2-}$ or (2,4), are seen in the pH range 3-6. Their relative proportions, measured as [Mo], are close to unity when the overall [Mo] is 0.3 M. However, both anions undergo a single protonation with pK_a close to 2.0, and the neutral monomer can also bind a chloride or sulfate ion, if these are present. The terminal oxygen resonances of both monomer and dimer have very similar chemical shifts (833 and 834 ppm respectively) implying similar Mo coordination. The monomer also has a resonance of equal area at 99 ppm, implying a rather long bond to an OH ligand, and the

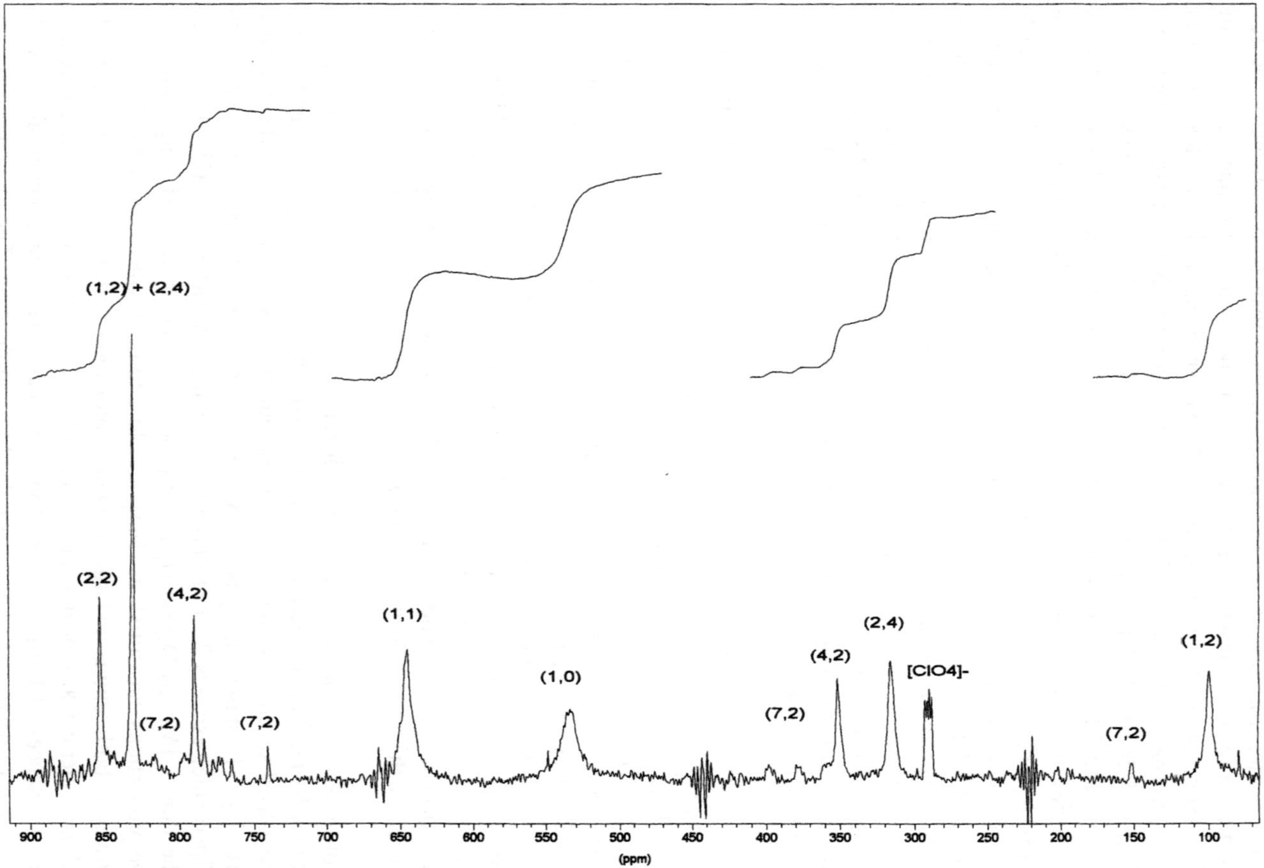

Fig. 1 Oxygen-17 NMR spectrum of a solution 300 mM overall in Mo, 350 mM in peroxide, pH 5.85, 298 K. 20 000 transients and 3% isotopic enrichment were used. Seven species are identified as (q,r), *i.e.* anions based on $Mo_q(O_2)_r$.

dimer has an additional resonance of half the area at 315 ppm. This bridging dimer O resonance shows complex behaviour upon deprotonation at 273 K. It splits into two peaks, one of which remains fixed in shift but becomes small as the pH rises. The other stays larger but shifts downwards in frequency, indicating a pK_a of approximately 8.5. Thus the dimer seems to have two distinct structures in equilibrium. In fact, a second structure has already been identified by Carpentier [24], along with a triperoxo dimer, and so the solution diperoxo dimer seems in fact to consist of two species in relatively rapid equilibrium. Possible structures are indicated in Fig. 2.

Higher oligomers are also seen, when the [Mo]/[peroxide] ratio is lower. One species is like the dimer in also having an area ratio for terminal O/bridging O of 2, but its chemical shifts are somewhat different and no protonation steps are seen. Equilibrium calculations suggest strongly that it is a (4,2) species, for which a possible candidate is the ion $[Mo_4O_{12}(O_2)_2]^{4-}$ reported by Stomberg [19]. Some uncertainty remains about the exact protonation state of this species.

Reference [11] reported two simple monomeric monoperoxo (1,1) species, MoX^- and MoX (pK_a = 2.7). In confirmation, ^{17}O NMR shows the presence of $[HMoO_3(O_2)]^-$, above pH 4 and at [Mo]/[peroxide] ≈ 1. Its shift is constant with pH, but it broadens into invisibility below pH 4, and its continuing presence at pH<4 and also possibly its protonation is not inconsistent with the NMR data. Unlike the other peroxomolybdate anions, it is clearly in chemical exchange with $[MoO_4]^{2-}$, especially at pH < 6 and at higher temperatures, where both resonances become very broad. Furthermore, a narrow resonance also appears at 411 ppm, in addition to a much larger and broader resonance at a similar shift, when the spectra are obtained using ^{17}O-enriched peroxide. At least part of this narrow resonance may be attributable to (1,1). Both these observations are explained if $[HMoO_3(O_2)]^-$ has monodentate peroxide, although the site of protonation remains unclear. Linear coordination will narrow the resonances of bound peroxide, and coordinative unsaturation at Mo will aid ligand exchange. Furthermore, the oxygen shift of (1,1) also implies tetrahedrally coordinated metal. It is 645 ppm, cp. 532 ppm for molybdate and *ca.* 800 ppm for higher coordination numbers at the metal, and it thus fits a relationship that was reported earlier to hold for tetrahedral coordination [25].

The formation of a bis(monoperoxo) (2,2,2) dimer $[(MoO_2(O_2))_2O(H_2O)_n]^{2-}$ is also indicated by equilibrium calculations based on the NMR data and by a peak at 852 ppm, presumably from the four terminal oxygens. This shift probably indicates an increase in coordination number relative to the monomer. The single bridging oxygen is not fully resolved, but probably contributes to a complex peak at 314 ppm.

At still lower peroxide concentrations, many new resonances are seen, with the correct area ratios for the eight species $[H_nMo_7O_{23}(O_2)]^{(6-n)-}$ and $[H_nMo_7O_{22}(O_2)_2]^{(6-n)-}$ (n = 0 to 3). Although the two diperoxo anions with n = 0 and 1 have already been studied by

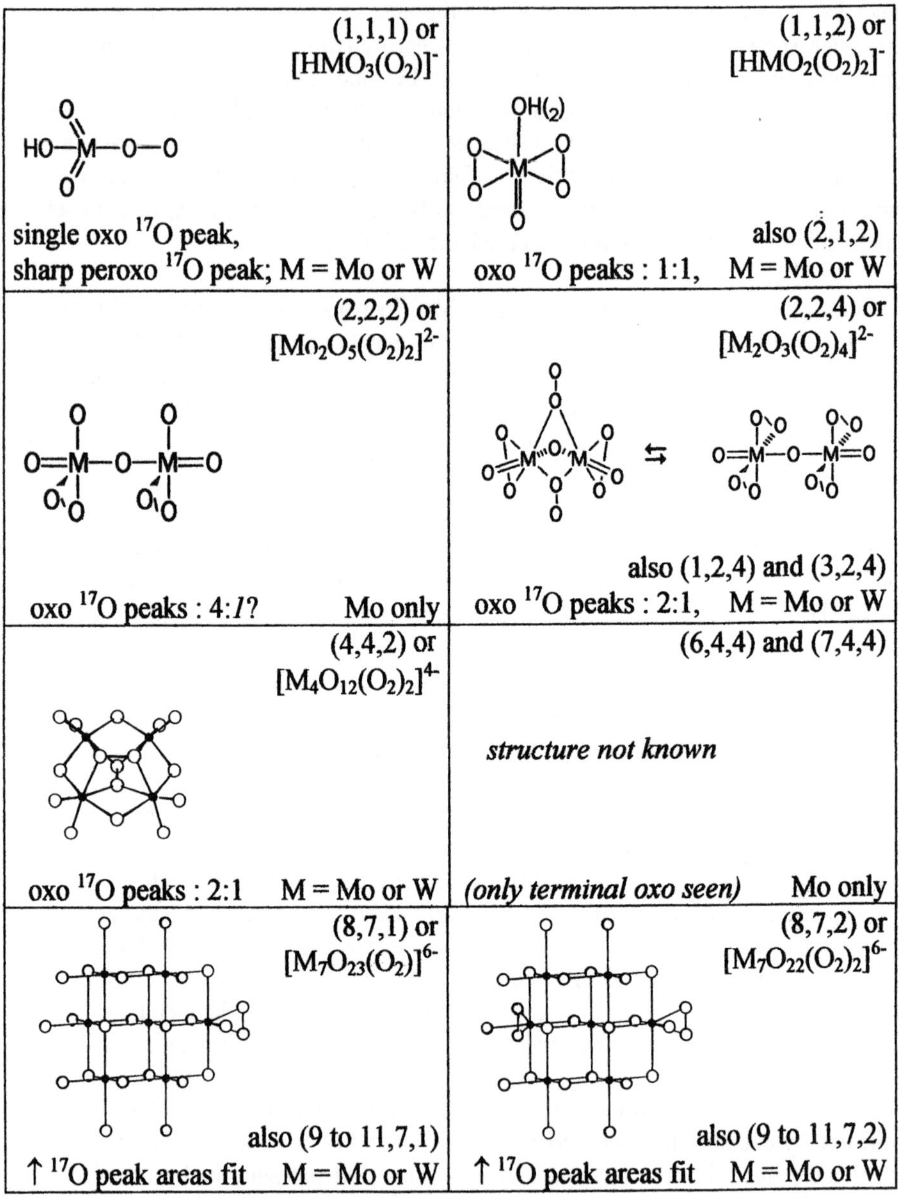

Fig. 2 Molecular formulas and proposed, simplified structures for eight peroxomolybdate species and six peroxotungstates. The (p,q,r) formulae indicate the number of protons p that must be added to q metal atoms and r peroxides for formation of the anion in question. Additional, coordinated but labile water molecules may also be present in some of these structures.

crystallography [17, 19] the monoperoxo anions have only been detected previously by potentiometry[11]. Fig. 3 shows how the chemical shifts vary with pH. These curves are not able in themselves to prove the presence of all four states of protonation, because of the closeness of the pK_a values. However, the equilibrium calculations shown in Fig. 4a require all these states to be present. Protonation clearly occurs at the cO oxygens, *i.e.* the edge sites nearest to the peroxo ligand, and this confirms the observation by Stomberg [17]. The spectra also confirm that peroxidation occurs solely at the corner g sites, and they show that this substitution of $[O_2]^{2-}$ for O^{2-} has the effect of preventing the internal O exchange process observed previously with $[H_nMo_7O_{24}]^{(6-n)-}$[12].

A few more minor resonances are also seen. A sharp but transient peak appears at 599 ppm, at pH *ca.* 10 and when free peroxide is also present. Peroxide decomposition is quite rapid in this region, and the peak may arise from a (1,3) triperoxo anion, for its W analogue has been implicated in the production of $^1O_2$ [1]. An even more transient, deep orange colour seen under similar conditions could arise from the tetraperoxo (0,4) anion[23]. At the other pH extreme, < 3, a broad terminal oxygen resonance grows rapidly with decreasing pH, as shown in Fig. 4b. The variations of its integral and shift with pH and concentration are consistent with the formulations (7,4,4) and (8,4,4), but the presence of several other species, such as Stomberg's $[Mo_{10}O_{22}(O_2)_{12}]^{8-}$ anion [22], would also be possible if their resonances were broadened and merged by exchange.

4. Tungsten^{VI}

Several peroxotungstates have been identified by crystallography, such as $[W_4O_8(O_2)_6(CO_3)]^{6-}$ [27], $[W(O_2)_4]^{2-}$[28] and $[W_2O_3(O_2)_4]^{2-}$[29]. However, they have proved difficult to study in solution. In principle, the study of peroxotungstates should be easier than that of peroxomolybdates, because ^{183}W NMR gives narrow resonances. However, the sensitivity of this method is quite low, so that has only rarely been applied to such unstable species. A recent study at a few pH values gives results consistent with those described below [1]. Furthermore, tungstate equilibria are attained slowly, if at all, and a more complex range of polyoxoanions is generally found than with Mo or V [30]. These generalities are confirmed by a study of peroxotungstates in progress. Fig. 5 shows a typical spectrum.

Because of the problems of attaining equilibrium without losing peroxide, it is unlikely that potentiometric methods will prove useful. Fortunately, however, there is a simple relationship between ^{17}O-W shifts and ^{17}O-Mo shifts in isostructural anions, as shown in Table 1. For all such oxygens, having chemical shifts >250 ppm, the W/Mo shift ratio is close to 0.79. This observation, along with integral data, enables several peroxotungstates to be recognised easily. They are (1,1,1); (n,1,2) where n = 1,2; (n,2,4) where n = 1,2,3; (4,4,2); (8,7,1) and (8,7,2). The two heptatungstate anions do not undergo protonation, unlike their heptamolybdate analogues. Consistently with earlier

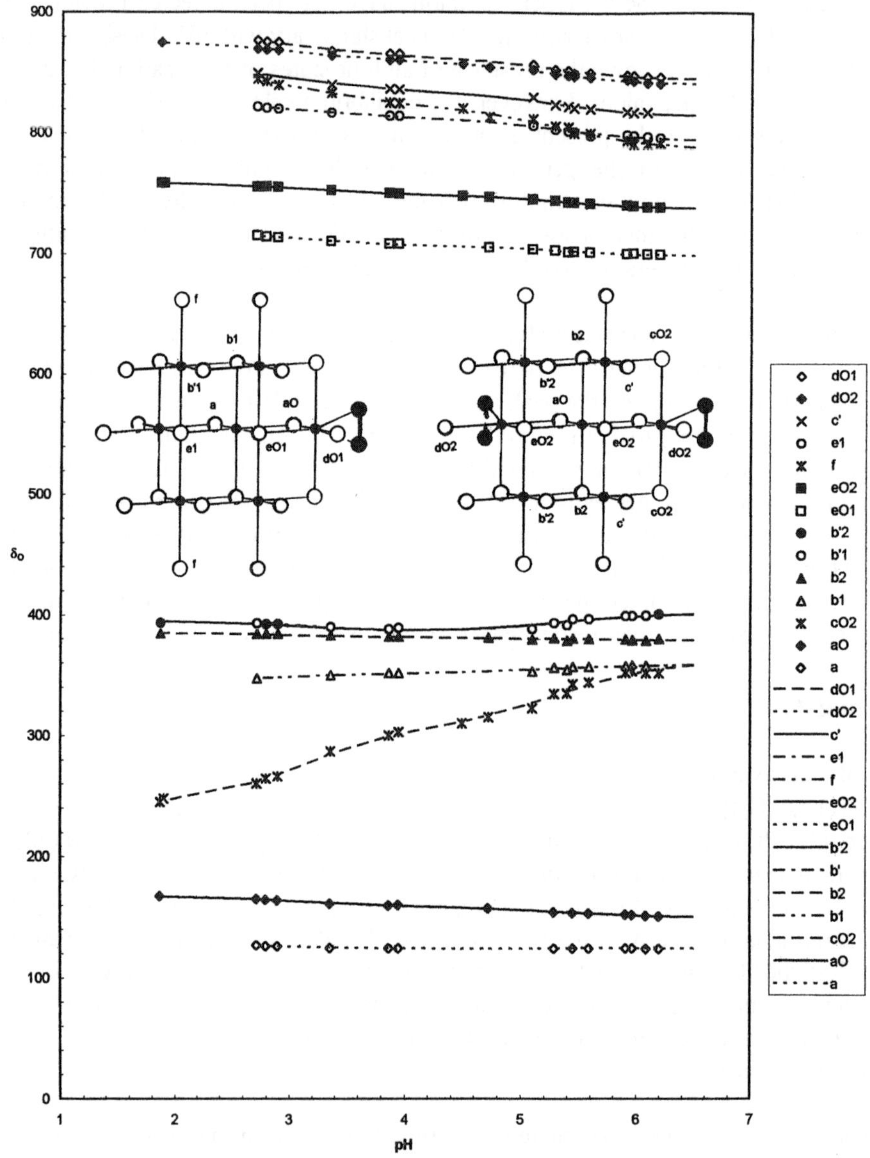

Fig. 3 Dependence of monoperoxo- and diperoxoheptamolybdate ^{17}O chemical shifts on pH. The lettering scheme is the standard one for heptamolybdate [29] but with an added O if at the peroxo-substituted end and a further '1' or '2' to indicate the number of peroxo ligands, where the (7,1) and (7,2) shifts are separable.

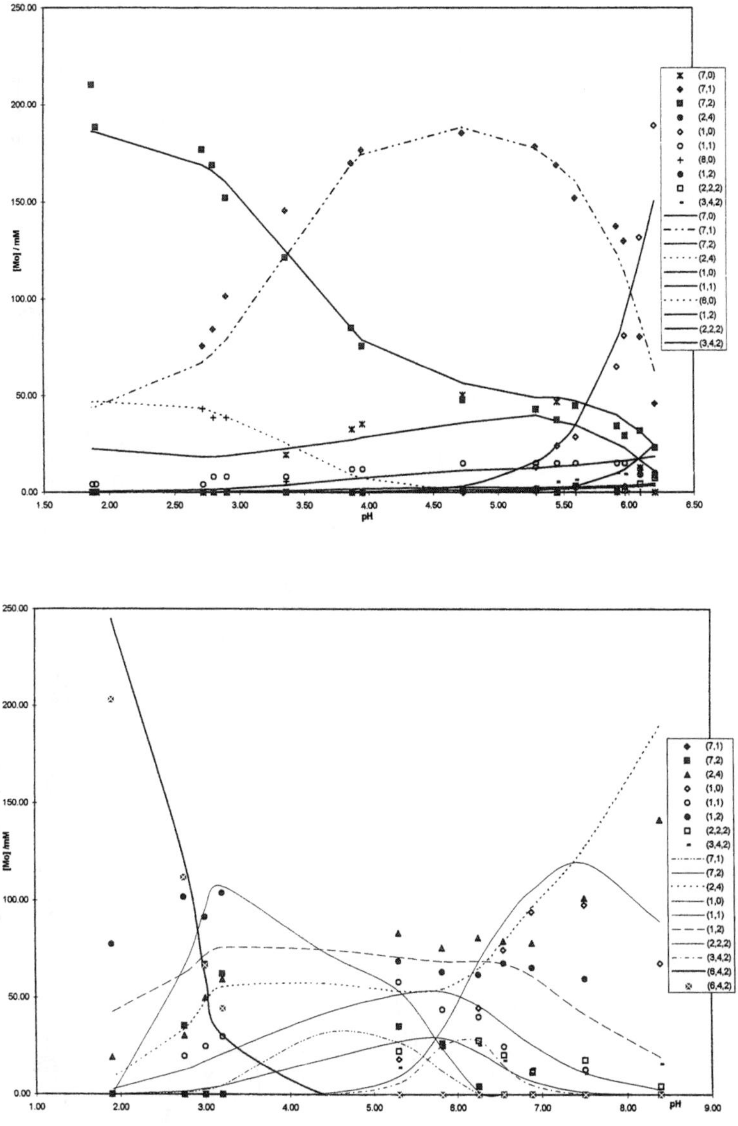

Fig. 4 Peroxomolybdate concentrations (mM) *vs.* pH. Experimental concentrations are as points and concentrations calculated according to the best-fit equilibrium model are lines. (a) 300 mM Mo and 60 mM peroxide; (b) 300 mM Mo and 350 mM peroxide. Species are identified by their (*q*,*r*) notations as above. The speciation model on which these figures are based is under further refinement.

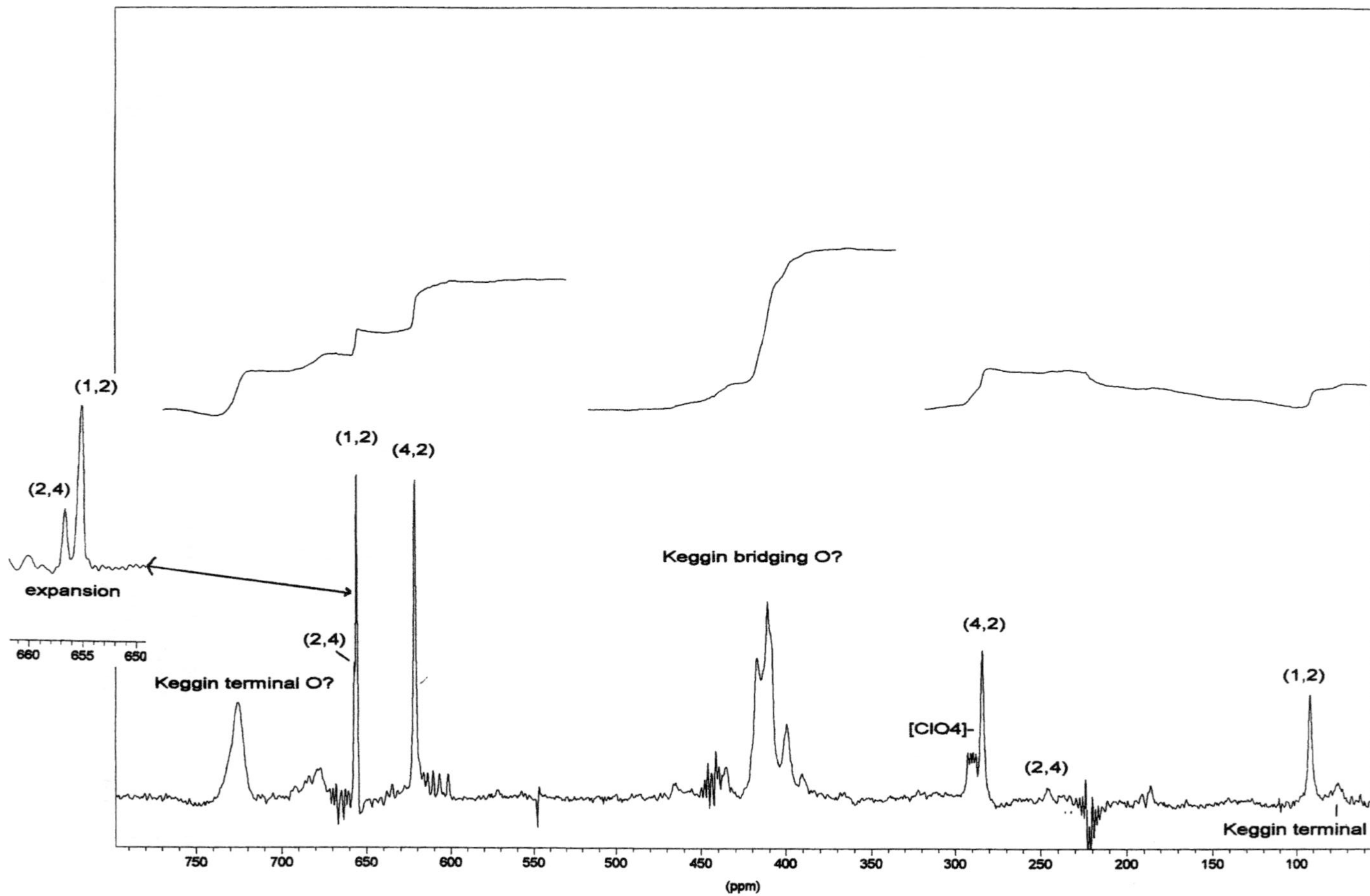

Fig. 5 Typical ^{17}O NMR spectrum of an aqueous peroxotungstate solution, with expansion at 640-650 ppm to show the very close dimeric and monomeric diperoxo species. Conditions as Fig. 1 except: 300 mM W, 165 mM peroxide, pH 4.72.

TABLE 1

Comparison of (peroxo)tungstate and (peroxo)molybdate ^{17}O NMR shifts

anion (main form)	M = W	M = Mo	$\delta_O(W)/\delta_O(Mo)$ as %
$[M_7O_{24}]^{6-}$	259, 298, 319, 590, 597, 626, 648, 648	339, 357, 399, 745, 745, 760, 812, 818	76, 72, 80, 79, 79, 84, 79, 80
$[M_7O_{23}(O_2)]^{6-}$	269, 289, 318, 556, 625, 627, 648, 648, 690	358, 362, 400, 700, 797, 791, 817, 818, 847	75, 80, 80, 79, 79, 79, 79, 79, 81
$[M_7O_{22}(O_2)_2]^{6-}$	285, 284, 318, 585, 627, 648, 690	379, 362, 358, 739, 791, 817, 842	75, 81, 84, 79, 79, 79, 82
$[MO_4]^{2-}$	422.7	533.2	79
$[HMO_3(O_2)]^-$	502	646	78
$[HMO_2(O_2)_2]^-$	657 (pK_a at 0.0)	834 (pK_a at 2.0)	79
$[M_2O_3(O_2)_4]^{2-}$	658 (pK_a at 8.0, 0)	833 (pK_a at 6.7, 2.1)	79
$[M_4O_{12}(O_2)_2]^{4-*}$	621, 284	790, 348	79, 82

* possibly a 5- anion

observations, most of the tungstate pK_a values are about 2 units lower than the corresponding molybdate ones [31].

As anticipated, many other species are also observed. One set of resonances occurs over most of the pH range 0-9, although not always as the major species present. Suggestively, the ^{17}O chemical shifts are all quite close to those observed previously for tungstate anions with α-Keggin structures. A second series of peaks is only seen from pH 4-9. These may relate to the peroxohexatungstate anions discussed below. A few other resonances are also found.

5. Niobium^{V}

The aqueous chemistry of Nb^{V} has received little attention, because it is dominated by the precipitation of Nb_2O_5. Also, ^{93}Nb NMR gives rather broad resonances, unsuited to species that are almost insoluble. However, the presence of peroxide should lessen the problems of aggregation. In a preliminary study, we have detected three sets of broad resonances, each with shifts that depend somewhat on pH. A set at *ca.* -700 ppm, relative to $NbOCl_3$, probably corresponds to oxoanions; a set at -800 to -900 ppm to monoperoxo anions and a set at *ca.* -1150 ppm to diperoxo anions. These shifts are reasonably consistent with the ^{51}V shifts in peroxovanadates.

6. ESI-MS spectra

Electrospray mass spectrometry has been shown to be viable for the investigation of dilute ($\leq 10^{-3}$ M), purely aqueous solutions [32]. This makes it an attractive possibility for the study of solutions at physiological or catalytic concentrations. However, it also presents some problems. The solutions must briefly be heated to 80 °C, just before the spraying process. Also, the complete loss of solvent can lead to confusing processes such as kinetically controlled aggregation [33], unexpected protonation or metallation and loss of bound H_2O. In general, the vapour state does not permit high anionic charges, and the vaporisation process does not permit very high concentrations of peroxide. Fortunately, kinetically controlled aggregation, when it occurs, yields a regular series of anions of monotonically increasing mass. A study of tungstate solutions gives spectra of a different type, as shown in Fig. 6. The presence of peroxoanions such as (1,1,1) and (1,1,2) is evident, along with other oligomers that may aggregate under NMR conditions. Some different spectra also reveal the presence of hexatungstate species with up to four substitutions of peroxo for oxo. These may also be contributing to the solution spectra above. Of course, the ESI-MS spectra show a single peak cluster for all isomers of a given anion, and are thus simpler than the NMR spectra. Also, they reveal the anionic charge, both through the isotope pattern and the mass change upon substitution by peroxide. Thus the method is usefully complementary to NMR, at least for aqueous systems in relatively slow exchange.

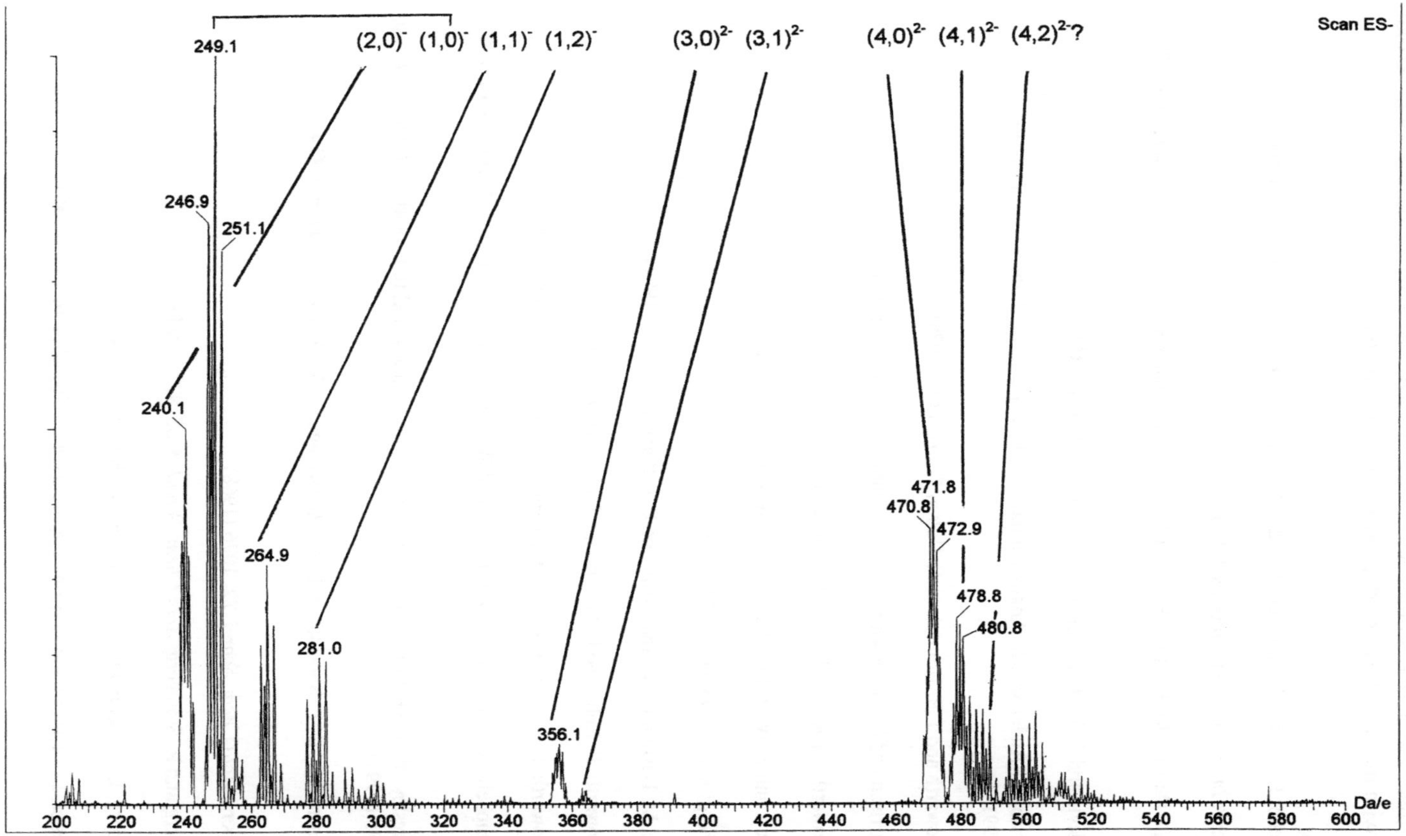

Fig. 6 Anion ESI-MS spectrum of *ca.* millimolar aqueous tungstate, derived by the addition of peroxide to dodecatungstate followed by boiling and dilution. Some species correspond to known solution anions, but others are oligomers that condense further at higher concentrations.

Taken together, the above studies show that V, Mo, W and Nb have a rich peroxo chemistry, with many features held approximately in common.

References

1. V. Nardello, J. Marko, G. Vermeersch and J. M. Aubry: *Inorg. Chem.* **37**, 5418 (1998).
2. V. Conte: *J. Inorg. Biochem.* in press(1999)
3. B. Sels, D. De Vos, M. Buntinx, F. Pierard, A. K.-D. Mesmaeker and P. Jacobs: *Nature* **400**, 855 (1999).
4. C. Djordevic and G. L. Wampler: *J. Inorg. Bioch.* **25**, 51 (1985).
5. L. Pettersson, I. Andersson, S. Angus-Dunne and O. W. Howarth: *J. Inorg. Biochem.* in press(1999)
6. O. W. Howarth and J. R. Hunt: *J. C. S. Dalton Trans*: 1388 (1979).
7. A. T. Harrison and O. W. Howarth: *J. C. S. Dalton Trans.* 1173 (1985).
8. O. W. Howarth: *Prog. NMR Spectroscopy* **22**, 453 (1991).
9. R. E. Drew and F. W. B. Einstein: *Inorg. Chem.* **11**, 829 (1972)
10. R. Stomberg: *Acta Chem. Scand.* **23**, 2755 (1969).
11. F. Taube, M. Hashimoto, I. Andersson and L. Pettersson: in press (1999)
12. O. W. Howarth, P. Kelly and L. Pettersson: *J. C. S. Dalton Trans.* 81 (1990).
13. O. W. Howarth, L. Pettersson and I. Andersson: *J. C. S. Dalton Trans.* 1799 (1991).
14. N. J. Campbell, A. C. Dengel, C. J. Edwards and W. P. Griffith: *J. Chem. Soc. Dalton Trans.* 1203 (1989).
15. N. M. Gresley, W. P. Griffith, A. C. Laemmel, H. I. S. Noguiera and B. C. Parkin: *J. Mol. Catalysis* ***A*117**, 185 (1997).
16. C. B. Shoemaker, D. P. Shoemaker, L. V. McAfee and C. W. DeKock: *Acta Cryst.* **C 41**, 347 (1985).
17. R. Stomberg: *Acta Chem. Scand.* **22**, 1076 (1968).
18. L. Trysberg and R. Stomberg: *Acta Chem. Scand.* **A 35**, 823 (1981).
19. R. Stomberg, L. Trysberg and I. Larking: *Acta Chem. Scand.* **24**, 2678 (1970).
20. I. Persdotter, L. Trysberg and R. Stomberg: *Acta Chem. Scand.* **A 40**, 335 (1986).

21. I. Persdotter, L. Trysberg and R. Stomberg: *Acta Chem. Scand.* **A 40**, 1 (1986).

22. I. Persdotter, L. Trysberg and R. Stomberg: *Acta Chem. Scand.* **A 40**, 83 (1986).

23. R. Stomberg: *Acta Chem. Scand.* **23**, 2755 (1969).

24. J.-M. Le Carpentier, A. Mitchler and R. Weiss: *Acta Cryst.* **B 28**, 1288 (1972).

25. E. Heath and O. W. Howarth: *J. C. S. Dalton Trans.* 1105 (1981).

26. R. Stomberg: *Acta Chem. Scand.* **A 39**, 507 (1985).

27. R. Stomberg: *J. Less Common Metals.* **143**, 363 (1988).

28. F. W. B. Einstein and B. R. Penfold: *Acta Cryst.* **A35**, 16 (supplement) (1963).

29. O. W. Howarth and P. Kelly: *J .C .S .Chem. Comm.* 1236 (1988).

30. O. W. Howarth and J. J. Hastings: *J. C. S. Dalton Trans.* 209 (1992).

31. O. W. Howarth, I. Andersson, J. J. Hastings and L. Pettersson: *J. Chem. Soc. Dalton Trans.* 2705 (1996).

32. M. J. Deery, O. W. Howarth and K. R. Jennings: *J. C. S. Dalton Trans.* 4783 (1997).

33. M. J. Deery, T. Fernandez, O. W. Howarth and K. R. Jennings: *J. C. S., Dalton Trans.* 2177 (1998).

Molybdate speciation in systems related to the bleaching of kraft pulp[#]

F. TAUBE, I. ANDERSSON, AND L. PETTERSSON*
Chemistry Department, Inorganic Chemistry, Umeå University, SE-901 87 Umeå, Sweden

Abstract. Peroxomolybdates have shown to be efficient selective agents in the degradation of lignin in non-chlorine based bleach processes of kraft pulp. Furthermore, the process can be improved when anions such as phosphate are present. To clarify the chemistry in aqueous solution, fundamental speciation studies of relevant systems have been made under conditions similar to those in the bleaching step.

In this article the equilibrium speciation in the system $pH^+ + qMoO_4^{2-} \rightleftarrows (H^+)_p(MoO_4^{2-})_q$ in 0.300 M $Na_2(SO_4)$ medium at 25 °C has been studied using potentiometric data in the range $2.5 \leq pH \leq 6.0$, $1.25 \leq Mo_{tot}/mM \leq 20$. The speciation was found to consist of the monomers MoO_4^{2-}(0,1), $HMoO_4^-$(1,1), H_2MoO_4(2,1) and the heptamers $Mo_7O_{24}^{6-}$ (8,7), $HMo_7O_{24}^{5-}$(9,7), $H_2Mo_7O_{24}^{4-}$(10,7) and $H_3Mo_7O_{24}^{3-}$(11,7) (numbers in parentheses refer to the values of p and q in the general reaction above). The following formation constants and 3σ were obtained: $\log \beta_{1,1} = 3.40 \pm 0.10$, $\log \beta_{2,1} = 7.79 \pm 0.06$ ($pK_a = 4.39$), $\log \beta_{8,7} = 52.43 \pm 0.04$, $\log \beta_{9,7} = 57.42 \pm 0.03$ ($pK_a = 4.99$), $\log \beta_{10,7} = 61.24 \pm 0.04$ ($pK_a = 3.82$) and $\log \beta_{11,7} = 63.90 \pm 0.10$ ($pK_a = 2.66$). The pK_a value for HSO_4^- was determined to 1.27 ± 0.01. The effects of different ionic media on this system are discussed.

Finally, this article presents some preliminary results of the ongoing speciation studies in the $H^+ - MoO_4^{2-} - H_2O_2$ and $H^+ - MoO_4^{2-} - H_2O_2 - H_2PO_4^-$ systems.

Key words: Molybdates, peroxomolybdates, equilibria, speciation, potentiometry.

1. Introduction

The catalytic properties of polyoxometallates (POMs) are well known. Recently some of these compounds have been used in the selective degradation of lignin in the bleach process of kraft pulp. A highly selective delignification has been obtained when adding molybdate to weakly acidic solutions in the presence of excess of hydrogen peroxide [1]. Furthermore, the delignification has proven to be more effective in the presence of phosphates [2]. Provided that the molybdate can be recycled, it can be used for a selective and efficient non-chlorine process, suitable for a closed pulp system. A key for understanding the chemistry of the molybdate in the bleaching step is to know the speciation and behavior of the species formed. This requires fundamental speciation studies of the $H^+ - MoO_4^{2-}$, $H^+ - MoO_4^{2-} - H_2O_2$ and $H^+ - MoO_4^{2-} - H_2O_2 - H_2PO_4^-$ systems, under conditions similar to those in the bleaching step. Because Na^+ and SO_4^{2-} are the most common ions in the bleaching process, the studies have been performed in

M.T. Pope and A. Müller (eds.), Polyoxometalate Chemistry, 161–173.

a sodium sulfate medium. The 0.300 M sulfate medium used is denoted $Na_2(SO_4)$, implying that $[Na^+]$ has been kept constant at 0.600 M and that $[SO_4^{2-}]$ can vary somewhat. Since SO_4^{2-} protonates at low pH it was necessary first to determine the pK_a value of HSO_4^- in 0.300 M $Na_2(SO_4)$ medium. The present paper will concern the speciation study of the H^+ - MoO_4^{2-} system in the range $2.5 \leq pH \leq 6.0$, together with a brief overview of the ongoing speciation studies in the H^+ - MoO_4^{2-} - H_2O_2 and H^+ - MoO_4^{2-} - H_2O_2 - $H_2PO_4^-$ systems.

There have been a number of investigations of equilibrium speciation in aqueous molybdate solutions. Several of the equilibrium analyses have been carried out by means of potentiometric titrations in a medium of constant ionic strength [3-12]. In addition to the monomeric species MoO_4^{2-} (0,1), $HMoO_4^-$ (1,1) and H_2MoO_4 (2,1), Sasaki and Sillén [5], as well as Farkas *et al.* [12], proposed that a series of heptamolybdates predominates in 3.0 M $Na(ClO_4)$ medium and 0.2 M K(Cl) medium respectively, namely $Mo_7O_{24}^{6-}$ (8,7), $HMo_7O_{24}^{5-}$(9,7), $H_2Mo_7O_{24}^{4-}$(10,7), $H_3Mo_7O_{24}^{3-}$ (11,7). The numbers in parentheses refer to the values of p and q in the general reaction $pH^+ + qMoO_4^{2-} \leftrightarrow (H^+)_p(MoO_4^{2-})_q$. From investigations in Na(Cl) media, Yagasaki *et al.* [11] (0.600 M Na(Cl)) and Cruywagen *et al.* [8] (1.0 M Na(Cl)), found that the octameric ion $Mo_8O_{26}^{4-}$(12,8) is formed instead of (11,7).

Earlier studies of molybdate speciation in $Na_2(SO_4)$ medium have been carried out by various methods but, to our best knowledge, not by potentiometry. In a study on polyanions by Glauber's salt cryoscopy, Jain *et al.* [13] proposed a model containing (8,7) and (9,7) together with the monomeric complexes (0,1) and (1,1) within the range $3.0 \leq pH \leq 6.0$. An-Pong *et al.* [14] found the (8,7) to predominate within $4.56 \leq pH \leq 6.79$ in a study based on the salt-ice point method. As in most POM systems, the species formed in the molybdate system have high nuclearities and high negative charges. The speciation is therefore very sensitive to the ionic medium, especially to the cation. The Na^+ concentration in the 0.300 M $Na_2(SO_4)$ medium is the same as in the 0.600 M Na(Cl) medium, which has been commonly used in previous POM studies at our department. The difference in speciation between the two media will be discussed.

Studies on peroxomolybdates have been reviewed by Connor and Ebsworth [15], Dickman and Pope [16], and also in Gmelin [17]. Most of the reviewed works include mono-, di- and, in some cases, tri- and tetraperoxomonomolybdate complexes. Furthermore, solid peroxide-poor (H_2O_2/Mo < 1) species have been isolated from solutions of polymolybdates mixed with small amounts of hydrogen peroxide. Evidence for hydrogen sulfate interaction with peroxomolybdates in solution is rare, but has been reported in strongly acidic media [18,19,20].

Equilibrium studies on possible peroxomolybdate species participating in the bleach process have, to our best knowledge, not been performed previously. In the present potentiometric study, a limitation of $pH \leq 5.5$ was necessary since the decomposition of hydrogen peroxide in peroxomolybdate solutions at higher pH values was substantial.

Only a limited number of peroxomolybdophosphate studies have been reported. Recently, it has been found that P- and As-containing compounds of the type $R_3[XO_4\{MO(O_2)_2\}_4]^{3-}$ where M = Mo or W; X = P or As, and R is a bulky cation, *e.g.* $N(C_6H_{13})_4^+$, are good catalysts for oxidative reactions by peroxide [21]. Crystals of $((C_6H_{13})_4N)_3[PMo_4O_{24}]$ have been isolated by Salles *et al.* [22]. Such compounds were shown to epoxidise both 1-octene and (R)-(+)-limonene in the presence of peroxide. Other peroxomolybdophosphates reported include the 2-aminopyridine salt of $[PO_2(Mo_2O_{11})_2]^{3-}$ [23], along with the structural characterisations of the mono- and bis-diperoxomolybdate species $[NH_4][(Ph_2PO_2)\{MoO(O_2)_2(H_2O)\}]$ and $[NMe_4][(Ph_2PO_2)\{MoO(O_2)_2\}_2]$ [24], and the tris-diperoxomolybdate species $[NMe_4][(PhPO_3)\{MoO(O_2)_2\}_2\{MoO(O_2)_2(H_2O)\}]$ [25]. The latter three species all show some catalytic activity for alkene epoxidation.

2. Experimental

2.1. CHEMICALS AND ANALYSES

Molybdate stock solutions were prepared by dissolving crystalline disodium molybdate, $Na_2MoO_4 \cdot 2H_2O$ (E. Merck *p.a.*) which had been recrystallized once. The concentration of molybdate was determined by evaporating water from a known amount of stock solution, drying the residue at 110 °C and then weighting as anhydrous Na_2MoO_4. Disodium sulfate, Na_2SO_4, (E. Merck *p.a.*) was dried at 80 °C for at least 24 hours and used without further purification. Solutions of sulfuric acid were standardized against tris(hydroxomethyl)-aminomethane (Tris, Sigma Chemical Co.). Diluted sodium hydroxide was prepared from a concentrated (50% H_2O and 50% NaOH) solution and standardized against sulfuric acid. The sodium hydroxide solutions were stored in plastic bottles. Hydrogen peroxide stock solutions were prepared from 30 % (9.7 M) hydrogen peroxide (E. Merck *p.a.*), standardized against potassium permanganate, and stored in black plastic bottles at ~ 4 °C. Sodium dihydrogenphosphate monohydrate $NaH_2PO_4 \cdot H_2O$ (E. Merck *p.a.*) was used as received, and solutions were standardised gravimetrically by evaporation at 120 °C to leave anhydrous NaH_2PO_4. In all preparation of solutions boiled and distilled (Milli-Q plus 185) water was used. Alkaline and neutral solutions were protected from CO_2(g) by the use of argon gas.

2.2. POTENTIOMETRIC MEASUREMENTS

The EMF measurements in the different systems were carried out as a series of potentiometric titrations in 0.300 M $Na_2(SO_4)$ medium at 25 °C (± 0.05 °C , thermostatted oil bath) with an automated, computer controlled potentiometric titrator.

As measuring electrodes two Ingold 201-NS glass electrodes were used. The free hydrogen concentration was determined by measuring the EMF of the cell:

- Ag, AgCl(s) | 0.010 M NaCl, 0.295 M Na_2SO_4 || 0.300 M $Na_2(SO_4)$ || equilibrium solution | glass electrode +

Under the assumption of constant activity coefficients, the measured EMF (in mV) may be written as $E = E_0 + 59.157 \log[H^+] + E_j$. E_j is the liquid junction potential at the 0.300 M $Na_2(SO_4)$ || equilibrium solution interface and is, for our experimental set-up, given by equation $E_j/\text{mV} = -76[H^+]$.

In ionic media where the medium anions do not react with H^+, *e.g.* in the frequently used 0.600 M Na(Cl) medium, the apparatus constant E_0 can be determined in titrations of acidified medium solutions, or simply by measuring the E value in solutions with known total concentration of H^+(H). In a sulfate medium this is not feasible since the pK_a value of HSO_4^- and the E_0 value are too strongly interdependent. Instead, an Ingold U402-M6-S7/100 combination electrode was calibrated against a phosphate buffer ($[HPO_4^{2-}] = [H_2PO_4^-]$ = 10 mM) in 0.600 M Na(Cl) medium with a known pK_a-value [26]. The pH-value of the same buffer in 0.300 M $Na_2(SO_4)$ medium was then measured. In order to determine the E_0-values for the glass electrodes, coulometric titrations were then made on 10 mM NaH_2PO_4 solutions in 0.300 M $Na_2(SO_4)$ medium. In the calculation of these titration data, the pK_a-value obtained from the combination electrode was used. In close connection, an acidified 0.300 M $Na_2(SO_4)$ solution with known H was coulometrically titrated with the same experimental set-up. By knowledge of the E_0 values of the glass electrodes, it was possible to determine the free proton concentration in the acidified solution and thereby the pK_a value for HSO_4^-. This solution was then also used for determining E_0 before and after each titration in the molybdate, peroxomolybdate and peroxomolybdophosphate systems. The titration data were considered acceptable if the difference between these E_0 values was ≤ 0.5 mV. The average of the two values was used as the E_0 value for the titration.

Due to the strong interdependence between the pK_a value for HSO_4^- and the E_j and E_0 values, accurate determination of E_j in 0.300 M $Na_2(SO_4)$ medium was not possible. Instead, the E_j value determined in 0.600 M Na(Cl) medium was used.

For pH determination of the NMR solutions, an Ingold U402-M6-S7/100 combination electrode was used and calibrated against buffer solutions of known $[H^+]$.

2.3. NMR MEASUREMENTS

^{31}P NMR spectra were recorded at 202.5 MHz on a Bruker AMX 500 MHz spectrometer at 25 ± 1 °C. Field-frequency stabilisation was achieved by placing the 8 mm sample tube into a 10 mm tube containing D_2O. All chemical shifts are reported relative to the external reference 85% H_3PO_4, assigned to 0 ppm. Spectra were quantitatively integrated after baseline correction. The deconvolution subroutine of the software program 1D WINNMR was used to obtain more precise integral values in the case of overlapped peaks.

2.4. CALCULATIONS

The EMF and quantitative ^{31}P NMR data were evaluated using the least squares program LAKE [27]. Modeling and construction of distribution diagrams were performed using the program SOLGASWATER [28].

3. Results and Discussion

3.1. H^+ - SO_4^{2-} SYSTEM

The pK_a value for HSO_4^- was determined to be 1.27 $\pm$ 0.01 (the error expressed as 3σ) from two coulometric titrations (51 points) within the pH range $1.7 \leq pH \leq 4.2$.

3.2. H^+ - MoO_4^{2-} SYSTEM

The equilibria are written with the components H^+ and MoO_4^{2-} according to the general equation:

$$pH^+ + qMoO_4^{2-} \rightleftarrows (H^+)_p(MoO_4^{2-})_q$$

The formation constants are denoted $\beta_{p,q}$ and the complexes are given in the notation (p,q). The total concentration of each component is given by equations (1) and (2):

$$H = h + [HSO_4^-] + \Sigma\Sigma p\beta_{p,q}h^p b^q \quad (1)$$

$$B = b + \Sigma\Sigma q\beta_{p,q}h^p b^q \quad (2)$$

H is the total concentration of H^+ over the zero level of H_2O, MoO_4^{2-} and SO_4^{2-}, B is the total concentration of molybdate, and h and b are the corresponding free concentrations of H^+ and MoO_4^{2-}.

The formation constants for the $(H^+)_p(MoO_4^{2-})_q$ species in the system were determined from 17 titrations (238 points) in the range $2.5 \leq pH \leq 6.0$ and $1.25 \leq B$ /mM ≤ 20.00. Generally, equilibrium was reached within 30 minutes. For a given concentration of B two titrations were made, one beginning at pH 2.5 and the other at pH 6.0. The endpoints of the two titrations should overlap. Together, they give the complete titration curve for the given concentration of B. The titration curves (Z versus

-log[H^+]) are illustrated in Figure1. Z is the average uptake of protons per molybdate ion, defined by:

$$Z = (H - h - [HSO_4^-]) / B \tag{3}$$

In order to check the reliability of the titrations some were performed as "constant Z" titrations. Instead of keeping B constant with varying Z values, as in the ordinary titrations, B is varied in the range $0.00 \leq B/mM \leq 20.00$, while Z is kept approximately constant. "Constant Z" titrations were performed for $Z \sim 0.5$, 1.14 and 1.4. For simplicity, these titrations are not illustrated in the figure.

Formation constants for arbitrary but systematically chosen complexes $(H^+)_p(MoO_4^{2-})_q$ are varied in LAKE, so that the sum of error squares,

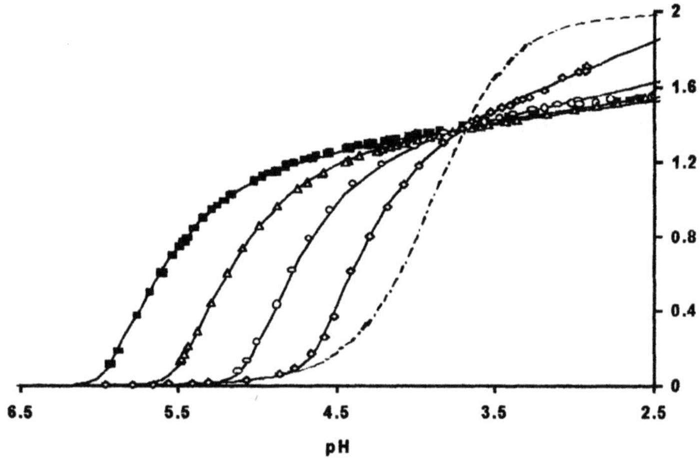

Fig. 1: Potentiometric data plotted as Z versus -log h (pH). Z is calculated as $Z = (H - h - HSO_4^-)/[Mo]_{tot}$, where H is the total concentration of protons and h is the "free" concentration of protons. Symbols represent experimental data points, ◇= 1.25 mM, o= 5.00 mM, Δ = 20.00 mM and ■ = 80.00 mM $[Mo]_{tot}$. The full curves were calculated using the model given in table 1a, the dashed curve represent the mononuclear wall, valid for $[Mo]_{tot} \leq$ 0.13 mM.

$$U = \Sigma(H_{calc} - H_{exp})^2 \tag{4}$$

or

$$U = \Sigma(H_{calc} - H_{exp})^2 / B \tag{5}$$

is minimized. H_{calc} is the calculated H for one specific titration point when using a certain model in LAKE, while H_{exp} is the analytically determined H in that point. The set of complexes giving the lowest U-value forms the model, which best explains the experimental data. Since all errors were placed on eq. (1), *i.e.* H^+, the molybdate concentration, B, is assumed correct.

The final model is presented in Table 1a, together with the 0.600 M Na(Cl) model [11] in Table 1b. As can be seen, the speciation in 0.600 M Na(Cl) contains an octamer (12,8), while the speciation in 0.300 M $Na_2(SO_4)$ consist entirely of heptamers plus monomers. An attempt to explain the experimental data with a model containing (12,8), instead of (11,7), led to a U value twice as high. Moreover, we observed a threefold increase in the 3 σ value for log $\beta_{10,7}$. When co-varying (12,8) and (11,7) the octameric (12,8) species was always rejected, indicating that this is a minor species at $[Mo]_{tot} \leq 20$ mM, if present at all.

Table 1. Equilibrium constants for molybdate complexes in two different media, (a): 0.300 M $Na_2(SO_4)$ medium, (b): 0.600 M Na(Cl) medium, (c):optimization result in 0.300 M $Na_2(SO_4)$ medium when using weight on *DA* instead of on *DA/B*, as in (a).

(p,q)	formula	log β ± 3 σ	pK_n
(a)			
(0,1)	MoO_4^{2-}	0.00	-----
(1,1)	$HMoO_4^-$	3.40 ± 0.10	3.40
(2,1)	H_2MoO_4	7.79 ± 0.06	4.39
(8,7)	$Mo_7O_{24}^{6-}$	52.43 ± 0.04	-----
(9,7)	$HMo_7O_{24}^{5-}$	57.42 ± 0.03	4.99
(10,7)	$H_2Mo_7O_{24}^{4-}$	61.24 ± 0.04	3.82
(11,7)	$H_3Mo_7O_{24}^{3-}$	63.90 ± 0.10	2.66
(b)			
(0,1)		0.00	-----
(1,1)		3.39 ± 0.10	3.39
(2,1)		7.35 ± 0.01	3.98
(8,7)		52.42 ± 0.02	-----
(9,7)		57.23 ± 0.03	4.77
(10,7)		60.78 ± 0.05	3.57
(12,8)	$Mo_8O_{26}^{4-}$	71.62 ± 0.06	-----
(c)			
(0,1)		0.00	-----
(1,1)		3.56 ± 0.34	3.56
(2,1)		7.83 ± 0.07	4.27
(8,7)		52.41 ± 0.04	-----
(9,7)		57.41 ± 0.03	5.00
(10,7)		61.22 ± 0.04	3.81
(11,7)		63.87 ± 0.11	2.65

One way to confirm the relevance of the model is to compare the results of optimization when using different residual weightings. The monomeric complexes, (1,1) and (2,1) predominate in solutions where B is low. When all weighting is set on DA, defined as $H_{calc} - H_{exp}$, data points with high B gives the highest contribution, since H is comparably high in these points, while data points with low B contribute very little to the U value, especially those data points in the neutral region. This will lead to an optimization result with larger 3 σ values on the formation constants for the monomeric complexes, particularly (1,1), compared to the heptameric ones.
Therefore, a common procedure in a two component system is to put all weighting on DA/B, *i.e.* $(H_{calc} - H_{exp}) / B$. If the optimized log β values are more or less independent of the weightings, this indicates that the model explains the experimental data well. Since the sulfate component in our system is proton active only at low pH values, we can consider our system as a two component system in the neutral pH range, *i.e.* where the effects of different weightings on the monomeric complexes are noticeable. In our case the optimized log β values were close to one another in the two different optimizations but, as expected, with a much higher 3 σ value for log $\beta_{1,1}$ when the weight was set on DA (Table 1c).

The relatively high 3σ for the two most acidic complexes, (2,1) and (11,7), in the proposed models, compared to the corresponding complexes, (2,1) and (12,8) in the 0.600 M Na(Cl) model (cf. Table 1a and b), is an effect of the relatively large amount of H^+ bound to HSO_4^- at low pH. Even a small error in the total concentration of sulfate, or in the pK_a value of HSO_4^-, will result in comparatively high DA-values for the acidic data points.

Although the interpretation of the influence of ionic medium and ionic strength upon isopolyanion equilibria is still controversial, the effect of the ionic medium is often ascribed to complexation between medium cations and highly charged polyoxometalate anions, thus stabilizing the anions [3]. A comparison between the model obtained in the present study and that in 0.600 M Na(Cl) by Yagasaki *et al.*[11] (Figures 2a-d) shows that the monomeric (2,1) complex is substantially stronger and the predominance of heptamolybdates less pronounced in the 0.300 M $Na_2(SO_4)$ medium. This effect would be expected in a weaker ionic medium, or when less Na^+ is available for stabilizing the heptamolybdates. One possible explanation in the present case of two apparently equal Na^+ concentrations is that Na^+ has a higher affinity towards SO_4^{2-} than Cl^-. This is supported by a study on different total activity coefficients for the sodium ion in different media by Elgqvist *et al.* [29]. They found that the ionic activity coefficient for Na^+ was lower in sulfate solutions than in chloride solutions, and that this effect could be partly ascribed to ion pair formation in sulfate media.

The stabilizing effect of Na^+ on heptamolybdates can be noticed by comparing a study in 1.00 M Na(Cl) medium [3] with the study in 0.600 M Na(Cl) medium. It becomes even more obvious in 3.00 M $Na(ClO_4)$ medium [4], where the heptamolybdates, especially the six minus charged (8,7) complex, are much more predominant than in 0.600 M Na(Cl) medium. Furthermore, the increasing strength of the (2,1) complex with decreasing sodium ion concentration can also be seen from these studies. The existence of the $Mo_8O_{26}^{4-}$ (12,8) complex in different sodium media seems less straightforward and not mainly governed by the sodium ion concentration. This species is not present in 3.00 M $Na(ClO_4)$ and 0.300 M $Na_2(SO_4)$ media but does exist in 0.600 M Na(Cl) medium, in which the ionic activity for Na^+ should be somewhat higher than in the sulfate medium. Of course, the different ionic strength in 0.300 M $Na_2(SO_4)$ compared to 0.600 M Na(Cl) medium, 0.9 and 0.6 respectively, may also contribute to changes in the speciation.

To check if the different speciation in these two media could be attributed to any sulfate interactions with molybdate complexes, FTIR-spectroscopic measurements were performed. No signs of such interactions were found.

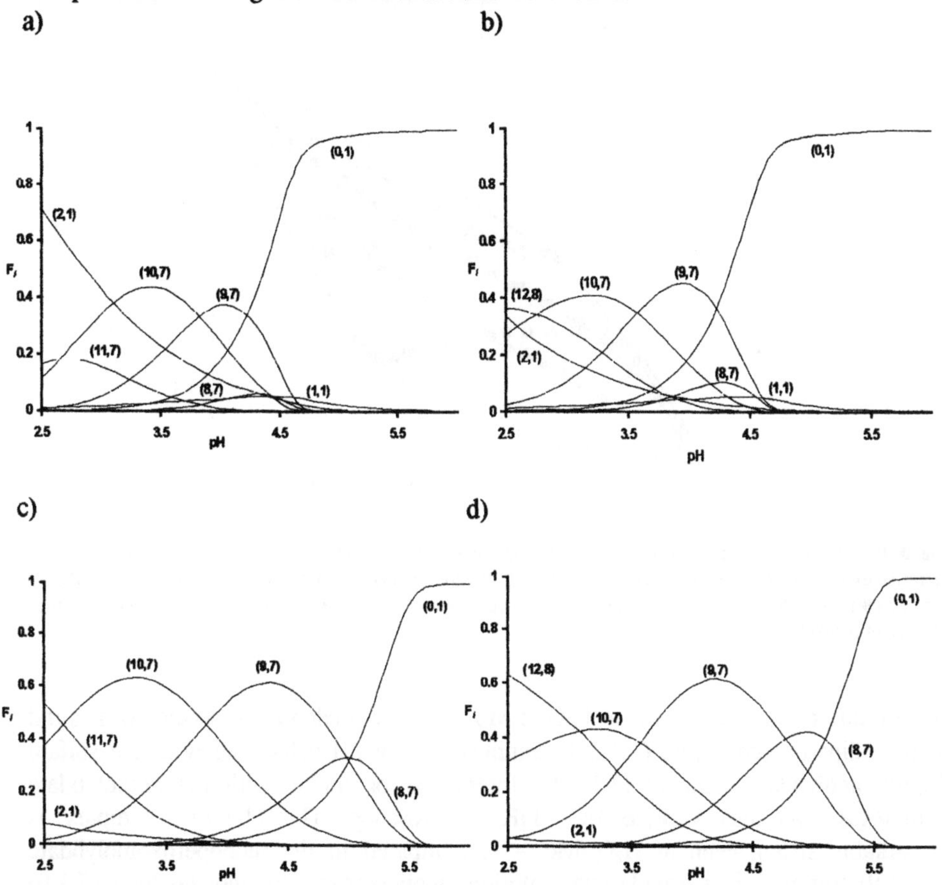

Figs. 2a-d, a: Distribution diagram at $[Mo]_{tot}$ = 1.25 mM in 0.300 M $Na_2(SO_4)$ medium plotted as F_i versus pH. F_i is defined as the ratio between [Mo] in a species and $[Mo]_{tot}$ in solution, **b:** $[Mo]_{tot}$ = 1.25 mM in 0.600 M Na(Cl) medium, **c:** $[Mo]_{tot}$ = 20.0 mM in 0.300 M $Na_2(SO_4)$ medium and **d:** $[Mo]_{tot}$ = 20.0 mM in 0.600 M Na(Cl) medium.

3.3. $H^+ - MoO_4^{2-} - H_2O_2 - SO_4^{2-}$ SYSTEM

This system is highly relevant to the industrial process where hydrogen peroxide is present in excess. To ascertain the speciation in such solutions, it was found necessary to perform a complete equilibrium analysis of the system, including low H_2O_2/Mo ratio data. Complexes are formed according to the general equation:

$$pH^+ + qMoO_4^{2-} + rH_2O_2 + sSO_4^{2-} \rightleftharpoons (H^+)_p(MoO_4^{2-})_q(H_2O_2)_r(SO_4^{2-})_s.$$

The titration curves (Z versus -log[H^+]) for B = 20.00 mM are illustrated in Figure 3.

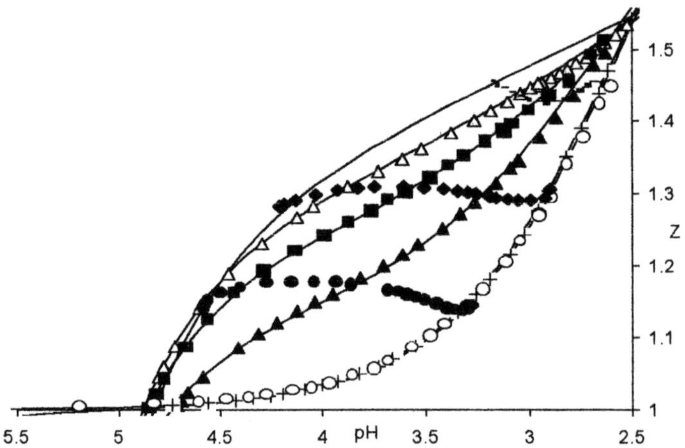

Fig. 3: Potentiometric data plotted as Z versus pH. Z is calculated as $Z = (H - h - [HSO_4^-]) / [Mo]_{tot}$. Symbols represent experimental data points. $[H_2O_2] / [Mo]_{tot}$ = 0.25 (Δ), 0.5 (■), 1.0 (▲), 2.3 (+), 3.9 (o), $Z \approx 1.15$ (●), $Z \approx 1.30$ (◆), $Z \approx 1.45$ (–). The curve without symbols represents the model curve for a molybdate solution in the absence of peroxide.

In peroxide-rich solutions, *i.e.* at H_2O_2 : Mo > 1 , the predominant species was found to be a diperoxomolybdate (1,1,2,0) complex and, at pH values below 2.3, a sulfato diperoxomolybdate complex (2,1,2,1). Interaction of sulfate with peroxomolybdate complexes was also verified by FTIR spectroscopy. The (1,1,2,0) complex is remarkably strong even at H_2O_2/Mo = 1, compared to the monoperoxomolybdate (1,1,1,0) complex. In peroxide-poor solutions monoperoxo heptamolybdates (p,7,1,0) was found to predominate, although the speciation in such solutions needs further investigation. In comparison with the $H^+ - MoO_4^{2-}$ system, the polymerization of monomolybdates into heptamolybdates is strongly suppressed in excess of peroxide owing to formation of the strong diperoxo complexes (1,1,2,0), (2,1,2,0) and (2,1,2,1). Furthermore, a dimeric diperoxo complex (2,2,4,0) is formed, although this species was found to be weak compared to the monomeric complexes at the moderate molybdate concentrations studied in the present work.

3.4. H^+ - MoO_4^{2-} - H_2O_2 - HPO_4^{2-} - SO_4^{2-} SYSTEM

Since the delignification of kraft pulp has proven to be more effective in the presence of phosphates [2], phosphate was included in the system. The resulting five component system has been studied by potentiometric titrations and ^{31}P NMR, but only at excess of hydrogen peroxide ($H_2O_2/Mo > 2$). Complexes are formed according to the general equation:

$$pH^+ + qMoO_4^{2-} + rH_2O_2 + sH_2PO_4^- + tSO_4^{2-} \rightleftharpoons (H^+)_p(MoO_4^{2-})_q(H_2O_2)_r(H_2PO_4^-)_s(SO_4^{2-})_t{}^{p-2q-s-2t}$$

Formation constants are denoted $\beta_{p,q,r,s,t}$, and complexes are given the notation (p,q,r,s,t) or $Mo_qX_rP_sS_t^{n-}$. X is used instead of the peroxo ligand to shorten the formulae. The total concentrations of molybdate, hydrogen peroxide, phosphate and sulfate are denoted *Mo*, H_2O_2, *P* and *S*. No mixed ligand phosphate-sulfate species could be detected. Therefore, the new species that are formed contain no sulfate ($t = 0$) so that the simplified notations $(p,q,r,s,)$ or $Mo_qX_rP_s^{n-}$ can also be used.

The interaction between molybdate, hydrogen peroxide, and phosphate gives rise to three pH-dependent ^{31}P NMR resonances, whose relative intensities are strongly dependent on the Mo/P ratio. This is clearly illustrated in Figure 4, showing three ^{31}P NMR spectra at $H_2O_2/Mo = 3$ (pH ~ 2.5) with Mo/P = 8 (top), 4 (middle), and 2 (bottom). The resonance to the left originates from species having the highest Mo/P ratio, and the calculations showed it to have the composition Mo_3X_6P. The other two resonances were found to originate from Mo_2X_4P and MoX_2P species, respectively. As can be seen from the spectra, the complexation of phosphate to peroxomolybdates is weak and an appreciable amount of monomeric phosphate, resonance P, is present even at Mo/P = 8.

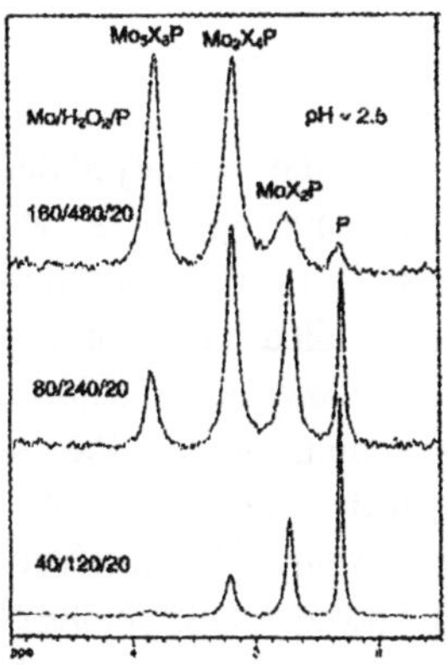

Fig. 4: ^{31}P NMR spectra of three solutions at H_2O_2 /Mo = 3 and Mo/P = 8, 4 and 2 respectively.

Figure 5 illustrates the ^{31}P chemical shifts as a function of pH for all the MoX_2P^{n-} (n=1, 2, 3), $Mo_2X_4P^{n-}$ (n=2, 3), and $Mo_3X_6P^{n-}$ (n=2, 3) complexes and also for the monomeric phosphate species. Such a plot shows the pH range of existence for each species and also the species that undergo protonation. The vertical dashed lines enclose the pH range that could be used for equilibrium calculations (potentiometric titration and ^{31}P NMR integral data). The chemical shift data outside this "equilibrium range" were used to determine the pK_a value of Mo_3X_6P and the second pK_a of MoX_2P.

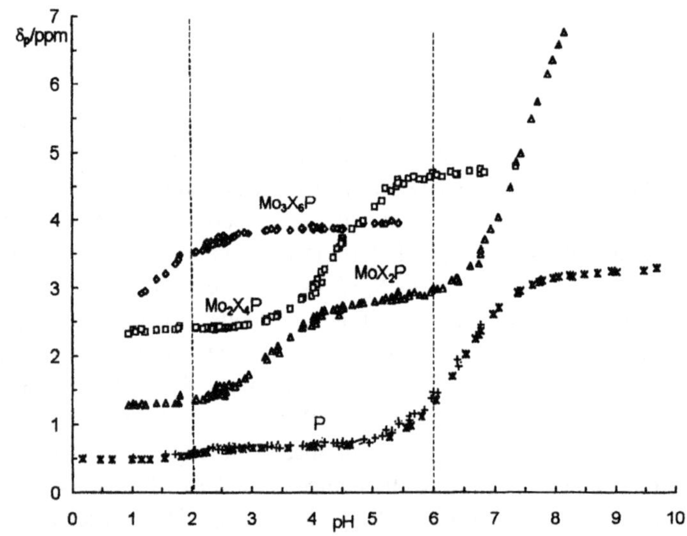

Fig. 5: ^{31}P NMR chemical shifts as a function of pH. The symbols represent experimental NMR points. The vertical dashed lines enclose the pH range used for equilibrium calculations.

4. CONCLUDING REMARKS

The speciation studies of the molybdate-, peroxomolybdate- and peroxomolybdophosphate systems have given a wider insight into species of pronounced interest for catalysis. With the quantitative speciations known, other aspects of the systems can be handled more precisely. The finding of a novel peroxomolybdosulfate complex in weakly acidic solutions introduce the possibility of complexation by other media anions under the same conditions. Indeed, the corresponding peroxomolybdochloride complex in 0.600 M Na(Cl) medium has in fact already been found, in an ongoing study. The catalytic properties of such complexes have not yet been evaluated. In the peroxomolybdophosphate system, the broadening of ^{31}P NMR resonances (the monomeric phosphate peak included) indicates that the species are in exchange and a dynamic study is in progress. Crystallisation experiments to obtain single crystals on these species and the peroxomolybdosulfate species, and ^{17}O NMR studies on concentrated solutions are in progress as well.

Acknowledgement

This work has been financially supported by The Strategic Foundation (SSF) and the Swedish Natural Science Research Council (NFR). We would like to thank Dr Oliver Howarth for valuable comments and linguistic corrections.

References

1 R. Agnemo, *9th ISWPC, Montréal, Canada* (1997) D2-1. ISBN 1-896742-14-9.
2 R. Agnemo, Personal communication.
3 J. Aveston, E.W. Anacker and J.S. Johnson: *Inorg. Chem.* **3**, 735 (1964).
4 Y. Sasaki, and L.G. Sillén: *Acta Chem. Scand.* **18**, 1014 (1964).
5 Y. Sasaki, and L.G. Sillén: *Ark. Kemi*, **29**, 253 (1967).
6 L.G. Sillén: *Pure Appl. Chem.*, **17**, 55 (1968).
7 L.G. Sillén, in *Coordination Chemistry*, ed. A.E. Martell, Van Nordstrand Reinhold, New York, 1971, vol. 1, pp. 491-541.
8 J.J. Cruywagen and J. B. B. Heyns: *Inorg. Chem.*, **26**, 2569 (1987).
9 K.H. Tytko, G. Baethe, E.R. Hirschfeld, K. Memhke and D.Z. Stellhorn: *Anorg. Allgem. Chem.*, **503**, 43 (1983).
10 K.H. Tytko, G. Baethe and J.J. Cruywagen: *Inorg. Chem.*, **24**, 3132 (1985).
11 A. Yagasaki, I. Andersson and L. Pettersson: *Inorg. Chem.*, **26**, 3926 (1987).
12 E. Farkas, H. Csóka, G. Micera and A. Dessi: *J. Inorg. Biochem.*, **65**, 281 (1997).
13 D.V.S. Jain and C.M. Jain: *Indian J. Chem.*, **12**, 178 (1974).
14 T. An-Pong, H. Shu-Hsun and T. Ch'ing-Ping: *K'o Hsueh T'ung*, **17**, 541 (1966).
15 J. A.Connor and E. A. V. Ebsworth: *Adv. Inorg. Chem. Radiochem.* **6**, 279 (1964).
16 M. H. Dickman and M. T. Pope: *Chem. Rev.* **94**, 569 (1994).
17 Gmelin Handbook of Inorganic Chemistry, Mo Suppl. Vol. 3b (1989).
18 F. Chauveau, P. Souchay and G. Tridot: *Bull. Soc. Chim. France, 1519* (1955).
19 F. C. Palilla, N. Adler and C. F. Hiskey: *Anal. Chem.* **25**, 926 (1953).
20 Y. Schaeppi and W. D. Treadwell: *Helv. Chim. Acta* **31**, 577 (1948).
21 A.C. Dengel, W.P. Griffith and B.C. Parkin: *J. Chem. Soc. Dalton Trans.* 2683 (1993).
22. L. Salles, C. Aubry, F. Robert, G. Chottard, R. Thouvenot, H. Ledon and J-M. Bregault: *New. J. Chem.* **17**, 367 (1993).
23. R. G. Beiles, Z. E. Rozmanova and O. B. Andreeva: *Russ. J. Chem.* **14**, 1122 (1969).
24 N. M. Gresley, W. P. Griffith, B. C. Parkin, A. J. P. White and D. J. Williams: *J. Chem. Soc. Dalton Trans.* 2039 (1996).
25. W. P. Griffith, B. C. Parkin, A. J. P. White and D. J. Williams: *J. Chem. Soc. Dalton Trans.* 3131 (1995).
26 A. Selling, I. Andersson, L. Pettersson, C.M. Schramm, S.L. Downey and J.H. Grate, *Inorg. Chem.*, **33**, 3141 (1994).
27 N. Ingri, I. Andersson, L. Pettersson, A. Yagasaki, L. Andersson and K. Holmström, *Acta Chem. Scand.*, **50**, 717 (1996).
28 G. Eriksson, *Anal. Chim. Acta*, **112**, 375 (1979).
29 B. Elgquist and M. Wedborg, *Marine Chemistry*, **2**, 1 (1974).

NMR Studies of Various Ligands Coordinated to Paramagnetic Polyoxometalates

BYUNG AHN KIM AND HYUNSOO SO
Department of Chemistry, Sogang University, Seoul 121-742, Korea

(Received: 3 October 1999)

Abstract. ^{1}H NMR spectroscopy was used to study various ligands coordinated to some paramagnetic polyoxometalates (POMs). Pure signals of the complexes were observed, which indicates that ligand exchange is slow on the NMR time scale. Mono- and diprotonated species of $[CoW_{11}CoO_{39}]^{8-}$ and $[SiW_9Cu_3O_{37}]^{10-}$ were detected from the spectra of pyridine coordinated to these POMs. 2-Aminopyridine binds to $[SiW_{11}CoO_{39}]^{6-}$ ($SiW_{11}Co$), whereas 2-methylpyridine does not. This indicates that hydrogen bonding between the amine group and a bridging oxygen atom on $SiW_{11}Co$ plays an important role in complex formation. 4-Aminopyrimidine forms two linkage isomers, **a** and **b**, binding to $SiW_{11}Co$ via N(1) and N(3), respectively. The relative amount of isomer **b** increases, when D_2O is replaced by DMF, indicating that hydrogen bonding between the amine group and $SiW_{11}Co$ is more favorable in DMF than in D_2O. 3,3-Dimethylpiperidine undergoes rapid chair-chair interconversion at room temperature. When it is coordinated to $SiW_{11}Co$ in D_2O, the conformation is frozen even at room temperature. When DMSO is added to a D_2O solution, the spectral change indicates that another conformation is stabilized in DMSO.

Key words: Paramagnetic NMR, polyoxometalates, protonation, hydrogen bonding, linkage isomers, conformations

1. Introduction

Polyoxometalates (POMs) are molecular analogs of extended oxide lattices, and ligands coordinated to POMs therefore are good models for substrates chemically adsorbed at metal oxide surfaces. Since polyoxometalate complexes are soluble in water and polar organic solvents, they are amenable to solution NMR spectroscopy. While diamagnetic POMs have only minor effects on the NMR spectra of the ligands, paramagnetic POMs cause dramatic shifts on NMR lines of the ligands. Hence, paramagnetic NMR spectroscopy [1] can be a powerful tool for studying polyoxometalate complexes.

Early in the history of paramagnetic NMR spectroscopy pyridine-type ligands coordinated to bis(acetylacetonato)cobalt(II), $Co(acac)_2$, were studied extensively [2]. Since the ligand exchange at the cobalt site is fast on the NMR time scale, an average spectrum of the free ligand and the complex was observed. The ligand exchange is slowed down when the cobalt ion is incorporated into a POM. Thus the ^{1}H NMR spectrum of a D_2O or DMSO-d_7 solution containing pyridine and $[SiW_{11}CoO_{39}]^{6-}$ ($SiW_{11}Co$) shows separate lines for the complex and the free ligand [3].

M.T. Pope and A. Müller (eds.), Polyoxometalate Chemistry, 175–186.

The ligand exchange is slow enough to produce a separate NMR spectrum for the complex, yet fast enough for saturation transfer between the signals from the complex and those from the free ligand. Therefore saturation transfer technique is useful in assigning the lines from the complexes. Two-dimensional techniques such as EXSY may be used to get the same information [4].

Slow ligand exchange makes paramagnetic NMR spectroscopy a useful technique for studying polyoxometalate complexes. Some ligands may be used as probes in studying formation, isomerization, and degradation of polyoxometalates containing paramagnetic metal ions. Conversely, some polyoxometalates may be used as probes to study conformation and conformational change of ligands coordinated to them.

2. Useful Polyoxometalates

Paramagnetic compounds have different electronic relaxation times. If the electronic relaxation is slow, good ESR spectra can be measured. Oxovanadium, manganese, and cupric complexes are such cases. If the electronic relaxation is fast, ESR spectra can be observed only at very low temperatures. But good NMR spectra are observed for ligands coordinated to the metal ions. Low-spin ferric, high-spin ferrous, cobalt(II), nickel(II), ruthenium(III), and lanthanide complexes are important examples [5].

Most Keggin and Dawson anions containing Co^{2+} or Ni^{2+} are useful POMs for NMR study. These transition metal ions carry an aqua ligand, which can be replaced readily by other ligands. Rates of ligand exchange for most of these POMs are slow on the NMR time scale. Notable exceptions are $[BW_{11}Co(L)O_{39}]^{7-}$ and $[AsW_{11}Co(L)O_{39}]^{5-}$, for which average NMR spectra are observed for the free ligand and the complex.

The isotropic NMR shifts in a paramagnetic system contain contributions from contact and pseudocontact shifts. When the ligand is rotating fast about the metal-ligand atom bond, the pseudocontact shifts are proportional to geometric factors, $(3\cos^2\theta - 1)/r^3$. Therefore, one can get structural information from NMR data, if the isotropic shifts can be separated into contact and pseudocontact contributions. This can be accomplished by measuring NMR spectra of the same ligand coordinated to $SiW_{11}Co$ and $SiW_{11}Ni$. The isotropic shifts contain contact and pseudocontact contributions for a $SiW_{11}Co$ complex, but contact shifts only for a $SiW_{11}Ni$ complex. Using the isotropic shifts in the $SiW_{11}Ni$ complex and calculated geometric factors, one can separate the isotropic shifts in the $SiW_{11}Co$ complex into contact and pseudocontact shifts.

No NMR spectrum is observed for ligands coordinated to $[SiW_{11}Fe^{III}O_{39}]^{5-}$, because the electronic relaxation of the ferric ion is not sufficiently fast. But good NMR spectra are observed for ligands coordinated to $[Co^{II}W_{11}Fe^{III}O_{39}]^{7-}$, indicating that the electronic relaxation time is shortened when the ferric ion is coupled with the fast-relaxing cobalt(II) ion.

Ordinary copper(II) complexes show no NMR spectra, because the electronic relaxation of the Cu^{2+} ion is very slow. However, good NMR spectra were observed for various ligands coordinated to $[SiW_9Cu_3O_{37}]^{10-}$ (SiW_9Cu_3), in which the three copper ions form an equilateral triangle [6]. The fast electronic relaxation may be attributed to spin frustration [7]. SiW_9Cu_3 is the only polyoxometalate which amino acids bind to.

TABLE 1

Paramagnetic polyoxometalates used for NMR studies

Paramagnetic atom	Polyoxometalates	Refs.
Co^{2+}	$SiW_{11}Co$, $PW_{11}Co$, $CoW_{11}Co$, $FeW_{11}Co$, $P_2W_{17}Co$	[3]
Ni^{2+}	$SiW_{11}Ni$, $[Ni_3(PW_9O_{34})_2]^{12-}$	[3, 8]
Fe^{3+}	$CoW_{11}Fe$	this work
Cu^{2+}	$[SiW_9Cu_3O_{37}]^{10-}$	[7]

3. Competition between Ligands and Solvent Molecules

Ligands compete with solvent molecules to bind to the metal ions in POMs. Heterocyclic compounds such as pyridine, imidazole, and pyrrolidine are readily coordinated to $SiW_{11}Co$ and $SiW_{11}Ni$ via the nitrogen atom in water. Aliphatic amines bind to these POMs only in polar organic solvents such as DMSO and DMF, but not in water. Amino acids are not soluble in these organic solvents. Amino acid esters, which are soluble in DMSO and DMF, bind to these POMs.

Various POMs can be transferred into nonpolar solvents by using tetraheptylammonium bromide as a phase-transfer agent [9]. Alcohols, ketones, and ethers bind to these POMs in toluene and other non-coordinating solvents. Recently Kozik et al. reported even carbon dioxide bind to $SiW_{11}Co$ in toluene [10].

TABLE 2

Ligands which bind to paramagnetic POMs

Polyoxometalates	Solvents	Ligands	Refs.
$SiW_{11}Co$, $SiW_{11}Ni$	D_2O	pyridine, imidazole, purine, aziridine, pyrrolidine, piperidine, quinuclidine	[3, 11, 12, 13, 14]
$SiW_{11}Co$, $SiW_{11}Ni$	DMSO, DMF	amines, amino acid esters	[15]
$SiW_{11}Co$, $SiW_{11}Ni$	toluene	alcohol, ether, ketone, carbon dioxide	[10]
SiW_9Cu_3	D_2O	pyridine, amino acid	[7]

4. Protonated Species of POMs

Evidence for protonation of POMs is found in a wide variety of experimental data including X-ray crystal structure and EPR spectra of mixed valence compounds [16, 17]. We have detected mono- and diprotonated species of several POMs by paramagnetic NMR spectroscopy.

4.1. $[Co^{II}W_{11}Co^{II}O_{39}]^{8-}$

Although the ^1H NMR spectrum of pyridine coordinated to $[CoW_{11}CoO_{39}]^{8-}$ shows many lines (Figure 1), the lines from α-, β-, and γ-protons are readily identified. Each group consists of two or three lines. The relative intensity of the strongest line decreases with increasing pH, indicating that this line originates from the monoprotonated species. In a similar way, the medium-intensity line and the weak line may be attributed to unprotonated and diprotonated species, respectively.

When a $[W{=}O]^{4+}$ group is replaced by a Co^{2+} ion, the bridging oxygen atoms between the cobalt and tungsten atoms are expected to be the most basic sites [18] and one or two of these sites must be protonated. Existence of three species indicates that intermolecular proton transfer is slow on the NMR time scale.

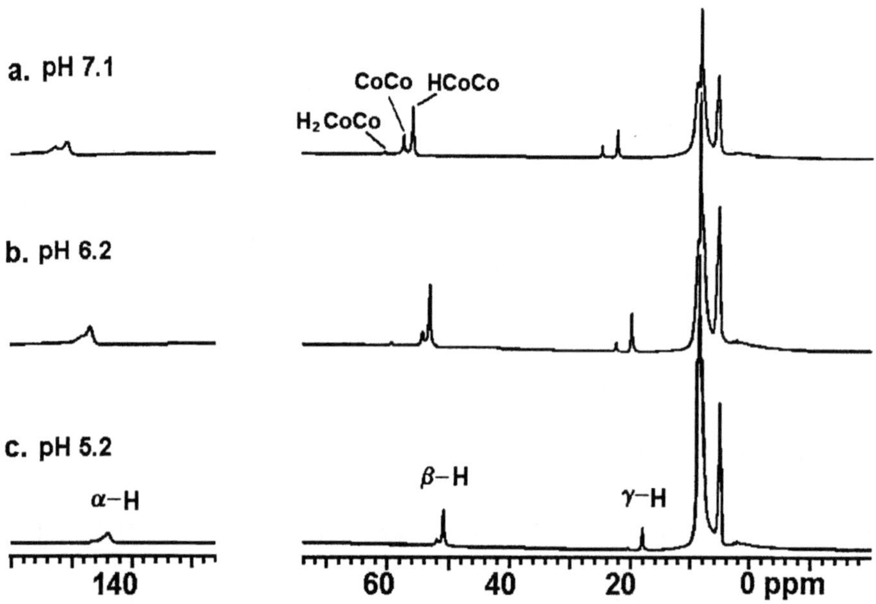

Fig. 1. ^1H NMR spectra of D_2O solutions containing $[Co^{II}W_{11}Co^{II}O_{39}]^{8-}$ and pyridine at pH (a) 7.1, (b) 6.2, and (c) 5.2. CoCo, HCoCo, and H_2CoCo represent un-, mono-, and diprotonated species, respectively.

4.2. $[SiW_9Cu_3O_{37}]^{10-}$

Protonated species were also detected for $[SiW_9Cu_3O_{37}]^{10-}$. A D_2O solution containing SiW_9Cu_3 and pyridine shows a very complex NMR spectrum (Figure 2). But, on comparing with the spectrum of $SiW_{11}Co(py)$, we could readily identify lines originating from α-, β- and γ-protons. Then, by varying the relative amounts of the POM and pyridine, we identified lines from tri-, di-, and monopyridine species. Among the lines originating from α-proton, those designated by A, B, and C come from di- and tripyridine species.

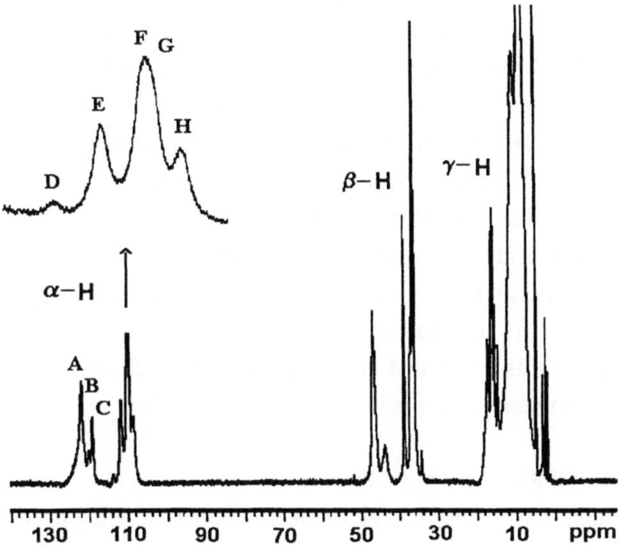

Fig. 2. ^1H NMR spectrum of a D_2O solution containing $[SiW_9Cu_3O_{37}]^{10-}$ and pyridine

The monopyridine species alone shows five lines designated by D, E, F, G, and H. These lines may be assigned to five different copper sites (see the scheme below). The three copper sites in the unprotonated species are equivalent. For the mono- and diprotonated species, there are two different copper sites each. If it is assumed that triprotonated species is not formed above pH 6, there are five copper sites for a monopyridine complex. Finally, based on their pH dependence and relative intensities the lines were assigned as follows: D, II2; E, II1; F+G, 00+I0; H, I1.

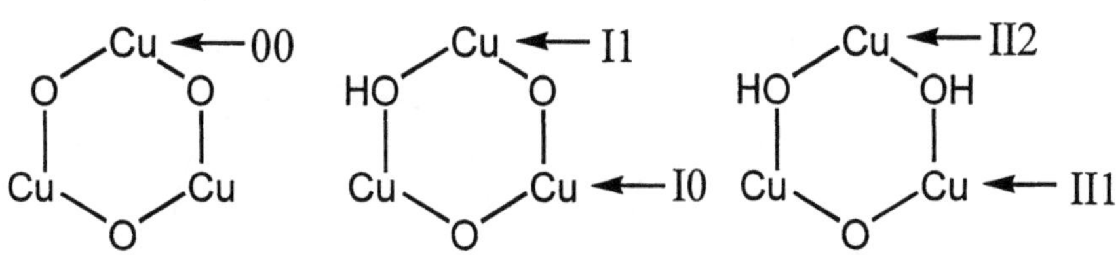

5. Roles of Hydrogen Bonding in Complex Formation

NMR studies of some ligands coordinated to POMs indicate that hydrogen bonding plays important roles in complex formation. 2-Aminopyridine, 4-aminopyrimidine, adenine, adenosine, etc. are good candidates to form hydrogen bonding with POMs.

5.1. 2-METHYLPYRIDINE vs. 2-AMINOPYRIDINE

When 2-methylpyridine and $SiW_{11}Co$ are mixed in D_2O, no NMR signal from the complex is detected [19]. On the other hand, 2-aminopyridine binds to $SiW_{11}Co$ readily. If 2-methylpyridine were coordinated to $SiW_{11}Co$, one proton in the methyl group should come within 2.3 Å from a bridging oxygen atom on $SiW_{11}Co$. This is shorter than the 2.6 Å suggested by Pauling for O...H van der Waals contact distance [20]. The methyl group, which cannot form a hydrogen bond, will hamper complex formation.

On the other hand, 2-aminopyridine can form a hydrogen bond and seems to have right geometry to form a hydrogen bond with $SiW_{11}Co$. We have calculated the distances between a bridging oxygen atom on $SiW_{11}Co$ and the amine group in the $SiW_{11}Co$ complex, using the structural data for 2-aminopyridine [21] and a Co–N distance of 2.20 Å [22]. The shortest distance between the amine nitrogen atom and a bridging oxygen atom is 2.8 Å and one of the protons is displaced 9° from the N...O vector. Thus, the geometry is favorable for hydrogen bonding.

5.2. LINKAGE ISOMERS OF 4-AMINOPYRIMIDINE

4-Aminopyrimidine forms two linkage isomers, **a** and **b**, binding to $SiW_{11}Co$ via N(1) and N(3), respectively. The ratio of isomers **a** and **b** depends upon the solvent: 5 : 1 in D_2O and 5 : 3.4 in DMF. It is probable that the POM and solvent molecules compete to form hydrogen bonding with the ligand. Since the competition is less serious in DMF than in D_2O, the relative amount of isomer **b** may increase in DMF.

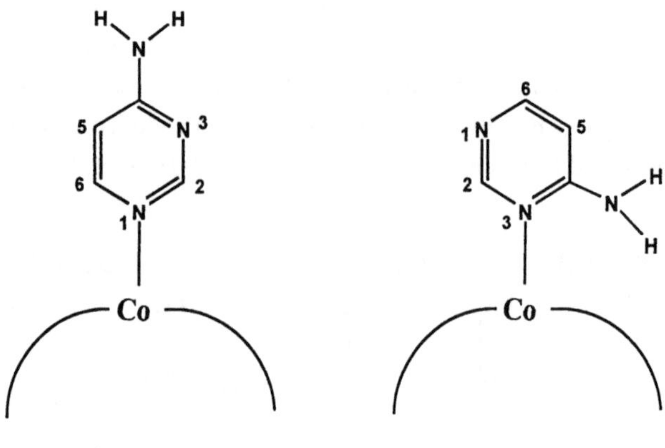

isomer **a** isomer **b**

The ^1H NMR spectrum of 4-aminopyrimidine coordinated to $SiW_{11}Co$ in DMF-d_7 at 25 °C is shown in Figure 3. All lines from the complexes were assigned by saturation transfer technique. The signal from the amine group in isomer **a** is split into two lines at –3.70 and –5.87 ppm, indicating that the internal rotation of the amine group is slow even at room temperature. These lines show temperature dependence that is characteristic of the two-site exchange problem, merging at 40 °C.

The line at –22.3 ppm comes from the amine group in isomer **b**. Its intensity corresponds to one proton, and this line may be attributed to the proton not involved in hydrogen bonding. The amine proton involved in hydrogen bonding is at 2.6 Å from the cobalt ion. Since the line width due to the dipolar relaxation is inversely proportional to the sixth power of the metal-proton distance, and the contribution of chemical exchange to the line width is the same for all protons in the ligand, the width of the missing line can be estimated. The estimated half width is at least 1100 Hz, indicating that the signal is too broad to be observed.

The line at –22.3 ppm is broadened much more rapidly than the other lines of isomer **b** as temperature is raised; it is barely seen at 35 °C. This indicates that the amine group in isomer **b** also undergoes internal rotation at high temperatures.

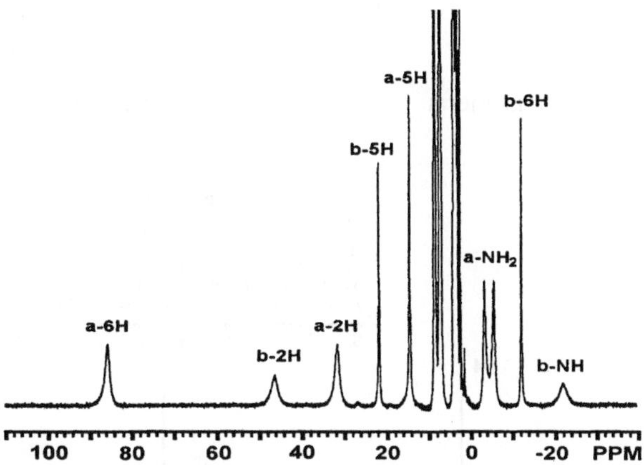

Fig. 3. ^1H NMR spectrum of a DMF-d_7 solution containing $[SiW_{11}CoO_{39}]^{6-}$ and 4-aminopyrimidine at 25 °C

6. Interaction of Amino Acids with POMs

Interactions of amino acids with polyoxometalates are of considerable interest. Histidine binds to $SiW_{11}Co$ via imidazole ring, forming two linkage isomers. Proline is coordinated to $SiW_{11}Co$ via the ring nitrogen at high pH. Simple amino acids such as glycine, alanine, etc. are not coordinated to $SiW_{11}Co$ in water, indicating that they cannot compete with water in binding to the Co^{2+} ion. Amino acids are not soluble in DMSO or DMF. Amino acid esters, which are soluble in these solvents, bind to $SiW_{11}Co$.

6.1. HISTIDINE

The ^1H NMR spectrum of a D_2O solution containing histidine and $SiW_{11}Co$ is shown in Figure 4 [11]. The NMR spectrum contains lines from two linkage isomers, **a** and **b**. An interesting feature is that the two lines originating from the CH_2 group of isomer **a** are separated by more than 20 ppm. If the CH_2 group is rotating freely around the C-C bond, the two protons, although diastereotopic, should have similar chemical shifts. This is evidenced by a small chemical shift difference for the two diastereotopic protons in isomer **b**. The large difference in their chemical shifts for isomer **a** suggests that the rotation around the C-C bond is hampered by the POM moiety.

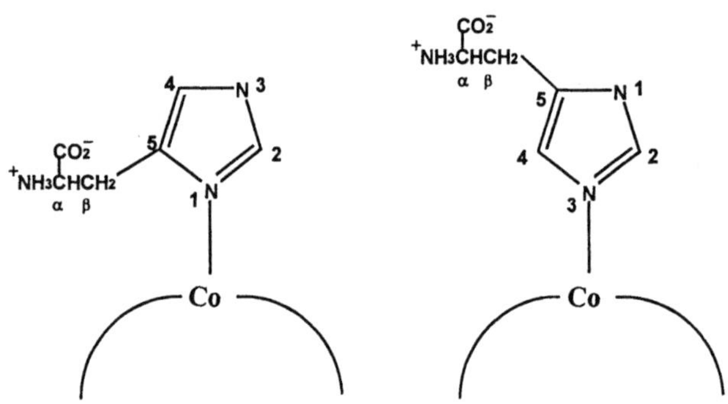

isomer **a** isomer **b**

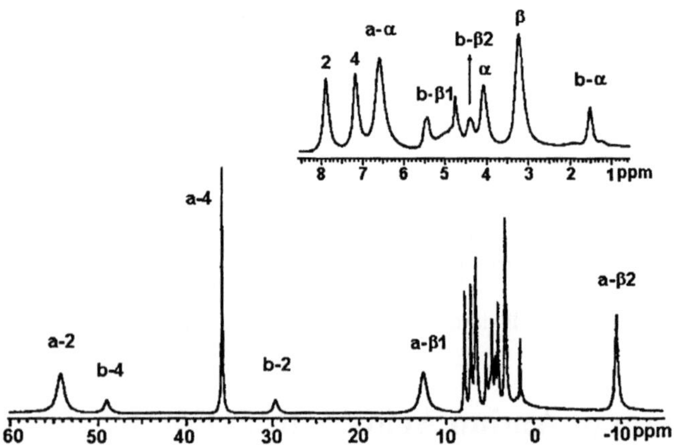

Fig. 4. ^1H NMR spectrum of a D_2O solution containing $[SiW_{11}CoO_{39}]^{6-}$ and L-histidine in a 1:1 molar ratio at pH 7.7.

6.2. GLYCINE AND N-METHYLGLYCINE

Amino acids bind to SiW_9Cu_3 in water. ^1H NMR spectra of glycine and N-methylglycine coordinated to SiW_9Cu_3 are shown in Figure 5. The largest isotropic shift is observed for the N-methyl group, indicating that N-methylglycine binds to SiW_9Cu_3

via the nitrogen atom. The two protons of the CH_2 group in N-methylglycine become diastereotopic in the complex, showing two separate lines at 63.3 and 53.1 ppm. The chemical shift of the CH_2 group in glycine is similar to those of the CH_2 group in N-methylglycine, indicating that glycine also binds to SiW_9Cu_3 via the nitrogen atom.

Although the two diastereotopic protons in the N-methylglycine complex may have different chemical shifts, the separation is much larger than those observed for diamagnetic molecules. The observed isotropic shifts may be interpreted in terms of three staggered conformers. The large separation can be explained only when the lowest energy conformer has the carboxylate group in the trans position with respect to the methyl group. Since the two protons occupy trans and gauche positions with respect to the copper atom in this conformer, they can have quite different isotropic shifts.

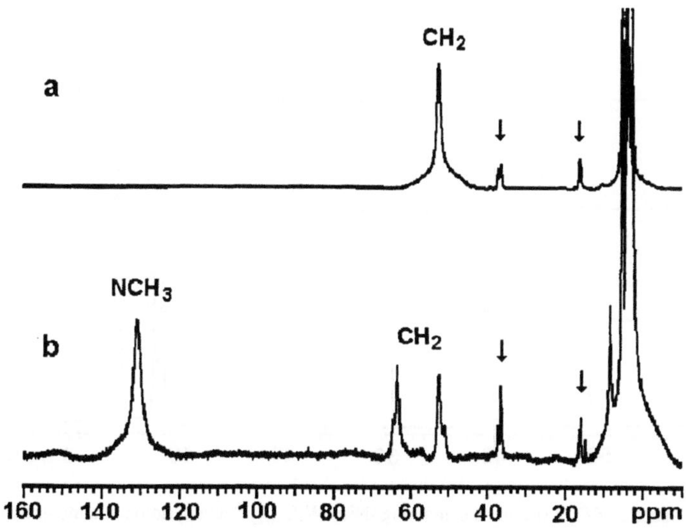

Fig. 5. ^1H NMR spectra of D_2O solutions containing $[SiW_9Cu_3O_{37}]^{10-}$ and (a) glycine and (b) N-methylglycine at pH 8.3. Pyridine-d_5 was added to mask the copper sites partially. Weak lines designated by arrows come from residual pyridine.

7. Transformation of POMs

Paramagnetic NMR spectroscopy can be used to follow slow transformations of POMs such as formation, isomerization and degradation.

7.1. ISOMERIZATION OF $[FeW_{11}CoO_{39}]^{7-}$

The ^1H NMR spectrum of pyridine coordinated to $[FeW_{11}CoO_{39}]^{7-}$ exhibits two sets of lines, the relative intensities of which are time-dependent. Equilibrium is reached in about 10 hours. It is probable that an isomerization reaction occurs slowly.

7.2. DEGRADATION OF $[Ni_3(PW_9O_{34})_2]^{12-}$

Degradation of $[Ni_3(PW_9O_{34})_2]^{12-}$ can be followed using paramagnetic NMR spectroscopy [8]. In this POM three nickel ions are sandwiched between two PW_9O_{34} groups. The ^1H NMR spectrum of pyridine coordinated to $[Ni_3(PW_9O_{34})_2]^{12-}$ shows two sets of lines (Figure 6). The stronger set is attributed to the unprotonated species and the weaker set to the monoprotonated species based on their pH dependence. The strong signal for the β-proton contains three lines ascribable to mono-, di-, and tripyridine complexes based on their concentration dependence.

Additional lines appeared below pH 5, and they were attributed to a degradation product, in which one nickel ion was replaced by a tungsten atom, $[Ni_2(WO_2)(PW_9O_{34})_2]^{12-}$ [23].

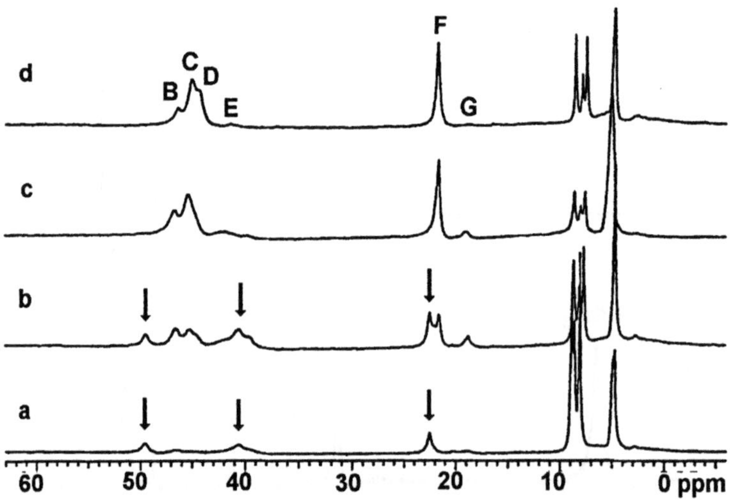

Fig. 6. ^1H NMR spectra of D_2O solutions containing $[Ni_3(PW_9O_{34})_2]^{12-}$ and pyridine in a molar ratio of 1 at pH (a) 4.1, (b) 5.0, (c) 6.2, and (d) 8.7. A broad line at 150 ppm originating from α-H is not shown. The lines designated by arrows come from $[Ni_2(WO_2)(PW_9O_{34})_2]^{12-}$.

8. Conformations of Piperidines

Conformations of six-membered ring compounds such as cyclohexane have been studied extensively. We have studied some piperidines coordinated to POMs. Piperidine and 3,3-dimethylpiperidine undergo rapid chair-chair interconversion at room temperature [13].

When coordinated to $SiW_{11}Co$, piperidine still undergoes conformational change. But the ^1H NMR spectrum of 3,3-dimethylpiperidine coordinated to $SiW_{11}Co$ shows separate lines for the axial and equatorial protons, indicating that the chair-chair interconversion does not occur even at room temperature (Figure 7).

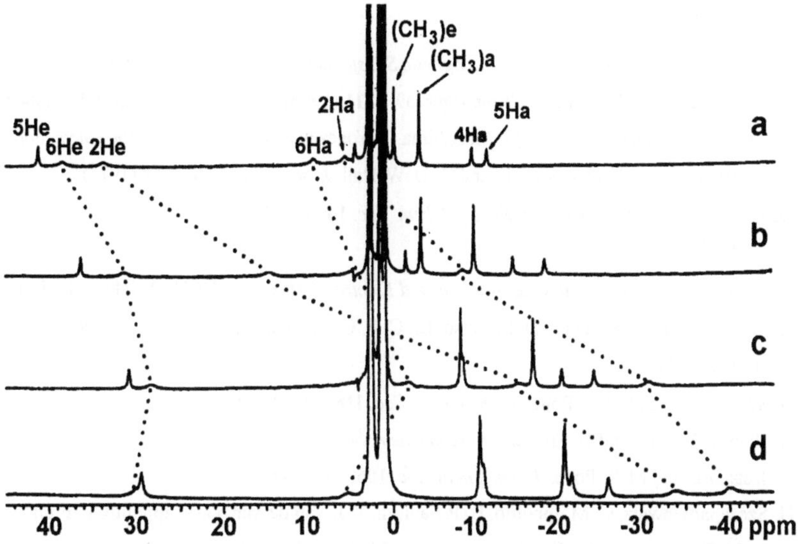

Fig. 7. ^{1}H NMR spectra of (a) D_2O, (b) 40% (by volume) DMSO-d_6, (c) 80% DMSO-d_6, and (d) DMSO-d_6 solutions containing $[SiW_{11}CoO_{39}]^{6-}$ and 3,3-dimethylpiperidine in a 1:1 molar ratio. The lines originating from the complex are labeled.

The equatorial protons at positions 2 and 6, $2H_e$ and $6H_e$, have similar chemical shifts in D_2O, which is in accordance with a chair conformation. On replacing D_2O by DMSO-d_7, the line from $2H_e$ is shifted upfield by 68 ppm, whereas the line from $6H_e$ is shifted upfield only by 9 ppm. Now that the lines from $2H_e$ and $6H_e$ are far apart, the conformer in DMSO cannot be a chair form. It is probable that a twisted form is stabilized in DMSO. It is suggested that a chair form and a twisted form are at equilibrium in the mixed solvent.

In summary, we have shown that ^{1}H NMR spectra of some ligands coordinated to paramagnetic polyoxometalates provide detailed information about the complexes. It is straightforward to use paramagnetic NMR spectroscopy in studying formation, isomerization, and degradation of POMs containing Co^{2+} or Ni^{2+}. On the other hand, more work is needed to develop paramagnetic NMR spectroscopy as a useful tool for determining the conformations of ligands coordinated to POMs.

Acknowledgment

Financial assistance from Sogang University Research Grant (1997) is gratefully acknowledged.

References

1. I. Bertini and C. Luchinat: *NMR of Paramagnetic Substances,* Elsevier, Amsterdam (1996).
2. (a) J. A. Happe and R. L. Ward: *J. Chem. Phys.* **39**, 1211 (1963). (b) D. Doddrell and J. D. Roberts: *J. Am. Chem. Soc.* **92**, 6651 (1970). (c) I. Morishima, T. Yonezawa, and K. Goto: *J. Am. Chem. Soc.* **92,** 6839 (1970). (d) W. deW. Horrocks, Jr. and D. DeW. Hall: *Inorg. Chem.* **10,** 2368 (1971).
3. M. Ko, G. I. Rhyu, and H. So: *Bull. Korean Chem. Soc.* **14,** 500 (1993).
4. Ref. 1, p 187.
5. L. Banci, I. Bertini, and C. Luchinat: *Nuclear and Electron Relaxation*, VCH, Weinheim (1991), p 88.
6. J. Liu, F. Ortega, P. Sethuraman, D. E. Katsoulis, C. E. Costello, and M. T. Pope: *J. Chem. Soc. Dalton Trans.* 1901 (1992).
7. H. Y. Woo, H. So, and M. T. Pope: *J. Am. Chem. Soc.* **118,** 621 (1970).
8. H. Y. Woo, J. Y. Kim, and H. So: *Bull. Korean Chem. Soc.* **16,** 1176 (1995).
9. D. E. Katsoulis and M. T. Pope: *J. Am. Chem. Soc.* **106,** 737 (1984).
10. S. H. Szczepankiewicz, C. M. Ippolito, B. P. Santora, T. J. Van de Ven, G. A. Ippolito, L. Fronckowiak, F. Wiatrowski, T. Power, and M. Kozik: *Inorg. Chem.* **37**, 4344 (1998).
11. M. Ko, G. I. Rhyu, and H. So: *Bull. Korean Chem. Soc.* **15,** 673 (1994).
12. S. M. Park and H. So: *Bull. Korean Chem. Soc.* **18,** 1002 (1997).
13. J. Hyun and H. So: *Bull. Korean Chem. Soc.* **18,** 961 (1997).
14. J. Hyun, S. M. Park, and H. So: *Bull. Korean Chem. Soc.* **18,** 1090 (1997).
15. B. A. Kim and H. So: *Bull. Korean Chem. Soc.* 1999 (in press).
16. V. W. Day and W. G. Klemperer: in *Polyoxometalates: From Platonic Solids to Anti-Retroviral Activity*, eds. M. T. Pope and A. Müller, Kluwer Academic Publishers, Dordrecht, the Netherlands, p. 100 (1994).
17. (a) M. M. Mossoba, C. J. O'Connor, M. T. Pope, E. Sinn, G. Hervé, and A. Tézé: *J. Am. Chem. Soc.* **102,** 6866 (1980). (b) S. P. Harmalker, M. A. Leparulo, and M. T. Pope: *J. Am. Chem. Soc.* **105,** 4286 (1983).
18. V. W. Day and W. G. Klemperer: *Science* **228,** 533 (1985).
19. Small formation constants, 3.6 and 1.0 L mol^{-1}, based on visible spectra were reported for 2-methylpyridine coordinated to $PW_{11}Co$ and $BW_{11}Co$, respectively. See T. J. Weakley: *J. Chem. Soc. Dalton Trans.* 341 (1973).
20. L. Pauling: *The Nature of Chemical Bond*, Cornell University Press, Ithaca, New York (1960).
21. M. Barfield and P. Fagerness: *J. Am. Chem. Soc.* **119,** 8699 (1997).
22. P. O. Lumme: *Polyhedron* **14,** 1553 (1995).
23. W. H. Knoth, P. J. Domaille and R. L. Harlow: *Inorg. Chem.* **25,** 1577 (1986).

Molecular Aspect of Energy Transfer from Tb^{3+} to Eu^{3+} in the Polyoxometalate Lattices: An Approach for Molecular Design of Rare-Earth Metal-Oxide Phosphors

TOSHIHIRO YAMASE

Research Laboratory of Resources Utilization, Tokyo Institute of Technology, 4259 Nagatsuta, Midori-ku, Yokohama 226-8503, Japan

Abstract. The exploitation of mixed heteronuclear rare-earth-element-containing polyoxometalates to probe the multipolar nature of heteronuclear rare-earth interactions is imaginative. It appears that polyoxometallolanthanoates are ideal for this type of investigation. Three structural types of Tb^{3+}/Eu^{3+} heterolanthanide-multinuclear polyoxometalates, $K_{15}H_3[Tb_{1.4}Eu_{1.6}(H_2O)_3(SbW_9O_{33})(W_5O_{18})_3]\cdot25.5H_2O$, $Na_7H_{19}[Tb_{4.3}Eu_{1.7}O_2(OH)_6(H_2O)_6Al_2(Nb_6O_{19})_5]\cdot47H_2O$, and $[NH_4]_{12}H_2[Tb_{3.1}Eu_{0.9}(MoO_4)(H_2O)_{16}(Mo_7O_{24})_4]\cdot13H_2O$ are studied by crystal structures, emission and excitation spectra, and emission decay dynamics. The excitation of the Tb^{3+} $^7F_6\rightarrow{}^5D_4$ transitions produces not only the emission lines of Tb^{3+}, but also those of Eu^{3+}, accompanied by nonexponential rise and decay curves of the emission from Tb^{3+} and Eu^{3+}. There is no significant exchange interaction between the lanthanide ions, as a result of the coordination of aqua and/or hydroxo ligands to the lanthanide ions. The mechanism of the $Tb^{3+}\rightarrow Eu^{3+}$ energy transfer is identified as a Förster-Dexter-type energy transfer from Tb^{3+} (donor) to Eu^{3+} (acceptor). The nearest-neighbor energy-transfer rates by electric dipole-dipole interactions between a Tb-Eu pair at 4.2K are estimated to be 4.5×10^4, 4.7×10^5, and 4.9×10^3 s^{-1} and the critical radii at 4.2 K are 10.3, 10.0, and 6.17 Å for $K_{15}H_3[Tb_{1.4}Eu_{1.6}(H_2O)_3(SbW_9O_{33})(W_5O_{18})_3]\cdot25.5H_2O$ (with Tb-Eu separation of 5.05 Å), $Na_7H_{19}[Tb_{4.3}Eu_{1.7}O_2(OH)_6(H_2O)_6Al_2(Nb_6O_{19})_5]\cdot47H_2O$ (with 3.76 Å separation), and $[NH_4]_{12}H_2[Tb_{3.1}Eu_{0.9}(MoO_4)(H_2O)_{16}(Mo_7O_{24})_4]\cdot13H_2O$ (with 6.17 Å separation), respectively. The low symmetry (C_s or C_1) of the LnO_8 (Ln=Tb and Eu) coordination polyhedra allows the nonvanishing electric-dipole transition probability for the $^7F_J\leftrightarrow{}^5D_0$ (J=0,1) transitions which leads to a faster transter rate at high temperatures. The photoexcitation of the host lattices (tungstate, niobate, and molybdate) induced the energy transfer from the oxygen-to-metal charge-transfer {O→M(=Nb, Mo, W) lmct} triplet states to Tb^{3+} ($^7F_6\rightarrow{}^5D_4$) and Eu^{3+} ($^7F_{0,1,2}\rightarrow{}^5D_{0,1}$). In the case of $[NH_4]_{12}H_2[Tb_{3.1}Eu_{0.9}(MoO_4)(H_2O)_{16}(Mo_7O_{24})_4]\cdot13H_2O$ this transfer is not complete and the O→Mo lmct triplet emission of molybdates is observed to provide the rate constant for the

M.T. Pope and A. Müller (eds.), Polyoxometalate Chemistry, 187–203.

energy transfer to Tb^{3+}/Eu^{3+} sites with 4.4 x 10^6 s^{-1}.

Key words: Polyoxometallolanthanoates, photoluminescence, Förster-Dexter-type energy transfer, Rare-earth metal-oxide phosphors, Molecular design

1. Introduction

We have investigated the intramolecular transfer of the oxygen-to-metal charge-transfer {O→M(=Nb, Mo, W) lmct} energy to the Eu^{3+} site in the polyoxometalloeuropate lattices for understanding the molecular insight into the transfer of the excitation energy of the host lattice to luminecence centers in the Eu^{3+}-doped metal oxide phosphors such as $Gd_2(WO_4)_3$:Eu, Y_2WO_6:Eu, $YNbO_4$:Eu, and $CaMoO_4$:Eu [1], and discussed the photoluminescence properties of the polyoxometalloeuropates in terms of both the energy transfer from the O→M lmct triplet states and the nonradiative relaxation of the $^5D_0$ state of Eu^{3+} [2]. Of particular interest for the energy transport phenomena among lanthanide centers in the oxide lattices is Eu-multinuclear polyoxometalate complexes, as exemplified by $K_{15}H_3[Eu_3(H_2O)_3(SbW_9O_{33})(W_5O_{18})_3]$•25.5$H_2O$, $Na_7H_{19}\{[Eu_3O(OH)_3(H_2O)_3]_2Al_2(Nb_6O_{19})_5\}$•47$H_2O$, and $[NH_4]_{12}H_2[Eu_4(MoO_4)(H_2O)_{16}(Mo_7O_{24})_4]$•13$H_2O$ [3-5]. Figures 1(a)-(c) show the structural features of the central Eu^{3+} aggregates for $K_{15}H_3[Eu_3(H_2O)_3(SbW_9O_{33})(W_5O_{18})_3]$•25.5$H_2O$, $Na_7H_{19}\{[Eu_3O(OH)_3(H_2O)_3]_2Al_2(Nb_6O_{19})_5\}$•47$H_2O$, and $[NH_4]_{12}H_2[Eu_4(MoO_4)(H_2O)_{16}(Mo_7O_{24})_4]$•13$H_2O$ respectively. In the anion of $K_{15}H_3[Eu_3(H_2O)_3(SbW_9O_{33})(W_5O_{18})_3]$•25.5$H_2O$, a central trinuclear $Eu_3(H_2O)_3$ core tetrahedrally arranges one B-α type SbW_9O_{33} and three W_5O_{18} ligands, giving an approximate point-symmetry of C_{3v}. Each Eu^{3+} in the $Eu_3(H_2O)_3$ core achieves square-antiprismatically eightfold coordination (with an approximate symmetry of C_s) by attachment to four O atoms from one W_5O_{18}, two O atoms from one SbW_9O_{33}, and two aqua-O atoms. The nearest Eu•••Eu distances in the trimeric $[Eu_3(H_2O)_3]^{9+}$ core singly bridged by water O atoms are 5.015(5)-5.067(4) Å [average, 5.050(3) Å] [3].

The anion of $Na_7H_{19}\{[Eu_3O(OH)_3(H_2O)_3]_2Al_2(Nb_6O_{19})_5\}$•47$H_2O$ consists of two $[Eu_3O(OH)_3(H_2O)_3]^{4+}$ clusters, two Al^{3+} cations, and five $[Nb_6O_{19}]^{8-}$ (apically two and equatorially three) anions. Each Eu^{3+} achieves a bicapped-trigonal-prismatically eightfold coordination (with a C_1 symmetry) via: one μ_3-O, two μ_3-OH^- ions, one terminal water O, and four O atoms belonging to the equatorial $[Nb_6O_{19}]^{8-}$ groups. The nearest Eu•••Eu distances in the $[Eu_3O(OH)_3(H_2O)_3]^{4+}$ half-core are 3.740(7)-3.777(5) Å [average, 3.756(2) Å]. Two half-cores are linked by three μ_4-O atoms, each of which belongs to each of three equatorial $[Nb_6O_{19}]^{8-}$ ligands, with the Eu•••Eu distances of 4.609(6)-4.763(6) Å [average, 4.69(2) Å] [4].

In $[NH_4]_{12}H_2[Eu_4(MoO_4)(H_2O)_{16}(Mo_7O_{24})_4]$•13$H_2O$ the anion contains a central $\{[Eu(H_2O)_4]_4(MoO_4)\}^{10+}$ core and four $[Mo_7O_{24}]^{6-}$ anions with configuration of a point symmetry of D_{2d}. Each Eu^{3+} achieves a tricapped trigonal-prismatically

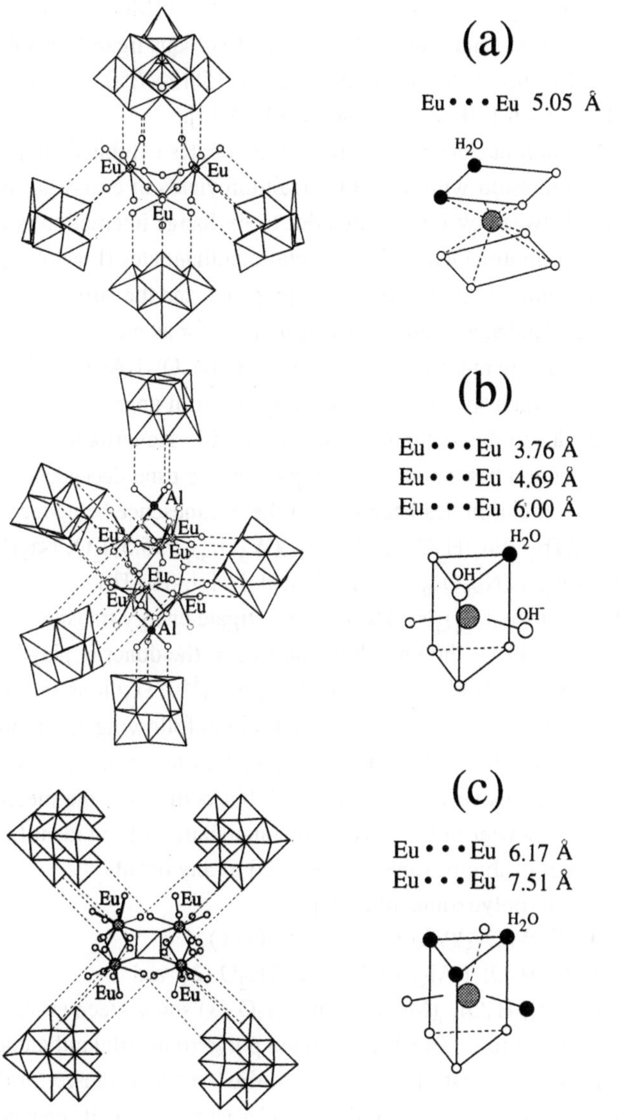

Fig. 1. Schematic representations and Eu•••Eu distances of the anions of $K_{15}H_3[Eu_3(H_2O)_3(SbW_9O_{33})(W_5O_{18})_3]\cdot 25.5H_2O$ (a), $Na_7H_{19}\{[Eu_3O(OH)_3(H_2O)_3]_2Al_2(Nb_6O_{19})_5\}\cdot 47H_2O$ (b), and $[NH_4]_{12}H_2[Eu_4(MoO_4)(H_2O)_{16}(Mo_7O_{24})_4]\cdot 13H_2O$ (c). A schematic representation of EuO_8 or EuO_9 coordination geometry for each anion is also shown.

ninefold coordination by attachment of one O atom from $[MoO_4]^{2-}$, four O atoms from three $[Mo_7O_{24}]^{6-}$ anions, and four O atoms from four aqua ligands (resulted in a C_s symmetry). The Eu•••Eu distances in the anion (6.158(2)-7.505(2) Å) are divided into two kinds of distances 6.170 (average) and 7.505 Å [5].

Such Eu-multinuclear polyoxometalloeuropates exhibited simple exponential decays of the $^5D_0$ emission with moderate quantum yields of emission, implying that the energy migration between two Eu^{3+} sites due to exchange interaction is negligible [2b]. A variety of the Ln-multinuclear polyoxometallolanthanoates (Ln=Er, Lu, and so on) with the same structure of anions have been prepared, and the structural change by the nucleation of heterolanthanide cations was minor, as far as we compared the crystallographic structure among three $\{[Ln_3O(OH)_3(H_2O)_3]_2Al_2(Nb_6O_{19})_5\}^{16-}$ (Ln=Eu, Er, and Lu) anions which showed a small change (within approximately 0.1 Å) in Ln•••Ln distances due to the lanthanide contraction [6]. Nevertheless, the Er/Eu- or Tb/Eu-mixed nucleation has a noticeable impact on the time dependence of the Eu^{3+} emissions, and our attention has been paid to the luminescnce behavior of Tb/Eu-mixed complexes $K_{15}H_3[Tb_{3-x}Eu_x(H_2O)_3(SbW_9O_{33})(W_5O_{18})_3]$•$25.5H_2O$, $Na_7H_{19}\{[Tb_{6-x}Eu_xO_2(OH)_6(H_2O)_6Al_2(Nb_6O_{19})_5\}$•$47H_2O$, and $[NH_4]_{12}H_2[Tb_{4-x}Eu_x(MoO_4)(H_2O)_{16}(Mo_7O_{24})_4]$•$13H_2O$ to investigate energy transfer processes between Tb^{3+} and Eu^{3+}. As will be shown below, the role of the donor is played by Tb^{3+}, that of the acceptor by Eu^{3+}. Both the Tb^{3+} ($^5D_4$) and Eu^{3+} ($^5D_0$) ions in the Tb/Eu-mixed complexes emit, where Eu^{3+} ($^5D_0$) excitation buildup following excitation of Tb^{3+} ($^5D_4$) provides strong evidence for Tb^{3+} to Eu^{3+} energy transfer. Energy transfer in solutions and solids has been extensively investigated [7], but neither has the mechanism of the $Tb^{3+}\rightarrow Eu^{3+}$ energy transfer in the polyoxometalate lattices been identified up to now, nor has the dynamics of this process been studied in any detail.

Tb^{3+} ions in polyoxometalloterbates $K_{15}H_3[Tb_3(H_2O)_3(SbW_9O_{33})(W_5O_{18})_3]$•$25.5H_2O$ (**1**), $Na_7H_{19}\{[Tb_3O(OH)_3(H_2O)_3]_2Al_2(Nb_6O_{19})_5\}$•$47H_2O$ (**2**), and $[NH_4]_{12}H_2[Tb_4(MoO_4)(H_2O)_{16}(Mo_7O_{24})_4]$•$13H_2O$ (**3**) show green emissions due to the $^5D_4\rightarrow{}^7F_J$ (J=6-0) transition. The Tb/Eu-mixed polyoxometallolanthanoates provide a favarable system for investigation of the $Tb^{3+}\rightarrow Eu^{3+}$ energy transfer in the oxide lattices, because the emission lines of donor and acceptor are well separated and can be measured without much interference by each other. For this reason, the $Tb^{3+}\rightarrow Eu^{3+}$ energy migration in three kinds (tungstate, niobate, and molybdate) of Tb/Eu-multinuclear lattices, $K_{15}H_3[Tb_{1.4}Eu_{1.6}(H_2O)_3(SbW_9O_{33})(W_5O_{18})_3]$•$25.5H_2O$ (**4**), $Na_7H_{19}\{[Tb_{4.3}Eu_{1.7}O_2(OH)_6(H_2O)_6Al_2(Nb_6O_{19})_5\}$•$47H_2O$ (**5**), and $[NH_4]_{12}H_2[Tb_{3.1}Eu_{0.9}(MoO_4)(H_2O)_{16}(Mo_7O_{24})_4]$•$13H_2O$ (**6**) is presented here in order to determine the energy transfer channels and to identify the type of the interaction between the partners involved in the transfer, together with photoluminescence properties for Tb^{3+} in **1**-**3**. **4** is regarded as a disordered mixture of Tb/Eu=1/2 and 2/1, **5** as the one of Tb/Eu=4/2 and 5/1, and **6** as the one of Tb/Eu=2/2 and 3/1. Thereby, it is reasonable to assume that the energy transfer from Tb^{3+} to Eu^{3+} for **4-6** occurs exclusively at the shortest distance of Tb•••Eu which is close to the shortest Eu•••Eu (5.05 Å for **4**, 3.76 Å for **5**, and 6.17 Å for **6**) for the coresponding pure Eu complex, although it is difficult to distinguish x-ray crystallographically between Tb and Eu.

2 Experimental

The pure Tb complexes **1-3** were prepared by replacing $Eu(NO_3)_3{\bullet}6H_2O$ as a starting material with $Tb(NO_3)_3{\bullet}6H_2O$ in our preparation procedures for the pure Eu complexes [3-5]. Identification was done by the agreement of their IR spectra with those of the pure Eu complexes. Interlanthanide substitutions for Tb/Eu-mixed complexes **4-6** were easily accomplished, and the resulting Tb/Eu-mixed complexes have stoichiometries which were, conveniently, close to the original composition of $Tb(NO_3)_3/Eu(NO_3)_3$ mixture. Energy dispersive X-ray (EDX) spectrometry analysis was performed for the determination of the atomic ratio of Tb/Eu on a JEOL JSX-3200 instrument. The IR spectra of **4-6** were consistent with those of the pure Eu complexes, too.

Diffuse reflectance and IR spectra were recorded on Hitachi 330 and JASCO FT/IR-5000 spectrophotometers at room temperature, respectively. Luminescence and excitation spectra of the sample powder pellet were obtained using a lock-in (NF L1-574) technique. The sample pellet (with a thickness of about 1 mm and a diameter of 10 mm) was prepared by pressing the sample powder under 3 x 10^7 Pa. The light source for the photoluminescence measurements was a Continuum 90300 YAG (355 nm, 400 mJ per pulse) laser, a 500-W xenon lamp (in a combination with a Nikon G-25 grating monochromator), or a LDL 20505 LAS dye laser (with LDC 480 dye, for the 459-510 nm wavelengths) pumped by a Questek 2320 XeCl (308 nm, 50 mJ per pulse) laser. The 488-nm light excitation of the Tb/Eu-mixed complexes was carried out by a 50 ns pulse of 5 mJ photons from the above dye laser. The luminescence was collected at an angle of 90° to the exciting light and focussed onto the entrance slit of a Spex 750M spectrometer (for high-resolution) or a Nikon G-25 grating monochromator which was equipped with Hamamatsu Photonix R636 photomultiplier tube. An absolute wave-number accuracy of ±2 cm^{-1} for the high-resolution luminescence spectra was estimated from the dye-laser allignment. Luminescence at low temperatures was measured using an Oxford Instruments CF 204 cryostat. The time profiles of the luminescence was measured on a LeCloy 9361 digital storage oscilloscope. No observable part of the original intensity of the incident light was transmitted through the sample pellet.

3. Luminescence Spectra

The luminescence spectra of the pure Tb complexes **1- 3** under 355-nm light irradiation consist of transitions of $^5D_4 \rightarrow ^7F_J$ around 488, 545, 584, 624, 650, 668, and 682 nm for J=6, 5, 4, 3, 2, 1, and 0 respectively, and the excitation spectra for the strongest $^5D_4 \rightarrow ^7F_5$ lines of the Tb^{3+} emission consist of $^7F_6 \rightarrow ^5D_4$ lines (around 488 nm), complicated $^7F_6 \rightarrow ^5D_{3,2}$ and $^5L_{10}$ lines (in the range 320-380 nm), and O→M (M=W,

Nb, or Mo) lmct bands (at <330 nm). Emission and excitation spectra at 4.2 and 300 K for **2** are shown in Figure 2, where the 300-K spectra are not fully resolved.

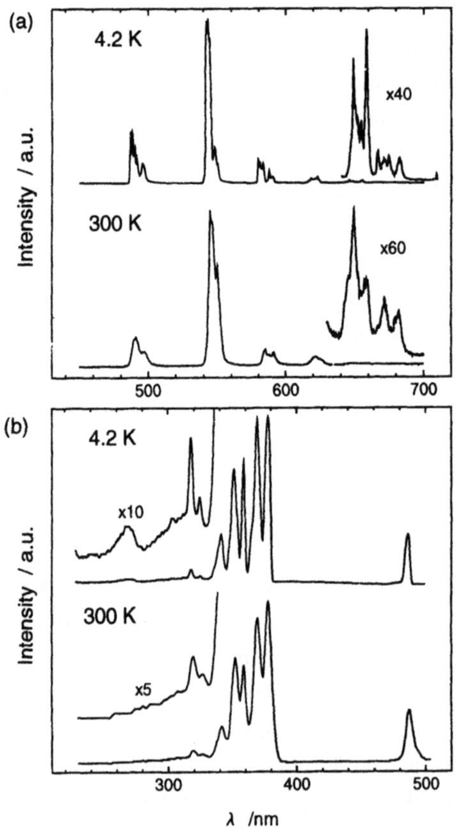

Fig. 2. Photoluminescence spectra (a) observed under the 355-nm light irradiation and excitation spectra (b) for the $^5D_4 \rightarrow ^7F_5$ transition (at 545 nm) of $Na_7H_{19}\{[Tb_3O(OH)_3(H_2O)_3]_2Al_2(Nb_6O_{19})_5\} \cdot 47H_2O$ (**2**) at 4.2 and 300 K.

The relative intensity ratio among f-f transitions for **1**-**3** hardly depends on temperature. The most remarkable difference in the luminescence spectra among **1**-**3** is a small distribution of both the $^7F_6 \rightarrow ^5D_4$ lines and the broad O→Nb lmct bands (at 250-350 nm) in the excitation spectra of **2**. The small contribution of the O→M lmct states to the excitation spectrum for **2** indicates a low yield of the intramolecular energy transfer from O→Nb lmct bands to Tb^{3+} in **2**, due to the small disparity between the electronic configurations of the excited and ground states of the highly symmetric $[Nb_6O_{19}]^{8-}$ ligands as discussed previously for the pure Eu complexes [2b]. In the excitation spectra of the pure Eu complexes the $^7F_1 \rightarrow ^5D_{0,1,2,3}$ lines due to the thermal population of $^7F_1$ state were developed with an accompanying reduction of the intensity of the $^7F_0 \rightarrow ^5D_{0,1,2,3}$ lines at high temperatues T>77K.[2b,3] In contrast to the pure Eu complexes none of excitation spectra of **1-3** exhibits hot lines ($^7F_5 \rightarrow ^5D_{3,2}$ and $^5L_{10}$) of the $^7F_5$ state of Tb^{3+}, since the large energy gap (about 2000 cm^{-1}) between the $^7F_6$ and $^7F_5$ states for Tb^{3+} compared to the case (about 300 cm^{-1} between the $^7F_0$ and $^7F_1$ states) for Eu^{3+} makes the thermal population (at T<300 K) of the $^7F_5$ state prohibitive.

Upon Tb^{3+} excitation (by the $^7F_6 \rightarrow ^5D_4$ transition) at 488 nm, the Tb/Eu-mixed complexes **4-6** exhibited both Tb^{3+} $^5D_4 \rightarrow ^7F_J$ (J=5-0) and Eu^{3+} $^5D_0 \rightarrow ^7F_J$ (J=0-4) lines in the region 540-720 nm. There was no observation of the Eu^{3+} $^5D_1 \rightarrow ^7F_J$ lines under this excitation at 488 nm. Figure 3 exemplifies the excitation spectra of **4** at 4.2 and 300 K for the Tb^{3+} $^5D_4 \rightarrow ^7F_5$ and Eu^{3+} $^5D_0 \rightarrow ^7F_2$ emissions.

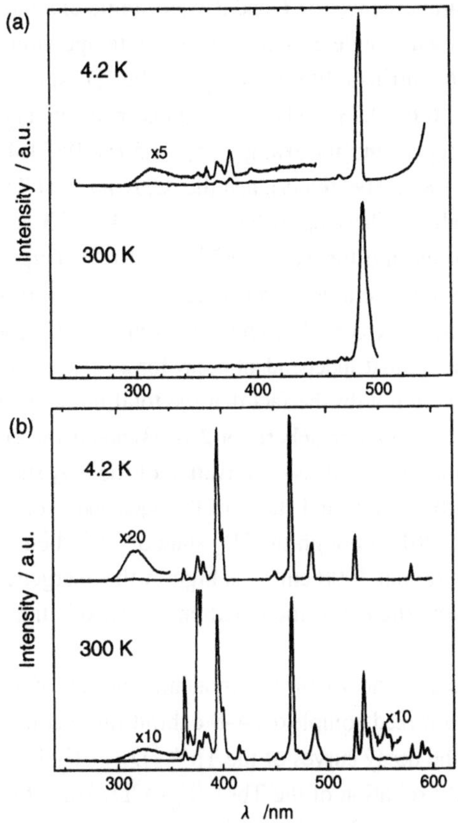

Fig. 3. Excitation spectra of $K_{15}H_3[Tb_{1.4}Eu_{1.6}(H_2O)_3(SbW_9O_{33})(W_5O_{18})_3]\cdot 25.5H_2O$ (**4**) at 4.2 and 300 K for the Tb^{3+} $^5D_4 \rightarrow ^7F_5$ (a) and Eu^{3+} $^5D_0 \rightarrow ^7F_2$ (b) emissions.

The excitation spectrum for the Eu^{3+} emission indicates the contribution of both the O→M (M=W, Nb, or Mo) lmct bands and the f-f transitions of Tb^{3+} in addition to the direct f-f transitions of Eu^{3+}, while the excitation spectrum for the Tb^{3+} emission indicates little contribution of the Eu^{3+} f-f transitions. This proves that there is energy transfer from Tb^{3+} to Eu^{3+} not from Eu^{3+} to Tb^{3+} in the polyoxometallolanthanoate lattices. The 300-K excitation spectrum of the Eu^{3+} $^5D_0 \rightarrow ^7F_2$ emission shows the very weak $^7F_2 \rightarrow ^5D_1$ lines at 554 nm, which were little observed at T≤200 K, in addition to the $^7F_1 \rightarrow ^5D_{0,1,2,3}$ lines at 589.5 and 594.5, 534.0 and 539.0, 471.0 and 474.5, and 415.0 and 418.5 nm. The observation of the hot lines due to the population of the $^7F_2$ state is ascribed to the relatively small gap (about 1000 cm^{-1}) between the $^7F_2$ and $^7F_0$ states.

4. Luminescence Decay Dynamics

The decay patterns of the $^5D_4 \rightarrow ^7F_J$ luminescence for **1** and **2** were single exponentials at all temperatures, while the decay for **3** was single exponential at low temperatures T≤50 K but nonexponential at T>50 K. The intrinsic lifetimes (τ_D) of the $^5D_4$ state for **1-3** at T≤50 K are in the range 1.45–1.64, 0.63-0.73, and 0.46-0.50 ms, respectively. The lifetime for **1** is slightly temperature-dependent [for example, τ_D=1.59±0.05, 1.42±0.05, and 1.30±0.06 ms, at 4.2, 77, and 300 K, respectively], and the decay for **3** at T>50 K strongly increases with nonexponentiality. These decay behaviors for **1** and **3** are associated with the nonradiative transition into the $Tb^{3+} \rightarrow W^{6+}$ (or Mo^{6+}) charge-transfer state, as demonstrated for the significant temperature dependence of the $^5D_4 \rightarrow ^7F_J$ decay for $K_3Na_4H_2[Tb(W_5O_{18})_2] \cdot 20H_2O$ (with τ_D=3.8, 2.7, and 1.7 ms at 4.5, 77, and 300 K, respectively) [8,9] in contrast to the almost independent decay for **2**. τ_D value for **2** and **3** is smaller than for **1**. In conjunction with the fact that the total number (*n*) of aqua and hydroxo ligands in the Tb^{3+} coordination sphere for **2** (*n*=3) and **3** (*n*=4) is larger than for **1**, this is predicted by the radiationless deactivation of the $^5D_4$ state through weak vibronic coupling with the vibrational states of the aqua and hydroxo ligands' high-frequency OH oscillators [2b], although the $^5D_4$ state of Tb^{3+} being approximately one OH-phonon energy (about 3000 cm^{-1}) as high in the energetic position as the $^5D_0$ state of Eu^{3+} is less efficient in the radiationless deexcitation by the OH oscillators [10].

While the pure Tb complexes **1-3** show simple exponential decays of Tb^{3+} emission (at T<50 K for **3**), the Tb/Eu mixed complexes **4-6** exhibit more complicated decay patterns. Figure 4 exemplifies the decay curves of the Tb^{3+} $^5D_4 \rightarrow ^7F_5$ line and the Eu^{3+} $^5D_0 \rightarrow ^7F_2$ line for **4** after pulsed excitation of the Tb^{3+} $^7F_6 \rightarrow ^5D_4$ transition (at 488 nm) at 4.2 K. The emission of Eu^{3+} at 4.2 K shows clearly a slow rise from zero to a maximum and then decays, which can be rationalized as a population of the $^5D_0$ state of Eu^{3+} by energy transfer from the $^5D_4 \rightarrow ^7F_J$ emission of Tb^{3+}, the $^5D_0$ state decay being now governed by the time behavior of the transfer process. At long times after the excitation pulse the decay curve of Eu^{3+} for **4** becomes exponential with

1.3±0.1 ms which is in agreement with the value (1.1±0.2) for the $^5D_0 \rightarrow ^7F_J$ emission of Eu^{3+} for the pure Eu complex.[2b]. The decay of the $^5D_4$ state for the Tb/Eu-mixed complexes **4-6** are nonexponential at all temperatures and increases with an accompanying faster buildup of the Eu^{3+} $^5D_0$ state when the temperature increases. Figure 5 shows the semilogarithmic decay curves of the $^5D_4 \rightarrow ^7F_5$ line for 4 after 488-nm light pulse irradiation. The decay curves presented in Figure 5 gives the most convincing demonstration that the Tb-Eu transfer rate increases with increasing temperature.

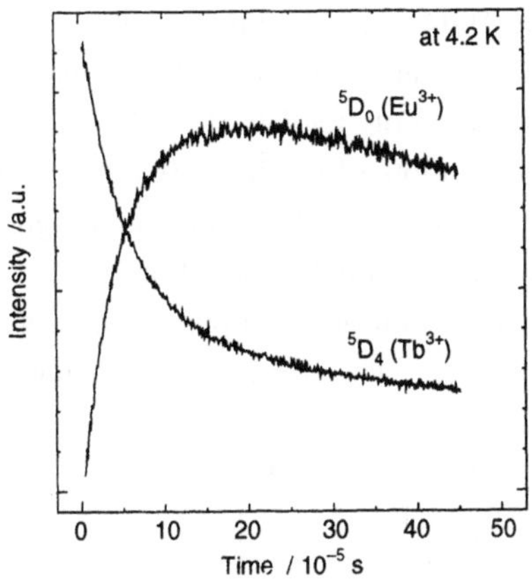

Fig. 4. Decay curves of the Tb^{3+} $^5D_4$ state and the Eu^{3+} $^5D_0$ state for $K_{15}H_3[Tb_{1.4}Eu_{1.6}(H_2O)_3(SbW_9O_{33})(W_5O_{18})_3]{\cdot}25.5H_2O$ (**4**) after pulsed excitation of the Tb^{3+} $^7F_6 \rightarrow ^5D_4$ transition at 4.2 K.

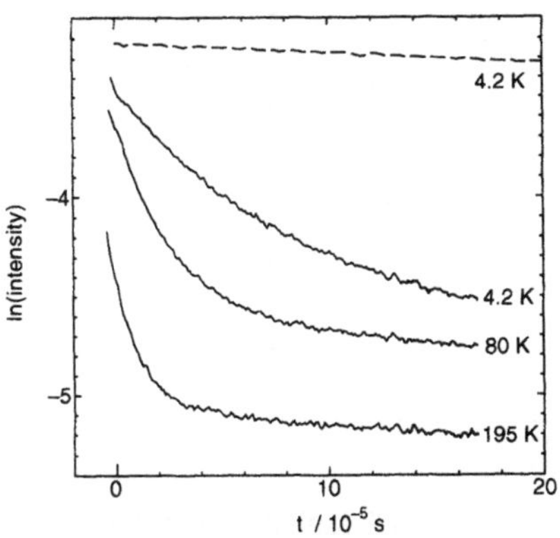

Fig. 5. Semilogarithmic decay curves of the Tb^{3+} $^5D_4 \rightarrow ^7F_5$ emission after pulsed excitation at 488 nm at a variety of temperatures for $K_{15}H_3[Tb_{1.4}Eu_{1.6}(H_2O)_3(SbW_9O_{33})(W_5O_{18})_3]{\cdot}25.5H_2O$ (**4**). Dashed line indicates the exponential decays for $K_{15}H_3[Tb_3(H_2O)_3(SbW_9O_{33})(W_5O_{18})_3]{\cdot}25.5H_2O$ (**1**).

5. Förster-Dexter-type Energy Transfer from Tb^{3+} to Eu^{3+}

The above results enable to be analyzed in terms of the Förster-Dexter theory for multipolar interaction between donor and acceptor [11]. Since both the Tb-Tb transfer and the back transfer from Eu^{3+} to Tb^{3+} at this temperature are negligible, the intensity $I(t)$ of the Tb^{3+} emission after pulsed excitation can be described by Inokuti-Hirayama model [12,13],

$$\ln I(t) + \frac{t}{\tau_D} = -\frac{4}{3}\pi^{3/2} n_A r_{DA}{}^3 (k_{ET} t)^{3/s} \qquad (1)$$

where n_A is the density of Eu^{3+} and k_{ET} is the nearest-neighbor transfer rate between a Tb-Eu pair at central cavity of the Tb/Eu-mixed complex, with separation r_{DA}. Also, s is 6, 8, or 10 depending on the multipolar nature of the Tb-Eu interaction, and τ_D (=1.59±0.05, 0.68±0.05, and 0.51±0.05 ms for **4**-**6** at 4.2K, respectively) is again the intrinsic decay time. Figure 6 shows plots of the $^5D_4 \rightarrow {}^7F_5$ emission decays of **4**-**6** in the forms of $\ln I(t) + t/\tau_D$ against $t^{3/s}$ for s=6 (which is used throughout this study), corresponding to an electric dipole-electric dipole interaction.

The approximate straight-line behavior indicates that our assumptions above are correct and that the choice of s=6 as a valid one. The effect of the lanthanide contraction on the

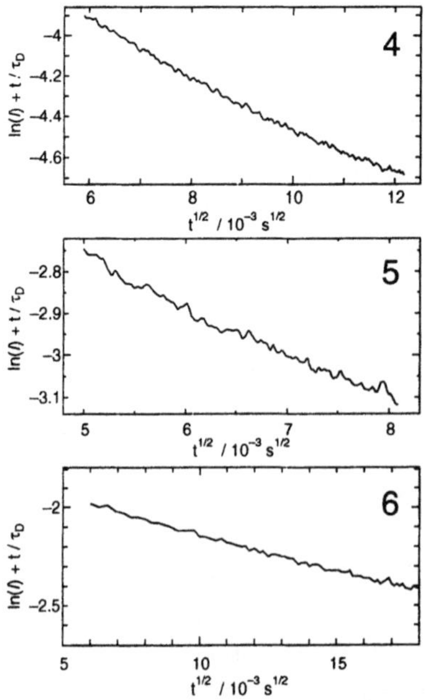

Fig. 6. The Tb^{3+} emission-decay data at 4.2 K plotted according to the Inokuti-Hirayama model for the electric dipole-dipole Tb-Eu interaction for $K_{15}H_3[Tb_{1.4}Eu_{1.6}(H_2O)_3(SbW_9O_{33})(W_5O_{18})_3]\cdot 25.5H_2O$ (**4**), $Na_7H_{19}\{[Tb_{4.3}Eu_{1.7}O_2(OH)_6(H_2O)_6Al_2(Nb_6O_{19})_5\}\cdot 47H_2O$ (**5**), and $[NH_4]_{12}H_2[Tb_{3.1}Eu_{0.9}(MoO_4)(H_2O)_{16}(Mo_7O_{24})_4]\cdot 13H_2O$ (**6**).

intermetalic distances between the isostructural anion species of Tb and Eu complexes is small, as implied by the small difference in mean Ln-O distances (2.42 and 2.43 Å) between $K_3Na_4H_2[TbW_{10}O_{36}]\cdot 20H_2O$ and $Na_9[EuW_{10}O_{36}]\cdot 32H_2O$, respctively [14]. Therefore, the r_{DA} value for the Tb/Eu-mixed complex is estimated to be close to the nearest Eu•••Eu distance for the pure Eu complex. For **4**, at n_A=5.46 x 10^{-4} ions/Å^3 [=1.6 x 4 /11680, where 4 and 11680 are number (Z) of molecules in unit cell and cell volume (in Å^3), respectively),[3] r_{DA} = 5.05 Å, and from the slope (=110 $s^{-1/2}$) in Figure 6 we find k_{ET} =(4.49±0.81) x 10^4 s^{-1}. From this the donor-acceptor interaction parameter (α) is calculated to be (7.45±0.13) x 10^8 (=k_{ET}•5.05^6) Å^{6}s^{-1}, and the critical radius (R_0) for the Tb-Eu energy transfer by making use of $R_0=(\tau_D\cdot\alpha)^{1/6}$, at which energy transfer and radiative decay of the donor (=Tb^{3+}) have the same probability, is obtained as R_0=10.3±0.2 Å. More distant neighbors for Tb^{3+} are associated with the intermolecular distance at 11.2 Å (twice) [3]. R_0=10.3 Å for **4** indicates that Tb^{3+} at 4.2 K enables to transfer its excitation energy to the Eu^{3+} ions which occupy either site of the two nearest neighbors within a sphere of radius of 10.3 Å. As shown by the crystal structure of the pure Eu complex in Figures 1(b) and (c), there are five nearest neighbors at distance 3.76 Å (twice), 4.69 Å (twice), and 6.00 Å (one) in **5**, and three nearest neighbors at distance 6.17 Å (twice) and 7.51 Å (one) in **6**. A similar treatment for the shortest Tb•••Eu distance yields k_{ET}=(4.68±0.90) x 10^5 s^{-1}, α=(1.31±0.25) x 10^9 Å^6 s^{-1} (=k_{ET}•3.76^6), and R_0=10.0±0.3 Å. for **5** at 4.2 K, and k_{ET}=(4.90±0.90) x 10^3 s^{-1}, α=(2.70±0.51) x 10^8 Å^6 s^{-1} (=k_{ET}•6.17^6), and R_0=7.2±0.5 Å. for **6**. The R_0 values calculated for **5** and **6** indicate that the transfer of the Tb^{3+} excitation energy to the Eu^{3+} ions occurs almost within a central cavity of the molecule, if we consider the nearest neighboring intermolecular Tb•••Eu distances of 11.0Å (twice) [4] for **5**, and 8.61 Å for **6**. The next nearest Tb•••Eu distances are 4.69 Å in the central $[Tb_{4.3}Eu_{1.7}O_2(OH)_6(H_2O)_6]^{8+}$ core for **5**, and 7.51 Å in the central $[Tb_{3.1}Eu_{0.9}(MoO_4)(H_2O)_{16}]^{10+}$ core for **6**. This allows us to estimate the rate of the electric dipole-dipole energy transfer to next nearest neighbors to be approximately 1/4 [=$(3.76/4.69)^6$] and 3/10 [=$(6.17/7.51)^6$) smaller than for the shortest Tb••• Eu distances of 3.76 and 6.17 Å for **5** and **6**, respectively. It is possible to fit the decay curves of the $^5D_4\rightarrow{}^7F_5$ emission at a variety of temperatures to the formula (with s=6) (1). Figure 7 shows plots of the calculated values of k_{ET} against temperature for **4**-**6**. As the temperature increases, k_{ET} for **4**-**6** increases, and the increasing effect for **5** is the largest.

The basic requirement for a Förster-Dexter energy transfer to occur is the spectral overlap of donor emission and acceptor absorption lines [12]. Figure 8 exemplifies energy situations of the emission and excitation line-peaks of Tb^{3+} (left) and Eu^{3+} (right) in the range of less than 488-nm excitation energy for **5**, which are estimated by the 4.2 and 300 K lines. In Figure 8 the excitation lines due to the $^7F_1$ and $^7F_2$ states, which are involved in the energy transfer at high temperatures, are shaded. The linewidths of the Tb^{3+}-emission lines are in the range of 40 to 100 cm^{-1} at half

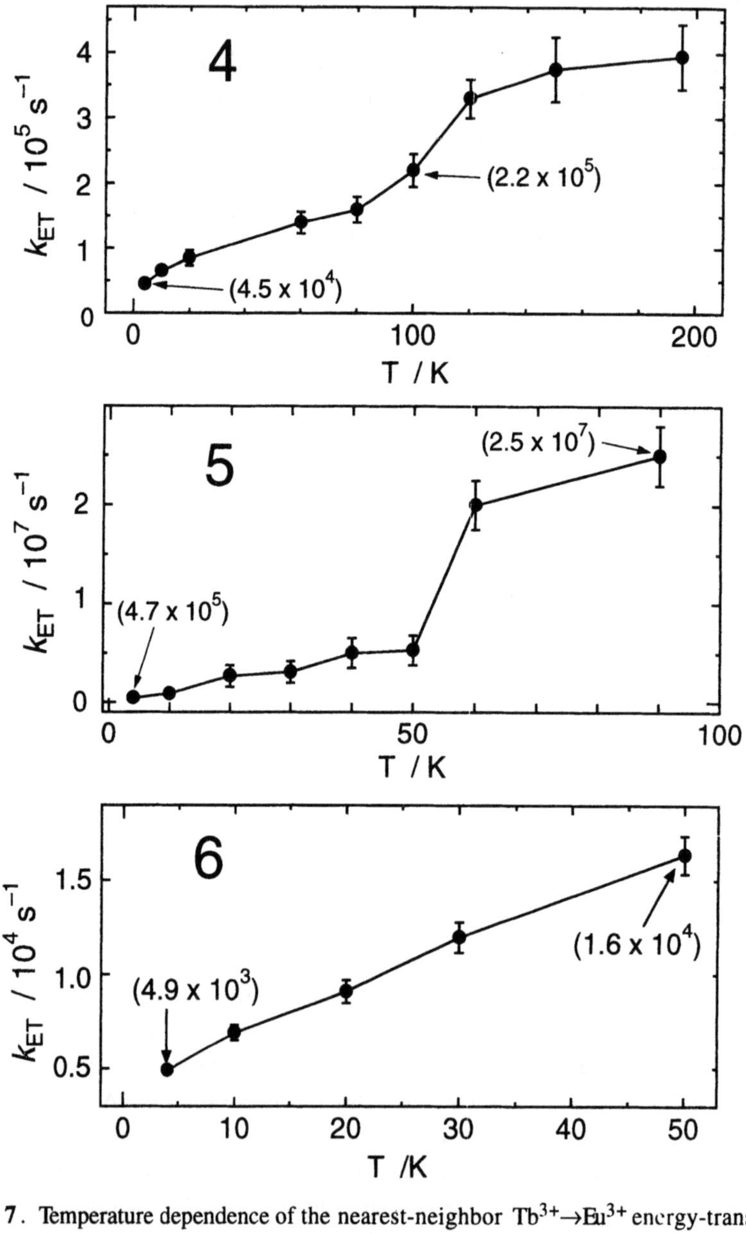

Fig. 7. Temperature dependence of the nearest-neighbor $Tb^{3+}\rightarrow Eu^{3+}$ energy-transfer rate (k_{ET}) for $K_{15}H_3[Tb_{1.4}Eu_{1.6}(H_2O)_3(SbW_9O_{33})(W_5O_{18})_3]\cdot 25.5H_2O$ (**4**). $Na_7H_{19}\{[Tb_{4.3}Eu_{1.7}O_2(OH)_6(H_2O)_6Al_2(Nb_6O_{19})_5\}\cdot 47H_2O$ (**5**), and $[NH_4]_{12}H_2[Tb_{3.1}Eu_{0.9}(MoO_4)(H_2O)_{16}(Mo_7O_{24})_4]\cdot 13H_2O$ (**6**). Numbers in parentheses indicate average values of k_{ET} at selected temperatures.

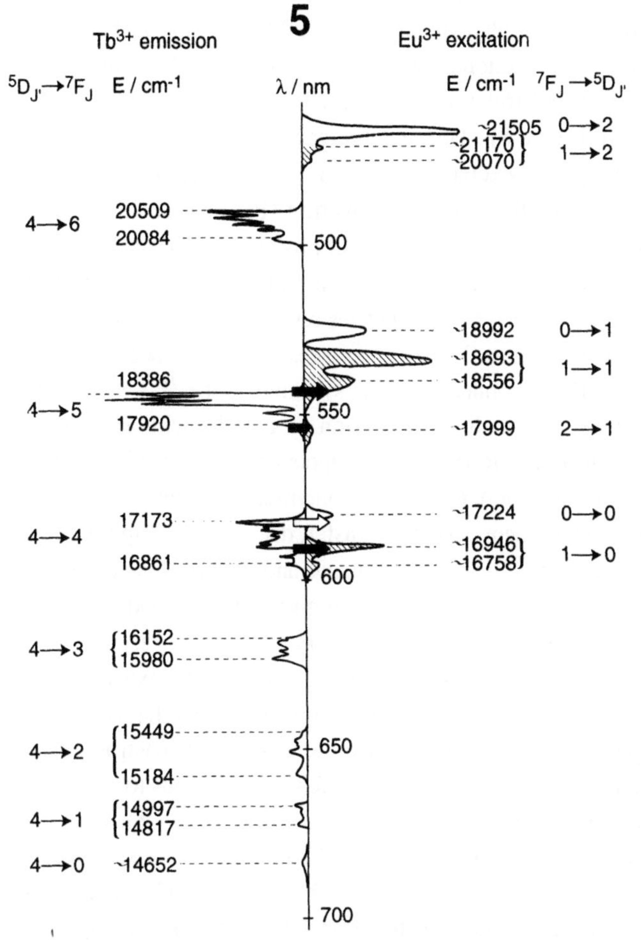

Fig. 8. Energetic comparison of the emission and excitation line-peaks for Tb^{3+} and Eu^{3+} in $Na_7H_{19}\{[Tb_{4.3}Eu_{1.7}O_2(OH)_6(H_2O)_6Al_2(Nb_6O_{19})_5\}\cdot 47H_2O$ (**5**). Excitation lines appearing at high temperature are shaded. The light arrow indicates energy transfer at low temperatures and the heavy arrows additional channels of energy transfer at high temperatures.

height. At low temperatures T<100 K where the population of the $^7F_2$ state is ignored, the population of the $^7F_1$ manifold of Eu^{3+} is low and the excitation transfer occurs mainly through the $^7F_0$ level. Therefore, it becomes evident that the condition for energy transfer is fulfilled around 17150 cm^{-1} (583 nm), where the $^5D_4 \rightarrow {}^7F_4$ line emission of Tb^{3+} around 17173 cm^{-1} overlaps with the $^7F_0 \rightarrow {}^5D_0$ excitation of Eu^{3+} around 17241 cm^{-1} (see a light arrow in Figure 9, symbolizing the utransfer channel). At high temperatures the transfer channels obviously increase due to the thermal population of $^7F_1$ and $^7F_2$ states of Eu^{3+}; an overlap between the $^5D_4 \rightarrow {}^7F_5$ line emission and the $^7F_1 \rightarrow {}^5D_1$ (and $^7F_2 \rightarrow {}^5D_1$ line around 17900 cm^{-1} only at T>200K) line excitation around 18600 cm^{-1} with addtional overlap of the $^5D_4 \rightarrow {}^7F_4$ line emission with the $^7F_1 \rightarrow {}^5D_0$ line excitation around 16960 cm^{-1} (see the heavy arrows as additional channels for energy transfer in Figure 8). Hence, large values of k_{ET} at high temperatures are expected after pulsed excitation of the $^7F_6 \rightarrow {}^5D_4$ line, as shown in Figure 7. The involvement of both the forbidden transition $^7F_0 \rightarrow {}^5D_0$ and the magnetic dipole transition $^7F_1 \rightarrow {}^5D_0$ in the Tb-Eu energy transfer process is associated with the low local symmetry at Tb^{3+} and Eu^{3+} sites in **4** -**6**. The LnO_8 coordination polyhedra for **4**-**6** approximately have C_s, C_1, and C_s symmetries respectively, because of distortions in the bond length and angles (Figure 1) as a result of the coordination of both aqua and hydroxo ligands [3-5]. Under such low symmetries the $^7F_J \leftrightarrow {}^5D_0$ (J=0,1) transitions have a nonvanishing electric-dipole transition probability with a resultant Förster-Dexter-type energy transfer due to the electric dipole-dipole interaction. Thereby, the decay behavior of Tb^{3+} emission in **4**-**6** is nonexponential, indicating that donor-donor transfer is much slower than donor-acceptor transfer, due to the disruption of the resonance energy transfer between Tb^{3+} ions by both aqua and hydroxo ligands [15,16]. Lower-symmetry crystal field of **5** enhances the transition probabilities and broadens the excitation and emission lines for both Tb^{3+} and Eu^{3+}. In addition, there seem to be spectroscopically inequivalent emission sites for **5**, as suggested by the existence of at least three Eu sites for the pure Eu complex [2b]. In conjunction with the fact that the pure Eu complexes corresponding to **4** and **6** indicated the presence of spectroscopically single Eu^{3+} site for each complex [2b,3], the structural feature of **5** leads to a faster donor-acceptor transfer rate compared to **4** and **6** (Figure 7). The linewidths of the Tb^{3+}-emission lines increase slightly with temperature. This will increase the spectral overlap between the Tb^{3+} $^5D_4 \rightarrow {}^7F_4$ emission lines and the Eu^{3+} $^7F_0 \rightarrow {}^5D_0$ excitation line, which is responsible for a faster donor-acceptor transfer rate with increasing temperature below 100 K.

6. Host Lattice Excitation and Energy Transfer

As above discussed, the $Tb^{3+} \rightarrow Eu^{3+}$ transfer channels at high temperatures indicate an increasing contribution of the Eu^{3+} $^5D_1$ state to the Eu^{3+} emission process. There was no observable $^5D_1 \rightarrow {}^7F_J$ emission for **4** and **5**, and the corresponding pure Eu complexes [2b,3]. Therefore, it is clear that in **4** and **5** the decay of the $^5D_1$ state is governed by fast nonradiative processes via a cross relaxation with a resultant deactivation to the $^5D_0$ state which leads to the $^5D_0 \rightarrow {}^7F_J$ emission. It should be noted that the $^5D_1 \rightarrow {}^7F_J$ emission is observed for **6** and its pure complex [17]. Figure 9 shows the spectrum (a) and decays (b) of the $^5D_1 \rightarrow {}^7F_J$ emission for **6** under the 355-nm light irradiation at 4.2 K. The $^5D_1 \rightarrow {}^7F_J$ emission spectrum at 4.2 K exhibits the $^5D_1 \rightarrow {}^7F_{0,1,2}$ lines around 526, 536, and 556 nm with very weak intensities. The decay of the $^5D_1 \rightarrow {}^7F_J$ emission was exponential, and the lifetime (13-8 μs) of the $^5D_1$ state decreased with increasing

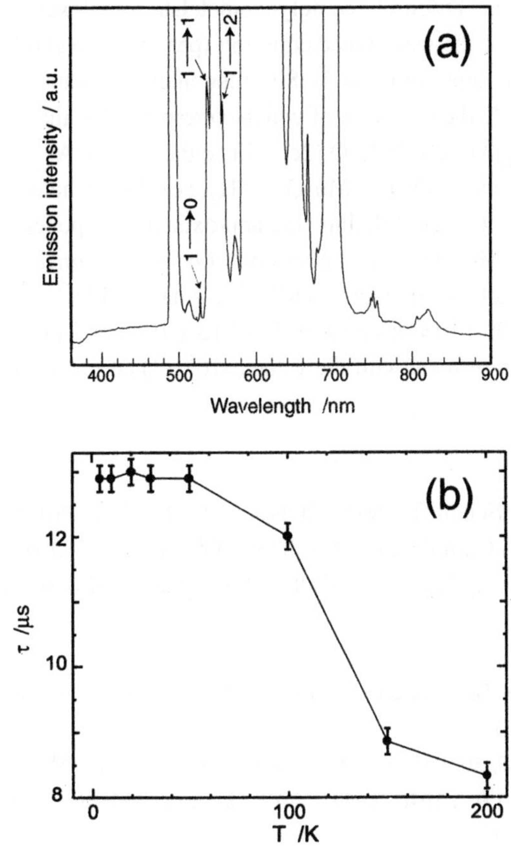

Fig. 9. Photoluminescence spectrum (a) and decays (b) of the $^5D_1\rightarrow^7F_J$ emission for $[NH_4]_{12}H_2[Tb_{3.1}Eu_{0.9}(MoO_4)(H_2O)_{16}(Mo_7O_{24})_4]\cdot13H_2O$ (**6**) under the 355-nm light irradiation at 4.2 K.

temperature in the 4.2-200 K range as shown in Figure 9(b) where the exponential decay of the $^5D_1\rightarrow^7F_1$ emission around 536 nm was measured as a function of the temperature. Since the lifetime of the $^5D_1$ state for **6** is identical with that for the pure Eu^{3+} complex [17], it is evident that the emission from the $^5D_1$ state for **6** is obtained mainly by exciting the O→Mo lmct bands: If excitation is transfered from a long-lived (0.5 ms at 4.2 K) manifold ($^5D_4$) of Tb^{3+} to a a short-lived (13 μs at 4.2 K) manifold ($^5D_1$) of Eu^{3+} with a resultant $^5D_1\rightarrow^7F_J$ emission, the $^5D_1$ lifetime value obtained by exciting the $^5D_4$ state should be practically identical to that of $^5D_4$. Thus, it is deduced that the O→M lmct excitation for the Tb/Eu-mixed complexes leads to the excitation of both Tb^{3+} $^5D_4$ and Eu^{3+} $^5D_1$ states, as a result of the Förster-Dexter resonance energy transfer, since the energy transfer from a broad band O→M lmct triplet emitter to a line absorber is possible for the present Tb/Eu-mixed polyoxometalate complexes. Figure 9(a) also shows the observation of the weak broad emission (peaking around 680 nm) of the O→M lmct triplet state with approximately 1/280 of the intensity against total f-f emissions of both Eu^{3+} and Tb^{3+}, the spectrum of which corresponded to the $^3T_{1u}$→$^1A_{1g}$ transition of $[NH_4]_6[Mo_7O_{24}]{\bullet}4H_2O$ [2c]. Since the $^3T_{1u}$→$^1A_{1g}$ emission due to the O→Mo lmct triplet state of $[NH_4]_6[Mo_7O_{24}]{\bullet}4H_2O$ exhibited the two exponential decays (6 + 15 μs at 4.2 K) [2c], thus, the approximately exponential decay (~0.23 μs) measured for the O→Mo lmct triplet emission of **6** let us estimate the energy transfer rate from the O→Mo lmct triplet state to Tb^{3+} (^7F_J→$^5D_4$) and Eu^{3+} ($^7F_0$→$^5D_{0,1}$) sites to be 4.3 x 10^6 s^{-1} [=1/(0.23 x 10^{-6}) - 1/(6 x 10^{-6}) or 1/(15 x 10^{-6})]. The estimated value is in good agreement with the value (>10^6 s^{-1}) proposed for the pure Eu complexes [2b].

Acknowledgement

I acknowledge Grants-in-Aid for Scientific Research, No. 06241104 (for "Priority Area, New Development of Rare Earth Complexes"), No. 09354009, and No. 10304055 from the Ministry of Education, Science, Sports, and Culture for support of this work.

References

1. (a) R. C. Powell and G. Blasse, *Structure and Bonding*, **42**, 43 (1980). (b) G. Blasse, *J. Chem. Phys.*, **45**, 2350 (1966).
2. (a) T. Yamase, *Chem. Rev.*, **98**, 307 (1998). (b) T. Yamase, T. Kobayashi, M. Sugeta, and H. Naruke, *J. Phys. Chem. A*, **101**, 5046 (1997). (c) T. Yamase and M. Sugeta, *J. Chem. Soc. Dalton Trans.*, 759 (1993).
3. T. Yamase, H. Naruke, and Y. Sasaki, *J. Chem. Soc. Dalton Trans.*, 1687 (1990).
4. T. Ozeki, T. Yamase, H. Naruke, and Y. Sasaki, *Inorg. Chem.*, **33**, 409 (1994).
5. H. Naruke, T. Ozeki, and T. Yamase, *Acta Crystallogr.*, **C47**, 489 (1991).
6. H. Naruke and T. Yamase, *J. Alloys Compd.*, **268**, 100 (1998). (b) H. Naruke and T. Yamase, *J. Alloys Compd.*, **255**, 183 (1997). (c) H. Naruke and T. Yamase, *Acta Crystallogr.*, **C52**, 2655 (1996).
7. See, e.g., (a) J.-C. G. Bünzli, In *Lnthanide Probes in Life, Chemical and Earth Sciences, Theory and Practice*; J.-C.-G. Bünzli and G. R. Choppin, Eds., Elsevier Publishing Co.: Amsterdam, 1989; chapter 7 and references therein. (b) G. Blasse and B. C. Grabmaier, In *Luminescent Materials*; Springer-Verlag: Berlin, Heidelberg, 1994; chapter 5 and references therein.
8. T. Ozeki and T. Yamase, *J. Alloys Compd.*, **192**, 28 (1993).
9. (a) G. Blasse, J. Dirksen, and Z. Zonnevijlle, *Chem. Phys. Lett.*, **83**, 449 (1981). (b) C. W. Stuck and W. H. Fonger, *J. Chem. Phys.*, **64**, 1784 (1976).
10. (a) W. D. Horrocks, Jr. and D. Sudnick, *Acc. Chem. Res.*, **14**, 384 (1981). (b) W. D. Horrocks, Jr. and D. Sudnick, *Science*, **206**, 1194 (1979).
11. T. Yamase and H. Naruke, *J. Phys. Chem. B*, **103** (1999), in press.

12. (a) T. Förster, *Ann. Phys.*, *2*, 55 (1948). (b) D. L. Dexter, *J. Chem. Phys.*, **21**, 836 (1953).

13. (a) H. Inokuti and F. Hirayama, *J. Chem. Phys.*, **43**, 1978 (1965). (b) J. Hegarty, GD. L. Huber, and W. M. Yen, *Phys. Rev. B*, **23**, 6271 (1981).

14. (a) T. Ozeki and T. Yamase, *Acta Crystallogr.*, **B50**, 128 (1994). (b) M. Sugeta and T. Yamase, *Bull. Chem. Soc. Jpn.*, **66**, 444 (1993). (c) T. Ozeki, M. Takahashi, and T. Yamase, *Acta Crystallogr.*, **C48**, 1370 (1992).

15. I. A. Kahwa, C. C. Parkes, and G. L. McPherson, *Phys. Rev. B*, **52**, 51 (1995).

16. (a) E. Moret, J.-C. G. Bünzli, and K. Schenk, *Inorg. Chim. Acta*, **178**, 83 (1990). (b) G. Blasse, *Inorg. Chim. Acta*, **169**, 33 (1990). (c) G. Blasse and L. H. Brixner, *Inorg. Chim. Acta*, **169**, 25 (1990).

17. H. Naruke and T. Yamase, *J. Lumin.*, **50**, 287 (1986).

CONDUCTING AND MAGNETIC ORGANIC / INORGANIC MOLECULAR MATERIALS BASED ON POLYOXOMETALATES.

Lahcène OUAHAB[a*], Stéphane GOLHEN[a], Smaïl TRIKI[b]

[a]Laboratoire de Chimie du Solide et Inorganique Moléculaire UMR CNRS 6511. Groupe Matériaux Moléculaires, Université de Rennes 1. Avenue du Général Leclerc, 35042 Rennes cedex, France.
[b]UMR CNRS 6521. Université de Bretagne Occidentale, 29285 Brest cedex, France

E-Mail : Ouahab@univ-rennes1.fr

Abstract:

The potentialities of the use of polyoxometalates as inorganic components in conducting and/or magnetic organic/inorganic hybrid molecular materials are illustrated with few examples of their chemical and electrochemical assemblies with organic donors derived from TTF and BEDT-TTF, nitronyl nitroxide and metallocenium radical cations.

Keywords: Polyoxometalates, electrical conductivity, magnetic properties, organic donor, inorganic acceptor, metallocenium, nitronyl nitroxide radicals cations.

1. Introduction

Polyoxometalates constitute a large class of compounds with a considerably rich topology associated with interesting chemical and physical properties. They are of great interests in different fields such as catalysis, medicine, material sciences...[1-5]. In this contribution we focus on the use of these molecular metal oxide clusters in conducting and magnetic molecular materials [6]. These materials are formed by chemical or electrochemical assemblies of polyoxometalates with organic and organometallic radical cations such as organic donors derived from TTF (tetrathiafulvalene), nitronyl nitroxide and metallocenium radical cations (see scheme 1).

The first report on TTF charge transfer salt containing a Keggin polyoxometalate as inorganic acceptor component was published by our group in 1988 [7]. In that publication we indicated that the aim of our work is to prepare materials with *mixed-valence* state on the organic and inorganic parts in order to obtain materials with conducting and magnetic properties. We presented TMTSF and BEDT-TTF materials containing such polyoxoanions at the NATO Advanced Study Institute meeting (Spetses

M.T. Pope and A. Müller (eds.), Polyoxometalate Chemistry, 205–229.

Island, Greece) in 1989. These compounds, namely $(TMTSF)_3[PW_{12}O_{40}]$ and $(BEDT\text{-}TTF)_3[PW_{12}O_{40}]$, were published in 1991 in the proceedings devoted to that meeting [6a]. We published the second report dealing with TTF derivative charge transfer salts with Lindquist polyoxoanions in 1989 [8]. In the same time, Davidson *et al.* reported the BEDT-TTF salt with a Keggin polyanion [9], namely $(BEDT\text{-}TTF)_8[SiW_{12}O_{40}]$. In 1990, Bellitto *et al.* published new TTF salts with Lindquist polyanion [10]. In 1993, Coronado *et al.* reported a TTF salt with $[Co_4(H_2O)_2(PW_9O_{34})_2]^{10-}$ [11]. In 1994, in collaboration with Delhaès from Bordeaux (France) and Coronado from Valencia (Spain), we investigated BEDT-TTF salts with paramagnetic Keggin polyanions [12].

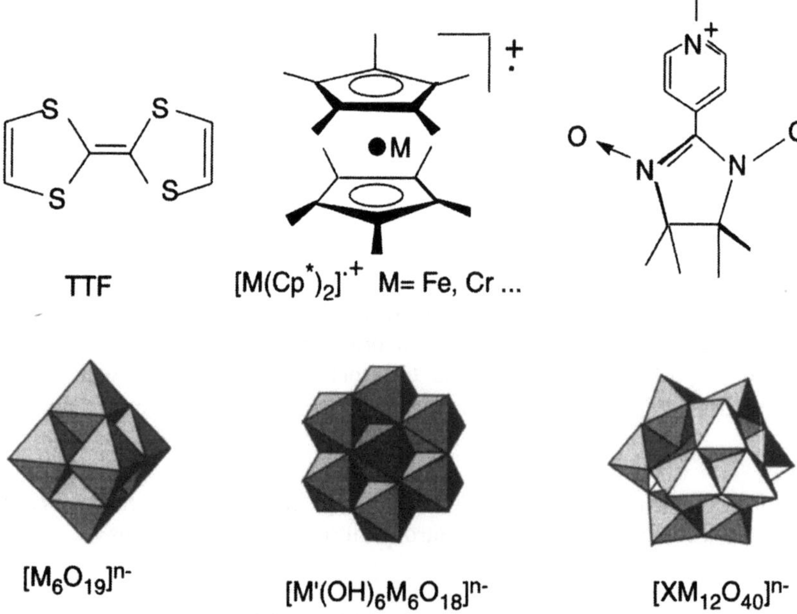

Scheme 1. Molecular precursors used.

Our interests in polyoxometalates as inorganic components in molecular materials are the following (i) their geometrical parameters (size and shape) can influence the dimensionality of their materials, (ii) the high negative electronic charge (from -2 to -16) can give rise to strong electrostatic interactions between the organic and inorganic networks, (iii) the presence of transition metals gives a flexibility in the sense that we can introduce paramagnetic centres and also modulate the above geometrical and electronic properties (points i and ii), thanks to the coordination chemistry, by changing the ligand on the metal, (iv) these large units are soluble and stable in organic solvent (CH_3CN, CH_2Cl_2...) which makes possible their molecular assemblies with organic and organometallic radical cations. Finally, the polyoxometalates are strong inorganic electron acceptors and it has been demonstrated that the Keggin polyoxometalate derivatives $[XM_{12}O_{40}]^{n-}$ can accept up to 32 electrons [13]. This offers the opportunity to prepare organic donor - inorganic acceptor materials with a mixed valence state on the organic and inorganic units [7,8]. The mixed valence state induces electron delocalisation on the organic part and spin localization on the inorganic part. This property was found in $(TTF)_6(Et_4N)[HPM_{12}O_{40}]$ (M=Mo, W) which constitute the first compounds containing paramagnetic polyoxometalates [7, 27] (see scheme 2).

The potentialities of polyoxometalates in material science [6] are displayed in scheme 2 through the properties of the Keggin polyoxometalates that are formulated as $[XM_{12}O_{40}]^{n-}$ or $[(XO_4)^{n-}M_{12}O_{36}]$, M=Mo, W. These polyanions can be diamagnetic or paramagnetic following reduction processes or by substitution of the central heteroatom or outer metal atoms by paramagnetic ones. Finally, it is possible to substitute terminal oxygen atoms by organic ligands [14].

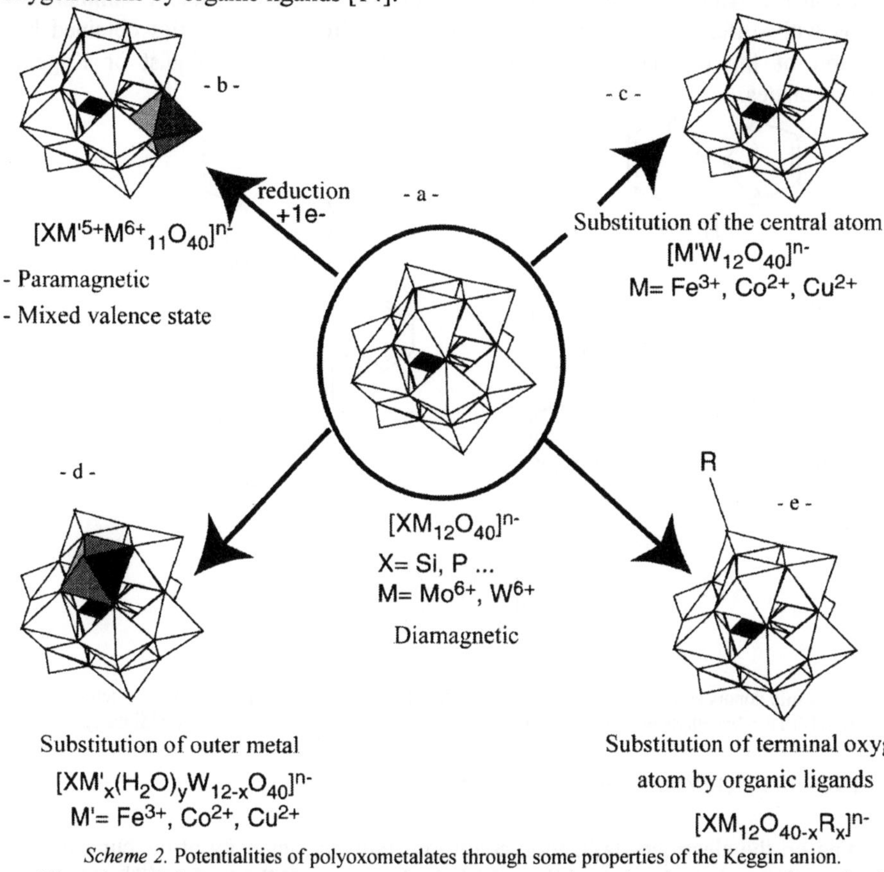

Scheme 2. Potentialities of polyoxometalates through some properties of the Keggin anion.

The aim of the use of paramagnetic polyoxometalates in charge transfer salts is to obtain ferromagnetic or antiferromagnetic coupling between localized spins in the inorganic sublattice through the mobile electrons of the organic sublattice thanks to the so called indirect exchange mechanism [15]. In this contribution we wish to present some results which account on the potentialities of polyoxometalates as precursors in molecular materials. We will limit ourselves to materials obtained in our laboratory.

2. TTF Derivatives Containing Materials

Most of the conducting and superconducting hybrid organic/inorganic radical cation salts are obtained by the well-known electrocrystallization technique which yields single crystals of good quality for physical measurements. The synthesis consists in the anodic oxidation of the organic donor in the presence of tetraalkylammonium salts of polyoxometalates as electrolyte [16]. In these materials the organic donors are packed as one-dimensional chains or two-dimensional layers separated by inorganic anions [16]. It is very difficult to control and to predict the solid state molecular organization and in some cases polymorphic compounds with different structures and properties are observed in the same batch for the same system donor / anion. Table 1 summarizes materials with different stoichiometries, crystal structures and properties obtained by association of the organic donors shown in scheme 3 and Lindquist or Keggin polyoxometalates.

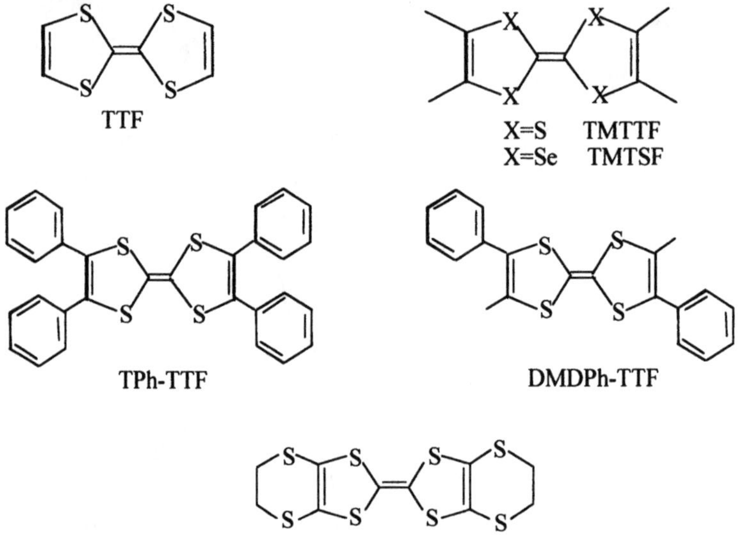

BEDT-TTF

Scheme 3. Organic donors associated with polyoxometalates. TTF = tetrathiafulvalene, TMTTF = tetramethyl-TTF, BEDT-TTF = bis-ethylene-TTF, TPhTTF = tetraphenyl-TTF, DMDPhTTF = dimethyldiphenyl-TTF

2. 1. LINQUIST POLYOXOMETALATES BASED SALTS

Depending on the electrocrystallization conditions (solvent, intensity of the current and polyoxometalate salts), the assemblies of Lindquist anions $[M'_nM_{6-n}O_{19}]^{(2+n)-}$ ($M=W^{VI}$, Mo^{VI} and $M'=Nb^V$ and V^V) with different donors yielded a rich polymorphism with ratio of 1:2. 1:3. 1:4, 1:5 and 1:6 for the anion:donor (see Table 1).

The 1:2 stoichiometry was observed for $[M_6O_{19}]^{2-}$ (M=W, Mo) and TTF [17], TMTTF (tetramethyl-TTF) [18], BEDT-TTF (bis-ethylene-TTF) [19], TPhTTF (tetraphenyl-TTF) [20] and DMDPhTTF (dimethyldiphenyl-TTF) [21]. These compounds were prepared using $(\alpha\text{-}Bu_4N)_2[M_6O_{19}]$ salts as electrolyte except TTF containing compound for which $(Et_4N)_2[M_6O_{19}]$ salt was used; the use of $(\alpha\text{-}Bu_4N)_2[M_6O_{19}]$ salt instead of the $(Et_4N)_2[M_6O_{19}]$ salt leads to the 1:3 phase described below. The crystal structures of these 1:2 salts consist of donor radical cations arranged as dimers and separated by

Table 1. Crystal data and room temperature conductivity for TTF derivative salts containing various polyoxometalates.

Compounds	a, (Å)	b, (Å)	c, (Å)	α, (°)	β, (°)	γ, (°)	SG	R^{i}	σ^{ii} ($S.cm^{-1}$)
$(TTF)_3W_6O_{19}$	9.965	10.503	10.634	71.93	78.63	63.3	$P\bar{1}$	0.020	$1.4\ 10^{-3}$
$(TTF)_3Mo_6O_{19}$	9.942	10.417	10.601	72.33	78.77	63.5	$P\bar{1}$	0.030	$1.4\ 10^{-4}$
$(TTF)_4W_6O_{19}(ACN)_{0.5}$	14.399	20.509	16.817	90.	100.74	90.	$P2_1/m$	0.033	ins
$(TTF)_7[\beta\text{-}Mo_8O_{26}]$	12.146	13.367	14.361	84.61	79.13	63.67	$P\bar{1}$	0.079	10^{-3}
$(TMTTF)_2[W_6O_{19}]$	8.469	13.337	16.776	94.35	95.68	98.95	$P\bar{1}$	0.035	ins
$(TMTTF)_2[Mo_6O_{19}]^{iv}$	8.421	13.268	16.740	94.54	95.69	99.03	$P\bar{1}$	------	ins
$(TPTTF)_2[W_6O_{19}]$	11.171	13.722	13.967	61.78	67.05	75.59	$P\bar{1}$	0.035	ins
$(DMDPTTF)_2[W_6O_{19}].CHCl_3$	11.336	24.225	20.422	90.	95.94	90.	C2/c	0.042	ins
$(DMDPTTF)_2[Mo_6O_{19}]$	9.620	11.051	13.017	113.10	96.80	90.26	$P\bar{1}$	0.032	ins
$(TMTSF)_3[W_6O_{19}]$. 2DMF	11.589	19.385	13.681	90.	99.53	90.	$P2_1/c$	0.049	1.3
$(TMTSF)_3[NbW_5O_{19}]$. 2DMF	11.599	19.469	13.429	90.	100.19	90.	$P2_1/c$	0.050	ins
$(ET)_2[Mo_6O_{19}]$	8.898	11.265	21.098	90.	96.34	90.	$P2_1/c$	0.036	ins
$(ET)_2[W_6O_{19}]^{iv}$	8.871	11.321	21.108	90.	96.33	90.	$P2_1/c$	----	ins
$(ET)_5[VW_5O_{19}].5H_2O$	19.770	11.598	41.340	90.	100.90	90.	$P2_1/c$	0.071	13.7
$(ET)_5[V_2W_4O_{19}].5H_2O$	19.827	11.908	40.876	90.	100.76	90.	$P2_1/c$	0.119	$7\ 10^{-3}$
$(ET)_4[Cr(OH)_6Mo_6O_{18}]^{iii}$	11.365	4.085	46.463	90.	90.	90.	$Pmn2_1$	—	—
$(TMTSF)_3[PW_{12}O_{40}]$	11.733	18.461	16.223	90.	98.05	90.	$P2_1/n$	0.053	ins
$(ET)_3[PW_{12}O_{40}]$ THF	12.715	13.612	13.511	77.98	62.77	74.07	$P\bar{1}$	0.064	ins
$(TTF)_6(Et_4N)[HXM_{12}O_{40}]^{v}$	15.383	19.284	14.088	90.	90.	90.	Cmmm	0.039	$5\ 10^{-2}$
$(TTF)_7[SiW_{12}O_{40}]$.2 ACN	12.742	13.164	14.197	91.76	104.75	89.98	$P\bar{1}$	0.13	—
$(TTF)_7[\beta\text{-}Mo_8O_{26}]$	12.146	13.367	14.361	84.61	79.13	63.67	$P\bar{1}$	0.079	—
$\alpha\text{-}(ET)_8\ [M'W_{12}O_{40}].solv^{vi}$	13.971	43.117	14.042	90.	107.25	90.	I2/m	0.068	10^{-2} - 10^{-3}
	16.66	43.42	12.14	90.	111.78	90.	I2/m	0.061	

^{i} Reliability R factor, ^{ii} Room temperature electrical conductivity, ins = insulator, ^{iii} non solved structure, ^{iv} Isostructural to previous compound; ^{v} X=P, Si; M=Mo, W, data correspond to X=P and M =Mo; ^{vi} M'=Co^{II}, Cu^{II}, Fe^{III}, B^{V}, H_2^{2+} data given for M'=Co^{II}, lines 1 and 2 correspond to α1 and α2 modifications, respectively, ACN = CH_3CN

anions (Figure 1). To balance the negative charge of the dianions, the organic molecules are fully oxidized, then these phases are rather ionic in agreement with their insulating and diamagnetic ground state.

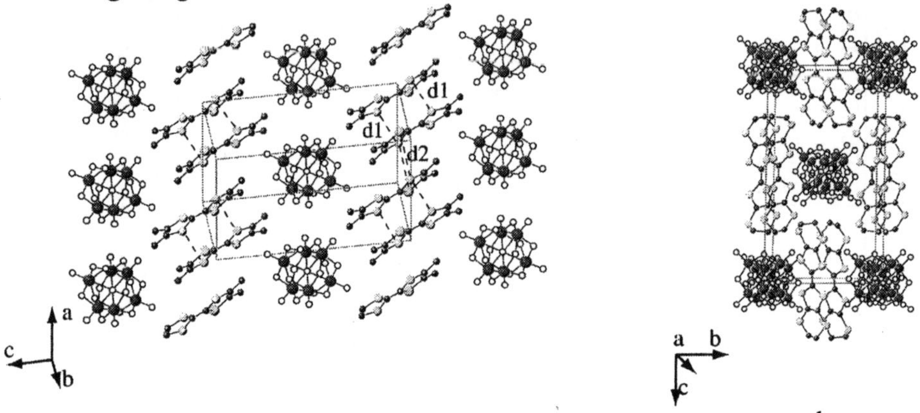

-a- -b-

Figure 1. Crystal structures of a) $(TMTTF)_2[M_6O_{19}]$ and b) $(BEDT\text{-}TTF)_2[M_6O_{19}]$ (M=W, Mo) d1=3.320Å, d2=4.024Å.

The 1:3 crystalline phases were obtained with $[M_6O_{19}]^{2-}$ (M=W, Mo) and TTF and TMTSF [8,22]. Using $(\alpha\text{-}Bu_4N)_2[M_6O_{19}]$ salts as electrolyte, additionally, two phases [23] were obtained with TMTSF and substituted Lindquist anions $[VW_5O_{19}]^{3-}$ and $[NbW_5O_{19}]^{3-}$. In all these compounds, the organic part is built up from two independent organic donors noted A and B, that are packed as infinite chains of trimers with the ...BAB...BAB...BAB... sequence (Figures 2 and 3) and lying in channels formed by the hexametalates. In $(TTF)_3[M_6O_{19}]$, M= W, Mo [8,22] salts (Figure 2), the TTF molecules exhibit a criss-cross overlap inside the trimer with intratrimer S...S contacts shorter than the Van der Walls (VdW) separations leading to strong overlap between the 3p sulfur orbitals. From the intramolecular bond length and the extended Hückel band structure calculations, the charge distribution on the organic molecules was estimated to be +0.5 on the B molecule and +1 on the A molecule. This is in agreement with the observation from the spectroscopic properties (IR-vis) in the near-IR region of the so-called A-band at 5000 cm^{-1} characteristic of a mixed valence state. The intertrimer S...S contacts are longer than the VdW separations. The trimerisation and the charge localization account on the weak electrical conductivity ($\sigma_{300K} = 10^{-3}$ $S.cm^{-1}$) and the diamagnetic behaviour of the salts.

The 1:3 TMTSF phases were obtained for both non-substituted and substituted Lindquist hexatungstate anions. All our efforts to obtain the corresponding salts containing substituted or non-substituted hexamolybdate anions are failed. The three crystalline phases, namely, $(TMTSF)_3[W_6O_{19}]$. 2DMF [8,22], $(TMTSF)_3[VW_5O_{19}]$. 2DMF [23] and $(TMTSF)_3[NbW_5O_{19}]$.2DMF [23] (DMF = N,N-Dimethylformamide) (Figure 3) are isostructural, the X-ray structures were solved for both salts containing non-substituted and nobium-substituted tungstate anions. The structrural data observed for the non-substituted anion $[W_6O_{19}]^{2-}$ are in good agreement and compare well with those found in the TTF salts described above and in other known Lindquist salts. In the substituted units, the Nb atoms are randomly distributed on the six metallic positions. In both structures, the organic molecules present slipped stacks with the cycle-to-double-bond intra- and intertrimer molecular overlaps identical to those usually observed in the $(TMTSF)_2X$ series, the so-called Bechgaard salts [24]. In both cases, the intratrimer Se...Se contacts are shorter than the corresponding VdW separations while the intertrimer separations are in the range of the VdW separations. The comparison of the partially oxidized chain $[(TMTSF)_3^{2+}]_n$ and the fully ionized one $[(TMTSF)_3^{3+}]_n$ shows weak variations for the Se...Se contacts; the intratrimer distances are shorter in $[(TMTSF)_3^{2+}]$ trimer than in the fully oxidized one, while, the intertrimer contacts are slightly longer [23].

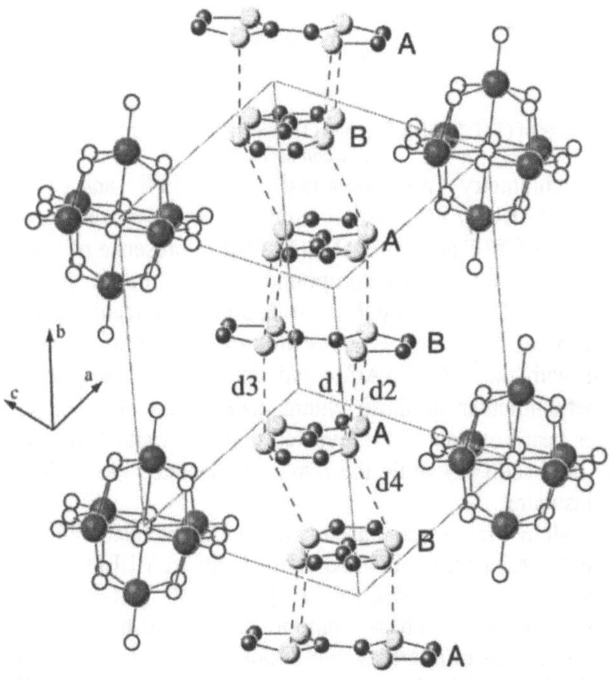

Figure 2. Crystal structure of $(TTF)_3M_6O_{19}$ showing criss-cross overlap between TTFs. Dashed lines indicate S...S contacts d1 = 3.32Å, d2 = 3.40Å, d3 = 4.10Å, d4 = 4.02Å.

These salts illustrate the possibility of controlling the electronic band filling by changing the charge on the polyoxoanion; for the salt containing the tungstate dianion, as in the above TTF_3 containing salts, the same charge distribution was found with however, a higher electrical conductivity (σ_{300K} = 1.3 $S.cm^{-1}$), in agreement with the presence of a A-band on the near-IR region (4000 cm^{-1}) and the diamagnetic ground state of the compound. For the salts containing substituted trianions $[VW_5O_{19}]^{3-}$ and $[NbW_5O_{19}]^{3-}$) the organic molecules bear the same charge of +1. The $\chi_M T$ product versus T shows at room temperature a value of 0.38 $emu.K.mol^{-1}$ and a continuous decrease on lowering the temperature to a value around 0.04 $emu.K.mol^{-1}$ indicating an antiferromagnetic coupling between the TMTSF units in the organic chain [23]. These results are in agreement with the paramagnetic and the insulating properties predicted by the molecular orbital and the band structure calculations performed on TTF and TMTSF tungstate salts [22].

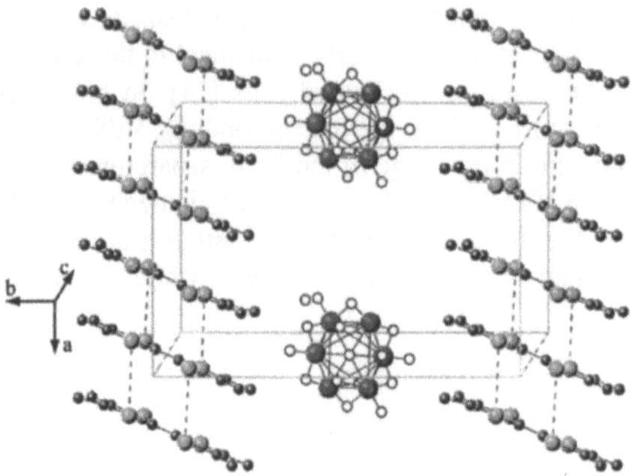

Figure 3. Crystal structure of $(TMTSF)_3[W_6O_{19}]$ and $(TMTST)_3[M'W_5O_{19}]$, M'= Nb, V. Dashed lines indicate Se...Se contacts < 4Å

The 1:4 stoichiometry was obtained with TTF and $[W_6O_{19}]^{2-}$, namely $(TTF)_4[W_6O_{19}]$. $0.5(CH_3CN)$ [25]. This compound was prepared by slow diffusion of acetonitrile solutions of TTF and $(\alpha\text{-}Bu_4N)_2[M_6O_{19}]$ in presence of few drops of sulfuric acid. The crystal structure of this compound contains up to six independent TTF molecules noted A, B, C, D, E and F. The A and D molecules form centrosymetric dimers interspersed by the C and E molecules to form infinite and parallel strongly dimerized chains with the ...AACAAC... and DDEDDE .... sequences. The B and F molecules are perpendicular to these chains (see Figure 4). The estimation of the electronic charges beard by each TTF revealed that the A and D molecules forming the dimers are fully oxidised (+1) while the remaining isolated molecules are neutral. The compound is an insulator.

Among the Lindquist molecular based salts, the more interesting from the physical point of view are the 1:5 phases obtained exclusively with BEDT-TTF and substituted Lindquist trianions, namely $(BEDT\text{-}TTF)_5[VW_5O_{19}]$. $5H_2O$ [26] and $(BEDT\text{-}TTF)_5[V_2W_4O_{19}]$. $5H_2O$ [23]. These compounds are isostructural and their crystal structure (figure 5) consists of alternate two-dimensional layers of organic and inorganic units as commonly observed in the BEDT-TTF containing materials. The organic layer is built up from five crystallographically independent BEDT-TTF molecules noted A, B, C, D and E.

The monosubstituted tungstate salt exhibits a metallic behaviour with a conductivity of 14 $S.cm^{-1}$ at room temperature to reach a maximum value of 30 $S.cm^{-1}$ at 250K. Bellow this temperature, a metal-insulator transition occurs. This behaviour is attributed to the cracks on the single crystals resulting from a loss of water molecules. The electrical measurements have been reproduced for several crystals but in each attempt the crystals crack. This first example of metallic material containing polyoxometalate demonstrates the possibility to stabilize the metallic states when passing from TTF to BEDT-TTF salt despite the presence of different organic carriers which favours inhomogeneous charge distribution and then poor electrical conductivities. $(BEDT\text{-}TTF)_6[M^VW_5O_{19}]$. $8H_2O$ (M^V =Nb and V) are the only 1:6 salt observed so far. The bad crystal quality of these two salts prevented their complete characterizations; the stoichiometries were checked by elemental and thermogravimetric analysis [23].

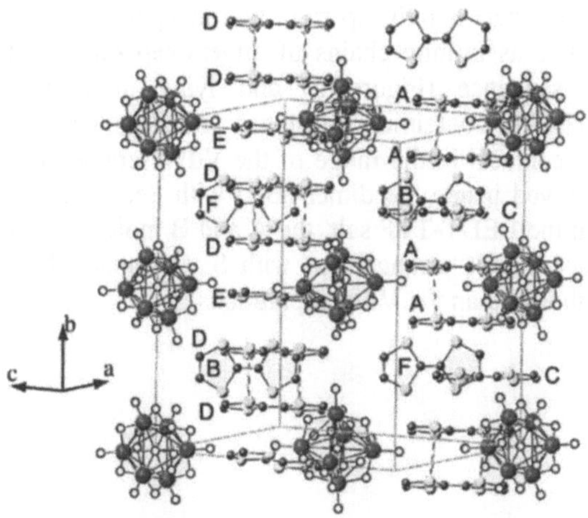

Figure 4. Crystal structure of $(TTF)_4[W_6O_{19}]$. Dashed lines indicate S…S contacts < 3.6Å

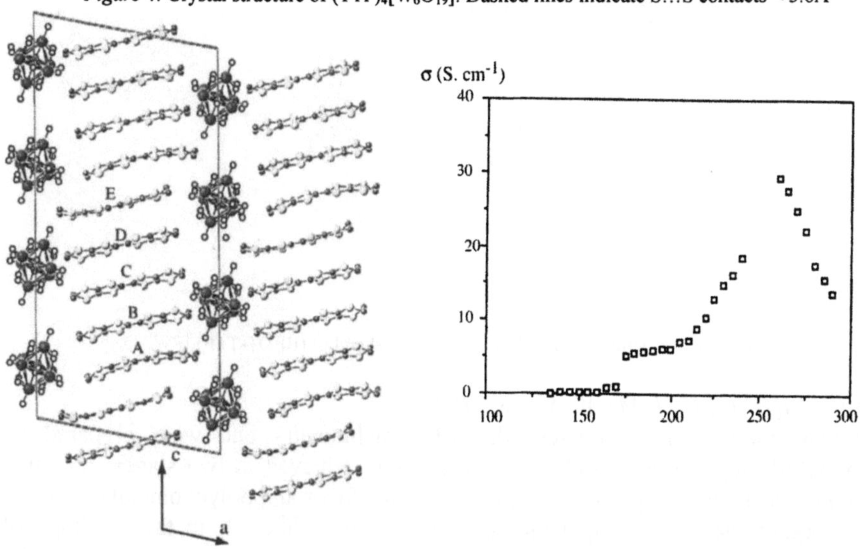

Figure 5. Crystal structure and plot of the electrical conductivity versus T for $(BEDT\text{-}TTF)_5[VW_5O_{19}]$. $5H_2O$.

2. 2. KEGGIN POLYOXOMETALATES BASED SALTS

The assemblies of Keggin polyoxometalates $[XM_{12}O_{40}]^{n-}$ (M=W, Mo and M' = P, Si, Fe^{3+}, Cu^{2+}, Co^{2+}, Zn^{2+}, H_2^{2+}) with different donors yielded crystalline phases with stoichiometries of 1:3. 1:6 , 1:7 and 1:8 (see Table 1).

TMTSF and BEDT-TTF Based Compounds

$(TMTSF)_3[PW_{12}O_{40}]$ [27] and $(BEDT\text{-}TTF)_3[PW_{12}O_{40}]$.THF (THF = tetrahydrofurane) [6a] are the two compounds known so far with the 1:3 stoichiometry with Keggin polyanion. With respect to the stoichiometry, the organic molecules are fully oxidized in agreement with the insulating character of the materials. In both compounds, the organic part is built up from two independent organic donors noted A and B, that are packed as infinite chains of dimers separated by monomers with the ...BB...A...BB...A... sequence (Figure 6) and lying in channels formed by the polyanions. In the TMTSF containing salt, the A and B molecules exhibit a criss-cross overlap with Se...Se contacts in the range of the VdW separations. A cycle-to-double-bond overlap is observed inside the dimer (BB) with Se...Se contacts shorter than the VdW separations. In the BEDT-TTF salt, the A and B molecules are perpendicular and present a unique face-to-side arrangement with S...S contacts between the A...B and B...B organic units shorter than the VdW separations.

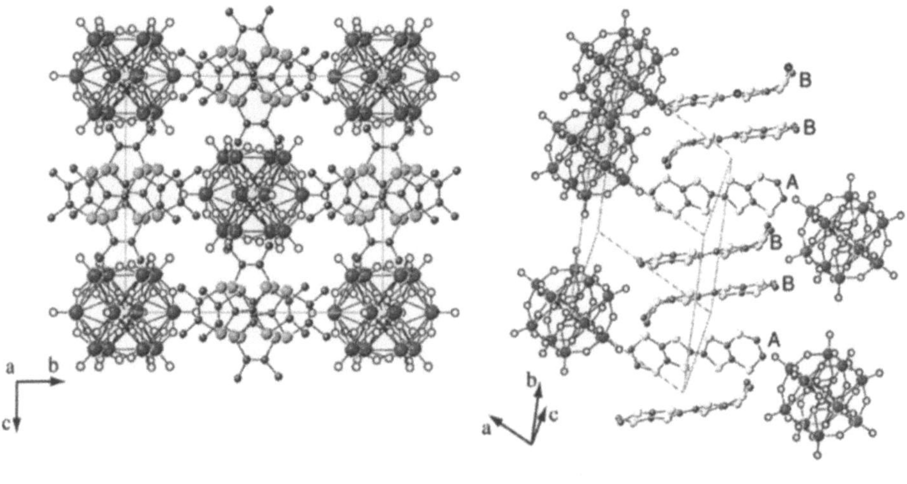

-a- -b-

Figure 6. Crystal structures of a) $(TMTSF)_3[PW_{12}O_{40}]$ and b) $(BEDT\text{-}TTF)_3[PW_{12}O_{40}]$.

$(TTF)_6(Et_4N)[HXM_{12}O_{40}]$, M= W, Mo; X= P, Si

These compounds [6,7,28] possess the 1:6 stoichiometry and were prepared by electrocrystallization. The crystal growth process is achieved in two stages. In a first step, redox reactions take place between the organic donor and polyoxometalates which act as inorganic acceptor components as indicated in equation 1. In fact, addition of acetonitrile solutions of TTF in electrochemical cells containing saturated acetronitrile solution of the oxidizing polyoxoanions, namely $[PMo_{12}O_{40}]^{3-}$, $[PW_{12}O_{40}]^{3-}$ or $[SiW_{12}O_{40}]^{4-}$, result in blue coloured solution characteristic of reduced polyanions (heteroblue). The solution containing the less oxidizing $[SiW_{12}O_{40}]^{4-}$ anion remains unchanged for several hours.

$$[XM_{12}O_{40}]^{n-} + 1e^- \longrightarrow [XM_{12}O_{40}]^{(n+1)-}$$
$$TTF \longrightarrow TTF^{+}. + 1e^-$$

In a second step, more neutral TTF is added in electrochemical cells and then the electrocrystallization process started under low constant current (I=1μA). Crystals were collected from the electrode after two weeks. Structural, physical (electrical conductivity, ESR, magnetic measurements) and extended Hückel band structure calculations revealed that in the phosphometalates containing salts, the anions reduced while it were not reduced in the case of the salts containing silicometalate anions. From these investigations the general formula was established as $(TTF)_6(Et_4N)$ $[HXM_{12}O_{40}]$ (see figure 7). The four compounds $(TTF)_6(Et_4N)[HPMo_{12}O_{40}]$ (1), $(TTF)_6(Et_4N)$ $[HSiMo_{12}O_{40}]$ (2), $(TTF)_6(Et_4N)$ $[HPW_{12}O_{40}]$ (3), $(TTF)_6(Et_4N)$ $[HSiW_{12}O_{40}]$ (4), are isostructural. The general crystal structure represented in Figure 7 is built from a polyoxoanion unit located at the origin of the C lattice, and three independent TTF molecules (labelled A, B and C). The A- and B-type molecules stack regularly with an eclipsed overlap along the [0 0 1] direction, with the sequence ...BAAB...BAAB..., in channels made of alternating polyanions and TTF molecules of the C-type (Figure 7). The disordered $(C_2H_5)_4N^+$ cation is located on the (0 0 ½) inversion centre. The intra-stack S...S contacts shorter than the corresponding VdW separations suggest a slight ...BAAB... tetramerization along the [001] direction with BAAB as a repeat unit.

All four salts present semi-conducting behaviour with weak room temperature conductivities ($\sigma_{300K} = 10^{-3}$ to 10^{-4} $S.cm^{-1}$). From the formula, without proton, obtained from the X-ray analysis, a simple ionic picture $(TTF)_2{}^0(TTF_4)^{2+}(Et_4N)^+[PM_{12}O_{40}]^{3-}$ for X=P and $(TTF)_2{}^0(TTF_4)^{3+}(Et_4N)^+[SiM_{12}O_{40}]^{4-}$ for X=Si suggests a diamagnetic behaviour for the phosphorus salts and a paramagnetic behaviour, due to one unpaired electron on each tetramer of the $(TTF_4)^{3+}$ sublattice, for the silicometalate ones. Surprisingly, while the silicometalate salts, (2) and (4), exhibit a diamagnetic behaviour; a temperature-dependent susceptibility was observed below 300K for the phosphometalate compounds (1) and (3). A Curie behaviour was observed over a large range of temperatures. The effective moments μ_{eff} deduced from the Curie constants correspond to 1.46 μ_B and 1.49 μ_B for compounds 1 and 3 respectively, characteristic of one unpaired electron. The reciprocal susceptibilities extrapolate to almost zero temperature showing that no magnetic interaction is occurring in these compounds. We did not detect any intrinsic ESR signal at room temperature in any salt, which could be associated with the organic sublattice. However, when the different salts were cooled down in liquid Helium at 4.2 K, we observed in the case of the phosphometalate salts (1), (3), an ESR signal which broadens and disappears quickly as the temperature increases. Based on the 4.2 K spectrum, the g-factor equals to 1.826 with a line width equal to 22 gauss for the compound 1. The corresponding values are 1.947 and 15 gauss for 3. These values are close to those observed in similar polyanions containing W^V and Mo^V (S = 1/2) species [29]. Consequently, we conclude that the phosphometalate anion in 1 and 3 has been reduced during the synthesis process and that the unpaired electron remains localized on a metallic site at low temperature and undergoes a rapid hopping delocalisation at room temperature. These different results allow us to state that the paramagnetic susceptibility determined from the magnetic measurements carried out on the phosphometalate salts 1 and 3 is due to W^{+5} and Mo^{+5} valence states. Moreover, the existence of only one unpaired electron in the phosphometalate salts 1 and 3 and the absence of any magnetic contribution in the silicometalate salts 2 and 4 lead us to propose a formal oxidation state of 2+ for the TTF tetramers BAAB forming the stack and 0 for the isolated TTF of C-type in every salt. In agreement with the spectroscopic and theoretical calculations [28] and in order to equilibrate the negative and positive charges we proposed that the appropriate formula for all of the four salts is $(TTF)_2{}^0(TTF_4)^{2+}(H)^+(Et_4N)^+[XM_{12}O_{40}]^{4-}$ assuming that a proton has been trapped around the inorganic cages during the synthesis of the salts [6,7,28].

The electron transfers observed in the phosphometalates containing salts 1 and 3 induce coexistence of mobile electrons on the TTF stacks and localized spins on the paramagnetic polyoxometalates. These two salts match very well the indirect exchange model but unfortunately, we have not observed the expected magnetic interactions. This is probably due to the weak conductivities of the compounds. Following these results, we decided to investigate materials with polyoxometalates containing paramagnetic transition metals [12] (see next section).

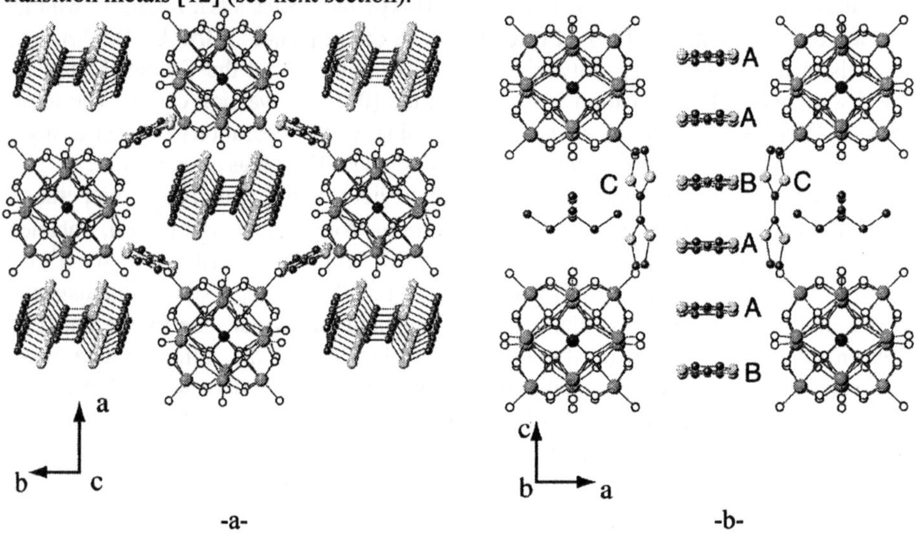

Figure 7. Views of the crystal structure of $(TTF)_6(Et_4N)[HXM_{12}O_{40}]$ in the *ab* and *ac* planes.

α-(BEDT-TTF)$_8$[M'W$_{12}$O$_{40}$]. n(solv) ; M'= Fe^{3+}, Co^{2+}, Cu^{2+} , B^{3+}, 2H^{+}, Zn^{2+} [12]

This work was performed with the contributions of three groups from Rennes 1, Bordeaux and Valencia (Spain). Depending on the electrocrystallization conditions, we have obtained up to three different crystallographic phases (noted as α1. α2 and X) [12] for nearly all the polyanions with the same stoichiometry 1:8 for the anion:donor. It is important to note the possibility of obtaining the same crystal structures in these series while we can change the anionic charge (-5 or -6) and the magnetic moment (S = 0 to S = 5/2). The charge -4 is found in $[SiW_{12}O_{40}]$ and the crystal structure of its BEDT-TTF salt [9] belongs to the α1 modification [16]. The crystals of the third phase X are twinned and prevented its crystal structure analysis. The crystal structures of both α1 and α2 series are very similar and consist in organic and inorganic layers alternating along the [0 1 0] direction (figure 8). The difference between the two phases concerns the packing in the *ac* plane, the *b* parameter being the same for the two phases. The organic network contains two different BEDT-TTF chains (see Figure 9), one regular and eclipsed chain (noted I in figure 8) and one dimerized chain (noted II in Figure 8).

From the spectroscopic properties and structural characteristics we assumed that in the eclipsed chain the BEDT-TTF molecules bear a mean charge of +1 while they are almost neutral in the dimerized chain. Here again, to balance the excess of negative charges on the polyanions, we need to consider the presence of protons in the crystal.

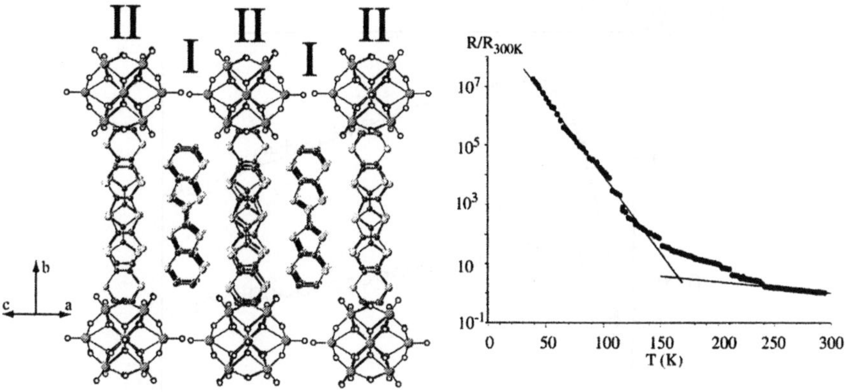

-a- -b-

Figure 8. a) General view of crystal structures of α- $(BEDT\text{-}TTF)_8[M'W_{12}O_{40}]$; b) Plot of the normalized resistance versus temperature

The room temperature measurements made on some single crystals show d.c. conductivities of about 10^{-1}-10^{-3} $S.cm^{-1}$ with a semiconducting behaviour in the temperature range 100-300 K. The magnetic properties of these salts have been investigated in the temperature range 4-300 K. For the salts containing a diamagnetic ion (M' = H_2^{2+}, Zn^{2+}, B^{3+}) a shoulder in the molar susceptibility around 60 K and a paramagnetic curie tail at lower temperatures has been observed (figure 10). The $\chi_M T$ versus T shows a decrease from values of 1.5 - 2 $emu.K.mol^{-1}$ to values of 0.1 - 0.2 $emu.K.mol^{-1}$ with the lowering of the temperature. This behaviour corresponds to strong antiferromagnetic interactions between the spins of the fully oxidized BEDT-TTF chain.

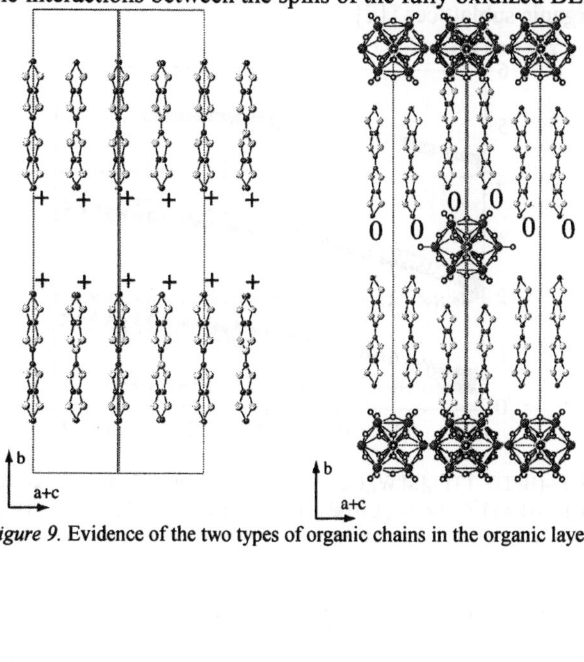

Figure 9. Evidence of the two types of organic chains in the organic layer.

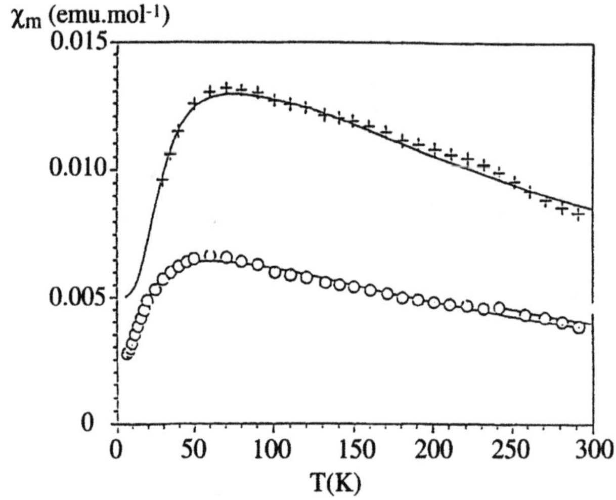

Figure 10. Plot of χ_M versus T for the salts containing diamagnetic ions: α- (BEDT-TTF)$_8$[M'W$_{12}$O$_{40}$]: M'= H$_2$ (O), M'= B (+). Continuous lines are the best fits.

In the salts containing paramagnetic ions (M' = Cu^{II}, Co^{II} and Fe^{III}), The χ_MT versus T plot (Figure 10) shows a decrease with the lowering of the temperature similar to that observed for the salts containing diamagnetic ions, reaching at low temperature, the values of the corresponding isolated paramagnetic polyoxometalates deduced from their tetraethylammonium salts (Figure 11). From these results we deduced that the same antiferromagnetic interactions occur between the spins of the BEDT-TTF molecules in the fully oxidized chains and also the absence of magnetic interactions between the organic and inorganic sublattices [12].

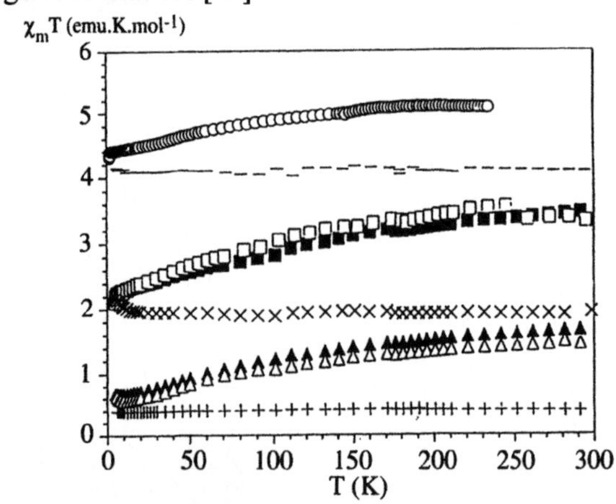

Figure 11. α1- and α2-(BEDT-TTF)$_8$[M'W$_{12}$O$_{40}$]: Plot of χ_MT versus T for the salts containing paramagnetic ions : M'= Cu (▲ for α1 and △ for α2), Co (■ for α1 and □ for α2), Fe (O for α1). The corresponding Bu$_4$N$^+$ salts of the polyanions are also shown: M'= Cu (+), Co (×), Fe (-).

3. Decamethylferrocenium Containing Materials

Since the discovery of bulk ferromagnetism in the charge transfer complex Fe(Cp*)$_2$-TCNE.CH$_3$CN [30], (Cp* = C$_5$(CH$_3$)$_5$ and TCNE = Tetracyanoethylene) a great interest was devoted to the organometallic decamethylferrocene donor and its derivatives in the field of molecular magnetism. In order to obtain magnetic interactions between localized spins through weak bonds, we are investigating molecular assemblies between polyoxometalates and metallocenium organometallic radicals [31-34]. Metallocene $M(Cp)_2$ possess easy accessible oxidation potentials which make them suitable molecular precursors for charge transfer complexes, it is possible to modulate their sizes and also their magnetic moments by changing the transition metal [(from S = 1/2 for Fe(III) to S = 3/2 for Cr(III)]. Table 2 gives the different salts obtained in this system.

3. 1. LINDQUIST POLYOXOMETALATES CONTAINING COMPOUNDS [34]

$[Fe(Cp^*)_2]_2[Mo_6O_{19}]$ was obtained by metathesis in DMSO using $(Bu_4N)_2[Mo_6O_{19}]$ and $Fe(Cp^*)_2BF_4$. The crystal structure (Figure 12) of this compound results from the packing in the [1 1 0] direction of bidimensional layers. Each layer consists in perpendicular chains, in the (1 0 2) plane, of decamethylferrocenium radical cations pairs separated by $[Mo_6O_{19}]^{2-}$ dianions. The chains are parallel to the [2 $\bar{3}$ $\bar{1}$] and [2 3 $\bar{1}$] directions. The distance between successive layers is equal to ca. 7.395Å and the shortest distance between Fe(III) ions is equal to ca. 7.763Å. As expected from these large distances between spin carriers, the magnetic properties of this salts showed a Curie law.

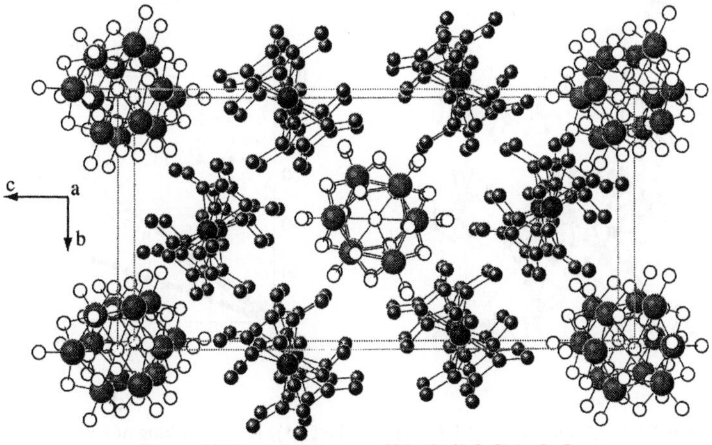

Figure 12. Crystal structure of $[Fe(Cp^*)_2]_2$ $[Mo_6O_{19}]$.

3. 2. KEGGIN POLYOXOMETALATES CONTAINING COMPOUNDS

These compounds were synthesized by metathesis or by redox reaction with both diamagnetic and paramagnetic Keggin polyanions [31-34]. The following anions were used $[PMo_{12}O_{40}]^{3-}$, $[PW_{12}O_{40}]^{3-}$, $[SiMo_{12}O_{40}]^{4-}$, $[SiW_{12}O_{40}]^{4-}$, $[FeW_{12}O_{40}]^{5-}$ and $[PCu(H_2O)W_{11}O_{39}]^{5-}$. Whatever the electronic negative charge of the Keggin polyanion (from -3 to -5), the same stoichiometry 1:4 has been observed; the general formula of these salts is $[Fe(Cp^*)_2]_4[XM'_nM_{12-n}O_{40}]$. n(solv); X= P, Si, Fe, M= W, Mo, M'=Cu. In many cases several phases with the same stoichiometry but different crystal structures were obtained in the same batch. Two types of crystal structures were observed (Figure 13): (i) 1-D structure where $[Fe(Cp^*)_2]^+$ cations form linear chains surrounded by columns of polyoxometalates; (ii) a 3-D structure where $[Fe(Cp^*)_2]^+$ occupy the vertices of a cuboctahedron (Figure 13). The 3-D structure results from the association of these cuboctahedra by sharing vertices in the *a* and *c* directions and sharing faces in the b direction. The magnetic properties of these compounds between 2-300 K indicate that all salts present very weak magnetic interactions obeying a Curie-Weiss law with small θ values.

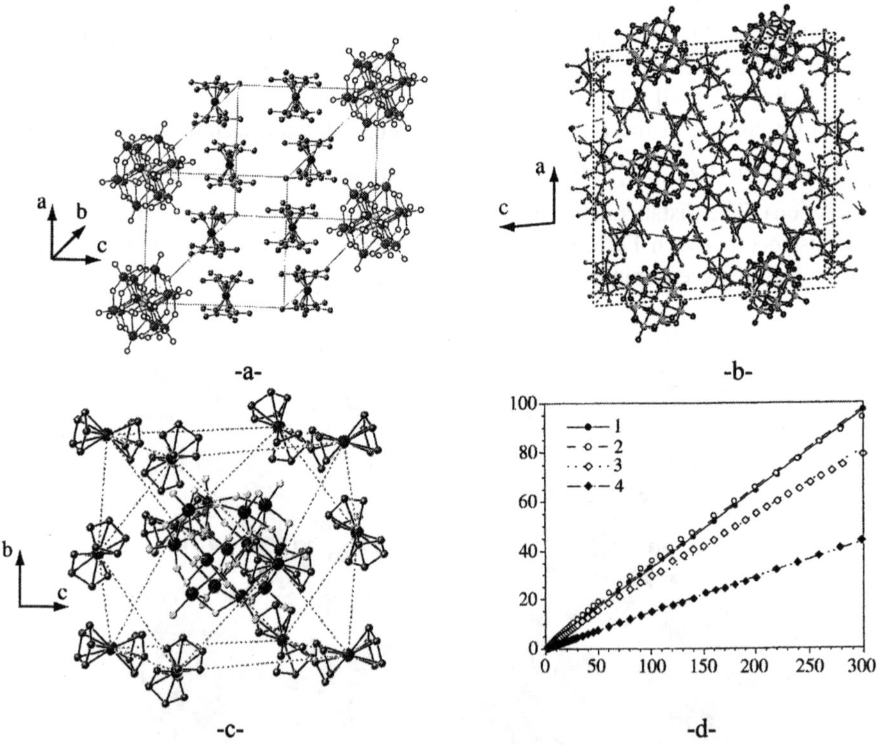

Figure 13. $[Fe(Cp^*)_2]_4$ $[XM_{12}O_{40}]$. *solv*: a) 1-D chains of $Fe(Cp^*)_2$. b) 3-D packing of $Fe(Cp^*)_2$. c) Evidence of $[Fe(Cp^*)_2]_{12}$ cuboctahedra encapsulating the polyoxometalate (Oxygen atoms are omitted for clarity), d) Magnetic properties. (1): $[Fe(Cp^*)_2]_4[SiMo_{12}O_{40}]$. DMF; (2) : $[Fe(Cp^*)_2]_4[SiW_{12}O_{40}]$. 2DMF; (3) : $[Fe(Cp^*)_2]_4[PMo_{12}O_{40}]$. 2DMF; (4) $[Fe(Cp^*)_2]_4[HFeW_{12}O_{40}]$. CH_3CN

3. 3. ANDERSON-EVANS POLYOXOMETALATES CONTAINING COMPOUNDS

The Anderson-Evans type of polyoxometalate $[M'(OH)_6M_6O_{18}]^{n-}$ is interesting for two important respects (i) this polyoxoanion is planar and (ii) it contains a paramagnetic ion (Cr, Ni, Fe,...) (see scheme 1) and therefore it can give rise to molecular solid state organization reminiscent to Miller's ferromagnet $Fe(Cp^*)_2$.TCNE.CH_3CN [30] with alternate stack of the $[Fe(Cp^*)_2]^+$ units and polyanions. Different crystalline phases of this with both decamethylferrocenium $Fe(C_5(CH_3)_5)_2^{+\cdot}$ and ferrocenium radical cations $Fe(C_5H_5)_2^{+\cdot}$ trianion were obtained by metathesis.

$[Fe(Cp^*)_2]_3\ [Cr(OH)_6Mo_6O_{18}]$. 20 H_2O salt.

The crystal structure of this compound [32, 34] represented in figure 14 is very unusual. It shows bidimensionnal layers containing orthogonal alternate stacks of the trianion and $[Fe(Cp^*)_2]^+$ centrosymmetric trimers with ...(ABA)POM(ABA)POM... sequence. In the trimer, the $[Fe(Cp^*)_2]^+$ units are perpendicular to each other. The distances between spin carriers (Fe-Fe distance of 7.889(2)Å and Cr-Fe distance of 8.887(2)Å are too large and consequently, we did not observed any magnetic interaction for this salt.

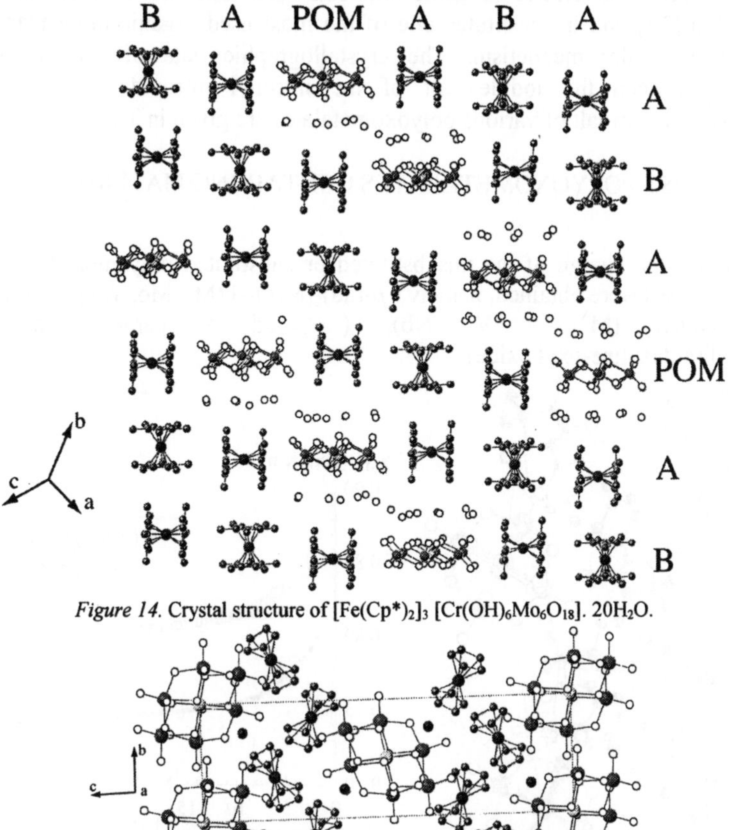

Figure 14. Crystal structure of $[Fe(Cp^*)_2]_3\ [Cr(OH)_6Mo_6O_{18}]$. $20H_2O$.

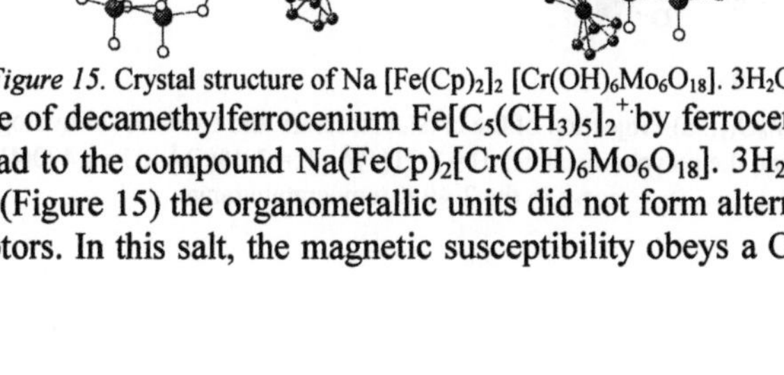

Figure 15. Crystal structure of Na $[Fe(Cp)_2]_2\ [Cr(OH)_6Mo_6O_{18}]$. $3H_2O$.

The exchange of decamethylferrocenium $Fe[C_5(CH_3)_5]_2^{+\cdot}$ by ferrocenium $Fe(C_5H_5)_2^{+\cdot}$ radical cation lead to the compound $Na(FeCp)_2[Cr(OH)_6Mo_6O_{18}]$. $3H_2O$ [31-34]. In its crystal structure (Figure 15) the organometallic units did not form alternate stacks of the donor and acceptors. In this salt, the magnetic susceptibility obeys a Curie law without any magnetic interaction. In order to initiate interactions in such materials, we try to decrease these separations between carriers by using smaller cations with higher magnetic moments like $Cr(Cp)_2^+$ (S = 3/2). However, while the Fe containing metallocenium are water and air stable, the Mn and Cr containing metallocenium are very sensitive and instable. The Anderson-Evans polyanions being stable only in water. So, as a prerequisite to this work, we synthesized the $(Cat^+)_3[Cr(OH)_6Mo_6O_{18}]$.6 DMSO salts ($cat^+$= Na^+ and Bu_4N^+) [32, 34] that are stable in dried organic solvent.

4. Nitronyl-nitroxide Containing Materials

Nitronyl-nitroxide derivative organic radicals are known to generate organic ferromagnets [35] and it constitutes one of the most used organic molecular building blocks in molecular magnetism. The crystallographic data of salts obtained by metathesis between the iodide salt of the nitronyl nitroxide radicals and the tetraalkylammonium salt of various polyoxometalates are given in Table 3.

4.1. LINDQUIST POLYOXOMETALATES CONTAINING MATERIALS

Depending on the charge of the unsubstituted or substituted polyanion, two different crystalline phases were obtained, namely (*p*-rad)$_2$[M_6O_{19}] (M= Mo, W) [36. 37] and (*p*-rad)$_3$[$M^VW_5O_{19}$] (M^V = V, Nb), (*p*-rad = para-methylpyridynium-tetramethylimidazoline-oxyl oxide).

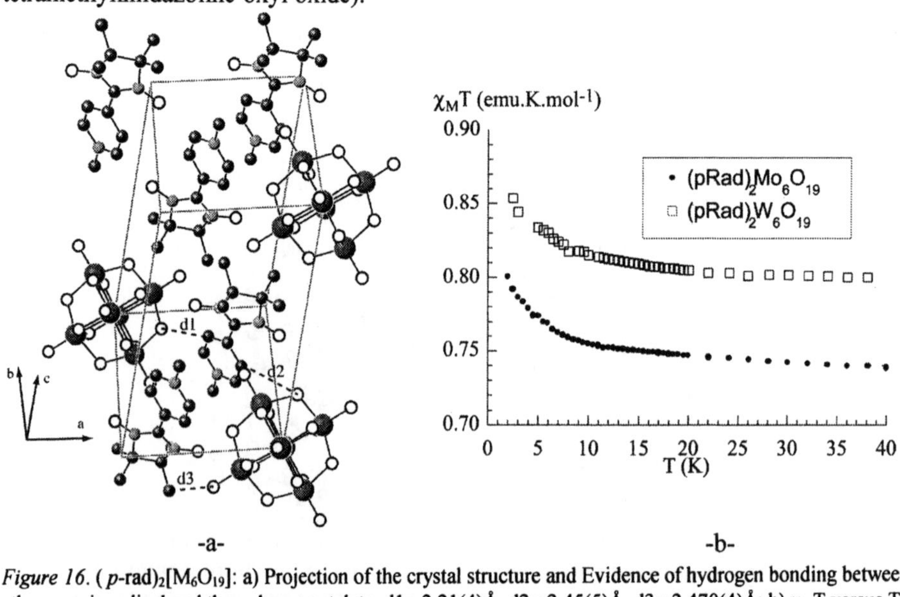

-a- -b-

Figure 16. (*p*-rad)$_2$[M_6O_{19}]: a) Projection of the crystal structure and Evidence of hydrogen bonding between the organic radical and the polyoxometalate. d1= 2.21(4)Å, d2= 2.45(5)Å, d3= 2.470(4)Å; b) χ_MT versus T curve in the 2-40 K temperature range.

Table 2. Crystal data for ferrocenium and decamethylferrocenium radical cation salts.

Compounds	a, (Å)	b, (Å)	c, (Å)	α, (°)	β, (°)	γ, (°)	SG	R[i]
$[Fe(Cp^*)_2]_2$ $[Mo_6O_{19}]$	9.906(4)	11.180(2)	23.11(2)	90.	91.92(2)	90.	$P2_1/c$	0.031
$[Fe(Cp^*)_2]_4$ $[SiMo_{12}O_{40}]$. 4 DMSO. 3 H_2O	28.817(4)	15.141(5)	31.432(6)	90.	116.93(1)	90.	C2/c	0.,083
$[Fe(Cp^*)_2]_4$ $[SiMo_{12}O_{40}]$. DMF	14.247(11)	15.004(4)	15.418(2)	63.26(2)	83.41(3)	69.77(4)	$P\bar{1}$	0.062
$[Fe(Cp^*)_2]_4$ $[SiW_{12}O_{40}]$. 2 DMF	14.234(11)	15.020(2)	15.410(3)	63.25(1)	83.92(3)	70.39(3)	$P\bar{1}$	-
$[Fe(Cp^*)_2]_4$ $[PMo_{12}O_{40}]$. 2 DMF	14.240(12)	15.011(3)	15.415(3)	63.25(2)	83.62(3)	70.12(3)	$P\bar{1}$	-
$[Fe(Cp^*)_2]_4[HFeW_{12}O_{40}]$. CH_3CN. 2 H_2O	29.495(9)	12.545(4)	29.482(24)	90.	94.39(4)	90.)	C2/c	0.045
$[Fe(Cp^*)_2]_4[HPCu(H_2O)W_{11}O_{39}]$. $6CH_3CN$	15.596(7)	15.138(3)	25.921(12)	90.	97.53(2)	90.	C2/c	0.074
$[Fe(Cp^*)_2]_3[Cr(OH)_6Mo_6O_{18}]$, $20H_2O$	12.489(3)	14.113(3)	15.662(4)	101.5(1)	105.82(2)	110.23(2)	$P\bar{1}$	0.037
Na $[Fe(Cp^*)_2]_2$ $[Cr(OH)_6Mo_6O_{18}]$. $3H_2O$	10.015(2)	7.722(7)	25.137(5)	90.	90.62(2)	90.	$P2_1/n$	0.051

[i] Reliability R factor,

Table 3. Crystal data for nitronyl nitroxide radical cation salts.

Compounds	a, (Å)	b, (Å)	c, (Å)	α, (°)	β, (°)	γ, (°)	SG	R[i]
(*p*-rad)$_2W_6O_{19}$(DMF)	8.268(1)	11.418(6)	12.076(2)	65.77(2)	79.27(2)	84.02(3)	$P\bar{1}$	0.019
(*p*-rad)$_2Mo_6O_{19}$(DMF)	8.240(8)	11.394(4)	12.059(1)	65.89(1)	79.64(3)	84.27(5)	$P\bar{1}$	0.022
(*p*-rad)$_3NbW_5O_{19}$	12.288(4)	14.122(5)	19.865(9)	98.03(2)	95.41(6)	110.06(2)	$P\bar{1}$	0.054
(*p*-rad)$_4$ $SiW_{12}O_{40}$	12.206(2)	14.835(2)	15.40(1)	85.67(4)	87.82(3)	85.59(4)	$P\bar{1}$	0.042
(*p*-rad)$_4$ $SiMo_{12}O_{40}$	12.19(1)	14.80(1)	15.51(2)	84.52(8)	86.65(7)	85.48(8)	$P\bar{1}$	0.080
(*m*-rad)$_5$ $SiW_{12}O_{40}$	25.43(2)	19.759(6)	23.04(4)	90.	91.12(4)	90.	C2/c	0.071

[i] Reliability R factor

In the crystal structure of the former salts (Figure 16a), the organic radicals cations form centrosymetric dimers with short intermolecular O...H distances between the radicals (~2.34Å). Another interesting feature in these materials concerns the interactions between the organic and inorganic units. In fact, short O...H contacts (~2.25Å) have been observed between the H atoms of the organic radical and the O atoms of the polyoxometalates indicating the existence of hydrogen bonding in these materials. The magnetic properties are represented in Figure 16b in the form of the $\chi_M T$ versus T plot. The $\chi_M T$ values observed at room temperature correspond to what are expected for two uncorrelated radical spins. Below 60K, these values increase as T is lowered and reach a maximum at 2K. These experimental data may be fitted with the Curie-Weiss law. Such behaviour reveals dominant ferromagnetic interactions between the spins of the organic radicals.

The crystal structure of the substituted polyanion salts [(*p*-rad)$_3$[$M^V W_5 O_{19}$] (M^V = V, Nb)] [37] (Figure 17) contains three independent organic radical cations noted A, B and C that form centrosymetric dimers separated by the polyanions with however, short intra and interdimers O...H contacts (2.25-2.60 Å). Here again, the magnetic susceptibility (Figure 17) revealed ferromagnetic interactions between the organic radicals as indicated by the increase of the $\chi_M T$ versus T plot below ~5K.

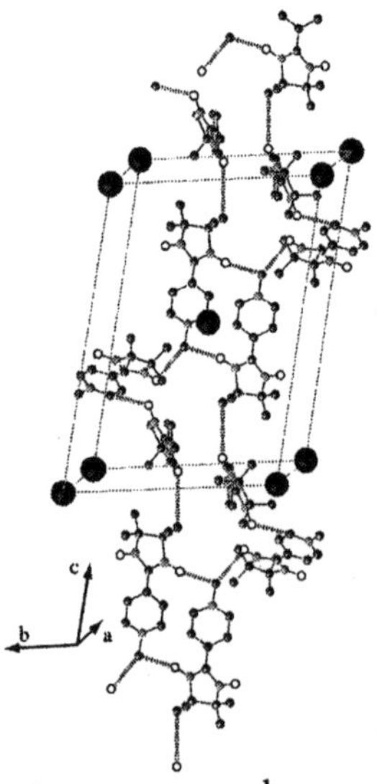

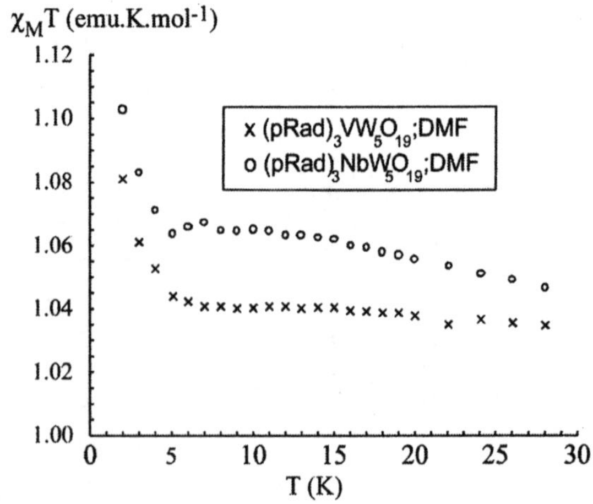

-a- -b-

Figure 17. (*p*-rad)$_3$[M'W_5O_{19}], M' = Nb, V: a)- Projection of the crystal structure, for clarity, the polyanions are represented by full circles b)- $\chi_M T$ versus T curve in the 2-30 K temperature range.

4.2. KEGGIN POLYOXOMETALATES CONTAINING MATERIALS

Two kinds of crystalline phases were observed with 1:4 and 1:5 stoichiometry for anion:cation ratio. The former stoichiometry was found in the isostructural (*p*-rad)$_4$ [$SiW_{12}O_{40}$] and (*p*-rad)$_4$ [$SiMo_{12}O_{40}$] salts [37]. Their crystal structures (Figure 18) contain two independent organic radicals noted A and B which generate centrosymmetric tetramers BAAB separated by polyoxometalates. Short intratetramer contacts (2.65-3.10 Å) were established between the oxygen atoms of the nitroxide group of one radical and the H atoms of the pyridinium ring of its adjacent radical. The magnetic properties obey a Curie-Weiss law with small negative θ values characteristic of weak antiferromagnetic interactions at low temperature.

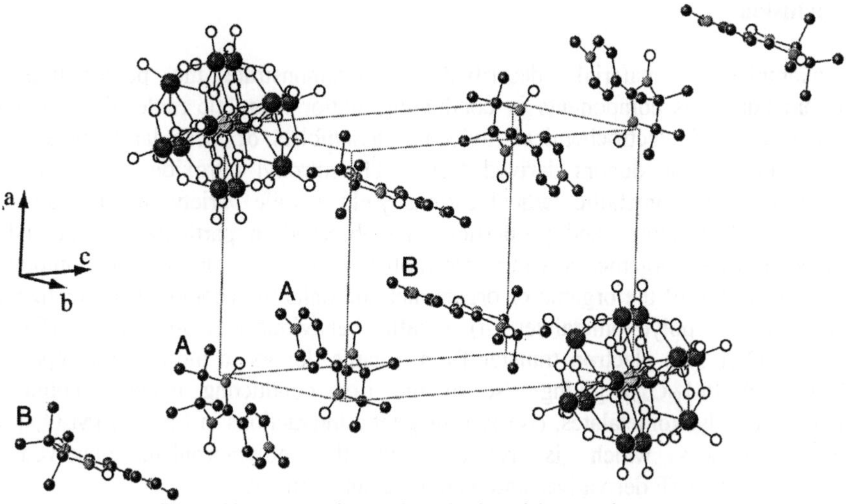

Figure 18. (*p*-rad)$_4$[SiW$_{12}$O$_{40}$]: Projection of the crystal structure

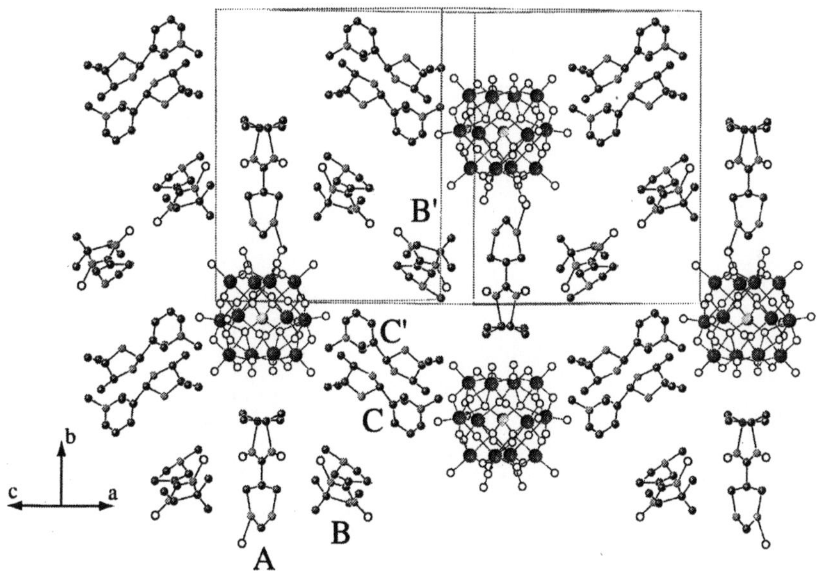

Figure 19. Crystal structure of (*m*-rad)$_5$[SiM$_{12}$O$_{40}$] (M=W, Mo).

The 1:5 stoichiometry was found in (*m*-rad)$_5$[SiW$_{12}$O$_{40}$] and (*m*-rad)$_5$[SiMo$_{12}$O$_{40}$] salts (*m*-rad = meta-methylpyridynium-tetramethylimidazoline-oxyl oxide) [37] containing one-electron reduced polyanions. This is a rare example of Keggin polyoxometalate salts containing more than four organic cations. The crystal structure (Figure 19) contains 2.5 independent meta radical units which form a spherical cavity fitted by the polyanion. More structural and magnetic investigations are underway for these two compounds.

5. Conclusion

The examples of materials described here demonstrate the potentialities of polyoxometalates as components in conducting and/or magnetic molecular materials. The chemical and electrochemical molecular assemblies of these large metal-oxide clusters with organic donors derived from TTF, nitronyl nitroxide derivatives and metallocinium organometallic radical cations yield a wide variety of materials with various crystal structures and properties. We observed in particular (i) the hybrid character of these organic/inorganic assemblies is reflected in all the compounds through the ability of the organic or organometallic units to accommodate the size and the shape of the polyoxometalates (ii) metallic behaviour was observed in (BEDT-TTF)$_5$[VW_5O_{19}], (iii) electron transfer from organic donors to inorganic acceptors in (TTF)$_6$(Et_4N)[$HPM_{12}O_{40}$] giving rise to the first conducting material containing paramagnetic polyoxometalates, (iv) ferromagnetic interactions in (*p*-rad)$_2$[M_6O_{19}], (v) structural richness which is reflected in the polymorphism observed in polyoxometalates/TTF derivatives and metallocenium systems.

References

[1] Souchay, P. (1969) *Ions Minéraux Condensés*, Masson, Paris.
[2] Pope, M.T. (1983) *Heteropoly and Isopoly Oxometalates*, Springer-Verlag New York.
[3] Evans, H.T. Jr. (1971) Perspectives in Structural Chemistry. 4, 1.
[4] Pope, M.T. and Müller, A. (1991) *Angew. Chem. Int. Ed. Engl.* 30, 34.
[5] Pope, M.T. and Müller, A., Eds, (1994) *Polyoxometalates: from Platonic Solids to Anti-retroviral activity*, Kluwer Academic Publishers, The Netherlands,.
[6] a)- Ouahab, L., Triki, S., Grandjean, D., Bencharif, M., Garrigou-Lagrange C., Delhaès P. (1991) in "*Lower-Dimensional Systems and Molecular Electronics*" Metzger, R.M., Day, P., Papavassillou, G.C., Eds, Plenum Press : NATO-ASI series, **B248**, 185, b) Ouahab, L. page 245 in Ref. 5. c) Coronado, E., Gomez-Garcia, C.J. page 233 in Ref. 5., d)- Ouahab, L. (1997) *J. Materials Chemistry*, **9.** 1909, e)- Gomez-Garcia, C.J., Coronado, E. (1995) *Comment Inorg. Chem.*, **17/5,** 255, f)- Coronado, E., Galan-Mascaros, J.R., Gimenez-Saiz, C., Gomez-Garcia, C.J. (1996) in "*Magnetism: A Supramolecular Function*" Kahn O., Ed., Kluwer Academic Publisher, 281 g)- Coronado, E. and Gomez-Garcia, C.J. (1998) *Chem. Rev*, **98,** 273, h)- Ouahab, L. (1998) *C. R. Acad. Sci. Paris*, **t.1. Serie IIc**, 369, i)- Ouahab, L. (1998) *Coord. Chem. Rev.*, **178-180,** 1501.
[7] Ouahab, L., Bencharif, M., and Grandjean, D. (1988) *C.R. Acad. Sci. Paris*, **Série II 307,** 749.
[8] Triki, S., Ouahab, L., Padiou, J., Grandjean, D. (1989) *J. Chem. Soc., Chem. Comm.*, 1068.
[9] Davidson, A., Boubekeur, K., Pénicaud, A., Auban, P., Lenoir, C., Batai,l P. and Hervé, G. (1989) *J. Chem. Soc. Chem. Commun.*, 1373.
[10] Bellitto, C., Attanazio, D., Bonamico, M., Fares, V., Imperatori, P., Patrizio, S., (1990) *Mat. Res. Soc. Pro.*, **173,** 143.
[11] Gomez-Garcia, C.J., Borras-Almenar, J.J., Coronado, E., Delhaès, P., Garrigou-Lagrange, C., Baker, L.C.W. (1993), *Synth. Met.*, **56,** 2023 .
[12] a)- Gomez-Garcia, C.J., Ouahab, L., Gimenez-Saiz, C., Triki, S., Coronado, E., Delhaès, P. (1994) *Angew. Chem. Int. Ed. Engl.*, **33-2,** 223, b)- Gomez-Garcia, C.J., Gimenez-Saiz, C., Triki, S., Coronado, E., Le Maguerès, P., Ouahab, L., Ducasse, L., Sourisseau, C., Delhaès (1995), P., *Inorg. Chem.,* **34,** 4139.
[13] Launay, J.P. (1976) *J. Inorg. Nucl. Chem.*, **18,** 807.
[14] Shaikh, S.N., Zubieta, J. (1986) *Inorg. Chem.*, **25,** 4613, b)- Proust, A., Fournier, M., Thouvenot, R., Gouzerh, P. (1994) *Inorg. Chim. Acta*, **215,** 61, c)- Chorghade, G.S., Pope, M.T. (1987) *J. Am. Chem. Soc.*, **109,** 5134, d)- Strong, J.B., Ostrander, R., Rheingold, A.L., Maatta, E. (1994), *J. Am. Chem. Soc.*, **116,** 3601, e)- Gouzerh, P. and Proust, A. (1998), *Chem. Rev.,* **98,** 77.
[15] a)- Elliott, J.R. (1965) *Magnetism*, Eds. Rado & Suhl, Acad. Press, New York Vol IIA, 385, b)- Heeger, A.J. (1969) *Solid State Physics*, B, 23, 283, c)- Delhaès, P. (1991), in NATO ASI *lower-Dimensional Systems and Molecular Electronics;* Eds Metzger, R.M., Day, P., Papavassillou, G.C., Plenum Press, New York, 43, d)- Ogawa, M.Y.,

Martinsen, J., Palmer, S.M., Stanton, J.L., Tanaka, J., Greene, R.L., Hoffman, B., Ibers, J.A (1987) *J. Am. Chem. Soc.* 109, 1115, e) Kahn, O. (1993) *Molecular Magnetism*; Verlag-Chemie: New York.

[16] Williams, J.M., Ferraro, J.R., Thorn, R.J., Carlson, K.D., Geiser, U., Wang, H.H., Kini, A.M., Whangbo, M.H. (1992) "*Organic Superconductors. Synthesis, Structure, Properties and Theory*", Grimes R.N., Ed., Prentice Hall, Englewood Cliffs, New Jersey.

[17] Bellitto, C., Attanazio, D., Bonamico, M., Fares, V., Imperatori, P., Patrizio, S. (1990) *Mat. Res. Soc. Pro.*, **173,** 143.

[18] Triki, S., Ouahab, L., Grandjean, D., Fabre, J.M. (1991) *Acta. Crystallogr.*, **C47,** 1371.

[19] Triki, S., Ouahab, L., Grandjean, D., Fabre, J.M. (1991)*Acta. Crystallogr.*, **C47,** 645.

[20] Triki, S., Ouahab, L., Grandjean, D., Amiell, J., Garrigou-Lagrange, C., Delahès, P. and Fabre, J.M. (1991) *Synt. Met.*, **41-43,** 2589.

[21] Triki, S., Ouahab, L., Fabre, J.M. (1994) *Acta. Crystallogr.*, **C50,** 219.

[22] Triki, S., Ouahab, L., Halet, J.F., Peña, O., Padiou, J., Grandjean, D., Garrigou-Lagrange, C., Delhaès P. *J. Chem. Soc. Dalton Trans.*, 1217 (1992).

[23] Triki, S. (1992) *PhD Thesis*, University of Rennes 1.

[24] Bechgaard, K., Jacobsen, S., Mortensen, K., Pedersen, H.J., Thorup N. (1980) *Solid. State Commun.,* **33,** 1119.

[25] Triki, S., Ouahab, L., Grandjean, D. (1993) *Acta. Crystallogr.*, **C49,** 132.

[26] Triki, S., Ouahab, L., Grandjean, D., Amiell, J., Garrigou-Lagrange, C., Delhaès, P. (1993) *Synt. Met.*, **56/1,** 1787.

[27] Ouahab, L., Grandjean, D., Bencharif, M. (1991) *Acta. Crystallogr.*, **C47,** 1980.

[28] a)- Ouahab, L., Bencharif, M., Mhanni, A., Pelloquin, D., Halet, J.F., Peña, O., Padiou, J., Grandjean, D., Garrigou-Lagrange, C., Amiell, J., Delhaès, P. (1992) *Chem. Mater.* **4,** 666, b)- Mhanni, A., Ouahab, L., Peña, O., Grandjean, D., Garrigou-Lagrange, C., Delhaès, P. (1991) *Synt. Met.*, **41-43,** 1703.

[29]- a) Prados, R.A., Pope, M.T. (1976) *Inorg. Chem.* 15, 2547, b) Launay, J.P., Fournier, M., Sanchez, C., Livage, J., Pope, M.T. (1980) *Inorg. Nucl. Chem. Lett.* 16. 257, c) Barrows, J.N., Pope, M.T. (1990) in *Electron Transfer in Biology and the Solid State*; Johnson, M.K., King, R.B., Kurtz, D.M., Kutal, C., Norton, M.L., Scott, R.A.. Eds, Advances in Chemistry Series 226. American Chemical Society: Washington, DC.

[30] Miller, J.S., Calabresse, J. C., Epstein, A.J., Bigelow, W., Zhang, J.H., Reiff, W.M. (1986) *J. Chem. Soc., Chem. Commun.*, 1026.

[31] Le Maguerès, P., Ouahab, L., Golhen, S., Peña, O., Gomez-Garcia, C.J., Delhaès, P. (1994) *Inorg. Chem*, 33, 5180.

[32] Golhen, S., Ouahab, L., Grandjean, D., Molinié, P. (1998) *Inorg. Chem.,* **7**, 1499.

[33] LeMaguerès, P. (1995) *PhD Thesis*, University of Rennes 1.

[34] Golhen, S. (1997) *PhD Thesis*, University of Rennes 1.

[35] a)- Iwamura, H., Koga, N. (1993), *Acc. Chem. Res.* **26.** 346, b)- Fujita, I., Teki, Y., Takui, T., Kinoshita, T., Itoh, K., Miko, F., Sawaki, Y, (1990) *J. Am. Chem. Soc.*, **112,** 4047, c)- Nakamura, N., Inoue, K., Iwamura, H., Fulioka, T., Sawaki, Y. (1992) *J. Am. Chem. Soc.*, **114,** 1484, d)- Iwamura, H. (1993) *Mol. Cryst. Liq. Cryst.*, **232,** 233, e)- Chiarelli, R., Novak, M.A., Rassat, A., Tholence, J.L. (1993) *Nature*, **363,** 147, f)-

Turek, P., Nozawa, K., Shiomi, D., Awaga, K., Inabe, T., Maruyama, Y., Kinoshita, M.,*Chem. Phys. Lett.* (1991) **180,** 327, g)- Tamura, M., Yakazawa, Y., Shiomi, D., Nozawa, K., Hosokoshi, Y., Ishikawa, M., Takahashi, M., Kinoshita, M. (1991) *Chem. Phys. Lett.*, **186,** 401, h)- Pei, Y., Kahn, O., Aebersold M.A., Ouahab, L., Le Berre, F., Pardi, L., Tholence, J.L. (1994) *Advanced Materials,* **6,** 681, i)- Lang, A., Pei, Y., Ouahab, L., Kahn, O. (1996) *Advanced Materials,* **8,** 60.

[36] Rimbaud, C., Ouahab, L., Sutter, J.P., Kahn, O. (1997) *Mol. Cryst. & Liq. Cryst.,* **306,** 67.

[37] Rimbaud, C. (1997) *PhD Thesis*, University of Rennes 1.

MOLECULAR MATERIALS FROM POLYOXOMETALATES

J. J. Borrás-Almenar, J. M. Clemente-Juan, M. Clemente-León, E. Coronado, J. R. Galán-Mascarós, C. J. Gómez-García

Dpto. Química Inorgánica. Univ. of Valencia. Dr. Moliner 50, 46100 Burjasot (Spain)
E-mail: eugenio.coronado@uv.es

Abstract

The present article highlights recent results and provide a perspective of the interest of polyoxometalates as inorganic component of molecular materials with active physical properties. Three different aspects will be presented: i) The interest of the magnetic and mixed valence clusters provided by polyoxometalate chemistry in molecular magnetism; ii) The use of these inorganic anions as magnetic component of crystalline conducting materials based on organic donor molecules; iii) The construction of well-organized films of polyoxometalate monolayers by using the Langmuir-Blodgett technique.

1. Introduction

The design of molecular materials with unusual magnetic, electrical and optical properties is an active focus of current research in materials science. In this contribution we will illustrate the possibilities provided by polyoxometalate chemistry in this area. Three different aspects will be presented: (i) the relevance of polyoxometalate chemistry in molecular magnetism. We will show that ideal examples of magnetic and mixed-valence clusters of increasing nuclearity can be obtained, wherein magnetic exchange interactions as well as electron transfer processes can be studied at the molecular level [1]. (ii) The use of these electron acceptors as inorganic component of crystalline radical salts based on organic donors which exhibit coexistence of localized magnetic moments and itinerant electrons [2]. (iii) The organization of these cluster anions as monolayers by using the Langmuir-Blodgett technique [3].

2. Magnetic Clusters

Molecular assemblies formed by a finite number of exchange-coupled paramagnetic centers, are currently receiving much attention in several active areas of research as molecular chemistry, magnetism or biochemistry. As they are in between the small molecular systems and the bulk state, the limited number of interacting centers often

M.T. Pope and A. Müller (eds.), Polyoxometalate Chemistry, 231–253.

allows us to model their properties with quantum mechanical approaches, avoiding the further approximations required to treat extended solids. On the other hand, when these clusters are large enough to give rise to ground states characterized by large spin and large magnetic anisotropy they may behave, below a given temperature, as magnets of nanometer size [4]. For example, the molecular cluster comprising twelve manganese ions, $[Mn_{12}O_{12}(CH_3COO)_{16}(H_2O)_4]$, shows below 4 K large magnetic hysteresis comparable to that observed in hard magnets. As they show a marked memory effect they may in principle be used in order to store information at the molecular level.

Polyoxometalates provide excellent examples of magnetic clusters. Recent reviews accounting for the state-of-the art in this area can be found in [5] and [6]. The ability of tungstates, and to a less extent of molybdates, for acting as ligands toward 3d-transition metal ions enables to obtain a variety of magnetic clusters possessing different spins and showing either ferromagnetic as well as antiferromagnetic exchange couplings. The bulky nonmagnetic polyoxometalate framework guarantees an effective magnetic isolation of the cluster, imposing at the same time its geometry. Furthermore, its chemistry allows the assembly of stable anion fragments into larger clusters. A chemical control of the magnetic nuclearity is therefore possible. The above characteristics make these complexes ideal candidates for modelling the magnetic exchange interactions in clusters of increasing nuclearities and definite topologies [1, 7]. In addition, they can be reversibly reduced to mixed-valence species (heteropoly "blues" and "browns") by injection of variable numbers of electrons. In the heteropoly blues these extra electrons are delocalized over a significantly large number of centers of the heteropoly framework [8]. The further introduction into these structures of paramagnetic metal atoms, leads to the creation of clusters in which localized magnetic moments and delocalized electrons can coexist and interact [9]. These features are motivating the development of new theoretical models in order to treat the problem of the electron transfer effects in large clusters, as well as to understand the interplay between electron delocalization, magnetic interactions and Coulomb repulsions in high nuclearity multielectronic systems [6].

A second important class of magnetic polyoxometalates are the polyoxovanadates (IV). As for the previous systems, polyoxovanadate chemistry provides unique examples to study the exchange interactions as well as the electron delocalization effects in large clusters [5, 10]. In this case, however, the exchange interactions occurs between the addenda vanadium atoms of the polyoxometalate anion. Thus, the polyoxovanadate cluster can be viewed as a fully magnetic cluster in which the $S = 1/2$ spins of the oxovanadium (IV) centers are antiferromagnetically coupled. As far as the electron delocalization is concerned, the vanadates exhibit remarkable examples of mixed-valence clusters in which the ratio between localized and delocalized spins can be chemically tuned. For example, in the $V_{18}O_{42}$ species one can partially oxidize the fully localized V_{18}^{IV} cluster by removing up to 8 electrons. This has a strong influence on the magnetic properties.

In this section we will present some relevant examples that illustrate the possibilities offered by polyoxometalate chemistry to study in detail magnetic exchange interactions as well as electron transfer processes in clusters. Two different systems will be presented: 1) The complexes based upon the trivacant tungstophosphate ligand $[PW_9O_{34}]^{9-}$. From the chemical point of view this family illustrates how the nuclearity of the cluster can be chemically controlled. From the magnetic point of view this family provides ideal examples of well-insulated magnetic clusters with predictable magnetic properties for which a detailed information on the nature of the magnetic exchange interactions can be extracted. 2) Mixed-valence polyoxometalates. We will try to understand how the interplay between electron delocalization and exchange interactions affect the magnetic properties of this class of high-nuclearity mixed-valence clusters.

2.1. INCREASING THE NUCLEARITY OF THE MAGNETIC CLUSTERS

By removal of a triad of octahedra sharing edges from the phosphotungstate anion α-$[PW_{12}O_{40}]^{3-}$ we end up with the trivacant ligand B-$[PW_9O_{34}]^{9-}$. Using this ligand various complexes of Ni(II) and Co(II) can be obtained. Thus, with Ni(II) up to three different complexes containing magnetic clusters of nuclearities 3, 4 and 9 can be isolated when one, two, or three phosphotungstate moieties are linked via their coordination to Ni(II) ions. Although none of the products is obtained exclusively, it is possible to find the best conditions to crystallize in high yield the salt of the desired nickel complex [7]. It has been found that the key factors that determine the nature of the final product are the pH and the temperature.

The Ni_3 cluster is present in the $[Ni_3(H_2O)_3(PW_{10}O_{39})H_2O]^{7-}$ polyoxoanion (Figure 1). This polyoxoanion may be viewed as a reconstituted Keggin-like structure $\{PNi_3W_9O_{40}\}$ wherein a W_3 triad has been substituted by Ni_3. Therefore, this triangular Ni_3 cluster is formed by three edge-sharing NiO_6 octahedra [11]. The Keggin structure is completed with an additional cap of WO_6 that produces a distorted $[Ni_3WO_4]$ cubane-type core with the Ni_3 cluster. Within the cluster each NiO_6 octahedron contains four oxide ions, a bridging oxygen from the central phosphate group, and a terminal oxygen from a coordinated water molecule. Recently a cluster showing this structure has been obtained wherein the additional cap of WO_6 has been replaced by a NiO_6 octahedra [12].

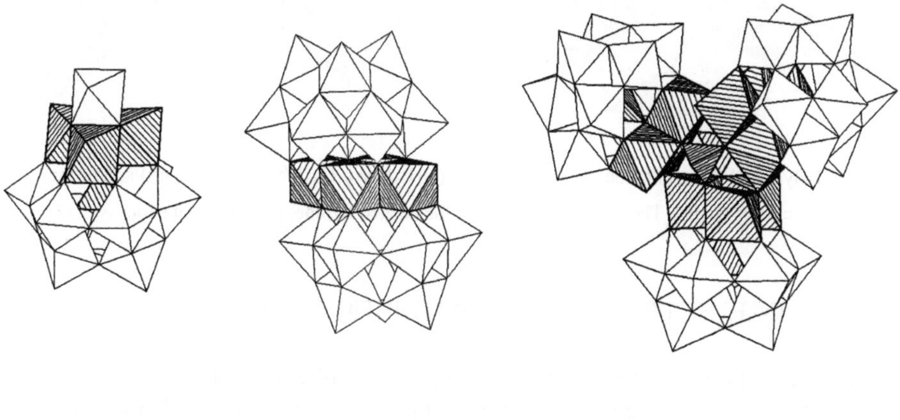

Figure 1. Polyhedral representation of the Ni_3 (left), Ni_4 (center) and Ni_9 (right) complexes. The Ni atoms are inside the shaded octahedra.

The Ni_4 complex exhibits the typical structure of the series $[M_4(H_2O)_2(PW_9O_{34})_2]^{10-}$ (M = Ni(II), Co(II), Mn(II), Cu(II), Fe(II)) first reported by Tourné et al. in 1973 for the cobalt derivative [13] and subsequently developed by others for the remaining metal atoms [14]. This structure consists of two heteropoly ligands $[PW_9O_{34}]^{9-}$ linked through coordination to a group of four divalent metal ions. This sandwiched central unit forms a centro-symmetric rhomb-like tetrameric cluster $\{M_4O_{16}\}$of four edge-sharing MO_6 octahedra.

Finally, the Ni_9 cluster is obtained by condensation of three reconstituted Keggin anions $\{PNi_3W_9O_{40}\}$. In the resulting giant polyoxometalate complex $[Ni_9(OH)_3(H_2O)_6(HPO_4)_2(PW_9O_{34})_3]^{16-}$ the magnetic cluster is formed by three triangular Ni_3O_{13} units sharing edges. These triangles are connected to each other by three OH^- bridging groups and two central HPO_4^{2-} groups in order to form a triangle of triangles (Figure 1). The periphery of this polyoxoanion is formed by three diamagnetic ligands $(PW_9O_{34})^{9-}$ which guarantee the insulation of the magnetic clusters. As for Ni_3 and Ni_4, in the Ni_9O_{36} unit the terminal oxygens of the NiO_6 octahedra belong to coordinated water molecules.

With cobalt (II) it has been possible to obtain the analogous tetranuclear [13] and nonanuclear clusters [15]. However, in the attempt to obtain the trinuclear cluster, a complex containing an heptanuclear Co_7 cluster has been obtained [16]. Its structure is unprecedented and is formed by two reconstituted Keggin units $\{PW_9Co_3\}$. These two units are linked together by a central unit formed by a tetrahedral Co site and six edge-sharing WO_6 octahedra (Figure 2).

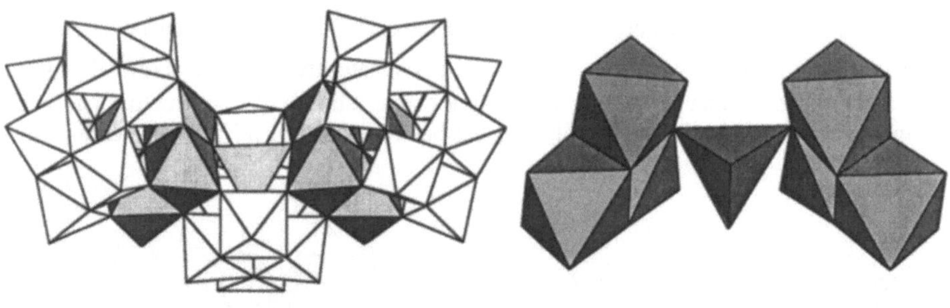

Figure 2. Polyhedral representation of the $[Co_7(H_2O)_2(OH)_2P_2W_{25}O_{94}]^{16-}$ polyanion (left) and of the Co_7 cluster (right). The Co atoms are inside the dashed tetrahedron and octahedra.

2.2 MODELING THE PROPERTIES OF MAGNETIC CLUSTERS

From the magnetic point of view these heteropoly complexes furnish various magnetic clusters of increasing nuclearities wherein the sign and nature of the exchange interactions can be directly correlated to the cluster structure and to the electronic ground states of the metal ions. As far as the sign is concerned, the presence in all cases of edge-sharing MO_6 octahedra with M-O-M angles close to 90° favors a ferromagnetic spin coupling due to the orthogonality of the magnetic orbitals. This can be seen in the magnetic curves of Ni_3, Ni_4, Ni_9 (figure 3) and Co_4 (figure 4) which show an increase of the magnetic moment when the temperature is decreased.

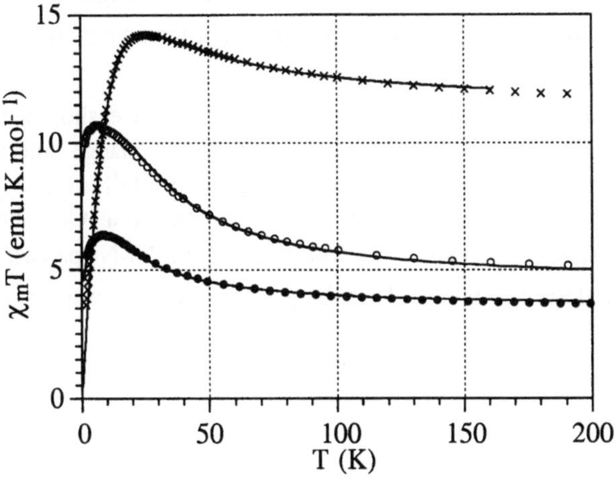

Figure 3. $\chi_m T$ vs temperature for the Ni_3 (filled circles), Ni_4 (empty circles) and Ni_9 (crosses) clusters. Solid lines are the best fit to the models (see text)

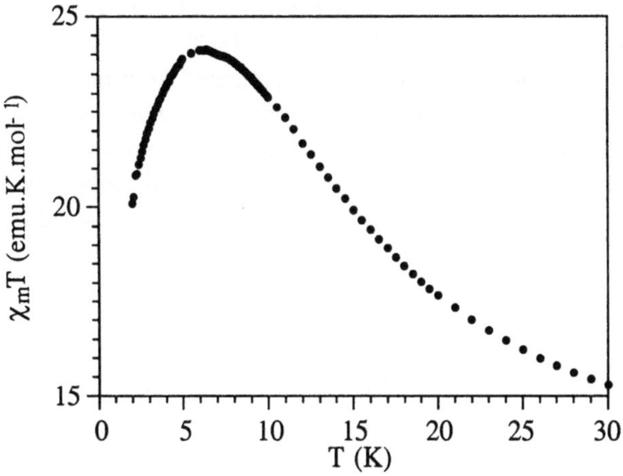

Figure 4. $\chi_m T$ vs temperature for the Co_4 cluster

On the other hand, in the particular case of the M_9 clusters the simultaneous presence of edge-sharing and corner-sharing octahedra with M-O-M angles of ca. 90° and

120°, respectively, favors a coexistence of ferromagnetic and antiferromagnetic exchange interactions. This coexistence accounts for the behavior of Ni_9 which shows an increase of the magnetic moment down to 30 K characteristic of dominant ferromagnetic interactions, followed by a sharp decrease at lower temperatures coming from the antiferromagnetic interactions which lead to a non magnetic S = 0 spin ground state.

This coexistence is also present in Co_9, although the magnetic moment shows a continuous decrease due to the presence of dominant antiferromagnetic interactions [17].

As far as the nature of the interaction is concerned, we should expect isotropic exchange interactions for Ni clusters and anisotropic exchange interactions for Co clusters. This is so because octahedral Ni^{II} possesses an orbitally non degenerate ground spin state ($^3A_2$ electronic term), whereas high-spin octahedral Co^{II} has an orbitally degenerate ground state ($^4T_1$ term). In the former case, the Ni-Ni pairwise interaction is expected to be well described by the Heisenberg Hamiltonian acting on the basis of the Ni spins S = 1:

$$H = -2J\, S_i S_j \tag{1}$$

In the second case, due to the combined effect of spin-orbit coupling and distortion of the octahedral site, the $^4T_1$ term is split into six Kramers doublets in such a way that the ground doublet is highly anisotropic and is separated from the first excited doublet by an energy gap of ca. 100 cm^{-1}. Thus, the magnetic moment associated to Co(II) can be described at low enough temperatures by an anisotropic spin S = 1/2. As a result the Co-Co pairwise interaction is expected to be conveniently described by an anisotropic exchange Hamiltonian acting on the basis of the effective spins S = 1/2:

$$H = -2J_x S_{ix} S_{jx} - 2J_y S_{iy} S_{jy} - 2J_z S_{iz} S_{jz} \tag{2}$$

We have used these two models to analyze the magnetic properties of these clusters. In general, these magnetic properties have provided valuable information on the sign and strength of the exchange interactions (Table 1).

TABLE 1. Magnetic characterization of magnetic clusters encapsulated by polyoxotungstate ligands. N and S_i are the magnetic nuclearities and the spin values of the interacting spins.

Anion	magnetic cluster	N	S_i	Magnetic parameters (cm^{-1})	comments	Ref.
$[Ni_3(H_2O)_3PW_{10}O_{39}(H_2O)]^{7-}$	Ni_3O_{13}	3	1	J = 3.9 D = 5.8	F exchange Heisenberg model + SIA[a]	11
$[Ni_4(H_2O)_2(PW_9O_{34})_2]^{10-}$	Ni_4O_{16}	4	1	J = 6.5 J' = 2.5 D = 4.5	F exchange Heisenberg model + SIA[a]	7
$[Ni_9(OH)_3(H_2O)_6(HPO_4)_2(PW_9O_{34})_3]^{16-}$	Ni_9O_{36}	9	1	J = 3.9 J' = -1.4	Coexistence of F and AF exchange Heisenberg model	7
$[Co_4(H_2O)_2(PW_9O_{34})_2]^{10-}$	Co_4O_{16}	4	"1/2"[b]	J_z = 12.2 J'_z = 3.7 J_x = 5.6 J'_x = 3.5 $J_x/J_y = J'_x/J'_y$ = 1.6	F exchange Anisotropic model	18
$[Co_9(OH)_3(H_2O)_6(HPO_4)_2(PW_9O_{34})_3]^{16-}$	Co_9O_{36}	9	"1/2"[b]	J_z = 8.4 J'_z = -12	Coexistence of F and AF exchange Ising model	17

[a] Single-Ion anisotropy
[b] Effective spin for octahedral Co(II)

However, when the complexity of the cluster increases (due to an increase in its nuclearity or to the presence of a large magnetic anisotropy), the number of magnetic parameters which have to be considered becomes too large. In such cases the information content of the magnetic susceptibility has shown to be insufficient for the determination of all the relevant magnetic parameters and other complementary techniques have been used in order to complete the information. An illustrative example is provided by Co_4. In this case the spin Hamiltonian appropriate to fit the magnetic susceptibility data should consider a minimum of 8 parameters namely:

(1) four exchange parameters J_z, J_{xy}, J_z', J_{xy}' to account for the two different exchange pathways J and J' (through the sides and through the diagonal of the rhomb), each one submitted to an axial anisotropy (Figure 5).

(2) four Landé parameters g_z, g_{xy}, g_z', g_{xy}' to account for the two different anisotropic Co^{II} sites.

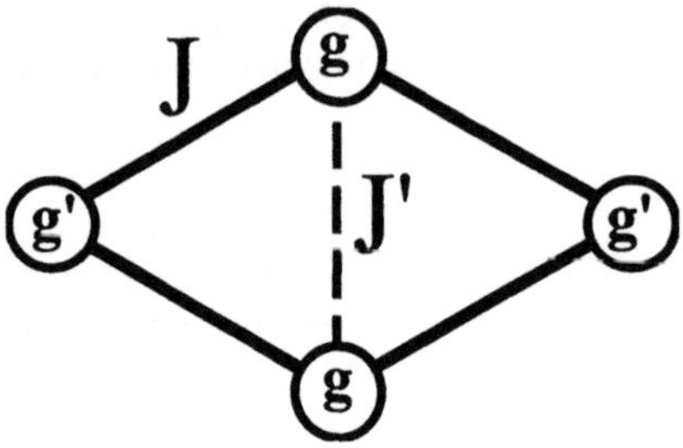

Figure 5. Exchange pathways for the M_4 clusters.

A fitting of the magnetic data to the above model indicates that this technique can only provide reliable information on the sign of the exchange interactions being almost insensitive to the presence of two different exchange pathways and to the amount of exchange anisotropy [19]. These data have been complemented with specific heat measurements and Inelastic Neutron Scattering measurements. The INS technique provides a direct access to the splitting of the low-lying energy levels of a cluster caused by exchange interactions. Therefore, a much deeper and more detailed insight into the nature of the magnetic coupling can be obtained with this spectroscopic technique than with the usual magnetic techniques [20]. At 2 K the INS spectrum of this compound shows up to six cold peaks which correspond to six magnetic excitations from the ground state to excited states of the tetramer (Figure 6). From these data an energy splitting pattern can be obtained (Figure 7) which can be closely reproduced by a fully anisotropic exchange model with the parameters J_z = 12.2 cm^{-1}, J_x = 5.6 cm^{-1}, J_z' =3.7 cm^{-1} , J_x' = 3.5 cm^{-1}, J_x/J_y = J_x'/J_y' = 1.6 [21]. With this spectroscopic technique evidence for an exchange-anisotropy in the Co_4 cluster is thus clearly demonstrated for the first time.

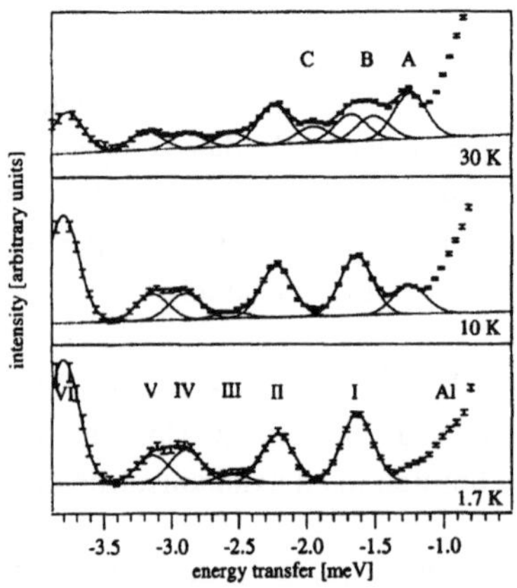

Figure 6. INS spectra of the deuterated salt $K_{10}[Co_4(D_2O)_2(PW_9O_{34})].22D_2O$ measured on the IN6 spectrometer with an incident neutron wavelength of 4.1 Å at 1.7, 10 and 30 K. Solid lines are the least-squares Gaussian fits with equal widths.

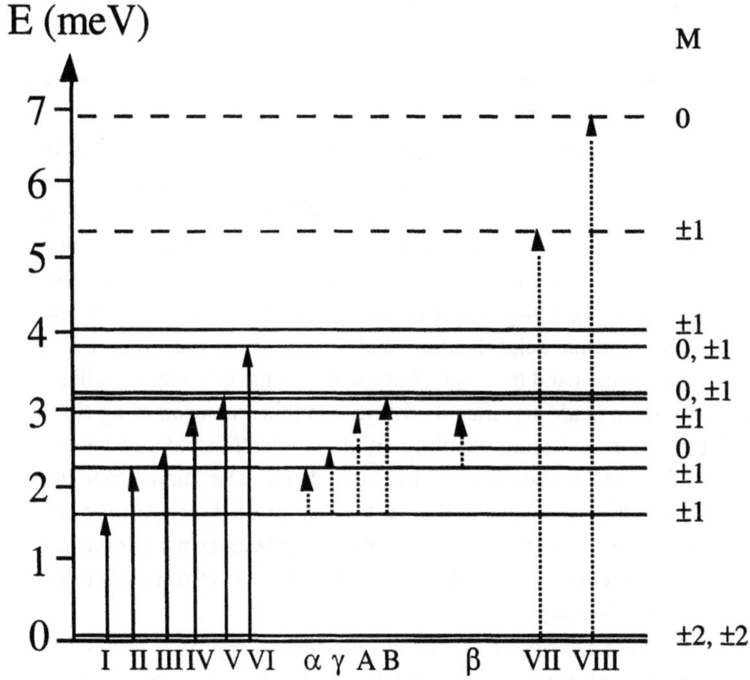

Figure 7. Energy level diagram of the Co_4 cluster-ground state derived from the INS experiments.

INS has also been used to complement the magnetic information in Ni_4. This cluster exhibits a ferromagnetic coupling between the Ni(II) ions, as evidenced by the magnetic susceptibility data (Figure 3). However, at low temperature the magnetic moment, instead of reaching the expected plateau for a ground spin state S = 4, exhibits a sharp decrease upon cooling down. This behavior is indicative of a splitting of this spin state caused by a single-ion anisotropy. As two different metal sites are present in the Ni_4 cluster, two different additional parameters, D and D', accounting for this local anisotropy should be taken into account. Again, the fit of the susceptibility involves a considerable number of parameters, so it is unable to provide reliable values of these two parameters. The information provided by INS has been able to distinguish between the two kinds of single-ion anisotropies. A larger value for the hydrated Ni site has been obtained (4.9 compared to 3.8 cm^{-1}), in good agreement with the larger distortion of this site.

To conclude this part it is important to emphasize that this kind of study has been possible thanks to the great versatility of polyoxometalate chemistry in providing examples of largely insulated magnetic clusters of increasing nuclearities, definite topologies and high symmetries, which can be easily deuterated in big amounts. From a fundamental point of view, the INS study of these clusters in combination with other magnetic techniques should provide a unique opportunity to progress in the understanding of the magnetic exchange interaction phenomenon in polynuclear metal complexes, as well as in its parametrization using effective Hamiltonians.

2.3. MODELING THE PROPERTIES OF MIXED-VALENCE CLUSTERS

Pursuing our interest in magnetic mixed-valence clusters we have explored the problems associated with the interplay between electron delocalization and spin interactions in polyoxometalate mixed-valence clusters. These problems are of current interest in magnetism due to the possibility of strongly stabilize a ferromagnetic coupling between the magnetic centers via a special kind of exchange coupling namely double-exchange [22, 23]. This mechanism is operative in a variety of mixed valence systems including the clusters of biological relevance [24] and the rare-earth manganates exhibiting giant magnetorresistance [25]. Until very recently, however, the interplay between electron delocalization and magnetic interactions was thoroughly treated for dimers, trimers and tetramers, but there was no treatment available for larger clusters [26]. We have made significant advances in this area examining the problem of electron delocalization and spin coupling in the 2-electron reduced heteropoly blues with Keggin [27] and Dawson-Wells structures [28]. The main conclusion of our model was that electron delocalization is at the origin of the experimentally observed spin pairing of the two electrons, as these transfer processes are operative even when the two delocalized electrons are widely separated in the polyoxometalate framework to minimize the Coulomb energy.

Recently we have developed a general computing approach that allows to treat much more complicated mixed-valence clusters [29]. By using this approach we have been able to model the magnetic properties of the mixed valence clusters $[V_{18}O_{42}]^{n-}$ reported by Müller et al. (Figure 8, left) [30]. In these clusters the V^{V}/V^{IV} ratios can be varied from the fully reduced single-valence cluster $\{V_{18}^{IV}\}$ to the partially oxidized mixed-valence cluster $\{V_{10}^{IV}V_8^{V}\}$. The magnetic properties indicate that the antiferromagnetic coupling is much larger in the latter (Figure 8, right).

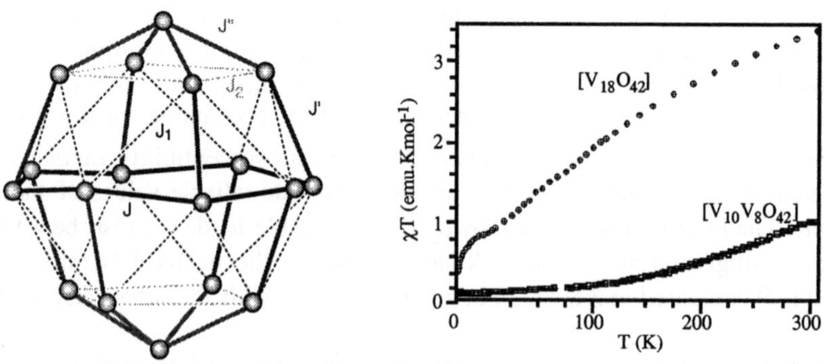

Figure 8. (left) Polyhedron describing the structure of the $[V_{18}O_{42}]$ species and scheme of the possible important exchange. (right) Experimental magnetic behaviour of the spin cluster $[V_{18}^{IV}O_{42}]$ and of the mixed-valence cluster $[V_{10}^{IV}V_8^{V}O_{42}]$.

This is a surprising result since the mixed-valence cluster comprises a smaller number of magnetic vanadium(IV) centers (10 compared to 18), and thus, one should expect a weakening in the V^{IV}-V^{IV} interactions on increasing the number of diamagnetic vanadium(V) centers. This stronger coupling has tentatively been attributed to electron-transfer effects. In order to confirm this assumption we have developed a model that considers the electron-transfer processes from V^{IV} to V^{V} centers, the exchange interactions between the V^{IV} centers, and the Coulomb repulsion parameter preventing two electrons to be on the same site [31]. This mixed-valence cluster involves 10 delocalized electrons undergoing electron-transfer processes among 18 metal sites, as well as an antiferromagnetic exchange coupling when they are on adjacent sites. Since the rigorous procedure to treat this system would require prohibitively large matrices to diagonalize and long computer times, we have considered a model system that comprises 10 metal sites and 6 delocalized electrons (Figure 9, left).

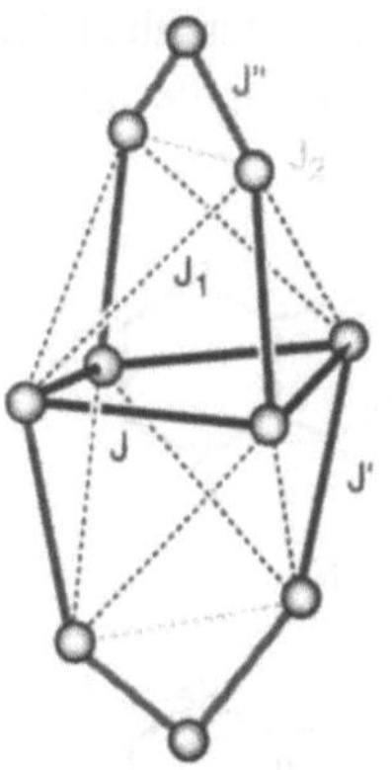

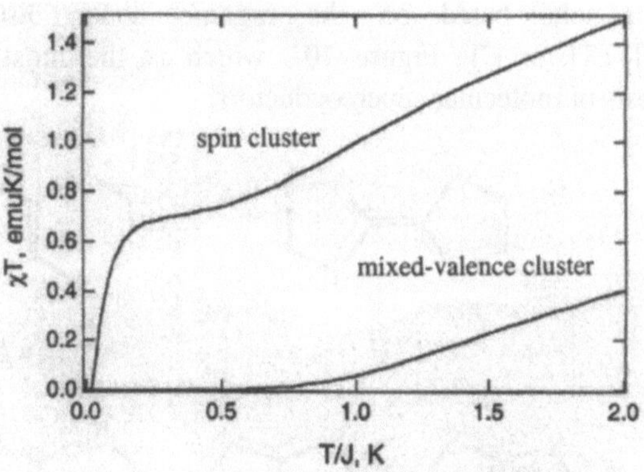

Figure 9. (left) Polyhedral model used to calculate the spin levels of the partially oxidized mixed-valence cluster $[V_{10}^{IV}V_8^{V}O_{42}]$. (b) Theoretical magnetic behavior of the spin cluster $[V_{18}^{IV}O_{42}]$ and of the mixed-valence cluster $[V_{10}^{IV}V_8^{V}O_{42}]$.

The evolution of the lower-lying spin levels with the transfer parameter t indicates that the role of the electron delocalization is that of stabilizing the antiferromagnetic ground state with respect to the fist excited spin-triplet level. Although the number of exchange and transfer pathways imposed by the geometry of the cluster is quite large (it involves 10 parameters) an attempt to quantitatively reproduce the experimental magnetic behavior has been made. The best set of parameters is (in units of K): J = -200, J' = -200; J" = -50, J_1 = -100, J_2 = -100, t = 400, t' = 400, t" = 1500, t_1 = -1000, t_2 = -1000. The corresponding theoretical curves are plotted in Figure 9 (right). As we can see, this analysis allows to reproduce the experimental trends, providing the first theoretical support to the magnetic findings.

An important point we want to stress is that polyoxometalates exhibit the largest nuclearities and most complex topologies never seen for clusters in the mixed-valence area. Therefore, the quantitative interpretation of their magnetic properties remains a challenging problem, as they are often too complex to be treated with the existing theories. Still, taking advantage of the high symmetry of these clusters, novel approaches have been developed in some cases that provide important hints on the role played by the electron transfer processes on the magnetic properties of these clusters.

3. Hybrid Salts based upon Organic π-Electron Donors and Magnetic Polyoxometalates

A current development in the area of molecular materials is to prepare hybrid materials formed by two molecular networks with the aim to combine in the same material two distinct physical properties such as electronic conductivity and magnetism. Examples include hybrid organic/inorganic materials formed by organic π–electron donors that furnish the pathway for conductivity, and inorganic metal complexes that act as structural or magnetic components. Polyoxometalates have been found to be extremely versatile building blocks of the aforementioned functionally active solids [2]. In 1994 we started the use of magnetic polyoxometalates as magnetic component of charge transfer salts based on the organic donor bis(ethylenedithio)tetrathiafulvalene (BEDT-TTF or ET; Figure 10), which is the most widely used molecule in the synthesis of molecular superconductors.

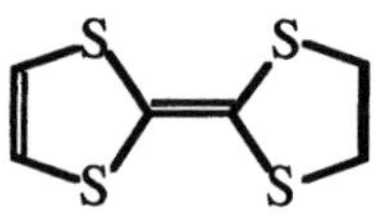

TTF

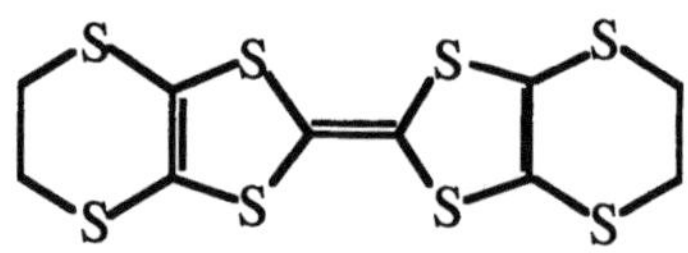

BEDT-TTF o ET

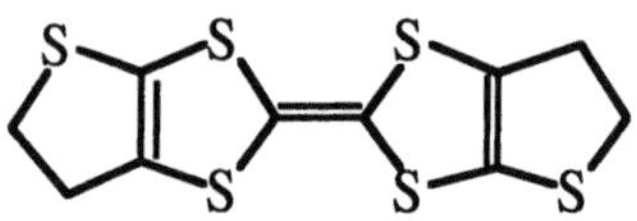

BEDT-TSeF

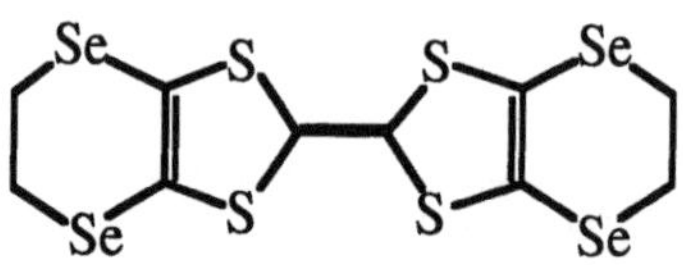

BEDSe-TTF

Figure 10. Some of the organic donors of the TTF type used with polyoxometalates to prepare hybrid organic/inorganic radical salts

Thus, we first reported, in collaboration with P. Delhaes and L. Ouahab groups, an extensive series of radical salts with the general formula $(BEDT\text{-}TTF)_8[XW_{12}O_{40}]$ [32] which contains Keggin polyoxoanions $[XW_{12}O_{40}]$ having paramagnetic ions in the tetrahedral site ($X = Fe^{III}$, Co^{II}, Cu^{II}) (Figure 11, left). All these materials showed *semiconducting* behaviors with $\sigma_{RT} \approx 10^{-1}\text{-}10^{-2}$ $S.cm^{-1}$ and $E_a \approx 100\text{-}150$ meV, and the magnetic properties indicated that the two sublattices were quasi-independent.

With the aim of increasing the magnetic coupling between the two components we used in a second step substituted Keggin anions having a magnetic ion in one of the peripheral octahedral sites (Figure 11, center). This resulted in the preparation of a new series of radical salts of formula $(BEDT\text{-}TTF)_8[XZ(H_2O)M_{11}O_{39}]$ ($X = P^V$, Si^{IV}; $Z = Cr^{III}$ Fe^{III}, Co^{II}, Ni^{II}, Cu^{II}, Zn^{II}; $M = W^{VI}$, Mo^{VI}) which maintains the structure of the above family: layers of BEDT-TTF molecules with an α packing mode alternating with layers of the inorganic Keggin anions (Figure 12).

Figure 11. The Keggin derivatives bearing a magnetic ion (shaded polyhedra) in the central tetrahedral cavity (left) and in the periphery (center). The Dawson-Wells anion (right).

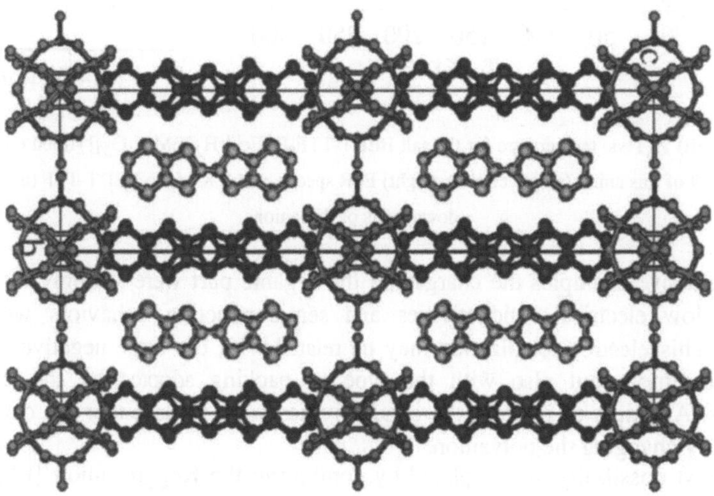

Figure 12. Structure of the series of radical salts BEDT-$TTF_8[XZ(H_2O)M_{11}O_{39}]$ ($X = P^V$, Si^{IV}; $Z = Cr^{III}$ Fe^{III}, Co^{II}, Ni^{II}, Cu^{II}, Zn^{II}; $M = W^{VI}$, Mo^{VI})

Interestingly, an unexpected arrangement of the Keggin anions was observed in the Mn derivatives (BEDT-TTF)$_8[PMn^{II}M_{11}O_{39}]$ (M = W, Mo). Here a related structure was found wherein the BEDT-TTFs were packed as in the other phase, but the Keggin units were linked by a bridging oxygen atom giving rise to an unprecedented chain of Keggin anions. The magnetic behavior of the Fe^{III} derivative is reported in figure 13 (left) and compared to that of the corresponding tetrabutylammonium salt. Upon cooling down, $\chi_m T$ shows a gradual decrease that is attributed to the strong antiferromagnetic coupling between the spins of the organic chains, approaching at low temperatures to the behavior of isolated Fe^{III}. On the other hand, the powder ESR spectrum (at 4.2 K) of the radical salt is merely the sum of the signals associated to Fe^{III} and BEDT-TTF cations (Figure 13.right). These results were general for all the other members of these series and demonstrated that also in this case the two molecular components behave independently.

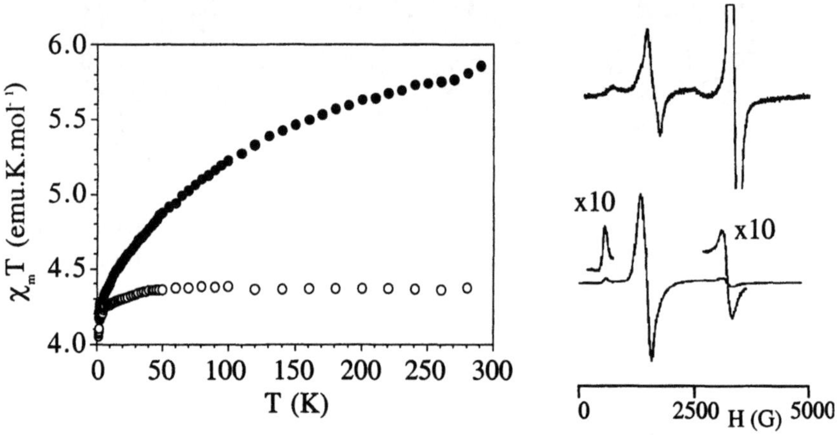

Figure 13. (left) $\chi_m T$ vs. temperature for the salt BEDT-$TTF_8[SiFe^{III}(H_2O)Mo_{11}O_{39}]$ (filled circles) and of the NBu_4^+ salt of this anion (empty circles). (right) ESR spectra at 4.2 K of the BEDT-TTF (up) and NBu_4^+ (down) salts of this anion

In the above examples the charges on the organic part were strongly localized and therefore, low electrical conductivities and semiconducting behaviors were always observed. This electron localization may be related with the large negative charges on the polyoxoanion, but also with the type of packing adopted by the BEDT-TTF molecules. Attempts to increase the conductivities were made either by changing the donor, or by changing the polyanion.

The first possibility was explored by combining the Keggin anion $[PMo_{12}O_{40}]^{3-}$ with BEDS-TTF (or BEST), a seleno-substituted molecule related with the BEDT-TTF [33] (see figure 10), in which the greater extended π-orbitals of the selenium are expected to enhance the donor-donor overlaps and therefore the electron delocalization. However, this small modification of the organic part (change of S by Se in the periphery) led to a different crystalline phase with stoichiometry 3:1 showing a different packing and thus, a different charge distribution on the organic molecules. What is interesting in this salt is that the Keggin phosphomolybdate is reduced by one electron giving rise to a paramagnetic mixed-valence cluster. However, from the electrical point of view, the organic component is formed by fully charged $(BEST)^{+\cdot}$ cations and therefore the salt is an insulator. From the structural point of view the most attractive aspect of this compound concerns the unprecedented packing of the organic molecules which instead of forming stacks in order to maximize the intermolecular contacts through the π molecular orbitals, are packed forming layers with side intermolecular contacts (Figure 14). Numerous short intermolecular distances involving the Se atoms are observed both within and between the layers, giving rise to a 3D organic network in which all the unpaired spins of the $(BEST)^{+\cdot}$ radicals are strongly antiferromagnetically coupled.

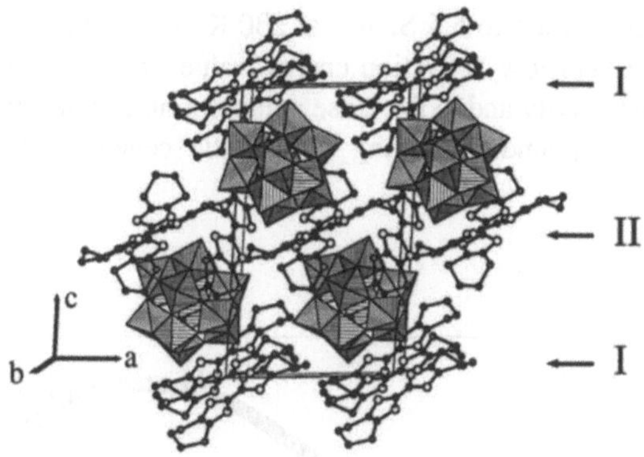

Figure 14. Structure of the radical salt $(BEST)_3H[PMo^{V}Mo^{VI}{}_{11}O_{40}]$ showing the two types of organic layers (I and II) alternating with the inorganic ones.

The second possibility was explored by using the Dawson-Wells anion $[P_2W_{18}O_{62}]^{6-}$ (Figure 11, right). With this polyanion we obtained the radical salt BEDT-$TTF_{11}[P_2W_{18}O_{62}].3H_2O$ [34]. The structural arrangement of the organic layer is formed by parallel chains of BEDT-TTF molecules. The organic molecules of neighboring chains are also parallel, leading to the so-called β–phase (Figure 15, left). What is remarkable in this structure is the presence of six crystallographically independent BEDT-TTF molecules (noted as A, B, C, D, E and F in figure 15.right) in such a way that each chain is formed by the repetition of groups of 11 BEDT-TTF molecules following the sequence ...ABCDEFEDCBA... stacked in an *exotic zigzag mode*.

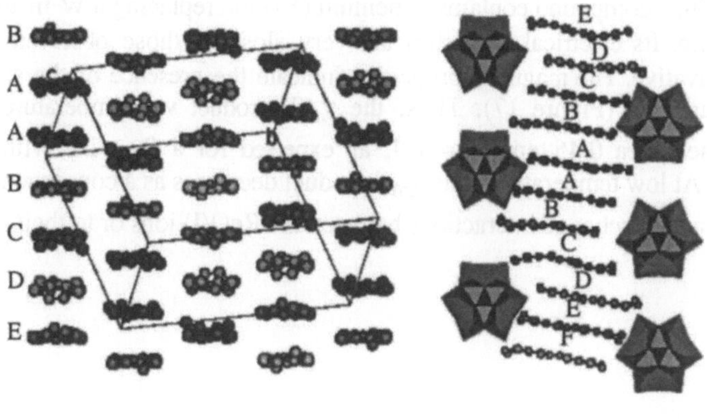

Figure 15. Structure of the organic layer in the radical salt β-BEDT-$TTF_{11}[P_2W_{18}O_{62}].2H_2O$ (left) and of the organic zigzag chains (right)

These unusual structural features illustrates well the ability of polyoxometalates to create new kinds of packing in the organic component. But the most attractive characteristic of this salt concerns its electrical conductivity that shows a *metallic-like behavior* in the region 230-300 K with an increase in the conductivity from ca. 5 $S.cm^{-1}$ at room temperature to 5.5 $S.cm^{-1}$ at 230 K. Below this T the salt becomes semiconductor with a very low activation energy value of 0.013 eV (Figure 16). In view of this high conductivity and to the close anion-cation contacts we have attempted to prepare a related compound containing a magnetic center on the surface of the Dawson-Wells polyanion.

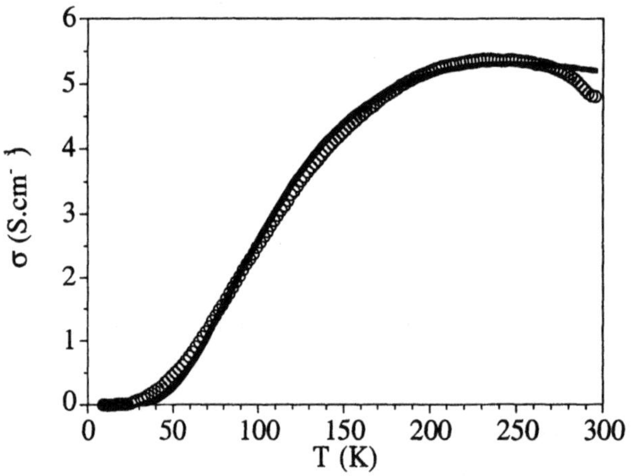

Figure 16. Thermal variation of the d.c. electrical conductivity in the ac plane for the radical salt β-BEDT-$TTF_{11}[P_2W_{18}O_{62}].2H_2O$. Solid line if the fit to a semiconductor model with a very low activation energy.

The obtained compound contains a rhenium (VI) ion replacing a W in the Dawson-Wells structure. Its electrical properties are very close to those observed in the non magnetic derivative. The magnetic properties indicate the presence of the spin S = 1/2 of the rhenium (VI) (Figure 17). Thus, the $\chi_m T$ product vs. temperature presents a constant value of ca 0.45 $emu.K.mol^{-1}$, as expected for a S = 1/2 with an orbital contribution. At low temperatures this $\chi_m T$ product decreases as a consequence of either antiferromagnetic exchange interactions between the Re(VI) ions or to their anisotropy.

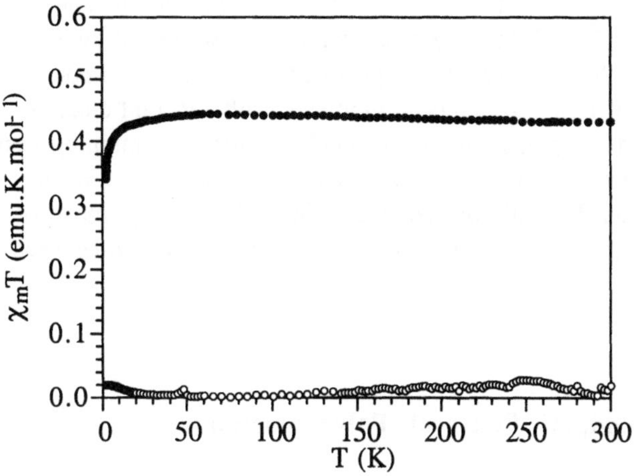

Figure 17. $\chi_m T$ vs temperature for the radical salt BEDT-TTF$_{11}$[$P_2ReW_{17}O_{62}$] (filled circes) and for the diamagnetic salt β-BEDT-TTF$_{11}$[$P_2W_{18}O_{62}$].2H_2O (empty circles).

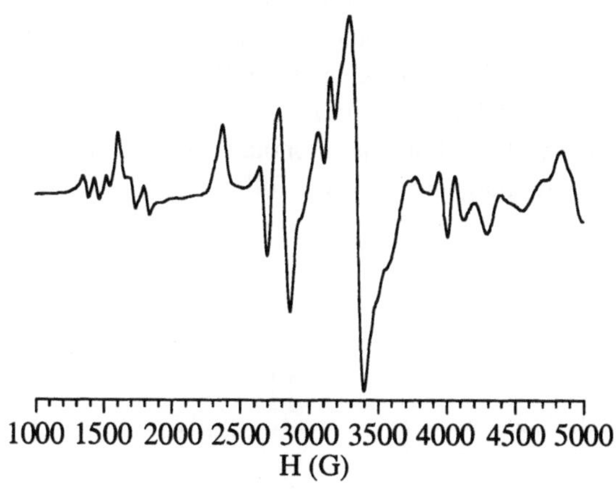

Figure 18. ESR spectrum at 4.2 K of the radical salt BEDT-TTF$_{11}$[$P_2ReW_{17}O_{62}$].

The ESR of this salt spectra confirm the presence of this ion (Figure 18). Thus, at low temperatures the signals coming from this ion are clearly observed together with that of the organic radicals (g = 2.0) coming from the presence of a small amount of isolated BEDT-TTF+ radicals due to crystal defects as is usually observed in this type of radical salts. This novel compound constitutes then an illustrative example of hybrid salt showing coexistence of highly delocalized electrons with localized magnetic moments.

To conclude this part we can say that the combination of magnetic polyoxometalates with TTF-type organic donors has witnessed rapid progress in the last few years. This hybrid approach has provided a variety of examples of radical salts with coexistence of localized magnetic moments with itinerant electrons. This is the critical step towards the preparation of molecular materials combining useful magnetic and electrical properties. Thus far, however, the weak nature of the cation/anion contacts as well as the still low conductivities exhibited by most of the reported materials have prevented the observation of an indirect interaction between the localized magnetic moments via the conducting electrons.

4. Langmuir-Blodgett Films of Polyoxometalates [35]

Another class of hybrid molecular materials we can prepare with polyoxometalates are the Langmuir-Blodgett films. Until very recently the LB technique was only used to organize amphiphilic organic molecules [36]. However, very recently it has been shown that inorganic complexes can also be organized as LB films. This technique allows to arrange molecules into well-organized multilayered films. In view of the multiple properties and electronic versatility of polyoxometalates an attractive possibility for us was to construct monolayers of these inorganic complexes. Thus, in collaboration with C. Mingotaud and P. Delhaes from the Centre de Recherches Paul Pascal we are exploring this possibility. By taking advantage of the adsorption properties of the polyanions along a positively charged organic monolayer spread in water we have been able to obtain the first organic-inorganic LB films following the method schematized in Figure 19.

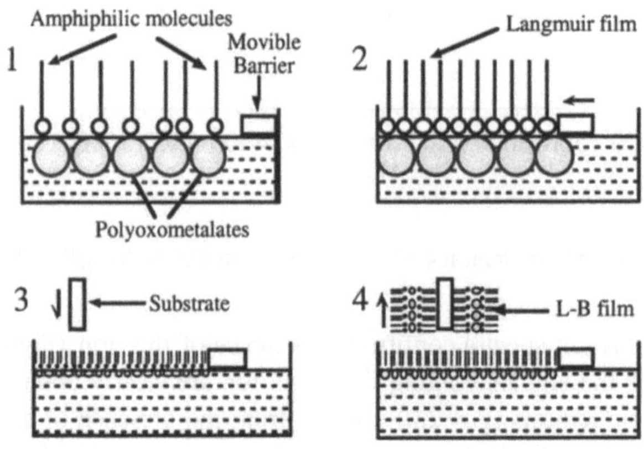

Figure 19. Method used for the formation of Langmuir films of polyoxometalates with amphiphilic molecules and their deposition as Langmuir-Blodgett films.

They are based on the chemically stable Keggin polyoxometalates $[X^{n+}W_{12}O_{40}]^{(8-n)-}$ (X^{n+} = 2(H^+), P^V, Si^{IV}, B^{III}, Co^{II}) and on the dimethyldioctadecyl-ammonium cation (DODA in Figure 20) [3, 37].

Figure 20. Molecular structure of the DODA cations.

The lamellar structure so obtained consists of monolayers of the polyoxoanions alternating with bilayers of the organic surfactant (Figure 21), as deduced from infrared linear dichroism and X-ray diffraction experiments.

Figure 21. Structure of the LB films obtained with $DODA^+$ and the Keggin polyoxometalates.

The aforementioned method is general and has been extended to organize in LB films magnetic polyoxometalates of increasing nuclearities [38]. We have used the polyoxometalates $[Co_4(H_2O)_2(PW_9O_{34})_2]^{10-}$ and $[Co_4(H_2O)_2(P_2W_{15}O_{62})_2]^{16-}$, as they encapsulate a Co_4O_{16} ferromagnetic cluster [1], and the giant polyoxometalates

$[Co_9(OH)_3(H_2O)_6(HPO_4)_2(PW_9O_{34})_3]^{16-}$ and $[Ni_9(OH)_3(H_2O)_6(HPO_4)_2(PW_9O_{34})_3]^{16-}$ that encapsulate a nonanuclear M_9O_{36} cluster (Figure 1). The structure of the two films containing the Co_4 cluster is similar to that reported for the Keggin anions. It consists of monolayers of polyoxometalates, which in this case are lying down over the lipid layer giving an interlayer separation of 10 Å, alternating with bilayers of $DODA^+$ lipids. As far as the magnetic properties of these films are concerned, they are very close to those obtained in the corresponding K^+ salts of the polyoxometalates, indicating that the magnetic cluster is maintained intact in the film. For example, we show in Figure 22 the magnetic properties of the LB film containing the Ni_9 cluster. As in the crystal the magnetic properties of this cluster show a coexistence of ferro and antiferromagnetic Ni-Ni pairwise interactions.

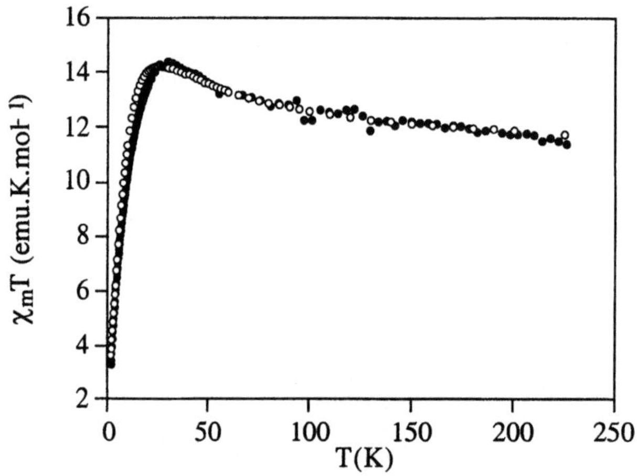

Figure 22. $\chi_m T$ vs. temperature for the LB film containing the $[Ni_9(OH)_3(H_2O)_6(HPO_4)_2(PW_9O_{34})_3]^{16-}$ polyanion (filled circles) and of the Na^+ salt of the same anion (empty circles).

The method we have described is general and can be extended to many kinds of polyoxoanions and other charged complexes, leading to new lamellar organized materials. In the present case the appropriate choice of polyoxometalates and lipid molecules should allow for ultrathin organized films incorporating the properties of the polyoxometalates to be constructed. Another interesting possibility we are also exploiting is to incorporate functional organic lipids furnishing a physical property (for example conductivity). This feature should allow the design of multifunctional LB films in the near future.

5. Concluding Remarks

We have shown in this article some relevant achievements in the field of the molecular materials that involve polyoxometalates. This quite recent application has witnessed rapid progress and some aspects are worth to mention:

1) In connection with the molecular magnetism, polyoxometalates are a rich class of compounds that provide ideal examples of magnetic and mixed-valence clusters of increasing nuclearities and topological and electronic complexities having prearranged geometries and often predictable exchange interactions.

2) In connection with the study of magnetic clusters, polyoxometalates offer a unique possibility of combining the routine magnetic techniques (magnetic susceptibility, magnetization, ESR) with Inelastic Neutron Scattering to get a thorough characterization of the ground state properties in a magnetic cluster. This is essential for in-depth understanding of the magnetic exchange interaction phenomenon. A relevant result obtained in this context has been the demonstration by INS of exchange-anisotropy in cobalt(II) clusters encapsulated by polyoxometalates. This constitutes the first direct evidence of exchange-anisotropy in cobalt(II) clusters.

3) In connection with the mixed valence clusters, the high symmetries of polyoxometalates have facilitated the development of exact quantum-mechanical models from which a clear picture of the relevant parameters involved in the electronic interactions can be extracted. Thus, despite the large nuclearities and extremely complex topologies of these clusters, models have been developed in some cases that provide important hints on the role played by electron transfer and exchange processes in molecular systems. For example, it has been shown that in these structures electron delocalization can result in a strong antiferromagnetic coupling between widely separated electrons.

4) In connection with the synthesis of novel classes of materials, these anions have been successfully used as magnetic component of molecular hybrids with interesting physical properties or combination of properties. Thus, they have been combined with organic donors of the tetrathiafulvalene (TTF) type to form radical salts with unprecedented organic packings and coexistence of localized magnetic moments with itinerant electrons. On the other hand, they have been incorporated into Langmuir-Blodgett films to produce organic-inorganic hybrids containing well-organized monolayers of polyoxometalates, separated by bilayers of amphiphilic lipid molecules.

6. Acknowledgments

Part of the results reported in this article has been developed in collaboration with C. Mingotaud, P. Delhaes and H. U. Güdel in the framework of the European COST action 518 (Project on Magnetic Properties of Molecular and Polymeric Materials). Financial support from the Spanish Ministerio de Educación y Cultura (Grants PB96-0862 and MAT98-0880) is gratefully acknowledged.

7. References

1. Coronado, E.; Gómez-García, C. J. *Comments Inorg. Chem.* **1995**, *17*, 255

2. Coronado, E.; Gómez.García, C. J. *Chem. Rev.* **1998**, *98*, 273.

3. Clemente-León, M.; Mingotaud, C.; Agricole, B.; Gómez-García, C. J.; Coronado, E.; Delhaes, P. *Angew. Chem. Int. Ed. Engl.* **1997**, *36*, 1114.

4. Gatteschi, D.; Caneschi, A.; Pardi, L.; Sessoli, R. *Science* **1994**, *265*, 1054.

5. Muller, A.; Peters, F.; Pope, M. T.; Gatteschi, D. *Chem. Rev.* **1998**, *98*, 239.

6. Coronado, E.; Clemente-Juan, J. M. *Coord. Chem. Rev.* (in press).

7. Clemente-Juan, J. M.; Coronado, E.; Galán-Mascarós, J. R.; Gómez-García, C. J. *Inorg. Chem.* **1999**, *38*, 55.

8. Pope, M. T. *Heteropoly and Isopoly Oxometalates*; Springer-Verlag: Berlin, **1983**.

9. Casañ-Pastor, N.; Baker, L. C. W. *J. Am. Chem. Soc.*, **1992**, *114*, 10384; Casañ-Pastor, N.; Baker, L. C. W. in M. T. Pope and A. Müller (eds.), *Polyoxometalates: From Platonic Solids to Anti-Retroviral Activity*, Kluwer Academic Publishers: Dordrecht, **1994**, p 203.

10. Gatteschi, D.; Pardi, L.; Barra A. L.; Müller, A. in M. T. Pope and A. Müller (eds.), *Polyoxometalates: From Platonic Solids to Anti-Retroviral Activity*, Kluwer Academic Publishers: Dordrecht, **1994**, p 219.

11. Gómez-García, C. J.; Coronado E.; Ouahab, L. *Angew. Chem., Int. Ed. Engl.* **1992**, *31*, 649.

12 Kortz, U.; Tézé, A.; Hervé, G. *Inorg. Chem.* **1999**, *38*, 2038.

13. Weakley, T.J. R.; Evans, H. T.; Showell, J. S.; Tourné G. F.; Tourné, C. M. *J. Chem. Soc. Chem. Comm.* **1973**, 139; Evans, H. T.; Tourné, C. M.; Tourné, G. F.; Weakley, T. J. R. *J. Chem. Soc., Dalton Trans.* **1986**, 2699.

14. (a) Finke, R. G.; Droege, M. W.; Domaille, P. J. *Inorg. Chem.* **1987**, *26*, 3886; (b) Weakley, T. J. R.; Finke, R. G. *Inorg. Chem.* **1990**, *29*, 1235; (c) Gómez-García, C. J.; Coronado, E.; Gómez-Romero, P.; Casañ-Pastor, N. *Inorg. Chem.* **1993**, *32*, 3378; (d) Zhang, X. Y.; Jameson, G. B.; O'Connor, C. J.; Pope, M. T. *Polyhedron* **1996**, *15*, 917; (e) Zhang, X. Y.; Chen, Q.; Duncan, D. C.; Lachiotte, R. J.; Hill, C. L. *Inorg. Chem.* **1997**, *36*, 4381.

15. Weakley, T. J. R. *J. Chem. Soc. Chem. Commun.* **1984**, 1406.

16. Coronado, E.; Forment-Aliaga, A.; Galán-Mascarós, J. R.; Gómez-García, C. J. (unpublished work)

17. Gómez-García, C. J.; Coronado, E.; Galán-Mascarós, J. R. *Adv. Mater.* **1994**, *6*, 221.

18 (a) Gómez-García, C. J.; Casañ-Pastor, N.; Coronado, E.; Baker, L.C.W.; Pourroy, G. *J. Appl. Phys.* **1990**, *67*, 5995; (b) Gómez-García, C. J.; Coronado, E.; Borrás-Almenar, J. J.; *Inorg. Chem.* **1992**, *31*, 1667; (c) Casañ-Pastor, N.; Bas, J.; Coronado, E.; Pourroy, G.; Baker, L. C. W. *J. Am. Chem. Soc.* **1992**, *114*, 10380; (d) Gómez-García, C. J.; Coronado, E.; Borrás-Almenar, J. J.; Aebersold, M.; Güdel, H. U.: Mutka, H.; *Physica B* **1992**, *180-181*, 238; (e) Clemente, J. M.; Andres, H.; Coronado, E.; Güdel, H. U.; Büttne, H.; Kearly, G. *Inorg. Chem.* **1997**, *36*, 2244; (f) Aebersold, M.; Andres, H. P.; Büttner, H.; Borrás-Almenar, J. J.; Clemente-Juan, J. M.; Coronado, E.; Güdel, H. U.; Kearley, D. *Physica B* **1997**, *234-236*, 764; (g) Andres, H.; Clemente-Juan, J. M.: Aebersold, M.; Güdel, H. U.; Coronado, E.; Büttner, H.; Kearly, G.; Melero, J.; Burriel, R. *J. Am. Chem. Soc.* **1999** (in press).

19. Casañ-Pastor, N.; Bas, J.; Coronado, E.; Pourroy, G.; Baker, L. C. W. *J. Am. Chem. Soc.* **1992**, *114*, 10380.

20. (a) Furrer, A.; Güdel, H. U. *Phys. Rev. Lett.* **1977**, *39*, 657; (b) Güdel, H. U.; Hauser, U.; Furrer, A. *Inorg. Chem.* **1979**, *10*, 2730.

21. Andres, H.; Clemente-Juan, J. M.; Aebersold, M.; Güdel, H. U.; Coronado, E.; Büttner, H.; Kearly, G.; Melero, J.; Burriel, R. *J. Am. Chem. Soc.* (in press).

22. Anderson, P. W.; Hasegawa, H. *Phys. Rev.* **1955**, *100*, 675.

23. Coronado, E.; Georges, R.; Tsukerblat B. S. in E. Coronado, P.Delhaes, D. Gatteschi and J. S. Miller (eds.). *Molecular Magnetism: From Molecular Assemblies to the Devices*, NATO ASI Series, Kluwer Academic Publishers, vol. E 321 **1996**, p. 65.

24. (a) Blondin, G.; Girerd, J. J. *Chem. Rev.* **1989**, *90*, 1359; (b) Christou, G. *Acc. Chem. Res.* **1989**, *22*, 328; (c) Christou, G. in ref. 20, p. 383.

25. Rao, C. N. R. *Chem. Eur. J.* **1996**, 2, 1499.

26. Borrás Almenar, J. J.; Coronado, E.; Georges, R.; Tsukerblat, B. S. in ref. 20, p. 105.

27. Borrás-Almenar, J. J.; Clemente, J. M.; Coronado, E.; Tsukerblat, B. S. *Chem. Phys.* **1995**, *195*, 1.

28. Borrás-Almenar, J. J.; Clemente, J. M.; Coronado, E.; Tsukerblat, B. S. *Chem. Phys.* **1995**, *195*, 16.

29. Borrás-Almenar, J. J.; Clemente-Juan, J. M.; Coronado, E.; Georges, R.; Palii, A. V.; Tsukerblat, B. S. *J. Chem. Phys.* **1996**, *105*, 6892.

30. Müller, A.; Sessoli, R.; Krickemeyer, E.; Bögge, H.; Meyer, J.; Gatteschi, D.; Pardi, L.; Westphal, J.; Hovemeier, K.; Rohlfing, R.; Döring, J.; Hellweg, I.; Beugholt, C.; Schmidtmann, M. *Inorg. Chem.* **1997**, *36*, 5239.

31. Coronado, E. et al., unpublished results.

32. (a) Gómez-García, C. J.; Ouahab, L.; Giménez-Saiz, C.; Triki, S.; Coronado, E.; Delhaes, P. *Angew. Chem., Int. Ed. Engl.* **1994**, *33*, 223; (b) Gómez-García, C. J.; Giménez-Saiz, C.; Triki, S.; Coronado, E.; Le Magueres, P.; Ouahab, L.; Ducasse, L.; Sourisseau, C.; Delhaes, P. *Inorg. Chem.* **1995**, *34*, 4139; (c) Galán-Mascarós, J. R.; Gimenez-Saiz, C.; Triki, S.; Gómez-García, C. J.; Coronado, E.; Ouahab, L. *Angew. Chem., Int. Ed. Engl.* **1995**, *34*, 1460.

33. Coronado, E.; Delhaes, P.; Falvello, L. W.; Giménez-Saiz, C.; Gómez-García, C. J. *Inorg. Chem.* **1998**, *37*, 2183.

34. Coronado, E.; Galan-Mascaros, J. R.; Gimenez-Saiz, C.; Gomez-Garcia, C. J.; Laukhin, V. N. *Adv. Mater.* **1996**, *8*, 801

35. Coronado, E.; Mingotaud, C., Adv. Mater. **1999**, *11*, 869.

36. Mingotaud, A. F.; Mingotaud, C.; Patterson, L. K. *Handbook of Monolayers;* 1st ed.; Academic Press: San Diego, **1993**.

37. Clemente-Leon, M.; Agricole, B.; Mingotaud, C.; Gómez-García, C. J.; Coronado, E.; Delhaes, P. *Langmuir* **1997**, *13*, 2340.

38. Clemente-León, M.; Mingotaud, C.; Gómez-García, C. J.; Coronado, E.; Delhaes, P. *Thin Solid Films* **1998**, *327-329*, 439.

Framework Materials Composed of Transition Metal Oxide Clusters

M. ISHAQUE KHAN
Department of Biological, Chemical, and Physical Sciences Illinois Institute of Technology Chicago, IL 60616, USA
Fax: Int. code + 312-567-3494 e-mail: Ishaque.Khan@.iit.edu

Abstract. Transition metal oxide clusters and their derivatives, which are important in such diverse fields as analytical chemistry, biochemical and geochemical processes, chemical sensing, catalysis, materials science, and medicine offer an unmatched variety of attractive building block units for the design and development of new materials whose properties could possibly be correlated with their constituent units at the molecular level. For example, these clusters may provide structural motifs for the rational synthesis of new metal oxide based catalysts and novel surfaces. Although the technique of bringing suitable transition metal oxide units together to generate new chemical systems with desirable properties remains underdeveloped, recent progress made in preparing new framework materials are encouraging. By adopting essentially one-pot synthetic approach, a series of novel framework materials, composed of well defined vanadium oxide clusters $\{V_{18}O_{42}(XO_4)\}$, of general formulation- $(Cat)_m[M_aL_{12}V_{18}H_bO_{42}(XO_4)]^{n-}.cH_2O$. (Cat = Li^+, $N_2H_5^+$; m = 0 , 2, 6; M = Mn, Fe, Co, Ni, Zn, Cd, etc.; a = 3; b = 0, 6; X = V, S; L = H_2O; n = 0, 2, 6; c = 24, 30) have been prepared and characterized. These are reviewed with reference to the synthesis, structure and physicochemical properties of the newly prepared representative framework solids.

Key Words: Crystal structure. framework solids. metal oxides. mixed-valent compounds. polyoxometalates. poloxovanadates. ion exchange. thermogravimetry.

1. Introduction

Transition metal oxide clusters or polyoxometalates and their derivatives are the subject of current investigation due to their relevance to and usefulness in analytical chemistry, catalysis, materials science, chemical sensing, biochemical and geochemical processes, and medicine [1]. The advancement in x-ray diffraction and other modern physicochemical techniques has enabled the structural characterization of a large number of transition metal oxide clusters which have been synthesized in recent years. Giant molecular clusters, containing up to 248 metal atoms, of molecular weights at par with proteins and their coordination compounds have been prepared and characterized in recent years [1e]. The structure and bonding patterns [2] in these supermolecules resemble the complex transition metal oxide surfaces which are employed as catalyst for industrial organic transformations [2d,3]. Many such catalysts are poorly understood because of their inaccessibility to the conventional physico-chemical techniques and, therefore, are not amenable to improvements in their performance. With their proven roles in catalysis [4] and in producing an unmatched variety of fascinating structural motifs [5,1b,1e], transition metal oxide clusters have potential to provide building units suitable for the rational synthesis of new metal oxide based catalysts and novel surfaces whose performance could possibly be modified and correlated with their constituent units at the molecular level. Therefore, the step-by-step expansion in the size of these systems,

M.T. Pope and A. Müller (eds.), Polyoxometalate Chemistry, 255–267.

going from molecular domain into infinite solids, by using well characterized building units to fashion materials with desirable properties is of current interest.

Although as compared to the polyoxomolybdates and polyoxotungstates the polyoxovanadates or vanadium oxide clusters appeared relatively recently on the polyoxometalates horizon, a decent pool of well characterized $\{V_xO_y\}$ type clusters with interesting physicochemical properties is now known. Representative members of this fascinating class of compounds include $[V_4O_{12}]^{4-}$ [6a], $[V_5O_{14}]^{3-}$ [6b], $[V_{10}O_{28}]^{6-}$ [6c], $[V_{10}O_{30}]^{11-}$ [6d], $[V_{12}O_{32}]^{4-}$ [6e], $[V_{13}O_{34}]^{3-}$ [6f], $[V_{15}O_{42}]^{9-}$ [6g], $[V_{15}O_{36}]^{5-}$ [6h], $[V_{18}O_{42}]^{12-}$ [7], $[V_{19}O_{49}]^{9-}$ [8], $[V_{22}O_{54}]^{6-}$ [9], and $[V_{34}O_{82}]^{10-}$ [10]. Many of these compounds, specially those of higher nuclearity (such as $[V_{12}O_{32}]^{4-}$, $[V_{15}O_{36}]^{5-}$, $[V_{18}O_{42}]^{12-}$, $[V_{19}O_{49}]^{9-}$, and $[V_{34}O_{82}]^{10-}$), may be formally regarded as V/O shells which may be assembled around some encapsulated (template) species [11]. The encapsulated species appear to exert structure directing (templating) effects in determining the shapes and sizes of the shells surrounding the templates. In this regard, the $\{V_{18}O_{42}\}$ shell has proven to be quite a versatile molecular container which could encapsulate molecules and ion of varying sizes and shapes [7a].

The structures of almost all of the prominent vanadium oxide clusters characterized to-date are related to and could be derived from the vanadium pentaoxide sheets which appear to be the mother of major polyvanadate based structures [2c]. The resemblance of the structure and bonding patterns of these molecular systems with the infinite metal oxide (e. g. V_2O_5) lattices- which play important roles in catalyzing many industrial chemical reactions [3a,3d], provides some incentives to explore the potential of polyoxovanadates as building blocks for tailor making new (e.g. zeolitic) materials. However, the technique of bringing suitable metal oxide building units together to generate true metal oxide surfaces and framework materials without incorporating additional conventional ligands into the structure is still in its infancy and has been limited to the synthesis of mainly one-dimensional chains [12].

We are currently studying the synthesis and characterization of new framework materials composed of well defined transition metal oxide motifs. By employing suitable precursors, we have prepared and characterized a series of novel framework materials based 100% on well defined metal oxide building blocks- $\{V_{18}O_{42}(XO_4)\}$ (X = S, V, Cl) without incorporating any conventional organic or inorganic (e.g. PO_4^{3-}, AsO_4^{3-}, SO_4^{2-}, etc.) ligands [13-18].

2. Synthesis

The $\{V_{18}O_{42}\}$ shell has been observed in a number of well characterized compounds which contain varying numbers of V^{IV} sites [7, 13-18]. The basic $\{V_{18}O_{42}\}$ shell in all these compounds is constructed from 18$\{VO_5\}$ square pyramids sharing edges through 24 μ_3-oxygen atoms. The shell could incorporate VO_4^{3-} and SO_4^{2-} [7a, 13-14] and encapsulate a variety of anions (Cl^-, Br^-, I^-, NO_2^-, SH^-, S^{2-}, $HCOO^-$) and neutral molecules of appropriate sizes [7a].

The $\{V_{18}O_{42}(XO_4)\}$ (X = V, S, etc.,) species could readily be generated in aqueous solution which may directly be used to react with a number of other metal salts to prepare a variety of materials with one-dimensional, two-dimensional, and three-dimensional framework structures. Thus, the reaction of hot (80 - 95 °C) aqueous solution of the appropriate vanadate, MVO_3 (M = Li, Na, K, NH_4), obtained from the reaction of the aqueous slurry of V_2O_5 with MOH, with hydrazinium sulfate or hydrazinium chloride results into a dark colored solution (pH = ~ 4.6). The dark solution reacts with a number of salts of transition- and main group metals, such as $MnCl_2{\cdot}4H_2O$, $KMnO_4$, $FeCl_2{\cdot}4H_2O$, $CoSO_4{\cdot}6H_2O$, $NiSO_4{\cdot}6H_2O$, $CdSO_4{\cdot}8H_2O$, and $MgSO_4{\cdot}7H_2O$, to give novel three-dimensional framework materials in high yield [13-16, 18].

By adopting this essentially one-pot synthetic method, we have prepared and characterized a series of novel framework materials composed of well defined vanadium oxide clusters, $\{V_{18}O_{42}(XO_4)\}$, of general formulation-

$(Cat)_m\,[M_aL_{12}V_{18}H_bO_{42}(XO_4)]^{n-}.cH_2O$. (Cat = Li^+, $N_2H_5^+$; m = 0 , 2, 6; M = Mn, Fe, Co, Ni, Zn, Cd, etc.; a = 3; b = 0, 6; X = V, S; L = H_2O; n = 0, 2, 6; c = 24, 30)

Here, we describe the following representative examples of this series of materials:

$[Mn^{II}_3(H_2O)_{12}H_6V^{IV}_{15}V^{V}_3O_{42}(V^{V}O_4)]\cdot 30H_2O$ (**1**)

$Li_6[Fe^{II}_3(H_2O)_{12}V^{IV}_{15}V^{V}_3O_{42}(V^{V}O_4)]\cdot 24H_2O$ (**2**)

$Li_6[Ni^{II}_3(H_2O)_{12}V^{IV}_{16}V^{V}_2O_{42}(SO_4)]\cdot 24H_2O$ (**3**)

$(N_2H_5)_2[Zn^{II}_3(H_2O)_{12}V^{IV}_{12}V^{V}_6O_{42}(SO_4)]\cdot 24H_2O$ (**4**)

$[M^{II}_3(H_2O)_{12}V_{18}O_{42}(XO_4)]\cdot 24H_2O$ (M = Fe, **5**; M = Co, **6**; X = V, S)

$Li_6[Cd^{II}_3(H_2O)_{12}V_{18}O_{42}(XO_4)]\cdot 24H_2O$ (X = V, S) (**7**)

$Li_6[Mn^{II}_3(H_2O)_{12}V^{IV}_{16}V^{V}_2O_{42}(XO_4)]\cdot 24H_2O$ (X = V, S) (**8**)

Black prism shaped crystals of the mixed-metal compound $[Mn_3(H_2O)_{12}V_{18}H_6O_{42}(VO_4)]\cdot 30H_2O$ (**1**) could be obtained (~ 60 % yield) by the reaction of $KMnO_4$ with the dark colored solution which is obtained by treating with hydrazinium sulfate the solution that results from the reaction of aqueous solution of $LiOH.H_2O$ with a slurry of V_2O_5 in water at 95 °C. Although **1** could be prepared over a wide range of different reaction temperatures and slightly different reaction stoichiometry and reactants, the optimum reaction temperature is ~ 95 ° C. Thus, **1** can also be prepared by using NaOH or KOH in place of $LiOH.H_2O$ or by directly employing the appropriate metal vanadates in the reaction medium. The reactions leading to the formation of **1** are, however, pH sensitive. The reactions carried out in neutral (pH=7.3) and basic (pH=12.3) media did not yield **1**, giving intractable orange brown amorphous materials. The reactions using $MnCl_2$ or $MnSO_4$ in place of $KMnO_4$ yielded a slightly different compound $Li_6[Mn^{II}_3(H_2O)_{12}V^{IV}_{16}V^{V}_2O_{42}(XO_4)]\cdot 24H_2O$ (X = V, S) (**8**) which crystallized in a different space group with more regular structure than what is observed in **1** (see below). Using iron(III) chloride in place of $KMnO_4$ resulted into the formation of $Li_6[Fe^{II}_3(H_2O)_{12}V_{18}O_{42}(VO_4)]\cdot 24H_2O$ (**2**) [18].

The reaction of aqueous solution of lithium vanadate with hydrazinium sulfate results into a dark colored solution which reacts with nickel(II) sulfate hexahydrate and zinc(II) sulfate heptahydrate to yield dark greenish-black prism shaped crystals of

$Li_6[Ni^{II}_3(H_2O)_{12}V_{18}O_{42}(SO_4)]\cdot 24H_2O$ (**3**) and $(N_2H_5)_2[Zn^{II}_3(H_2O)_{12}V_{18}O_{42}(SO_4)]\cdot 24H_2O$ (**4**), respectively. The presence of sulfate ion in the reaction mixture is essential for the syntheses of **3** and **4**. Thus, while it can also be prepared by using hydrazine chloride ($NH_2NH_2\cdot 2HCl$) in place of hydrazine sulfate or by substituting nickel sulfate with nickel chloride in the above described synthetic method, **3** could not be synthesized when both sulfate sources ($NiSO_4\cdot 6H_2O$ and $NH_2NH_2\cdot SO_4$) were replaced by $NiCl_2\cdot 6H_2O$ and $NH_2NH_2\cdot 2HCl$, respectively; a brown-black powder of a different material was obtained instead in the latter case which was not further characterized.

Black prism-shaped crystals of $[M^{II}_3(H_2O)_{12}V_{18}O_{42}(XO_4)]\cdot 24H_2O$ (M = Fe, **5**; M = Co, **6**; X = V, S) could be isolated in 50-60% yield from the dark colored solution obtained from the reaction of V_2O_5 with $LiOH\cdot H_2O$, hydrazinium sulfate, and $FeCl_2\cdot 4H_2O$ or $CoSO_4\cdot 6H_2O$ in water at 84-86 °C. The analogous reactions yield the greenish-black species $Li_6[Cd_3(H_2O)_{12}V_{18}O_{42}(XO_4)]\cdot 24H_2O$ (**7**) and $Li_6[Mn^{II}_3(H_2O)_{12}V_{18}O_{42}(XO_4)]\cdot 24H_2O$ (X = V, S) when, respectively, $CdSO_4\cdot 8H_2O$ and $MnCl_2\cdot 4H_2O$ or $MnSO_4$ are allowed to react with the reduced polyvanadate solution.

3. Spectral and Physical Properties

The dark colors of the materials prepared from the above described reactions are indicative of the reduced vanadium sites. This was confirmed by the manganometric titrations where the number of reduced (V^{IV}) centers in these compound were determined by titrating their acidic solutions against standardized $KMnO_4$ solution. The results of titration showed varying number V^{IV} centers per formula units in different compounds.

With few exceptions, these materials are stable in air for long time and are insoluble in common solvents. Thus, crystals of **1** are indefinitely stable in air, insoluble in cold water and common organic solvents, slightly soluble in warm water, moderately soluble in DMSO giving purple-black colored solution stable for several days at room temperature (solution in cold DMSO is bluish-black in color; dissolution of **1** in hot DMSO gave purple-black solution that retained its color for hours after being cooled to room temperature), and quite soluble in boiling water. The aqueous solution of **1** changed color from black to light green to yellow within few hours. When stored under nitrogen, most of these solids are indefinitely stable. Crystals of **3** undergo slow deterioration losing their shining faces which slowly develops cracks at room temperature. This, however, could be minimized when the reaction temperature and/or reaction time to prepare **3** is somewhat increased. Some of the crystals, notably those of **3**, **5**, and 7, undergo slow oxidation in air. Consequently, the number of reduced (V^{IV}) sites decreases with time.

Besides the vibration bands due to water, encapsulated XO_4 species, and any cations, the infrared spectra of these compounds exhibit strong IR bands in 1000 - 900 cm^{-1} region which is the characteristic region of $\upsilon(V{=}O)$. Multiple features attributable to the bridging V-O-V groups are also found in 840 - 400 cm^{-1} region.

4. Structure

Compounds **1-8** adopt highly symmetrical three-dimensional framework structures which are closely related to each other as described below.

The extended structure of the crystals of **1**, shown in Figure 1, is composed of the transition metal oxide building-block units given in Figure 2. The structure consists of "spheres" of $\{V_{18}O_{42}(VO_4)\}$ clusters each one linked, in three dimensional array, by six μ_2-$\{Mn(H_2O)_4\}$ bridges to six other neighboring $\{V_{18}O_{42}(VO_4)\}$ units generating a network of $[-\{V_{18}O_{42}(VO_4)\}-\mu_2-Mn(H_2O)_4-\{V_{18}O_{42}(VO_4)\}-]_\infty$ arrays running along three orthogonal directions.

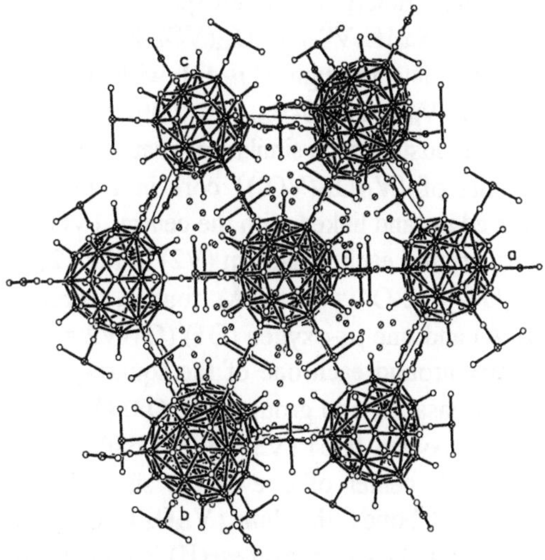

Figure 1. A view of the structure of $[Mn_3(H_2O)_{12}V_{18}H_6O_{42}(VO_4)]\cdot 30H_2O$ (**1**) showing arrays of $\{V_{18}O_{42}(VO_4)\}$'spheres' interconnected through $\{Mn(H_2O)_4\}$ bridging groups and channels occupied by the water molecules (striped circles) of crystallization. Hydrogen atoms are not shown

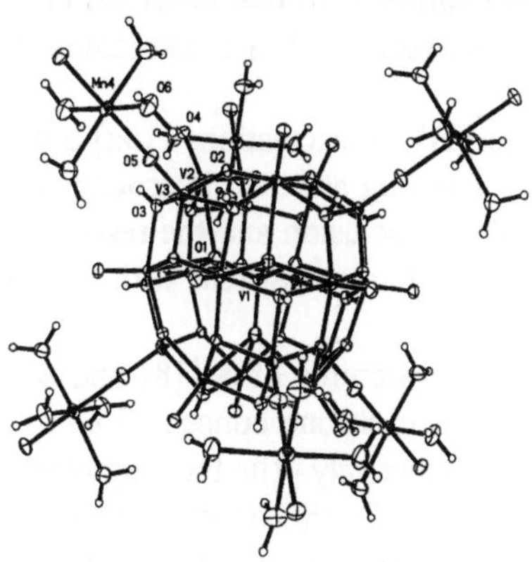

Figure 2. The building block units in the crystal structure of $[Mn_3(H_2O)_{12}V_{18}H_6O_{42}(VO_4)]\cdot 30H_2O$ (**1**) showing the atom labeling scheme in the asymmetric unit. Small open circles represent hydrogen atoms.

The constituent $\{V_{18}O_{42}(VO_4)\}$ cluster unit is constructed from the $\{V_{18}O_{42}\}$ shell encapsulating a tetrahedral $\{VO_4\}^{3-}$ group which interacts with the 12 V-centers of the shell, each oxygen (O1) of the $\{VO_4\}^{3-}$ unit (V1-O = 1.661 Å) interacting in μ_4-mode with three V(2)- centers, forming V(1)-O(1)-V(2)$_{shell}$ bonds (O1-V$_{shell}$= 2.440 Å) and forcing the local idealized tetrahedral symmetry upon the $\{V_{18}O_{42}(VO_4)\}$ unit.

The molecular container property of the$\{V_{18}O_{42}\}$ shell has neatly been illustrated by its encapsulation of a variety of anions and molecules. The shell may exist with different electronic populations in two closely related structural forms with different symmetries (T_d and D_{4d}) that are influenced by the stereochemical needs and the extent of interaction of the encapsulated moiety- the 'guest species', with the V-centers of the shell [7a, 14]. For example, in the $Na_6[H_7V^{IV}{}_{16}V^{V}{}_2O_{42}(VO_4)]\cdot 21H_2O$ cluster [7a], the $\{VO_4\}^{3-}$ (V-O = 1.71 Å) group interacts through 4$\{\mu_4$-O$\}$ with the 12 V-centers of the shell forming covalent (V-μ_4-O)-V$_{shell}$ (μ_4-O-V$_{shell}$ = 2.39 Å) bonds and conferring the tetrahedral symmetry to the cluster anion similar to what is observed in **1**.

The 12$\{VO_6\}$ units of $\{V_{18}O_{42}(VO_4)\}$ core in **1** [14] are fused with 6$\{VO_5\}$ groups through common edges and linked with the central $\{VO_4\}$ unit via corner sharing. The octahedral geometry around each vanadium (V2) in the 12$\{VO_6\}$ units is defined by a terminal oxo- group (O4) (V2-O4 = 1.609 Å), four μ_3-oxygens (O2, O3) of the shell (V2-O = 1.949 - 1.981 Å) and one μ_4-oxygen (O1) (O1-V2 = 2.440 Å) from the central $\{VO_4\}$ unit. The geometry around each one of the square-pyramidal vanadium (V3) in 6$\{VO_5\}$, is defined by four basal μ_3-oxo groups (O2, O3) (V3-O = 1.889 - 2.019 Å) from the shell and an apical μ_2-oxygen (O5) (V3-O5 = 1.640 Å) which in turn is linearly bonded to the manganese(II) center of one of the six $\{Mn(H_2O)_4\}$ bridges, forming $\{V(3)$-O(5)-$Mn(H_2O)_4$-O(5)-$\}$ bonds, that link $\{V_{18}O_{42}(VO_4)\}$ clusters with each other. The octahedral geometry around each manganese(II) is completed by four oxygen atoms (O6) from the aqua ligands (Mn-O(H$_2$) = 2.200 Å), each one disordered over three positions, and two trans- μ_2-oxo groups (O5) (Mn-O = 2.136 Å). The framework structure in **2** is essentially isomorphous to that observed in **1** with the exception that the constituent $\{V_{18}O_{42}(VO_4)\}$ clusters in **2** are interconnected by six μ_2-$\{Fe(H_2O)_4\}$ bridging groups.

The bond valence sum (BVS) calculations [19a] and metal-oxygen bond lengths identify twelve O3 groups in **1** having attached hydrogen, which refines with occupancy of 0.5, and O6 to be H_2O. This conclusion and the result of manganometric titration of V^{IV} sites (15 V^{IV} per formula unit) are consistent with the charge balance consideration and mixed-valence nature of **1**.

The analogous network structures of **3** [18] and **4** [13] (Figure 3) consists of arrays of $\{V_{18}O_{42}(SO_4)\}$ clusters each one connected to six others via $\{Ni(H_2O)_4\}$and $\{Zn(H_2O)_4\}$ bridging groups, respectively. The building block units in the structure of **4,** shown in Figure 4, consists of $\{V_{18}O_{42}(SO_4)\}$ cluster formed from the $\{V_{18}O_{42}\}$ shell[6] encapsulating a tetrahedral $\{SO_4\}^{2-}$ moiety with disordered oxygen atoms. Here, the host shell behaves as container for the $\{SO_4\}^{2-}$ group. The guest $\{SO_4\}^{2-}$ group which, unlike the encapsulated $\{VO_4\}^{3-}$ groups in **1** and **2,** is not an integral part of the shell, has

Figure 3. View of the extended structure of $(N_2H_5)_2[Zn_3(H_2O)_{12}V_{18}O_{42}(SO_4)]\cdot 24H_2O$ (**4**) showing interpenetrating nets of $\{V_{18}O_{42}(SO_4)\}$ clusters interconnected through $\{Zn(H_2O)_4\}$ bridging groups, and rectangular channels occupied by the hydrogen bonded water molecules (open circles) and cations (striped circles). Hydrogen atoms and the central cluster in the unit cell have been omitted.

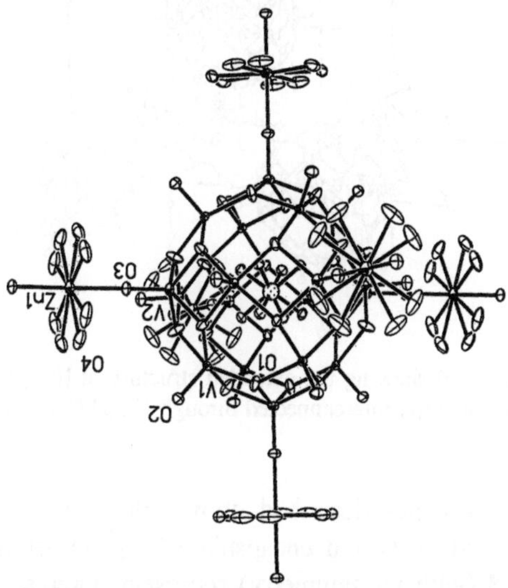

Figure 4. The building block units in the crystal structure of $(N_2H_5)_2[Zn_3(H_2O)_{12}V_{18}O_{42}(SO_4)]\cdot 24H_2O$ (**4**) showing the atom labeling scheme. Key: Central circle with regular dot pattern represents sulfur atom; Atoms bonded to the sulfur atom represent O5 atoms. Selected bond lengths: **4,** V(1)-O(2) = 1.588 Å; V(1)-O(1) = 1.9515 Å; V(2)-O(3) = 1.634 Å; V(2)-O(1) = 1.953 Å; Zn(1)-O(4)(H_2) = 2.063 Å; Zn(1)-O(3) = 2.122 Å; S(1)-O(5) = 1.472 Å; V(1)-V(2) = 2.9441 Å.

normal S-O distances (1.472 Å), and rattles inside the host shell. The Zn(1)-O(4) distance (2.063 Å) and the bond valence sum value (0.36) identify O(4) oxygen as H_2O. This in combination with the result of the manganometric titration of V^{IV} sites (12 V^{IV} per formula unit) requires two units of negative charge per $[Zn_3(H_2O)_{12}V^{IV}_{12}V^{V}_{6}O_{42}(SO_4)]$ unit which is balanced by the two $N_2H_5^+$-hydrazinium cations.

The mixed-valence compounds **5-8** [15,18] also exhibit highly symmetrical isomorphous structures (Figure 5). In each case, the structure consists of $\{V_{18}O_{42}\}$ cages with crystallographic m-3m (O_h) symmetry [20] linked by bridging $\{M(H_2O)_4\}$ groups (M = Fe, Co, Cd, Mn) into two interpenetrating three-dimensional networks- a consequence of body-centered symmetry [21]. Each cage hosts a two-fold disordered tetrahedral $\{XO_4\}$ group. Based on crystallographic, chemical, and spectroscopic evidences [22], the final refinement model assumed a disordered distribution of both (VO_4^{3-} and SO_4^{2-}) anions for the $\{XO_4\}$ group.

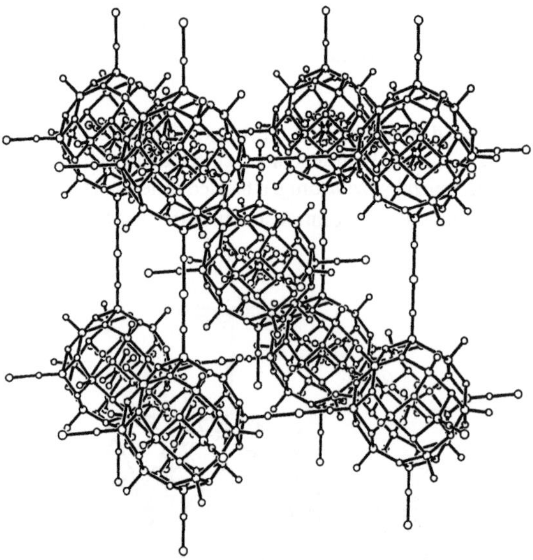

Figure 5. View of the unit cell showing the extended structure of $[Fe_3(H_2O)_{12}V_{18}O_{42}(XO_4)]\cdot 24H_2O$, **5**, containing arrays of $\{V_{18}O_{42}(XO_4)\}$ interconnected through $\{Fe(H_2O)_4\}$ bridging units. Water molecules have been omitted for clarity.

In the earlier examples described above, the host $\{V_{18}O_{42}\}$ shell incorporate VO_4^{3-} groups (compounds **1-2**) and encapsulate SO_4^{2-} moieties (compounds **3-4**). The $\{V_{18}O_{42}\}$ cages in **5-8** (with O_h symmetry) represent, interestingly, somewhat different situations illustrating, further, the structural flexibility of the $\{V_{18}O_{42}\}$ motif.

A view of the $[V_{18}O_{42}(XO_4)]$ cluster in **5** is shown in Figure 6. All bond distances in the cluster are within normal ranges. Similar to what has been observed in **1-4**, the geometry around each V1 in 12$\{VO_5\}$ groups is defined by a terminal oxo-(O2) group and four μ_3-oxygens (O1). The geometry around V2 in the remaining 6$\{VO_5\}$ units is

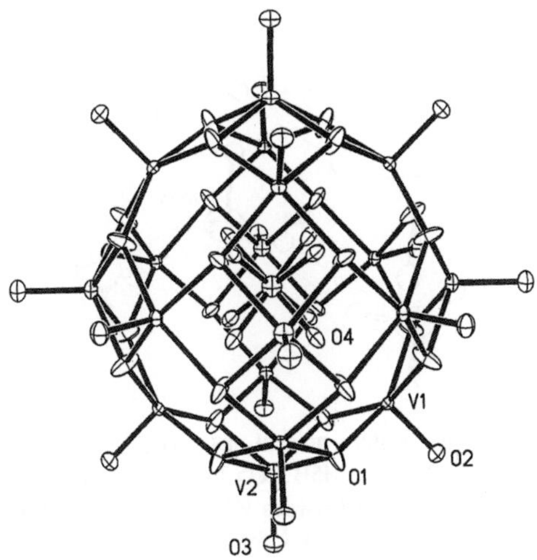

Figure 6. A view of the $[V_{18}O_{42}(XO_4)]$ cluster in the crystal structure of **5**, showing the atom labeling scheme in the asymmetric unit. The unlabeled central atom, bonded to O4 atoms, represents X. Selected bond lengths (Å) and angles (°): **5**; V1-O1 1.948(2), V1-O2 1.587(6), V2-O1 1.942(4), V2-O3 1.636(9), Fe1-O3 2.105(9), V1-O1-V2 98.16(14), V1-O1-V1 138.2(3). **6**; V1-O1 1.923(3), V1-O2 1.590(9), V2-O1 1.954(7), V2-O3 1.652(15), Co1-O3 2.118(14), V1-O1-V2 97.3(2), V1-O1-V1 138.7(5). **7**; V1-O1 1.960(14), V1-O2 1.589(4), V2-O3 1.619(7), Cd1-O3 2.230(6), V1-O1-V2 98.04(11), V1-O1-V1 138.2(2)

defined by four μ_3-oxo (O1) groups and an apical μ_2-oxygen (O3) which is also bonded to the M^{II} center of one of the six $\{M(H_2O)_4\}$ bridges that interlink $\{V_{18}O_{42}(XO_4)\}$ clusters. The coordination sphere of M^{II} is completed by four oxygen (O6) atoms from the aqua ligands ($M-O(H_2)$ = 2.089 Å for **5** and 2.047 Å for **6**), each one exhibiting a twofold disorder, and two trans- μ_2-(O3) groups. The elongated ellipsoid for O1, present in **5-6**, is probably a consequence of steric interactions arising from a short O…O contact (2.545(9) Å for **5** and 2.521(13) Å for **6**) between O1 and an O4 site that is only 50% occupied. The M(1)-O(6) distance and the BVS value (0.39) of O(6) group identify it as H_2O. This conclusion and the result of the manganometric titration of V^{IV} sites (~ 9.5V^{IV} per formula unit) [19b] correspond to the above formulation.

The packing in **1-8** (Figures 1,3,5) generate channels and cavities, defined by $\{V_{18}O_{42}(XO_4)\}$ and $\{M(H_2O)_4\}$ units, which are occupied by counter ions and/or lattice water molecules. The lattice water in the cavities are, in general, hydrogen bonded to the oxo- groups present on the surface of the $\{V_{18}O_{42}(XO_4)\}$ clusters. Moreover, the lattice water molecules are clustered together through intermolecular hydrogen bonds. A view of such clusters present, as $\{(H_2O)_6\}$ groups with disordered hydrogen atoms, in the cavities of $Li_6[Cd_3(H_2O)_{12}V_{18}O_{42}(XO_4)]\cdot 24H_2O$ (**7**) is shown in Figure 7. An edge-on view of the unit cell content in **7**, which contain an average of 15.5V^{IV} centers per formula unit, showing four of the eight $\{(H_2O)_6\}$clusters is also presented in Figure 8.

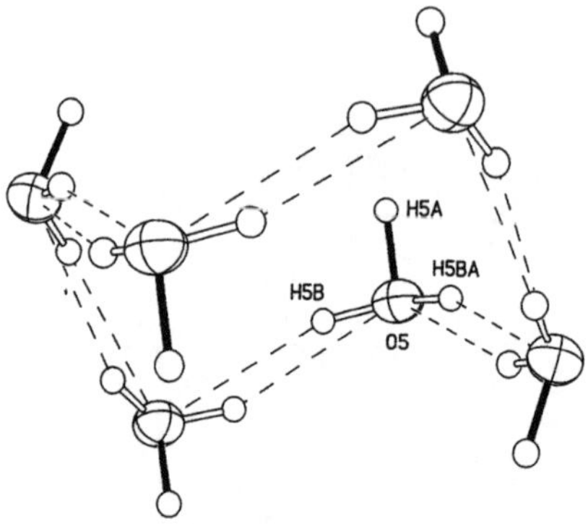

Figure 7. A view of the lattice water present as $\{(H_2O)_6\}$ clusters in the channels in $Li_6[Cd_3(H_2O)_{12}V_{18}O_{42}(XO_4)]\cdot 24H_2O$ (7). The H5B...O5 hydrogen bond distances are 1.94Å for 7 and 1.87 Å for 3. There is also a hydrogen bond between H5A and triply bridging oxo- group (O1) of the cage (H5A...O1 distances are 2.02Å for 7 and 1.94 Å for 3).

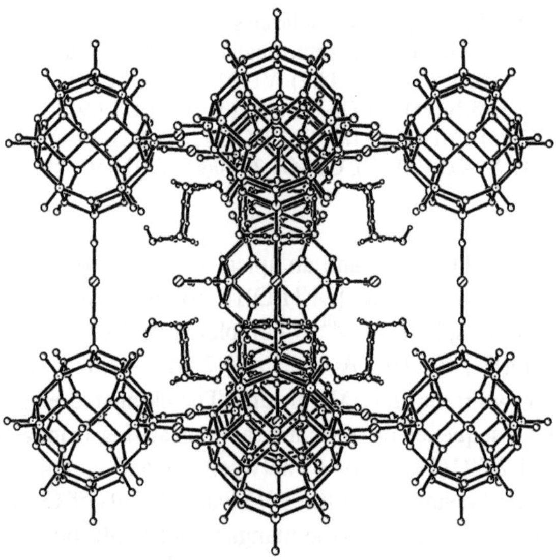

Figure 8. A edge-on view of the unit cell in 7 showing channels and four of the eight $\{(H_2O)_6\}$ clusters. The central XO_4 groups have been omitted.

5. Thermal and Ion Exchange Properties

The lattice water occupying the voids present in these materials are readily removable at relatively lower temperatures. The thermogravimetric analysis of a sample of **1** shows initial weight loss corresponding to the total removal of the lattice water at 70 °C followed by the weight loss due to the removal of coordinated water at ~ 257 °C. Further heating up to 500 °C yielded a reduced metal oxide phase. The dehydrated (at 120 °C) sample of **1** exhibited reversible water absorption as evidenced by IR spectral studies.

The cations present in the cavities of solids are readily exchangeable by other cations (NH_4^+, Na^+, K^+, etc.) without noticeable change in the framework structures. Thus, when soaked in 2M aqueous solutions of the electrolyte ECl (E = Na, K, or NH_4) containing the desired cations (Na^+, K^+, or NH_4^+ ions) for 24 hours with occasional stirring at room temperature, the crystals of **3** and **7** undergo ion exchange accompanied by a decrease in the pH of the electrolyte solutions by 1-1.5 units. Since neither the results of the bond valence sum calculations nor the Fourier difference map indicated the presence of any potentially exchangeable protons associated either with the constituent clusters, bridging groups or within the voids in the structure of **3** and **7**, this drop in pH may be attributed to the ion exchange which may release Li^+ ions from the solid into the solution with accompanying hydrolysis. The presence of any 'hidden' acidic groups in **3** and **7** is although less probable, especially in view of the satisfactory charge balance and the well resolved structure, but can not be ruled out.

6. Conclusion

The syntheses and characterization of the framework materials described here represent a step in the direction of the preparation of transition metal oxide based solids composed of suitable metal oxide motifs with controllable properties. These stable framework structures are composed of essentially transition metal oxide clusters linked through additional metal oxide units without incorporating any conventional ligand in the framework. In view of the rapidly expanding pool of the well characterized metal oxide clusters, this has potential to provide access to a variety of new synthetic materials. Given the proven role of polyoxometalates in catalysis [4] and in the development of new oxide supported transition metal catalysts [4f] their application in preparing new surfaces could be valuable. The heterometallic centers, such as Mn^{II}, Fe^{II}, Co^{II}, Ni^{II}, Cd^{II}, and Zn^{II}, located in the walls of the tunnels in these compounds may be incorporated in new solids to fashion materials that may also exhibit properties, e.g. catalytic activity, associated with the heterometallic sites. These compounds may also exhibit useful ion exchange, sorptive, and chemical sensing properties. Well characterizable solid surfaces may provide additional opportunities for theoretical studies of specific metal oxide surfaces to gain insight of their reactivities and surface dynamics [2d].

The chemistry associated with $\{V_{18}O_{42}(XO_4)\}$ cores is further exploitable for not only understanding the property of systems already prepared but also for synthesizing new surfaces, for example, functionalized derivatives, for which these clusters will provide good models. The replacement of H_2O from $\{M(H_2O)_4\}$ by other organic and inorganic groups may offer possibility of anchoring groups (e.g. metal ions, clusters, organometallic moities, and asymmetric units) deemed suitable for enhancing the reactivity of the surface.

References

1. (a) M. T. Pope, *Heteropoly and Isopoly Oxometalates,* Springer, Berlin, (1983).
   (b) *Polyoxometalates: From Platonic Solids To Anti-Retroviral Activity* (Eds.: M. T. Pope, A. Müller), Kluwer Academic, Dordrecht, (1994).
   (c) M. T. Pope, A. Müller, *Angew. Chem. Int. Ed. Engl.* **30,** 34 (1991).
   (d) Some 600 refereed publications and over 120 patents on the chemistry and technology related to polyxometalates in just one year; For more information see, D. E. Katsoulis, *Chem. Rev.* **98,** 359 (1998), and references.
   (e) For an up to-date account on some of the most fascinating supermolecular systems, see the following excellent review article: A. Müller, P. Kögerler, C. Kuhlmann, *J. Chem. Soc., Chem. Commun.* 1347 (1999), and references therein.
2. (a) L. C. W. Baker in *Advances in the Chemistry of Coordination Compounds,* (Ed.: S. Kirschner), Macmillan, New York, (1961), p. 608.
   (b) V. W. Day, W. G. Klemperer, *Science* (Washington, DC) **228,** 4699 (1985).
   (c) W. G. Klemperer, T. A. Marquart, O. M. Yaghi, *Angew. Chem. Int. Ed. Engl.* **31,** 49 (1992).
   (d) K. Isobe, A. Yagasaki, *Acc. Chem. Res.* **26,** 524 (1993).
3. (a) I. M. Campbell, *Catalysis at Surfaces,* Chapman and Hall, London, (1988).
   (b) H. Kung, *Transition Metal Oxides: Surface Chemistry and Catalysis,* Elsevier, New York, (1989).
   (c) R. K. Grasselli, J. D. Burrington, *Adv. Catal.* **30**, 133 (1981).
   (d) J. M. Thomas, W. J. Thomas, *Principles and Practice of Heterogeneous Catalysis,* VCH, Weinheim, (1997).
4. (a) N. Mizuno, M. Misono, *Chem. Rev.* **98,** 199 (1998).
   (b) Y. Izumi, K. Urabe, M. Onaka, *Zeolite, Clay, and Heteropoly Acid in Organic Reactions,* VCH, Weinheim, (1992).
   (c) A. Corma, *Chem. Rev.* **95**, 559 (1995).
   (d) C. L. Hill, *Coord. Chem.Rev.* **143**, 407 (1995).
   (e) I. V. Kozhevnikov, *Chem. Rev.* **98,** 171 (1998).
   (f) M. Pohl, D. K. Lyon, N. Mizuno, K. Nomiya, R. G. Finke, *Inorg. Chem.* **34**, 1413 (1995), and references therein.
5. A. Müller, F. Peters, M. T. Pope, D. Gatteschi, *Chem. Rev.* **98,** 239 (1998).
6. (a) J. Fucks, S. Mahjour, and J. Pickard, *Angew. Chem. Int. Ed. Engl.* **15**, 374 (1976).
   (b) V. W. day, W. G. Klemperer, and O. M. Yaghi, *J. Am. Chem. Soc.*, **111**, 4518 (1989).
   (c) H. T. Evans, Jr., *Inorg. Chem.*, **5**, 967 (1966); A. G. Swallow, F. R. Ahmed, and W. H. Barnes, *Acta Cryst.*, **21**, 397 (1966); V. W. day, W. G. Klemperer, and D. J. Maltbie, *J. Am. Chem. Soc.*, **109**, 2991 (1987); P. A. Durif, M. T. Everbuch-Pouchot, and J. C. Guitel, *Acta Cryst.* **B36**, 680 (1980).
   (d) Y. Zhang, R. C. Haushalter, and J. Zubieta, *Inorg. Chim. Acta,* **277**, 263 (1998).
   (e) V. W. day, W. G. Klemperer, and O. M. Yaghi, *J. Am. Chem. Soc.*, **111**, 5959 (1989).
   (f) D. Hou, K. S. Hagen, and C. L. Hill, *J. Am. Chem. Soc.*, **114**, 5864 (1992).
   (g) D. Hou, K. S. Hagen, and C. L. Hill, *J. Chem. Soc., Chem. Commun.*, 426 (1993).
   (h) A. Müller, E. Krickemeyer, M. Penk, H.-J. Wallberg, and H. Bögge, *Angew. Chem. Int. Ed. Engl.,* **1987**, 26, 1045; A. Müller, M. Penk, Ralf Rolfing, E. Krickemeyer, and J. Döring, *Angew. Chem. Int. Ed. Engl.* **29**, 926 (1990).
7. (a) Müller, A.; Sessoli, R.; Krickemeyer, E.; Boegge, H.; Meyer, J.; Gatteschi., D.; Pardi, L.; Westphal, J.; Hovemeier, K.; Rohlfing, R.; Doring, J.; Hellweg, F.; Beugholt, C.; Schmidtmann, M. *Inorg. Chem.* 36, 5239 (1997).
   (b) Johnson, G. K.; Schlemper, E. O. *J. Am. Chem. Soc.* **100,** 3645 (1978).
8. (a) L. Suber, M. Bonamico, and V. Fares, *Inorg. Chem.* **36**, 2030 (1997).

(b) A. Müller, M. Penk, E. Krickemeyer, H. Bögge, H.-J. Wallberg, *Angew. Chem. Int. Ed. Engl.* **27,** 1719 (1988).

9. A. Müller, E. Krickemeyer, M. Penk, Ralf Rolfing, A. Armatage, and H. Bögge, *Angew. Chem. Int. Ed. Engl.* **30**, 1674 (1991).

10. A. Müller, R. Rolfing, J. Doring, and M. Penk, *Angew. Chem. Int. Ed. Engl.* **30**, 588 (1991).

11. A. Müller, *Nature* **352,** 115 (1991); P. C. H. Mitchell, *Nature* **348**, 15 (1990).

12. (a) J. R. D. DeBord, R. C. Haushalter, L. M. Meyer, D. J. Rose, P. J. Zaf, J. Zubieta, *Inorg. Chim. Acta* **256**, 165 (1997).
(b) M. I. Khan, E. Yohannes, and R. J. Doedens, unpublished results.

13. M. Ishaque Khan, E. Yohannes, and D. Powell, *J. Chem. Soc., Chem. Com.*, 23 (1999).

14. M. Ishaque Khan, E. Yohannes, and D. Powell, *Inorg. Chem.* **38,** 212 (1999).

15. M. Ishaque Khan, E. Yohannes, and Robert J. Doedens, *Angew. Chemie. Int. Ed. Engl.* **38,** 1292 (1999).

16. M. Ishaque Khan, E. Yohannes, R. J. Doedens, S. Tabussum, S. Cevik, L. Manno, and D. Powell, *Cryst. Engineering* **2,** 000 (in press- 1999).

17. M. Ishaque Khan, S. Tabussum, C. Zheng, *Polyhedron* (under review).

18. M. Ishaque Khan et. al. unpublished results.

19. (a) I. D. Brown in *Structure and Bonding in Crystals, Vol II* (Ed.: M. O'Keefe, A. Navrotsky), Academic Press, New York, (1981), p.1.
(b) Crystals undergo oxidation in air. Consequently, the number on reduced sites decreases with time.

20. The O_h symmetry refers to the average of the two disordered components. The symmetry, for example, of $\{V_{19}O_{46}\}$ species composed of $\{V_{18}O_{42}\}$ cage encapsulating an ordered $\{VO_4\}$ group will be no higher than T_d.

21. S. R. Batten, R. Robson, *Angew. Chem. Int. Ed. Engl.* **37**, 1460 (1998).

22. (a) For example, the X-O distances (1.538(15) Å for **5** and 1.51(2) Å for **6**) are intermediate between the expected values for SO_4^{2-} and VO_4^{3-}. In the case of **5**, (for which the structural results are most precise), the O displacement ellipsoid is elongated along the X-O bond. The displacement parameter of X was higher than expected when it was refined as 100% V and unrealistically low when only S was included. In the final refinement, the displacement parameter of X was fixed at 0.015 $Å^2$ and the relative proportions of V and S were allowed to vary. This model converged to approximately equal proportions of V and S. b) IR spectra of these compounds exhibit bands attributable to SO_4^{2-} and VO_4^{3-}. Elemental analysis (%S) corresponds to the formulation of **5** and **6.** All attempts to prepare **5** and **6** without using sulfate have been unsuccessful.

Perspectives in the Solid State Coordination Chemistry of the Molybdenum Oxides

PAMELA J. HAGRMAN, DOUGLAS HAGRMAN, JON ZUBIETA
Department of Chemistry, Syracuse University, Syracuse, NY 13244

Abstract. The chemistry of polyoxomolybdate anions may be exploited in a building block approach to the synthesis of solid state oxide materials. One strategy controls the oxide microstructure by introducing a secondary metal-ligand subunit as a charge-compensating, space-filling and structure-directing component. The resulting organic-inorganic hybrid materials are representative of solid state coordination chemistry in which the oxide microstructure reflects the geometry of the ligand, the coordination preferences of the secondary metal center, and the synergistic interaction of the coordination complex cation with the polyoxomolybdate component.

Keywords: polyoxomolybdate, solid state coordination chemistry, organic-inorganic hybrid materials, hydrothermal synthesis.

1. Introduction

The ubiquitous materials of the metal oxide family possess a vast range of stoichiometries and structure types which endow them with a variety of useful physical properties and applications to materials as diverse as magnetic oxides, sensors, phosphors, ceramics, catalysts, ion exchanges and molecular sieves and even biomaterials, as listed in Table 1 [1-24]. In the specific case of the molybdenum oxides, the major subclasses of materials include binary and ternary oxides, bronzes, and isopoly- and heteropolyoxoanion clusters [25-29].

M.T. Pope and A. Müller (eds.), Polyoxometalate Chemistry, 269–300.

Table 1. Selected Examples and Applications of Inorganic Oxides[1-24]

Metal Class	*Examples*	*Applications*
Magnetic oxides	CrO_2, ferrites, spinels	Magnetic tapes, transformer cores, computer memories
Oxide sensors	$BaTiO_3$, $PbZrO_3$	Temperature sensors, piezo speakers
Phosphors	Doped Zn_2SiO_4, Y_2O_3	Oscilloscope tubes
Electronic materials	$BaTiO_3$, ZnO	Surge protectors, high T semiconductors
Ceramics	Al_2O_3, $PbZr_{1-x}Ti_xO_3$	Ferroelectrics
Catalysts	$(VO)_2P_2O_7$, $Bi_2Mo_2O_9$, V_2O_5/TiO_2, SiO_2-MgO_3	Selective oxidations, polymerization, oxidative dehydrogenation, cracking, alkylation
Ion exchange	NASICON, Zeolites	Fast ion conductors, detergents
Molecular sieves	Zeolites	Separating agents
Biomaterials	Organoapatites	Artificial bone

However, despite the practical significance of molybdenum oxides, their structural design, in the sense of controlled synthesis of metastable but persistent extended structures, remains a challenge in solid state chemistry [30]. A powerful tool for the design of novel oxide materials exploits the incorporation of organic molecules to alter the inorganic microstructure or to transmit structural information inherent in the coordination preferences of the metal centers. Examples of the structure-directing role of organic constituents on oxide microstructures include zeolites [31], biomineralized materials [32], mesoporous compounds of the MCM-41 class [33], and transition metal phosphates [34, 35].

A related approach, which we have developed for the synthesis of organically modified molybdenum and vanadium oxides, may be described as solid state coordination chemistry [36, 37]. One strategy relies on linking molecular cluster subunits from the vast family of chemically robust polyoxoanions, either through direct coordination to form oxo-bridged arrays of clusters or through secondary metal centers acting as inorganic bridging groups. The essential building blocks are the polyoxomolybdate anion and a coordination complex cation, that is, a secondary transition metal/ligand subunit. Since under specified conditions of pH and concentration, the polyoxometalate anions form by spontaneous self-assembly, the emphasis lies in the ligand geometry and the choice of the secondary metal center. Of course, variation of the secondary metal site can provide different coordination preferences and a range of photochemical, electrochemical and reaction properties.

The interplay of the coordination preferences of the secondary metal site and the geometric constraints imposed by the ligand provides considerable structural flexibility, as well as an effective subunit for the spatial transmission of structural information. Two classes of molybdenum oxide materials which have evolved from this variant of small molecule synthesis will be discussed: (a) polyoxomolybdate anions linked through molecular coordination complex cationic subunits [38-42] and (b) polyoxomolybdate anions encapsulated within polymeric coordination complex cations [43, 44].

1.1 GENERAL COMMENTS ON THE BUILDING BLOCK STRATEGY.

The secondary metal-ligand component may be introduced as an appropriate molecular subunit which coordinates to the peripheral oxo-groups of the polyoxomolybdate clusters as a bridging group. In this instance, the architecture of the resultant composite solid will reflect the preferred coordination geometry of the secondary metal, the coordination sites available at the secondary metal-ligand subunit, and the number of points of attachment engaged on the oxide cluster. The simplest case is that of bimodal linkage at the secondary metal subunit to produce a one-dimensional structure (Figure 1a). However, a coordination complex subunit capable of additional interactions may result in a tesselated architecture, such as that of Figure 1b. Of course, the variable coordination modes adopted by the cluster subunit may also influence the structure of the product. For example, a bimodal bridging complex may combine with an oxide substructure through four point, rather than two point, attachment to generate the structure of Figure 1c, rather than that of Figure 1a. Consequently, the oxide subunit must not be regarded as

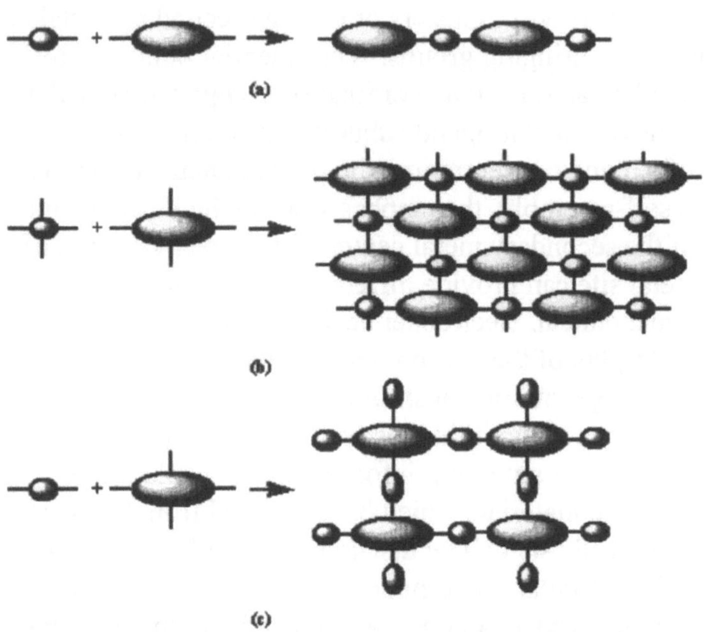

Figure 1. Schematic representations of various modes of linking of secondary metal-ligand subunits (smaller spheres) and polyoxomolybdate clusters (larger spheres) based on available attachment points.

structurally "innocent" clay to be molded by the secondary metal-ligand template, but rather as a synergistically active structure-directing component.

Furthermore, the ligand may be chosen so as to promote the self-assembly of cationic molecular cluster subunits. The linking of polyoxomolybdate anions through subunits of increasing nuclearities is illustrated schematically in Figure 2.

The cationic component need not be restricted to molecular subunits, but rather thorugh judicious choice of ligand may be introduced as a one-, two- or three-dimensional substructure. As illustrated in Figure 3, the appropriate rod-like bridging ligand in combination with the coordination preferences of the secondary metal site can provide a variety of cationic polymer scaffoldings for the entrainment and manipulation of oxide substructures. In a sense, the ligand provides a scaffold for the transmission of metal-based structural information in one-, two- or three-dimensions. This strategy for the molecular design of molybdenum oxide solids represents a conflation of polyoxomolybdate chemistry and of "crystal engineering" exploiting polyfunctional rod-like ligands as tethers for the construction of extended solids with diverse topologies [45, 46].

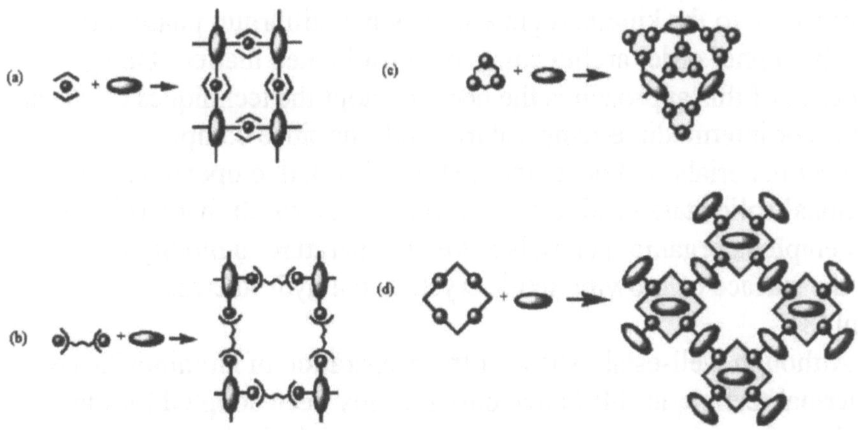

Figure 2. Schematic representations of the linking of secondary metal-ligand subunits Sf varying degrees of aggregation to polyoxomolybdate clusters.

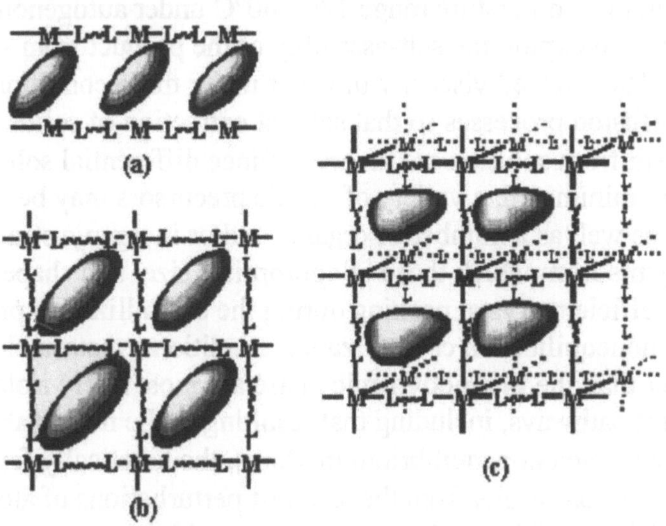

Figure 3. Schematic representations of the encapsulation of polyoxomolybdate clusters within the cavities of polymeric coordination complex cations.

2. Synthesis

Our approach to the synthesis of oxides exploits the synergism between organic and inorganic components in the sense that the organic component serves to imprint some structural information onto the inorganic network. There is thus a hierarchical coding of the chemistry and crystal growth. This approach requires a shift in synthetic paradigm from the thermodynamic to the kinetic regime in which equilibrium phases are replaced by higher order architectures of consolidated matter. One consequence of this approach is the need to adopt the techniques of *chemie douce* [47] or intermediate-temperature synthesis since complex hierarchical materials will decompose at the elevated temperatures of conventional solid state synthesis. The general approach that has been adopted employs organic materials at low temperature to modify or control the surface of growing oxide crystals in a hydrothermal medium[48].

Although well-established for the preparation of aluminosilicates, hydrothermal techniques [49] have only recently been adopted for the preparation of a wide variety of metastable materials, including transition metal phosphates, metal organophosphonates, and complex polyoxoalkoxometalates [29]. Hydrothermal reactions are typically carried out in the temperature range 120-260°C under autogenous pressure, so as to exploit the self-assembly of the product from soluble precursors. The reduced viscosity of water under these conditions enhances diffusion processes so that solvent extraction of solids and crystal growth from solution are favored. Since differential solubility problems are minimized, a variety of simple precursors may be introduced, as well as a number of organic and/or inorganic structure-directing agents from which those of appropriate size and shape may be selected for efficient crystal packing during the crystallization process. Under such nonequilibrium crystallization conditions, metastable kinetic phases rather than the thermodynamic phase are most likely isolated. While several pathways, including that resulting in the most stable phase, are available in such nonequilibrium mixtures, the kinetically favored structural evolution results from the smallest perturbations of atomic positions. Consequently, nucleation of a metastable phase may be favored.

The nature of the molybdenum oxide component can be directed to some extent by the choice of reaction conditions. Thus, while low pH (3 to 5), short reaction times (10-48 h), and lower temperature ranges (110-160° C) favor the formation of molybdate clusters as structural building blocks, conditions of higher pH (5 to 8) and more extreme temperatures and exposure times generally result in architectures in which the secondary metal is incorporated into a bimetallic oxide network and/or in which the molybdate cluster identity is lost upon fusing into one- or two-dimensional molybdenum oxide substructures.

3. Plyoxoanions Linked Through Molecular Cationic Subunits

Solid state materials constructed from components based on polyoxomolybdate anions and coordination complex cations are representative of the significant contemporary theme in solid state chemistry of molecularly derived solids with extended linkages [45, 46]. A variety of chain, layer, and three-dimensional frameworks based on polyoxometalate clusters, either directly fused through bridging oxo-groups or linked through heterometallic sites, have been described in recent years [45, 46]. Examples include the one-dimensional $[\{Co(H_2O)_4\}_2(H_2W_{12}O_{42})]^{6-}$ [50], the network solid $[\{Ni(en)_2\}_3(V_{18}O_{42}Cl)]^{-0.5}$ [39], and the framework materials $[Na_2Co_2H_2W_{12}O_{42}(H_2O)_{12}]^{4-}$ [51] and $[Zn_3V_{18}O_{42}(SO_4)(H_2O)_{12}]^{5-}$ [52].

The prototypical structure for the materials of this review is provided by the molecular cluster $[\{Cu(o\text{-phen})_2\}_2Mo_8O_{26}]$ **(MOXI-27)** which consists of an α-$[Mo_8O_{26}]^{4-}$ clusters decorated with two $\{Cu(o\text{-phen})_2\}^{2+}$ moieties covalently attached to peripheral oxo-groups, as shown in Figure 4. While this structure lacks extension into the polymeric domain, the ease of attachment of heterometal-diamine complexes to the surface of such clusters is established.

A number of observations on the structure of **MOXI-27** are relevant with respect to a synthetic approach to materials based on polyoxoanion and coordination complex cation components. (1) Extension of the structure beyond the molecular domain is apparently precluded by the steric demands of the *o*-phen ligands on the Cu(II) sites. A less sterically-demanding ligand set should allow expansion of the copper coordination sphere from "4+1" to "4+2" Jahn-Teller distorted geometry. Alternatively, reducing the stoichiometry of the ligand, so as to favor a $\{Cu(LL)\}^{2+}$ rather than the $\{Cu(LL)_2\}^{2+}$ subunit, would also

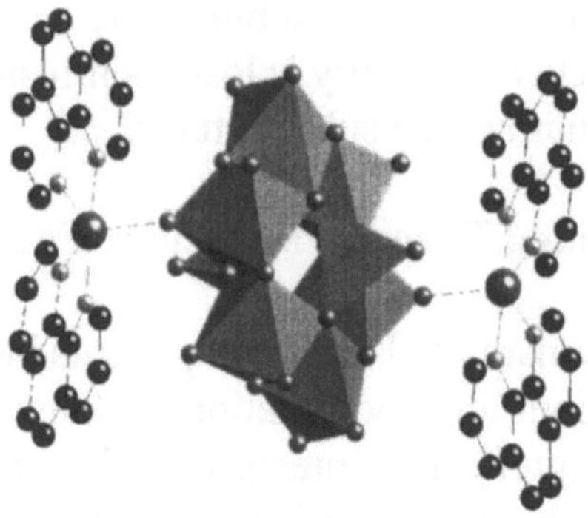

Figure 4. A view of the structure of $[\{Cu(o\text{-}phen)_2\}_2Mo_8O_{26}]$ (**MOXI-27**).

provide vacant coordination sites on the Cu(II) for bridging of polyoxoanions into extended arrays. (2) The geometric preferences of the secondary metal center also have a profound influence on the ability of the coordination complex subunit to bridge polyoxoanion sites. Thus, Cu(II) with a preference for "4+1" or "4+2" axically distorted geometries tends to provide rather weak interactions with the oxo-groups of the anions when present as the square-planar $\{Cu(LL)_2\}^{2+}$ moiety. Transition metals with a strong preference for more regular octahedral geometry, such as Ni(II) or Fe(III), may, therefore, be more effective as cationic linkers between polyoxoanions. (3) The ligand may be designed to provide specific linking modes at the secondary metal site or to promote aggregation of the coordination complex subunit into clusters or even polymers. Thus, 2,2':6':2"-terpyridine affords tridentate meridional coordination about the secondary metal, effectively restricting the available coordination sites to a meridional arrangement. Similarly, a ligand such as tetra-(4'-pyridyl)pyrazine (tppz) not only imposes meridional coordination, but also generates a binuclear cationic subunit $\{M_2(tppz)\}^{n+}$ with a designed spacer between the polyanion subunits.

The various influences of secondary metal coordination preferences, ligand geometry and cationic cluster subunits are manifested in the family of materials constructed from polyoxomolybdates and coordination complex cations. For example, the steric constraints imposed by the bulky *o*-phen ligands of $[\{Cu(o\text{-}phen)_2\}_2Mo_8O_{26}]$ are evidently relaxed in the ethylenediamine derivative $[\{Cu(en)_2\}_2Mo_8O_{26}]$ (**MOXI-11**).

As shown in Figure 5, the structure is constructed from δ-$[Mo_8O_{26}]^{4-}$ clusters linked by $\{Cu(en)_2\}^{2+}$ groups into the virtual two-dimensional sheet. It is noteworthy that the axial Cu-O distances of 2.46(1)Å and 2.98(1)Å are rather long and the subunits are consequently only weakly associated.

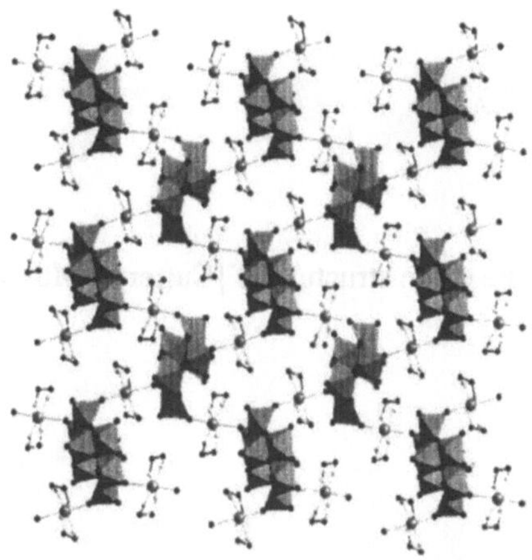

Figure 5. A view of the structure of $[\{Cu(en)_2\}_2Mo_8O_{26}]$ (**MOXI-11**).

Ligand influences are demonstrated by the structure of $[Cu(terpy)Mo_2O_7]$ (**MOXI-41**), shown in Figure 6a. The structure consists of a one-dimensional chain constructed from $\{Mo_4O_{14}\}^{4-}$ clusters linked through $\{Cu(terpy)\}^{2+}$ subunits. As shown in Figure 6b, the $\{Mo_4O_{14}\}^{4-}$ cluster consists of four edge- and corner-sharing $\{MoO_5\}$ distorted square pyramids. This $\{Mo_4O_{14}\}^{4-}$ cluster imbedded in the chain of **MOXI-41** is unusual in several respects. There is no precedent in the polyoxomolybdate chemistry for an isolated tetranuclear cluster, furthermore, the exclusively five coordinate geometry is unanticipated for a molybdate core. The Cu(II) center exhibits a $\{CuN_3O_2\}$ square pyramidal geometry with Cu-O bond distances of 1.893(2)Å and 2.189(2)Å for the basal and opical positions, respectively, reflecting the structural Jahn-Teller effect at the d^9 Cu(II) site.

It is instructive to compare the structure of **MOXI-41** to that of the Zn(II) anologue of identical stoichiometry $[Zn(terpy)Mo_2O_7]_1$ (**MOXI-42**), shown in Figure 7. The one-dimensional chain of **MOXI-42** consists of an undulating ribbon of edge- and corner-sharing $\{MoO_5\}$ square pyramids, decorated by peripheral $\{NiN_3O_2\}$ square pyramids. The molybdate chain may be described in terms of binuclear $\{Mo_2O_8\}$ subunits of edge-sharing square pyramids, linked through corner-sharing of oxo-groups in an *anti* disposition. The Ni(II) site links to the bridging oxo-groups of two adjacent binuclear molybdate subunits to form a cyclic $\{NiMo_2O_3\}$ moiety.

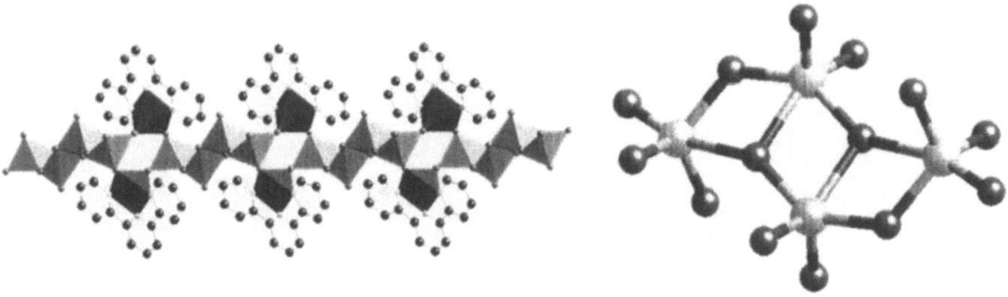

Figure 6. Left: a view of the chain structure of $[Cu(terpy)Mo_2O_7]$ (**MOXI-41**). Right: the $\{Mo_4O_{14}\}^{4-}$ cluster.

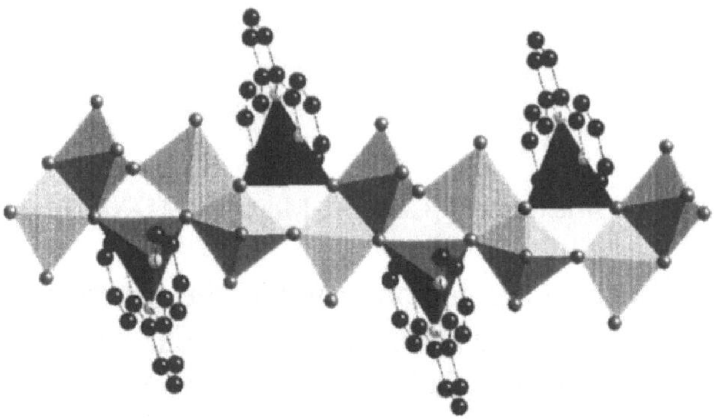

Figure 7. A view of the structure of $[Zn(terpy)Mo_2O_7]$ (**MOXI-42**).

The structures of **MOXI-41** and **MOXI-42** appear to reveal the structural influences of the secondary metal site. In the case of **MOXI-41**, the axial lengthening of the Cu-O bonds prevents the molybdate subunits from fusing into a one-dimensional chain and consequently the {Mo4O14}4- clusters retain their identity. In contrast, the short Zn-O distances in **MOXI-42** allow the fusion of the molybdate polyhedra into a one-dimensional substructure.

However, this naïve interpretation requires considerable refinement to take into account the subtle interplay of ligand geometry, secondary metal coordination geometry and extent of molybdate aggregation and interaction with the cationic subunits. A comparison of the structure of $[Ni(2,2'\text{-}bpy)_2Mo_4O_{13}]$ (**MOXI-10**) and $[Cu(2,2'\text{-}bpy)Mo_2O_7]$ (**MOXI-17**) and those of **MOXI-41** and **MOXI-42** reveals the structural versatility of this building block approach and the inherent complexity of the system. As shown in Figure 8, the structure of [Ni(2,2'-

bpy)$_2$Mo$_4$O$_{13}$] (**MOXI-10**) is constructed from β-$\{Mo_8O_{26}\}^{4-}$ clusters linked by $\{Ni(bpy)_2\}^{2+}$ bridges into a one-dimensional chain. In contrast, the structure of [Cu(2,2'-bpy)Mo_2O_7] (**MOXI-17**) consists of a buckled one-dimensional molybdenum oxide chain linked to peripheral $\{CuN_2O_4\}$ octahedra. The Cu(II) sites exhibit the usual "4+2" coordination mode, with two short Cu-O distances of 1.993(11)Å (ave) and two long Cu-O distances of 2.380(10)Å (ave).

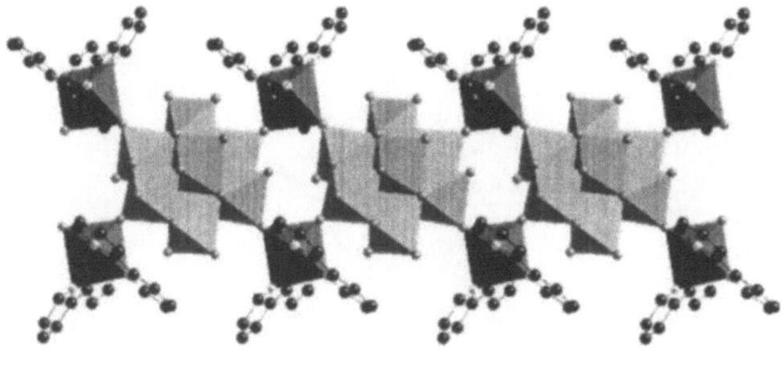

Figure 8. The structure of [Ni(2,2'-bpy)$_2$Mo$_4$O$_{13}$] (**MOXI-10**).

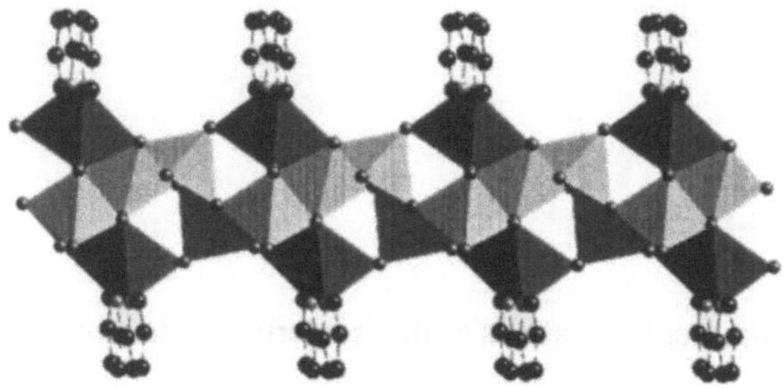

Figure 9. A view of [Cu(2,2'-bpy)Mo_2O_7] (**MOXI-17**).

The structures of **MOXI-41, MOXI-42, MOXI-10** and **MOXI-17** illustrate an interplay of structure-determining factors. It is noteworthy that of the [M(terpy)Mo_2O_7] set, **MOXI-41** and **MOXI-42**, the Cu(II) species exhibits discrete $\{Mo_4O_{14}\}^{4-}$ clusters while in the Ni(II) derivative the molybdate polyhedra fuse into a one-dimensional chain. This observation appears to reflect at least impart the coordination preferences of the $\{Cu(terpy)\}^{2+}$ and $\{Ni(terpy)\}^{2+}$ subunits, as noted above. However, in contrast to these latter structures, of the **MOXI-10** and **MOXI-17** pair it is the Ni(II) species which displays discrete molybdate clusters while the Cu(II) material exhibits a one-dimensional molybdate chain. However, in this latter case, the influences of the M(II) components cannot be directly compared as the subunits are present as $\{Ni(bpy)_2\}^{2+}$ and $\{Cu(bpy)\}^{2+}$. It would appear that the reduced steric hindrance associated with a single bidentate ligand at the $\{Cu(bpy)\}^{2+}$ site allows the close approach and fusing of the molybdate octahedra, while the sterically constrained $\{Ni(bpy)_2\}^{2+}$ subunits cannot be readily accommodated on the surface of a one-dimensional molybdate chain but rather require some separation into aggregate subunits.

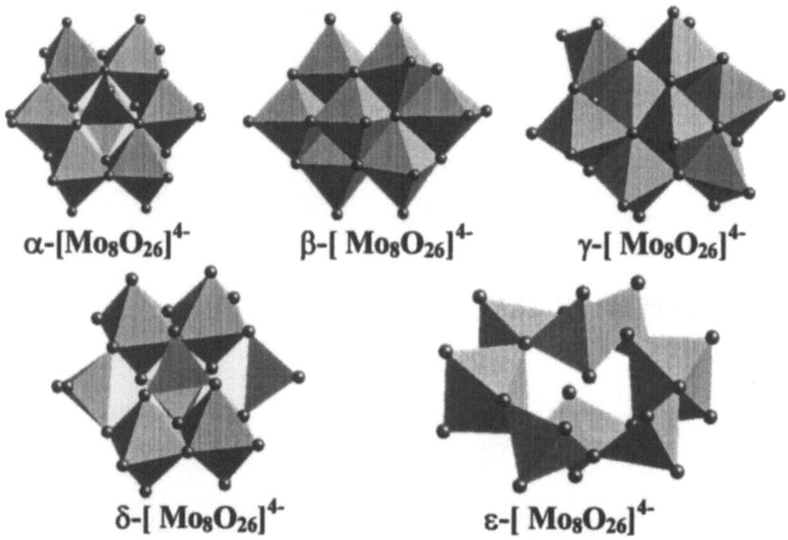

Figure 10. Polyhedral representations of the isomeric structures of $[Mo_8O_{26}]^{4-}$.

It is instructive to note that while the octamolybdate unit $[Mo_8O_{26}]^{4-}$ appears in $[\{Cu(o\text{-}phen)_2\}_2Mo_8O_{26}]$, $[\{Cu(en)_2\}_2Mo_8O_{26}]$ and $[Ni(2,2'\text{-}bpy)_2Mo_4O_{13}]$, it occurs in three distinct isomeric forms, α-, δ-, and β-, respectively. Five isomers of $[Mo_8O_{26}]^{4-}$ have now been described. The α, β, and γ isomers had been isolated in a number of salts, [28, 53] while the δ-form has been reported as the (α-δ) or (β-δ) "intermediate" structure [54]. The ε-form is thus far unique to the heterometal-diamine molybdenum oxide materials. As shown in Figure 10, the isomer structures differ in number, types and fusion modes of molybdenum polyhedra. The five forms are conceptually and chemically interrelated by minimal bond breaking through lengthening of axial interactions and polyhedral rotations. The occurrence of a particular isomer in the hydrothermal product is not predictable, a not unreasonable observation given the lack of significant energy differences between the structure types.

Although the structure of polyoxomolybdate component can not be predictably manipulated, the coordination complex cation may be fashioned through incorporation of ligands with appropriate geometric requirements. This aspect of the structural chemistry is most readily manifest in the construction of polynuclear coordination complex aggregates, as illustrated by the incorporation of binuclear, trinuclear and tetranuclear cationic subunits in $[\{Cu_2(tppz)(H_2O)_2\}Mo_8O_{26}]$ (**MOXI-44**), $[\{Cu_3(4,7\text{-phen})_3\}_2(Mo_8O_{26})(Mo_6O_{19})]$ (**MOXI-15**), and $[\{Cu(4,4'\text{-}bpyS)\}_4Mo_8O_{26}]$ (**MOXI-36**), respectively.

As noted, the structural complexity of the secondary metal-ligand subunit may be enhanced by appropriate ligand design. Thus, the tetraphyridylpyrazine ligand is anticipated to act as a binucleating group to provide a novel building block for structures of this class. As shown in Figure 11a, the structure of $[\{Cu_2(tpypz)(H_2O)_2\}(Mo_8O_{26}\}]$ (**MOXI-44**) consists of α-$[Mo_8O_{26}]^{4-}$ clusters linked by $\{Cu_2(tpypz)(H_2O)\}^{4+}$ units. The Cu(II) sites exhibit the characteristic "4+2" distorted octahedral geometry. Each copper center of a binuclear unit coordinates to three nitrogen donors of the tpypz ligand, two terminal oxo-groups from adjacent $[Mo_8O_{26}]^{4-}$ clusters, and an aqua ligand. Consequently, each $\{Cu_2(tpypz)(H_2O)_2\}^{4+}$ cluster links four α-molybdate subunits), while each α-$[Mo_8O_{26}]^{4-}$ cluster bonds to four binuclear copper units. The binuclear copper units are oriented with respect to the molybdate clusters in the two-dimensional network so as to project the aqua ligands into cavities within the layer.

The orientation of the ligand donor groups and the coordination tendencies of the metal have been exploited in constructing the trinuclear coordination complex cluster of $[\{Cu_3(4,7\text{-phen})_3\}_2(Mo_8O_{26})(Mo_6O_{19})]$ (**MOXI-15**), shown in Figure 12. The planar $\{Cu_3(4,7\text{-phen})_3\}^{3+}$ cluster exploits the incorporation of Cu(I) with the preference for digonal coordination to exploit the 60° angle between N-donors of the ligand in the fashioning of the ring structure of this cation.

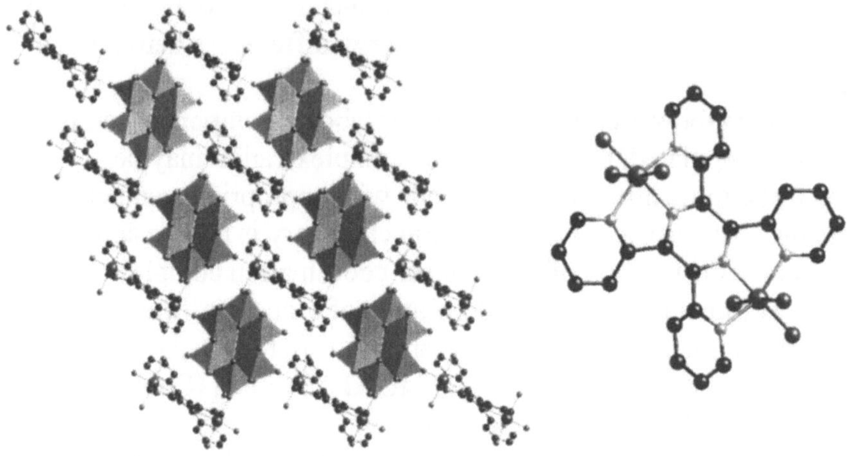

Figure 11. Left: A view of $[\{Cu_2(tpypz)(H_2O)_2\}Mo_8O_{26}]$ (**MOXI-44**). Right: The binuclear $\{Cu_2(tpypz)(H_2O)_2\}^{4+}$ subunit.

As shown in Figure 12, the structure of **MOXI-15** is constructed from the linking of three independent molecular clusters: $\{Cu_3(4,7\text{-phen})_3\}^{3+}$, β-$[Mo_8O_{26}]^{4-}$ and $[Mo_6O_{19}]^{2-}$. While the hexa- and octa-molybdate clusters are well known aggregates [28], the trinuclear copper(I) cluster cation provides a novel twenty-one membered cyclic structure. The extended structure is generated through the bonding of terminal and bridging oxo-groups of the clusters to the Cu(I) sites of the rings. One motif consists of linear chains of β-$[Mo_8O_{26}]^{4-}$ clusters linked through $\{Cu_3(4,7\text{-phen})_3\}^{3+}$ rings into a one-dimensional ribbon. Adjacent parallel ribbons are linked through $[Mo_6O_{19}]^{2-}$ clusters into a two-dimensional sheet structure. Each hexamolybdate cluster is sandwiched between two planar $\{Cu_3(4,7\text{-phen})_3\}^{3+}$ rings. The hexanuclear clusters are aligned with opposite Mo_3 triangular faces of the Mo_6 octahedron parallel to the Cu_3 rings. A curious feature of the structure is the presence of both $[Mo_6O_{19}]^{2-}$ and β-$[Mo_8O_{26}]^{4-}$ clusters as building blocks for the two-dimensional sheet.

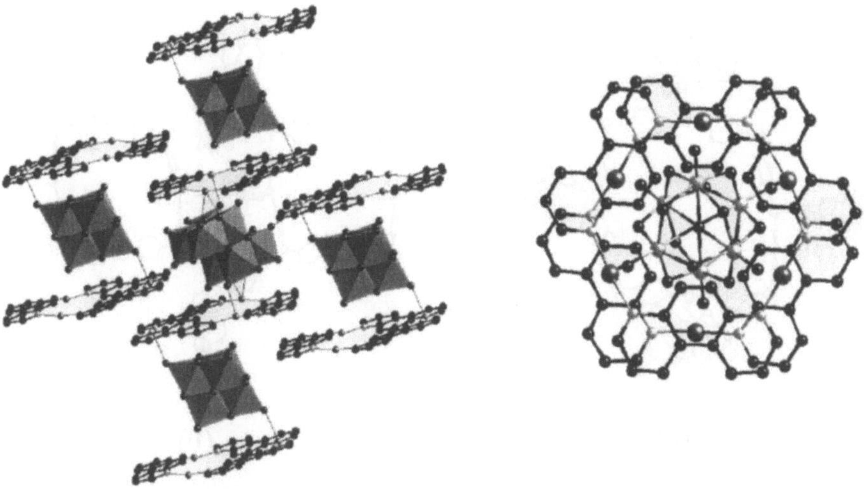

Figure 12. Left: A view of $[\{Cu_3(4,7\text{-phen})_3\}_2(Mo_8O_{26})(Mo_6O_{19})]$ (**MOXI-15**). Right: The sandwiching of the hexamolybdate subunit between two $\{Cu_3(4,7\text{-phen})_3\}^{3+}$ clusters.

The self-assembly of a tetranuclear coordination complex subunit, by analogy to the $\{Cu_3(4,7\text{-phen})_3\}^{3+}$ moiety of **MOXI-15**, may be accomplished in principle by exploiting a dipodal ligand with an interdonor angle of *ca.* 90° and a secondary metal site with a tendency toward linear coordination geometry, such as Cu(I) or Ag(I). It should be noted, however, that the dimensions of the resultant "square" $\{M_4(LL)_4\}^{4+}$ provide an interior cavity which may also accommodate an anionic cluster of significant dimensions. This general approach proved successful in the preparation of an unusual example of a two-fold interpenetrated framework structure, $[\{Cu(py_2\text{-}S)\}_4Mo_8O_{26}]$ (**MOXI-36**). As shown in Figure 13, the structure of **MOXI-36** is constructed from β-$[Mo_8O_{26}]^{4-}$ clusters and $[\{Cu(S\text{-}py_2)\}_4]^{4+}$ rings.

The cation forms a square of a *ca.* 12.7Å on an edge and 17.1Å across the digonal. The geometry is a consequence of the valence angle of *ca.* 100° at the sulfur bridge of the ligand, which orients the pyridyl nitrogen donors at an approximate right angle, and of the preference of Cu(I) for digonal coordination to pyridyl donors. That the dimensions of the square are such as to accommodate an interior cluster occupant has unusual structural consequences, as noted below.

As shown in Figure 13a, the four copper sites of the cationic square each bond to the peripheral oxygens of an anion cluster. Similarly, as shown in Figure 13b, each cluster links to four $[\{Cu(S\text{-}py_2)\}_4]^{4+}$ rings.

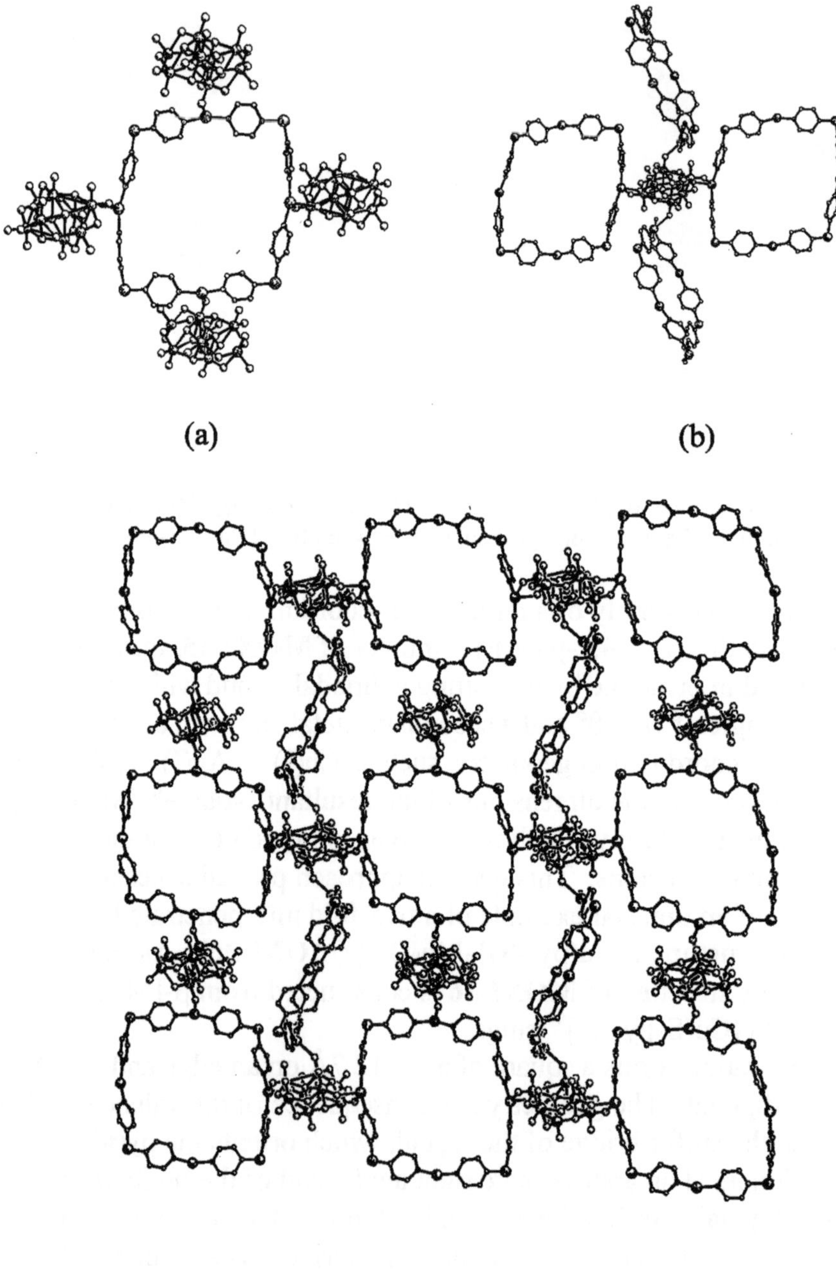

(a) (b)

(c)

Figure 13.

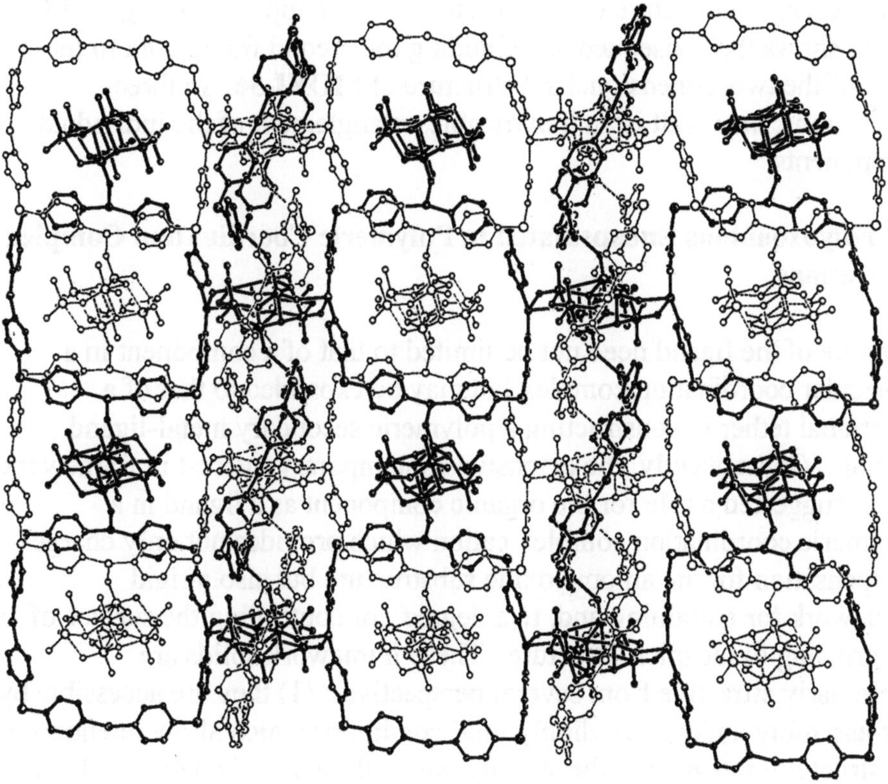

(d)

Figure 13. (a) The covalent attachment of one $\{Cu_4(py_2\text{-}S)_4\}^{4+}$ cluster of $[\{Cu(py_2\text{-}S)\}_4Mo_8O_{26}]$ (**MOXI-36**) to four polyanion clusters. (b)The linkage of one polyanion to four cationic clusters. (c) The propagation of the structure of one framework. (d) The consequences of interpenetration; one independent framework is highlighted, while the second is lightened.

As shown in Figure 13c, the structure is developed in two-dimensions through the linking of cation "squares" and molybdate clusters. However, when this two-dimensional motif is viewed edge-on as in Figure 13d, the four-fold connectivity at each of the component building blocks is seen to result in the projection of cationic and anionic clusters above and below the "stepped" layer. This pattern of covalent attachment produces a three-dimensional framework.

This single three-dimensional framework exhibits considerable void volume. In common with the majority of solids of this type [55], these voids are occupied by an independent framework, such that every cationic ring serves as a host to a molybdate cluster of the second framework. This feature of the structure is most apparent in Figure 13, which shows the consequences of adding the second framework to the view of the two-dimensional substructure of **MOXI-36**. In three-dimensions, the result is an inextricable entanglement of the individual components.

4. Polyoxoanions Encapsulated in Polymeric Coordination Complex Cations.

The role of the ligand need not be limited to that of a component in a molecular coordination complex but may be expanded to that of a functional tether in constructing a polymeric secondary metal-ligand subunit. Consequently, the extensive contemporary interest in framework solids suggested a role for the organic component as a ligand in a polymeric coordination complex cation which provides not only charge-compensation for the anionic oxide substructure but also a rigid framework for entraining and, to a degree, for controlling the surface of the growing oxide microstructure. These framework solids are particularly attractive from several perspectives: (1) they are accessible by self-assembly under exceedingly mild conditions which allow retention of the structural integrity of the starting materials and the isolation of kinetic, rather than exclusively thermodynamic, products; (2) they possess a remarkable chemical and structural diversity; and (3) the numerous examples described in recent years demonstrate that some control is achievable in defining their architectures. In the design of coordination polymers, the expectation is that the geometry of the metal will be propagated through the bridging ligands. The approach to the design of such coordination polymers has been to exploit organic ligands which through their geometries and coordination preferences impose a specific topology: chain (1-D), ladder (1-D), brick network (2-D), bilayer (2-D), and framework (3-D), for example. From the perspective of structural modification of molybdenum oxide phases, the secondary metal-organonitrogen polymer subunits may be regarded as cationic subunits which provide not only charge compensation and space filling requirements but also scaffolding of variable charge, topology, and channel dimensions for the incorporation of oxide subunits. In a naïve representation such as Figure 3, the structure of the oxide component will conform to the constraints of the cationic skeleton. While the emerging chemistry suggests a more complicated synergism between the oxide anionic and polymer cationic subunits, this naïve model provides a useful template for synthetic design.

The prototypical material of this class is [{Cu(4,4'-bpy)}$_4$Mo$_8$O$_{26}$] (**MOXI-12**), which consists of δ-$[Mo_8O_{26}]^{4-}$ clusters and $\{Cu(4,4'\text{-bpy})\}^+$ linear chains, as shown in Figure 14. The cation component $\{Cu(4,4'\text{-bpy})\}^+$ provides the scaffolding as chains of digonally coordinated Cu(I) centers. The relative orientation of the chains is unusual and results in cavities within the scaffolding that accommodate the $[Mo_8O_{26}]^{4-}$ units. There are two distinct chain environments forming a right angle or virtual square network. The chains associated with the Cu(1) site form pairs of parallel, face-to-face rods that run parallel to the crystallographic *a* axis. Propagation of this motif in the *ab* plane results in a bilayer of parallel $\{Cu(4,4'\text{-bpy})\}_n^{n+}$ strands. The parallel orientation of the aromatic rings within the chains allows a close approach of the paired chains within the bilayer. In contrast, the chains associated with the Cu(2) site exhibit a 44.3° twist between the rings of the bipyridyl units, which precludes close approach of a neighboring strand. The Cu(2) chains are at right angles to the Cu(1) bilayers, run parallel to the crystallographic *b* axis, and propagate as parallel strands in the *ab* plane.

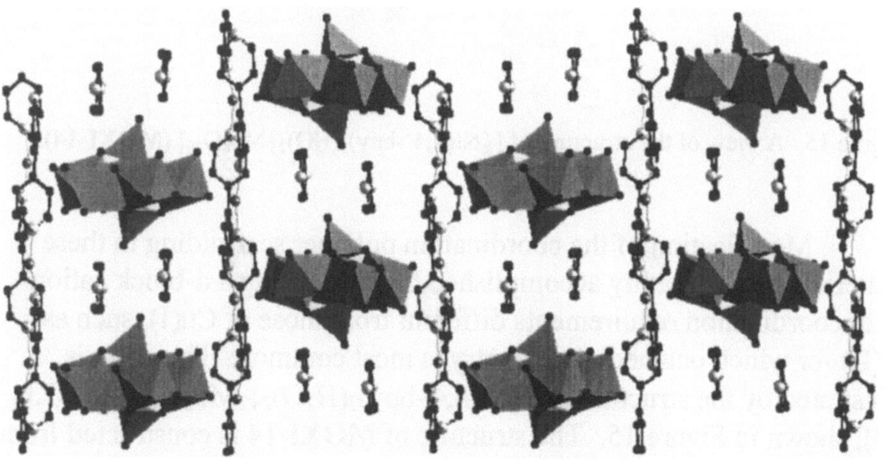

Figure 14. The structure of [{Cu(4,4'-bpy)}$_4$Mo$_8$O$_{26}$] (**MOXI-12**).

The $[Mo_8O_{26}]^{4-}$ clusters occupy cavities defined by two adjacent pairs of Cu(1) chains and one Cu(2) strand from each of two adjacent layers. The terminal oxo group of one tripodal Mo site exhibits weak long-range interactions of 2.555(5) and 2.691(4)Å with the Cu(1) sites, while oxo groups of tetrahedral and octahedral ring Mo sites display essentially non-bonding interactions with the Cu(2) sites. This long-range connectivity provides a virtual layer structure, in which the Cu(1) bilayers and $[Mo_8O_{26}]^{4-}$ subunits are sandwiched between the layers of parallel $\{Cu(2)(4,4'\text{-bpy})\}_n^{n+}$ strands. Adjacent layers are slotted together since alternate Cu(2) strands interact with $[Mo_8O_{26}]^{4-}$ clusters of an adjacent layer.

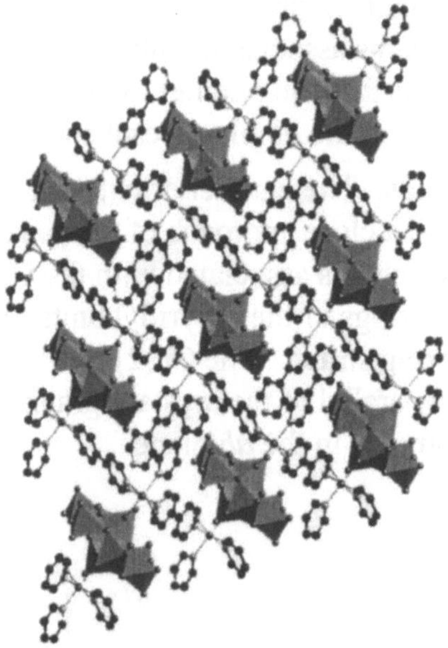

Figure 15. A view of the structure of $[\{Ni(4,4'\text{-bpy})_2(H_2O)_2\}Mo_8O_{26}]$ (**MOXI-14**).

Modification of the coordination polymer scaffolding in these materials can be readily accomplished by introducing a d-block cation with coordination requirements different from those of Cu(I), such as N(II) for which octahedral geometry is most common. The point is illustrated by the structure of $[\{Ni(4,4'\text{-bpy})_2(H_2O)_2\}_2Mo_8O_{26}]$ (**MOXI-14**), shown in Figure 15. The structure of **MOXI-14** is constructed from novel ε-$[Mo_8O_{26}]^{4-}$ clusters and $\{Ni(H_2O)_2(4,4'\text{-bpy})_2\}_n^{2n+}$ one-dimensional zig-zag chains. The molybdate cluster adopts an unprecedented structure consisting of six $\{MoO_5\}$ square pyramids and two $\{MoO_6\}$ octahedra linked by edge-sharing interactions into an ellipsoidal cluster, whose major axis is defined by the vector joining the octahedral sites. The geometry adopted by this ε-$[Mo_8O_{26}]^{4-}$ subunit reflects the steric constraints imposed by the polymer scaffolding of the $\{Ni(H_2O)_2(4,4'\text{-bpy})_2\}_n^{2n+}$ motif. The cationic component forms zig-zag chains of *fac*-$\{NiO_3N_3\}$ octahedra linked through bridging, bidentate 4,4'-bpy groups. Each Ni(II) site of the chain is coordinated to nitrogen donors from three 4,4'-bpy ligands, two aqua ligands, and a terminal oxo group from the molybdate cluster subunit. Since the 4,4'-bpy groups occupy *cis* positions on the Ni(II) sites, the $\{Ni(H_2O)_2(4,4'\text{-bpy})_2\}_n^{2n+}$ chain has a fold of about 90° at each Ni atom. It is noteworthy that the third bpy unit of each Ni center adopts monodentate coordination. The pendant bpy groups of adjacent $\{Ni(H_2O)_2(4,4'\text{-bpy})_2\}_n^{2n+}$ chains project into the interchain region to form π-stacked pairs, which define the boundaries of the interchain cavity.

The polymeric scaffolding of **MOXI-14** is connected to the polyoxoanion clusters by covalent bonding of the terminal oxo group of one Mo center to the Ni(II) site. This produces a 3-D covalent framework with "chains" of $[Mo_8O_{26}]^{4-}$ groups incorporated within the polymeric $\{Ni(H_2O)_2(4,4'\text{-bpy})_2\}_n^{2n+}$ backbone. The anionic subunits are confined within a rectangular channel defined by neighboring $\{Ni(H_2O)_2(4,4'\text{-bpy})_2\}_n^{2n+}$ strands and the pendant 4,4'-bpy groups that project from these strands. The ellipsoidal structure adopted by the $[Mo_8O_{26}]^{4-}$ moiety in **MOXI-14** conforms to the shape constraints thus imposed.

Structural modification may be readily accomplished by introducing variations in the tether length or geometry. For example, expansion of the tether length by introducing an ethene group in 4,4'-bipyridylethene (bpe) results in the related structure $[\{Cu(bpe)\}_4Mo_8O_{26}]$ (**MOXI-13**) shown in Figure 16. The α-form of the octamolybdate is observed with the capping tetrahedral sites involved in bonding to the parallel $\{Cu(bpe)\}^{+1}$ double chains. In contrast to the structure of **MOXI-11**, the interlamellar chains are parallel to the double chains.

It is noteworthy that under oxidizing conditions, so as to substitute Cu(II) for the Cu(I) in **MOXI-13**, the reaction of Cu(II), molybdate and

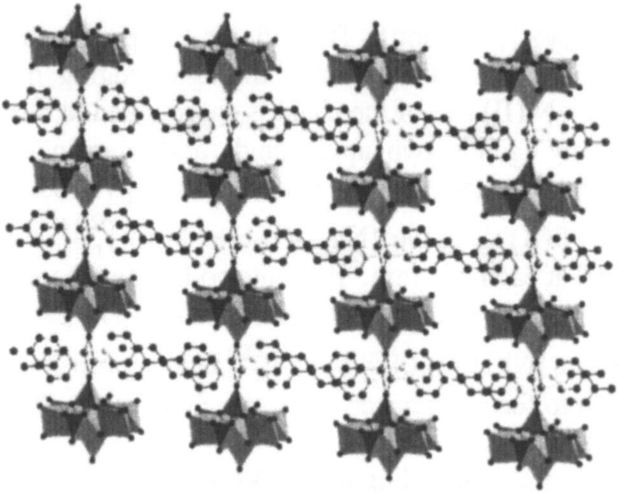

Figure 16. A view of [{Cu(bpe)}$_4$Mo$_8$O$_{26}$]. (**MOXI-13**).

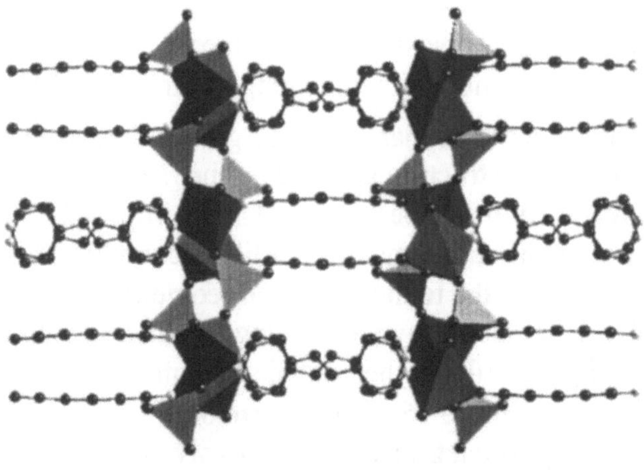

Figure 17. The 3-dimensionalcovalently linked structure of [Cu(bpe)MoO$_4$]. (**MOXI-13**).

bpe yields [Cu(bpe)MoO$_4$] (**MOXI-1**), shown in Figure 17. The pronounced differences in the coordination requirements of Cu(I) and Cu(II) are reflected in the structures of **MOXI-13** and **MOXI-1**. In the latter case, there are no distinct polyoxoanion subunits, but rather a bimetallic oxide layer constructed of corner-sharing {MoO$_4$} tetrahedra and trigonal bipyramidal {CuO$_3$N$_2$} sites. The layers are buttressed by bpe ligands occupying the interlamellar region to produce a three-dimensional covalent framework. **MOXI-1** provides the structural prototype for a class of materials based on integration of the secondary metal sites into the oxide substructure, with a concomitant absence of discrete polyanion and coordination complex cation subunits [56-61].

The ligand geometry may also be tuned by altering the relative dispositions of the donor groups, as is 3,3'-bipyridine. The structural consequences can be quite dramatic, as manifested in the comparison of **MOXI-14** in Figure 15 with $[\{Ni(3,3'\text{-bpy})_2\}_2Mo_4O_{14}]$ (**MOXI-48**), illustrated in Figure 18. The structure of **MOXI-48** consists of α-$[Mo_4O_{14}]^{4-}$ clusters imbedded in cavities of the $\{Ni(3,3'\text{-bpy})_2\}_n^{2n+}$ network. Alternatively, the structure may be viewed as a bimetallic oxide network $[Ni_2Mo_4O_{14}]$ with 3,3'-bpy units in the cavities.

Figure 18. A view of the network structure of $[\{Ni(3,3'\text{-bpy})_2\}_2Mo_4O_{14}]$ (**MOXI-48**).

Further elaboration of the ligand structure appears to provide more predictable cationic substructures and consequently structural constraints on the molybdenum oxide substructure. This influence of ligand architecture on oxide structure is demonstrated by the structure of [{Cu(tptz)}$_2$Mo$_6$O$_{19}$] (**MOXI-46**). The tridentate tri-4-pyridyltriazine ligand reacts with Cu(I) to provide a two-dimensional secondary metal-ligand substructure shown in Figure 19. While the Cu(I) sites of **MOXI-45** adopted the common digonal coordination mode, the coordination preferences of Cu(I) are also consistent with trigonal planar and tetrahedral geometries. The ligand geometry favors the planar coordination mode and bridging of copper sites to form the two-dimensional hexagonal grid of Figure 19.

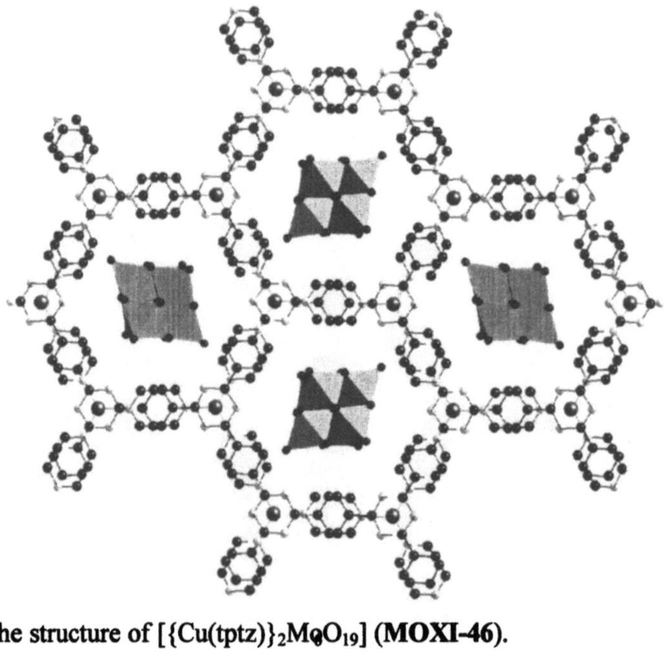

Figure 19. The structure of [{Cu(tptz)}$_2$Mo$_6$O$_{19}$] (**MOXI-46**).

This theme of structural elaboration of the ligand may be further developed to incorporate more complex geometries. The tptz ligand generated the anticipated "hexagonal" network substructure in **MOXI-46**. A ligand such as tetra-4-pyridylporphyrin may generate a three-dimensional framework through metal ligation at the pyridyl sites, as well as incorporating metal centers in the porphyrin cavity.

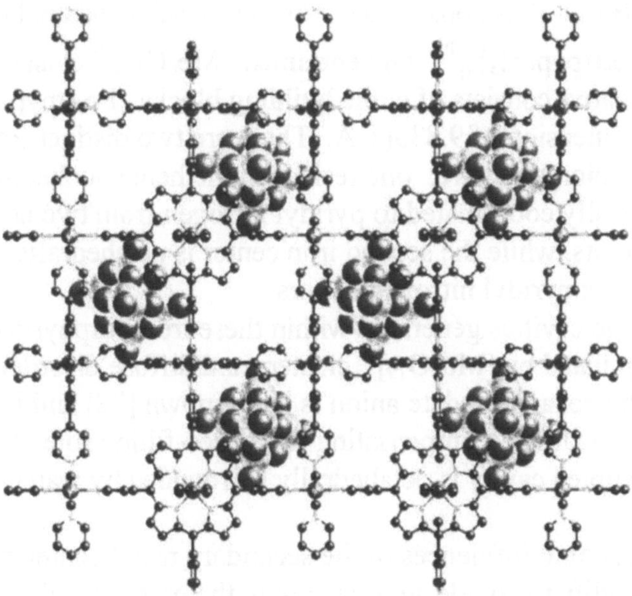

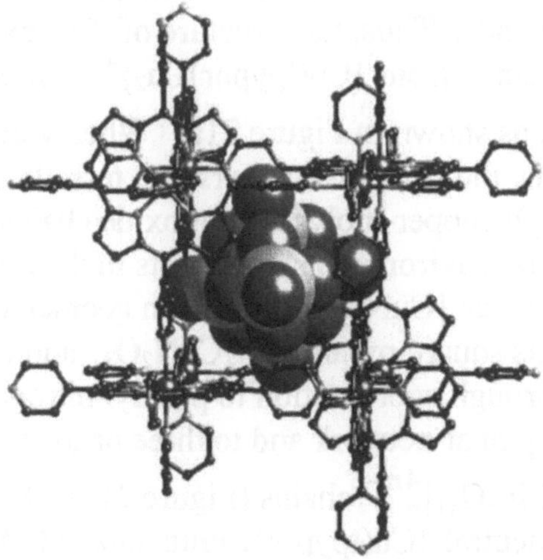

Figure 20. Above: A view of [{Fe(tpypor)}$_3$Fe(Mo$_6$O$_{19}$)] (**MOXI-39**). Below: The encapsulation of the hexamolybdate anion within the {$Fe_8(tpypor)_6$}$^{8+}$ cage.

In assessing the structure-directing role of the secondary or porphyrin/pyridyl group-ligated metal center, the isolation of [{Fe(tpypor)}$_3$Fe(Mo$_6$O$_{19}$)$_2$]•xH$_2$O (**MOXI-39**), a material constructed from a three-dimensional $[\{Fe(tpypor)\}_3Fe]_n^{4n+}$ framework and isolated $[Mo_6O_{19}]^{2-}$ polyoxo anions, was effected. As shown in Figure 20, the structure of **MOXI-39** is constructed from a three-dimensional cationic framework $\{Fe_4(tpypor)_3\}_n^{4n+}$ and entrained $\{Mo_6O_{19}\}^{2-}$ cluster anions. The cationic matrix consists of cubic building blocks $\{Fe_8(tpypor)_6\}^{8+}$ with an edge dimension of 9.833(1)Å. There are two distinct iron centers in the cationic microstructure: one resides in the heme pocket and is additionally axially coordinated to pyridyl nitrogen from two adjacent {Fe(tpypor)} units, while the second iron center is octahedrally coordinated to six pyridyl nitrogen donors.

The large cavities generated within these iron-porphyrin cubes are alternately populated by $[Mo_6O_{19}]^{2-}$ clusters and diffuse disordered water molecules. The hexamolybdate anion is well-known [28] and in this material adopts a charge-compensating and space-filling role. Each $[Mo_6O_{19}]^{2-}$ occupied cavity is octahedrally surrounded by water-filled cavities.

The structural influences of the secondary metal center may also be exploited to modify the oxide substructure in the presence of such structure-directing ligands. Thus, the structure of $[Cu(tpypor)Cu_2Mo_3O_{11}]$ (**MOXI-35**) is constructed from $\{Cu(tpypor)Cu_2\}^{4+}$ building blocks and $[Mo_3O_{11}]_n^{4n-}$ chains, as shown in Figure 21a. When viewed along the crystallographic *c* axis, the structure appears as a tessellated porphyrin network linked through copper-molybdenum oxide chains. There are three distinct copper(II) environments: one rests in the porphyrin pocket and displays square planar $\{CuN_4\}$ coordination geometry while the remaining two exhibits square pyramidal $\{CuN_2O_3\}$ and octahedral $\{CuN_2O_4\}$ ligation through coordination to pyridyl nitrogens from two tpypor groups of one planar network and to three or four oxo-groups, respectively, of the $[Mo_3O_{11}]_n^{4n-}$ chains (Figure 21b). Thus, the structure may be described as neutral {Cu(tpypor)} units linked to bimetallic $\{Cu_2Mo_3O_{11}\}$ oxide chains. These oxide chains are constructed from a molybdate ribbon of alternating $\{MoO_6\}$ octahedra and $\{MoO_4\}$ tetrahedra in a corner-sharing arrangement and decorated by $\{MoO_4\}$ tetrahedra sharing a single corner with a octahedral sites, and the square pyramidal and octahedral copper centers which share three and four vertices, respectively, with the molybdate ribbon. One of these copper sites bridges adjacent octahedral $\{MoO_6\}$ sites and one terminal $\{MoO_4\}$ tetrahedron in corner-sharing modes exclusively while the second engages in corner-

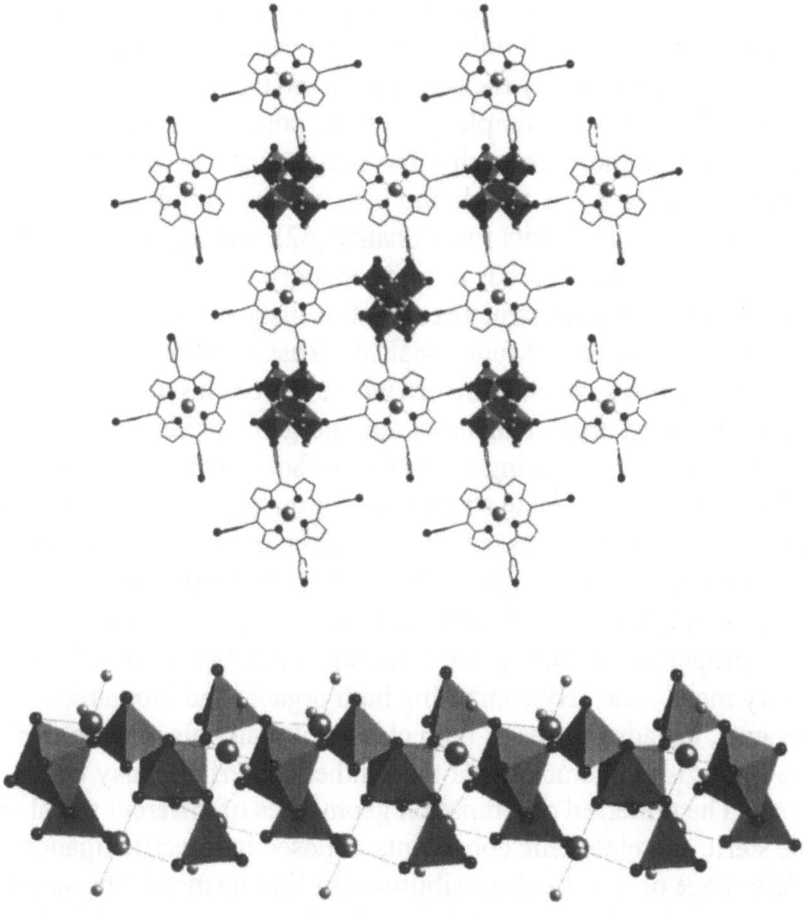

Figure 21. Above: A view of the tesselated pattern adopted by the porphyrin substructure of [Cu(tpypor)$Cu_2Mo_3O_{11}$] (**MOXI-35**). Below: The 1-D chain of **MOXI-35**; the Mo sites are polyhedra while the Cu sites are shown as ball and stick representations.

sharing interactions with a tetrahedral $\{MoO_4\}$ site and edge-sharing interactions with two adjacent $\{MoO_6\}$ octahedra. This polyhedral arrangement is quite distinct from that observed for "naked" copper-molybdate phases.

5. Conclusions

The vast range of solid state properties exhibited by the oxides is a result of their diversity of chemical composition and structure types. However, while many naturally occurring oxides and minerals possess complex crystal structures, the majority are of simple composition and have highly symmetrical structures with rather small unit cells. Most silicates, important ores, gems, many rocks and soils are examples of these materials. Although such simple oxides can possess unique and specific properties, such as piezoelectricity, ferromagnetism, or catalytic activity, as a general rule there is a correlation between the complexity of the structure of a material and its functionality [62]. One approach to the design of novel oxide materials mimics Nature's use of organic molecules to modify inorganic microstructures. In this instance, the inorganic oxide contributes to the increased functionality via assimilation as one component in a hierarchical structure where there is a synergistic interaction between organic material and inorganic oxide. Synthetic studies of materials possessing such an interface, coupled with the acquisition of the appropriate structural information, should contribute to the development of an increased understanding of methods to control the structure-property relationships within these hybrid materials.

The molybdenum oxides of this study exploit the structure-directing properties of the organic component functioning as a ligand to a secondary metal sites. By combining both organic and inorganic components, the advantages of both classes of materials may be exploited and the synergistic interactions at the synthetic interfaces may be explored. The preferred coordination geometries of different metal sites and the steric and electronic constraints imposed by specific ligands provide a range of structural possibilities for linking metal sites and for transmitting and replicating the structural information implicit in the metal coordination throughout the solid.

While a degree of design has been introduced into the cationic subunit of these structures by considering the coordination preferences of a secondary metal and the geometry and binding modes of the ligand, it is evident that the synergistic interactions between the oxide and organic or metal-organic substructures at the hydrothermal interface often result in unexpected structural types. Thus, the very complexity of these composite materials sets natural limits on the degree of predictability. If design requires total predictability, then such composite materials are rendered effectively undesignable due to their complexity and the inherent dynamism of the kinetic domain of their synthesis. However, a more positive viewpoint suggests that as the products of empirical development are unraveled, a reciprocity of structure-function relationships will emerge, such that control of desired properties may be achieved. Thus, while the syntheses may result in unanticipated structures, the practical imperative of exploiting organic components in the preparation of oxides with microporous properties and unusual magentic properties has been demonstrated. Furthermore, the equally compelling imperatives of curiosity and chemical aesthetics - a new and undeveloped chemistry, combined with stunning and, in the eyes of this beholder, beautiful structures - have also been satisfied.

Acknowledgements: The work described in this review was supported by NSF Grant No. CHE 9617232.

6. References

1. Reynolds, T.G.; Buchanan, R.C., in *Ceramic Materials for Electronics* 2nd ed., R.C. Buchanan, ed., Dekker, New York, **1991**, p. 207.
2. Büchner, W.; Schliebs, R.; Winter, G.; Büchel, K.H., *Industrial Inorganic Chemistry,* VCH, New York **1989**.
3. McCarroll, W.H., in *Encyclopedia of Inorganic Chemistry,* R.B. King, ed., John Wiley and Sons, New York, **1994**, *vol. 6,* p. 2903.
4. Haertling, G.H., in *Ceramic Materials for Electronics,* 2nd. ed., R. C. Buchanan, ed., Dekker, New York, **1991**, p. 129.
5. Leverenz, H.W., *Luminescence of Solids,* Wiley, New York, **1980**.
6. Einzinger, R., *Ann. Rev. Mater. Sci.,* **1987**, *17,* 299.
7. Tarascon, J.M.; Barboux, P.; Miceli, P.F.; Greene, L.H.; Hull, G.W.; Eibschutz, M.: Sunshine, S.A., *Phys. Rev.,* **1988**, *B37,* 7458.
8. Matkin, D.I., in *Modern Oxide Materials,* B. Cockayne and D.W. Jones ets., Academic Press, New York, **1972**, p. 235.
9. Rao, C.N.R.; Rao, K.J., in *Solid State Compounds,* A-K Cheetham and P. Day, eds., Clarendon Press, Oxford, **1992**, p. 281.
10. Bierlein, J.D.; Arweiler, C.B., *Appl. Phys. Lett.,* **1987**, *49,* 917.
11. Centi, G.; Trifuro, F.; Ebner, J.R.; Franchetti, V.M., *Chem. Rev.,* **1988**, *88,* 55.
12. Grasselli, R.K., *Appl. Catal.,* **1985**, *15,* 127.
13. Gasior, M.; Gasior, I.; Grzybowska, B., *Appl. Catal.,* **1984**, *10,* 87.
14. Okuhara, T.; Misono, M., in *Encyclopedia of Inorganic Chemistry,zi* R.R. King, ed., John Wiley and Sons, New York, **1994**, *vol. 6*, p. 2889.
15. Niiyama, H.; Echigoya, E., *Bull. Chem. Soc. Jpn.,* **1972**, *45*, 83.
16. Yamaguchi, T., *Appl. Catal.,* **1990**, *61,* 1.
17. Clearfield, A., *Chem. Rev.,* **1988**, *88*, 125.
18. Newsam, J.M., in *Solid State Compounds,* A.K. Cheetham and P. Day eds., Clarendon Press, Oxford, **1992**, p. 234.
19. Ruthven, D.M., *Principles of Absorption and Absorption Processes,* Wiley-Inerscience, New York, **1984**.

20. Szostak, R., *Molecules Sierves-Principles of Synthesis and Identification,* Van Nostrand Rhenihold, New York, **1988**.
21. Raleo, J.A., *Zeolite Chemistry and Catalysis,* ACS Monograph (7), American Chemical Society, Washington, DC, **1976**.
22. Murakami, Y.; Iijima, A.; Ward, J.W., eds., *New Developments in Zeolite Science,* Elsewier, Amsterdam, **1986**.
23. Vaughan, D.E.W., *Properties and Applications of Zeolites,* Chem. Soc. Special Publ. No. 33, R.P. Townsend, ed., The Chemical Society, London, 1979, p. 294. Cheetham, A.K., *Science,* **1994**, *264,* 794.
24. Venuto, P.B., *Microporous Materials,* **1994**, *2,* 297.
25. Braithwaite, E.R. Haber, J., eds., Molybdenum: An Outline of its Chemistry and Uses, Elsevier: Amsterdam, **1994**.
26. Robbinson, C.L., in *Comprehensive Inorganic Chemistry,* Pergamon Press: Oxford, **1973**, *3,* 623.
27. Pope, M.T., *Prog. Inorg. Chem.,* **1991**, *39,* 181.
28. Pope, M.T., *Heteropoly and Isopoly Oxometalates,* Springer Verlag: Berlin, **1983**.
29. Khan, M.I., Zubieta, J., *Prog. Inorg. Chem.,* **1995**, *43,* 1.
30. The difficulties and opportunities inherent in the designed syntheses of solids have been succinctly expressed by Roald Hoffmann, who noted in 1993: “But in two or three dimensions, it's a synthetic wasteland. The metholodogy for exercising control so that one can make unstable but persistent extended structures on demand is nearly absent. Or to put it in a positive way - this is a certain growth point of the chemistry of the future.” Both assertions are correct, and the explosive growth of “crystal engineering”, of organic templating of inorganic materials and of directed assembly of inorganic supramolecular assemblies suggest that the future is now. Hoffmann's insightful commentary combines an appreciation and anticipation of the current thrust of solid state chemistry with an elegant aesthetic sense for the beauty of chemical structures. Hoffmann, R. How chemists should think, *Sci. Amer.,* **1993**, February, 66.
31. (a) Smith, J.V., *Chem. Rev.,* **1988**, *88,* 149-182. (b) Occelli, M.L.; Robson, H.C., *Zeolite Synthesis,* American Chemical Society, Washington, DC, **1989**.
32. Mann, S., *Nature,* **1993**, *365,* 499.
33. Kresge, C.T.; Leonowicz, M.E.; Roth, W.J.; Vartuli, J.C.; Beck, J.S., *Nature,* **1992**, *359,* 710.
34. Haushalter, R.C.; Mundi, L.A., *Chem. Mater.,* **1992**, *4,* 31-48.
35. Khan, M.I.; Meyer, L.M.; Haushalter, R.C.; Schweitzer, C.L.; Zubieta, J.; Dye, J.L., *Chem. Mater.,* **1996**, *8,* 43.
36. Zhang, Y.; DeBord, J.R.D.; O'Connor, C.J.; Haushalter, R.C.; Clearfield, A.; Zubieta, J., *Angew. Chem., Int. Ed. Engl.,* **1996**, *35,* 989.
37. Hagrman, D.; Hagrman, P.J.; Zubieta, J., *Angew. Chem., Int. Ed. Engl.,* **1999**, *38,* 3165.
38. DeBord, J.R.D.; Haushalter, R.C.; Meyer, L.M.; Rose, P.J.; Zapf, P.; Zubieta, J., *Inorg. Chem. Acta,* **1997**, *256,* 165.
39. Hagrman, D.; Sangregorio, C.; O'Connor, C.J.; Zubieta, J., *J. Chem. Soc. Dalton Trans.,* **1999**, 3707.
40. Hagrman, D.; Zapf, P.J.; Zubieta, J., *Chem. Commun.,* **1998**, 1283.

41. LaDuca, R.L. Jr.; Rarig, R.S.; Zapf, P.; Zubieta, J., *Inorg. Chim. Acta,* **1999**, 0000.
42. Hagrman, D.; Zubieta, J., *Chem. Commun.*, **1998**, 2005.
43. Hagrman, D.; Zubieta, C.; Haushalter, R.C.; Zubieta, J., *Angew. Chem. Int. Ed. Engl.*, **1997**, *36,* 873.
44. Hagrman, D.; Zubieta, J., *Chem. Commun.*, **1998**, 2005.
45. "Crystal engineering" of metal-organic coordination 1-D, 2-D and 3-D solids has witnessed an explosive growth in the past decade. A particularly useful overivew in terms of interpenetration is provided in: Batten, S.R.; Robson, R, *Angew. Chem., Int. Ed. Engl.*, **1998**, *37,* 1460 and references therein.
46. Some representative references from leading researchers in the vast field: (a) Zaworotko, M.J., *Chem. Soc. Rev.*, **1994**, *23,* 283; (b) Gudbjartson, H.; Biradha, K.; Poirier, K.M.; Zaworotko, M.J., *J. Am. Chem. Soc.*, **1999**, *121,* 2599; (c) Abrahams, B.F.; Jackson, P.A.; Robson, R., *Angew. Chem., Int. Ed. Engl.*, **1998**, *37*, 2656; (d) Carlucci, L.; Ciani, G.; Prosperpio, D.M.; Sironi, A., *Inorg. Chem.*, **1998**, *37,* 5941; (3) Yaghi, O.M.; Li, H.; Davis, C.; Richardson, D.; Groy, T.L., *Acc. Chem. Res.*, **1998**, *31,* 474; (f) Lopez, S.; Keller, S.W., *Inorg. Chem.*, **1999**, *38,* 1883; (g) Kondo, M.; Okubo, T.; Asami, A.; Noro, S.; Yoshitomi, T.; Kitagawa, S.; Ishii, T.; Matsuzaka, H.; Seki, K., *Angew. Chem., Int. Ed. Engl.*, **1999**, *38,* 140; (h) Goodgame, D.M.L.; Grachvogel, D.A.; Williams, D.J., *Angew. Chem., Int. Ed. Engl.*, **1999**, *38,* 153; (i) Sharma, C.V.K.; Rogers, R.D., *Chem. Commun.*, **1998**. (j) Fujita, M.; Kwon, Y.J.; Washizu, S.; Ogura, K., *J. Am. Chem. Soc.*, **1994**, *116,* 1151; (k) Withersby, M.A.; Blake, A.J.; Champness, W.R.; Hubberstey, P.; Li, W.-S.; Schröder, M., *Angew. Chem., Int. Ed. Engl.*, **1997**, *37,* 2327; (l) Blake, A.J.; Mill, S.J.; Hubberstey, P.; Li, W.-S., *J. Chem. Soc., Dalton Trans.*, **1998**, 909; (m) Moore, J.S.; Lee, S., *Chem. Ind. (London)*, **1994**, *14,* 556; (n) Mayr, A.; Guo, J., *Inorg. Chem.*, **1999**, *38,* 921.
47. Gopalakrishnan, J., *Chem. Mater.*, **1995**, *7,* 1265.
48. A remarkably diverse range of structural types has been accessed through hydrothermal reactions of inorganic precursors in the presence of organic cations. See refs. 6-40 and also some recent examples: (a) Mann, S.; Burkett, S.L.; Davis, S.A.; Fowler, C.E.; Mendelson, N.H.; Sim, S.D.; Walsh, D.; Whilton, N.T, *Chem. Mater.*, **1997**, *9,* 2300; (b) Ekambaran, S.; Sevov, S.C., *Angew. Chem., Int. Ed. Engl.*, **1999**, *38,* 372; (c) Ayyappan, S.; Bu, X.; Cheetham, A.K.; Rao, C.N.R., *Chem. Mater.*, **1998**, *10,* 3308; (d) Bu, X.; Feng, P.; Stucky, G.D., *Science,* **1997**, *278,* 2080; (e) Lu, J.J.; Xu,Y.; Goh, N.K.; Chia, L.S., *Chem. Commun.*, **1998**, 1709; (f) Riou-Cavellic, M.; Grenèche, J.-M.; Riou, D.; Férey, G.; (g) Lii, K.-H.; Zima, V., *Chem. Mater.*, **1998**, *10,* 1914; (h) Bircsak, Z.; Harrison, T.A., $(CN_3H_6)_2(VO_2)_3(PO_4)(HPO_4)$, *Inorg. Chem.*, **1998**, *37,* 3204; (i) Clearfield, A., *Chem. Mater.*, **1998**, *10,* 2801; (j) Grohol, D.; Gingl, F.; Clearfield, A., *Inorg. Chem.*, **1999**, *38,* 751; (k) Lobachev, A.N., *Crystallization Processes under Hydrothermal Conditions,* Consultants Bureau, New York, **1973**.
49. For a recent commentary on rational synthesis of zeolites see: Akporiaye, D.E., *Angew. Chem., Int. Ed. Engl.*, **1998**, *37,* 2456.
50. Evans, H.T. Jr.; Weakley, T.J.R.; Jameson, G.B.; *J. Chem. Soc., Dalton Trans.*, **1996**, 2537.

519.Loose, I.; Bösing, M.; Klein, R.; Krebs, B.; Schulz, R.P.; Scharbert, B., *Inorg. Chim. Acta,* **1997**, *263,* 99.
52. Khan, M.I.; Yohannes, E.; Powell, D., *Chem. Commun.,* **1999**, 23.
53. Inoue, M.; Yamase, T., *Bull. Chem. Soc. Jpn.,* **1995**, *68,* 3055.
54. Xi, R.; Wang, B.; Isobe, K.; Nishioka, T.; Toriumi, K.; Ozawa, Y., *Inorg. Chem.,* **1994**, *33,* 833.
55. Batten, S.R.; Robson, R., *Angew. Chem., Int. Ed. Engl.,* **1998**, *37,* 1460.
56. Zapf, P.J.; Warren, C.J.; Haushalter, R.C.; Zubieta, J., *Chem. Commun.,* **1997**, 1543.
57. Hagrman, D.; Haushalter, R.C.; Zubieta, J., *Chem. Mater.,* **1998**, *10,* 361.
58. Hagrman, D.; Warren, C.J.; Haushalter, R.C.; Seip, C.; O'Connor, C.J.; Rarig, R.S. Jr.; Johnson, K.M. III; LaDuca, R.L. Jr.; Zubieta, J., *Chem. Mater.,* **1998**, *10,* 3294.
59. Zapf, P.J.; Haushalter, R.C.; Zubieta, J., *Chem. Mater.,* **1997**, *9,* 2019.
60. Zapf, P.J.; Hammond, R.P.; Haushalter, R.C.; Zubieta, J., *Chem. Mater.,* **1998**, *10,* 1366.
61. Zapf, P.J.; LaDuca, R.C. Jr.; Rarig, R.S.; Johnson, K.M.; Zubieta, J., *Inorg. Chem.,* **1998**, *37,* 3411.
62. Complexity is a subject of significant and general scientific interest. Complexity in chemistry refers to the description and manipulation of systems of molecules, as in living cells and materials. In the latter context, organic-inorganic hybrid structures partake of the chemical complexity of materials, with the attendant complications of predictability and rational design. See, for example, Whitesides, G.M.; Ismagilov, R.F., Complexity in chemistry, *Science,* **1999**, *284,* 89. The relationship between complexity and functionality is abundantly evident in biological systems. Chemists may learn from biology and make the creative leap to the design of inorganic materials whose structures are influenced by organic molecules.

Polyoxometalate Clusters in a Supramolecular SelfOrganized Environment: Steps Towards Functional Nanodevices and Thin Film Applications.

DIRK G. KURTH
Max-Planck-Institute of Colloids and Interfaces, D-14424 Potsdam, Germany

DIRK VOLKMER
Universität Bielefeld, Anorganische Chemie I, D-33501 Bielefield, Germany

Abstract. Polyoxometalates (POMs) represent a well-defined class of inorganic compounds with potential applications in fundamental and applied science. To implement POMs in functional materials and nanotechnological devices it will be of paramount importance to control the surface chemical properties of the primarily water-soluble clusters. Two different approaches toward this goal are presented here, which are primarily based on ion-exchange of POM counter cations with suitable surface-active compounds. In the first method, the POM counter cations are replaced by charged surfactants, leading to discrete hydrophobic *surfactant-encapsulated clusters* (SECs). These materials combine the physicochemical properties of the inorganic POM core with the diverse assets of organic compounds including wetting, adhesion, solubility as well as bio-compatibility. Using Langmuir-Blodgett techniques, it is feasible to deposit SEC mono- and multilayers and thus to engineer highly-ordered POM arrays that approach macroscopic length scales. In the second method, multi-layered thin films of POMs and polyelectrolytes are co-assembled on arbitrary substrates by a sequential deposition process. These films are robust and permeable, which renders them potentially attractive for applications in heterogeneous catalysis, electrochemical and molecular recognition devices, such as sensors.

Key Words: Polyoxometalates, surfactant-encapsulated clusters, supramolecular chemistry, thin films, layer-by-layer self-assembly, Langmuir and Langmuir-Blodgett films.

M.T. Pope and A. Müller (eds.), Polyoxometalate Chemistry, 301–318.

1. Introduction

Nanotechnology attempts to establish technical processes that are accomplished by devices and machinery of nanoscopic dimensions.[1] A logical consequence of this miniaturization is the development of *molecular devices*, i.e. functional supramolecular architectures that process a specific elementary step within a coherent sequence of matter/signal transformations. Potential fields of applications for molecular devices are diverse; currently investigated "hot topics" include (arrays of) chemosensors to derive electric signals from molecular recognition events ("nanosensors"), or molecular switches and logical elements for highly integrated circuits ("molecular electronics"). Self-assembling supramolecular systems represent a feasible and elegant synthetic approach to fabricate the required quantities of nanosized supramolecular architectures with novel structural and functional properties. The final supramolecular entity spontaneously evolves from a chemical library of suitably shaped molecular building blocks through a sequence of recognition, growth, and termination steps.[2] Discrete nano-sized supramolecular assemblies have thus been synthesized exploiting ligand-metal ion interactions,[3] π-π interactions,[4] or hydrogen-bonding mediated recognition processes.[5]

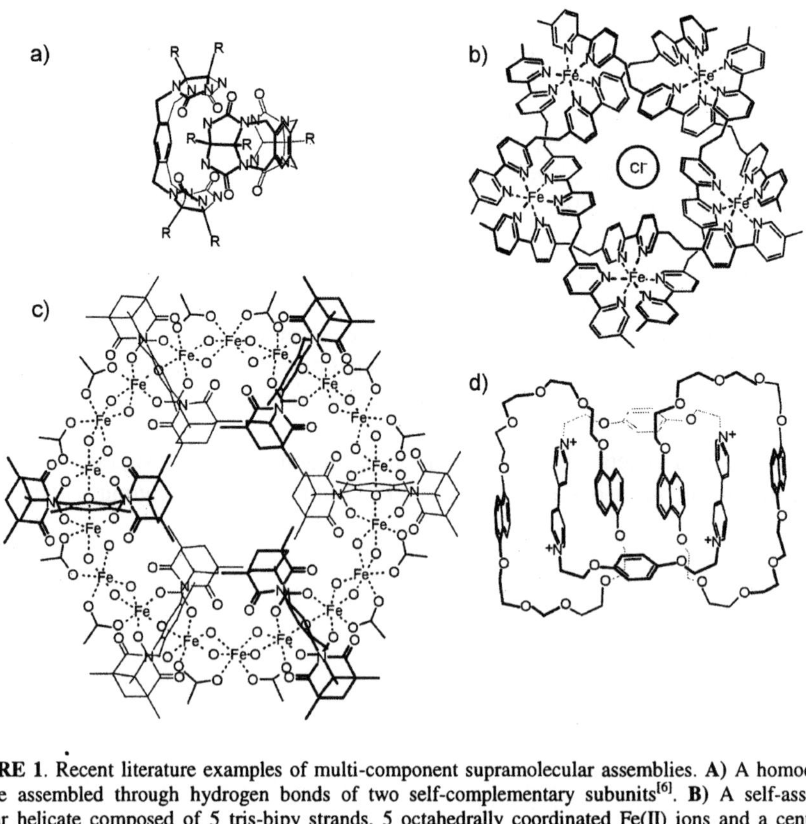

FIGURE 1. Recent literature examples of multi-component supramolecular assemblies. **A**) A homodimeric capsule assembled through hydrogen bonds of two self-complementary subunits[6]. **B**) A self-assembled circular helicate composed of 5 tris-bipy strands, 5 octahedrally coordinated Fe(II) ions and a central Cl^- anion[7]. **C**) A cyclic octadecanuclear Fe(III) complex (molecular 18-wheeler)[8]. **D**) A self-templated [3]catenane of interlocked macrocyclic rings [9].

Despite the increasing size and complexity, and the striking beauty of the aforementioned supramolecular structures, their possible fields of applications yet have to be explored. Transition metal polyoxometalates (POMs), in contrast, represent a well-known class of inorganic supramolecular structures with a huge variety of applications in fundamental and applied science such as catalysis,[10] electrochemistry,[11] electrooptics,[12] medicine,[13] corrosion protection, dyes/pigments, dopants in (non-)conductive polymers, dopants in sol-gel matrixes, bleaching of paper pulp, and analytical chemistry.[14] Their valuable physicochemical properties suggest using POMs as functional components in advanced materials. With the recent discovery of self-assembling, discrete and nano-sized polyoxomolybdates ("giant wheels", "Keplerates"), [15] novel inorganic components are now available, opening access to the construction of molecular devices such as nanoreactors or nanosensors.

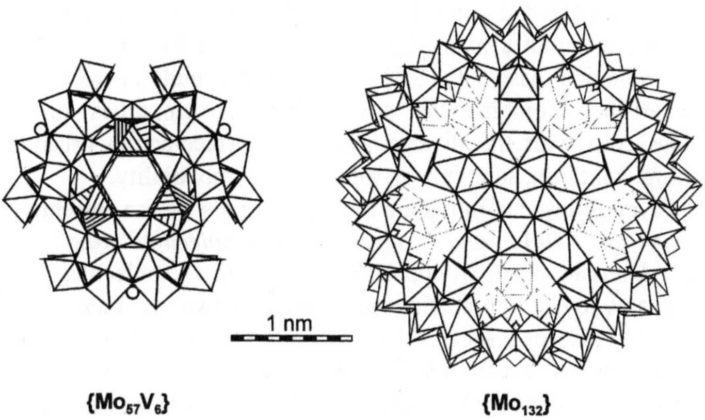

FIGURE 2. Recent examples of giant polyoxomolybdates as discussed in the text. **A)** Polyhedral representation of the cluster $[H_3Mo_{57}V_6(NO)_6O_{83}(H_2O)_{18}]^{21-}$ (abbreviated as **{Mo57V6}**). The cluster resembles a flattened ellipsoid with approximate dimensions of a = b = 2.2 nm, c = 1.2 nm.[16] **B)** Polyhedral representation of the novel type of isopolyoxomolybdate clusters ("Keplerates"), $[(H_2O)_n \subset Mo_{132}O_{372}(CH_3COO)_{30}(H_2O)_{72}]^{42-}$ (abbreviated as **{Mo132}**). The cluster has a nearly spherical Mo-O framework architecture with an outer diameter of approximately 3 nm and a solvent-filled cavity. The internal cavity is quite large: from the crystal structure of **{Mo132}** the void volume is estimated to be 1.5 (±0.3) nm^3. As many as 50 (±15) H_2O molecules can reside within the central cavity.[17] The framework has (Mo-O) 9-ring apertures with diameters ranging from 0.30 – 0.56 nm that may facilitate shape-selective loading of the central cavity. The nanocavity makes the Keplerate a promising candidate for the construction of molecular reactors for catalysis and sensing.

A central challenge of integrating POMs into molecular devices and advanced materials is to control the surface chemical properties of the primarily water-soluble clusters. Modifying the surface chemical properties of POMs opens avenues to mediate their contact with common technical substrates and junctions, to tailor their compatibility with organic materials and biological tissue, to manipulate their solubility properties and to engineer novel nano- and mesoscopic supramolecular architectures. Many possible applications involve thin films or layered materials, such as displays, sensors, or protective coatings. However, while the chemical routes that govern POM self-assembly have been extensively investigated in the past years, the self-organization of the discrete clusters into extended, highly-ordered ensembles represents the next milestone to the implementation of POMs into functional hierarchically structured materials. In order to exploit novel *collective properties* of POM ensembles it will be important to gain more precise control on the spatial arrangement of the clusters in their supramolecular self-organized environment. As an example, the three-dimensional periodic arrangement of POM clusters may cause perturbations in the electronic states of the isolated POM components and/or energy levels may become successively delocalized over the entire ensemble.[18] To achieve these goals it is, therefore, necessary to develop new and to improve existing methods of fabricating thin films of well-defined composition and dimensions with POMs as functional components.

Our current investigations pursue two different objectives. First, we are exploring the properties of *surfactant-encapsulated clusters* (SECs), i.e. complexes of suitable cationic surfactants and POMs. Their structural characterization by various analytical techniques (see CHAPTER 2) reveals a particular structural motif in which a single POM cluster resides within a hydrophobic shell of surfactant molecules, leading to a discrete, nearly spherical assembly. SECs are attractive building blocks for molecular devices because they combine the physicochemical properties of the inorganic polyoxometalate core with the diverse assets of surface-active organic compounds including wetting, adhesion, solubility as well as bio-compatibility.

The second type of organic-inorganic hybrid materials we are currently investigating are *thin films of POMs and charged macromolecules* (e.g. polyallylamino hydrochloride). Alternate layers of oppositely charged macromolecules can be adsorbed on polar substrates by a dip-coating procedure known as Layer-by-Layer (LbL) technology.[19] The adsorbed species form a uniform layer of molecular dimensions during each deposition cycle. Due to the electrostatic repulsion between the immobilized and the dissolved macromolecules, the adsorption process is self-terminating. The LbL method provides a facile synthetic route towards uniform POM-based multilayer films (see CHAPTER 3). The sequential deposition process permits the combination of POMs with many other functional components as part of the polycationic macromolecules. Since different species may be deposited within each deposition cycle, it is conceivable to create anisotropic multi-layer films, which display a gradient of physicochemical properties in the direction normal to the substrate. A further attractive feature of polyelectrolyte thin films is that they are permeable for small molecules, which renders them potentially attractive for a host of applications in photochemistry, electrochemistry, catalysis, and sensing.

2. Surfactant-Encapsulated Clusters (SECs)

The vast progress achieved in POM synthesis has so far not been complemented by a comparable development in POM surface chemistry. One possible strategy in surface modification is to exchange labile ligands, which are coordinatively bound to the peripheral metal atoms of the cluster core.[20] Due to the high stability of the terminal M–O bond in polyoxomolybdates and -tungstates and the relatively slow ligand exchange, especially for polyoxotungstates, chemical routes which target on covalent modifications of the POM surface often lead to rearranged cluster cores. The prolonged reaction times and the rather vigorous reaction conditions often lead to mixtures of chemical derivatives of the original clusters in only moderate or low yields.[21]

An alternative strategy described here relies on a self-assembly process: the counter cations from the hydration sphere of the POM anions are replaced by suitable alkylammonium surfactants, resulting in discrete supramolecular entities; due to the close packing of the long-chain alkylammonium amphiphiles on the surface of the POM we coined the term "surfactant-encapsulated clusters" (SECs) to emphasize the particular structural properties of these core-shell type of supramolecular assemblies. This colloid chemical approach has been successfully used in the past to stabilize a variety of semiconductor and precious metal nanoparticles,[22] but has never been applied convincingly to POM chemistry.[23]

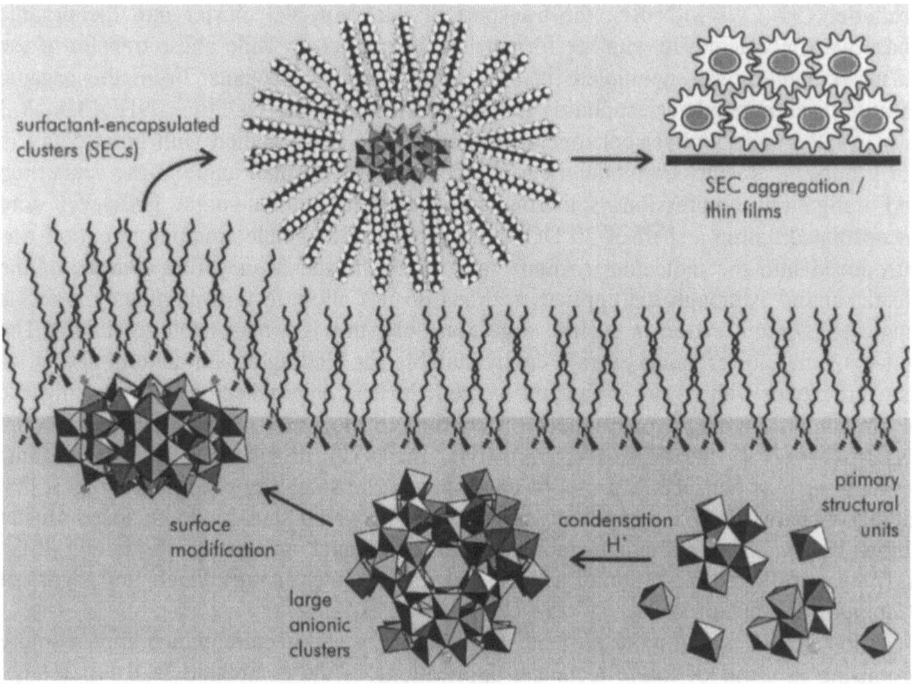

FIGURE 3. Sequential self-assembly and concomitant phase transfer of surfactant-encapsulated $\{Mo_{57}V_6\}$ clusters.

The SEC synthesis is achieved in a two-step procedure: first, a water-soluble salt of the envisaged POM component is prepared and its structure is characterized by standard analytical methods (e.g. single crystal X-ray structure analysis). In the second step, an aqueous solution of the POM salt is treated with a water-immiscible organic solvent containing an appropriate amount of a cationic surfactant. This preparation scheme permits characterization of the materials brought into use at each step. In the systems studied so far, we found that the cationic surfactants can displace the counter-cations, leading to an immediate transfer of the encapsulated clusters into the organic phase (FIGURE 3). To ensure that the SEC material isolated from organic solution is void of excess surfactant, a substoichiometric amount may be applied. The molecular composition of the SEC does not depend critically on the amount of surfactant used (e.g. a surfactant deficiency as high as 10% with respect to the charge of the cluster anion is tolerable), which indicates that the encapsulation indeed is a self-terminating process.

Experiments with different surfactant/POM combinations indicate that steric requirements for the packing of the surfactant alkyl chains and the molecular balance of the hydrophobic/hydrophilic properties play an important role in stabilizing the structure of the surfactant-encapsulated clusters. To give an example, aqueous solutions of the partially reduced POM $[H_3Mo_{57}V_6(NO)_6O_{83}(H_2O)_{18}]^{21-}$, $\{\mathbf{Mo_{57}V_6}\}$, were equally treated with stoichiometric quantities of the commercially available surfactants dioctadecyldimethylammonium bromide (DODA•Br, $(CH_3)_2(C_{18}H_{37})_2N{\bullet}Br$), octadecyl trimethylammonium bromide ($(CH_3)_3(C_{18}H_{37})N{\bullet}Br$), or trioctadecylmethylammonium bromide ($CH_3(C_{18}H_{37})_3N{\bullet}Br$); the transport of the $\{\mathbf{Mo_{57}V_6}\}$ cluster into the organic phase was only achieved with the former DODA surfactant while phase transfer of the cluster failed or was incomplete in case of the latter surfactants. From the organic solution, a compound corresponding to the empirical formula $(DODA)_{20}(NH_4)\{\mathbf{Mo_{57}V_6}\}$ was isolated and its physicochemical properties have been studied with great detail in our groups.[24] Results from analytical ultracentrifugation, small angle X-ray scattering, and Langmuir compression isotherms are consistent with a single $\{\mathbf{Mo_{57}V_6}\}$ core encapsulated within a shell of 20 DODA molecules. One single ammonium cation was introduced into the molecular formula to compensate the 21 negative charges of the cluster anion. Although the applied methods do not allow to unambiguously detect a single ammonium cation in such a large assembly, this is a reasonable postulate: The center of the cluster anion bears a cavity suitable for binding an ammonium cation. In fact, the cavity with its pre-organized oxygen electron pairs resembles the binding site of ammonium-binding crown ethers. In contrast to the water-soluble starting material $(NH_4)_{21}\{\mathbf{Mo_{57}V_6}\}$, the SEC $(DODA)_{20}(NH_4)\{\mathbf{Mo_{57}V_6}\}$ dissolves readily in organic solvents such as benzene, toluene, or chloroform. The solubility properties suggest that the alkyl chains form a compact hydrophobic shell which shields the enclosed cluster anion. To demonstrate the high stability of this compound, an aqueous dispersion of the SEC was refluxed and sonicated for several minutes upon which no signs of decomposition occurred.

Another SEC with a more intricate supramolecular architecture, which appears more promising in terms of future technical applications, is the compound $DODA_{40}$ $(NH_4)_2$ $[(H_2O)_n \subset Mo_{132}O_{372}(CH_3COO)_{30}(H_2O)_{72}]$ (= $DODA_{40}(NH_4)_2\{\mathbf{Mo_{132}}\}$).[25] This compound initially contains approximately 50 H_2O molecules which are clathrated in the central nano-sized cavity of the POM anion. (FIGURE 4). Based on elemental analysis, 40 DODA molecules encapsulate the cluster, leading to a discrete, nearly spherical particle with a molecular mass of approximately 43.900 g/mol. The solvent-accessible surface (SAS) of the encapsulated $\{\mathbf{Mo_{132}}\}$ cluster calculated for a probe radius of 0.28 nm is approximately 33.4 nm^2, which yields an average surfactant area of 0.84 nm^2. Within the given range of uncertainty, this corresponds reasonably well to the empirical value of 32 (±8) nm^2 for the complete surface area of 40 DODA molecules.[26]

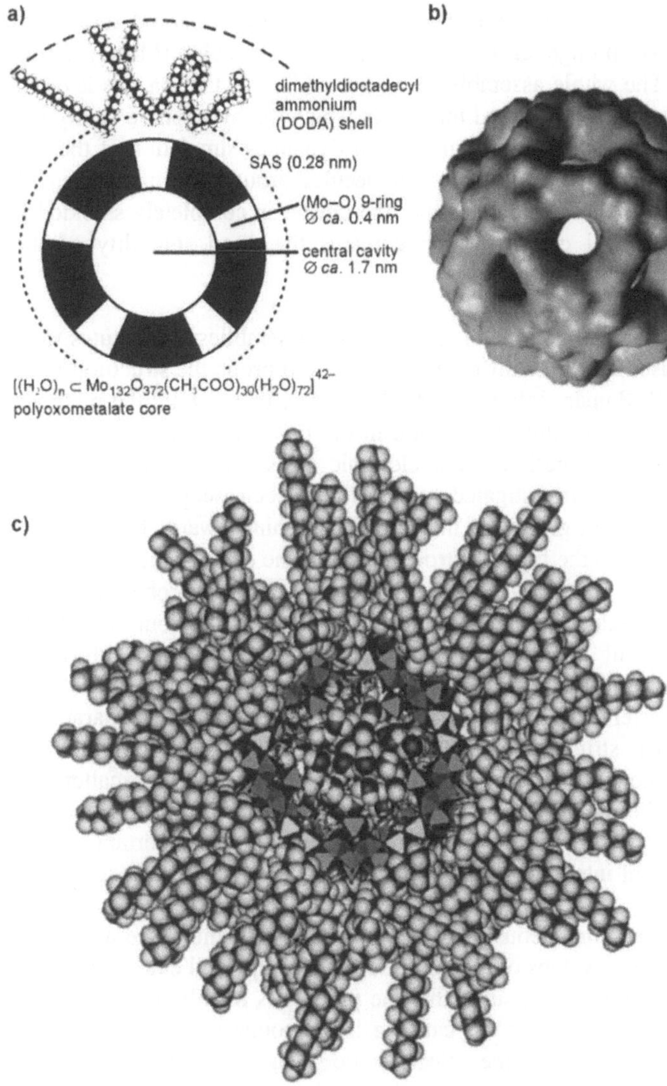

FIGURE 4. Supramolecular architecture of $(DODA)_{40}(NH_4)_2$**{Mo$_{132}$}**. A) Schematic representation of the core-shell structure of a single SEC. B) Solid rendered representation of the Connolly surface of the Keplerate cluster **{Mo$_{132}$}**. C) Cross-section through a SEC model showing the surfactant shell and the clathrated H_2O molecules in space-filling representations, while the Mo-O framework is displayed as polyhedral model.

Geometric matching of the two juxtaposed ionic surfaces may be a critical point for driving the self-encapsulation process to completeness. The surface charge density of **{Mo$_{132}$}** is such that all DODA molecules find sufficient space to form a single layer at a van der Waals distance to the cluster surface. The covered surface area of an 0.84 nm^2 / DODA molecule furthermore suggests a rather tight packing of the amphiphile at the cluster surface. The SAS of **{Mo$_{132}$}** for an 0.28 nm probe displays a continuous spherically shaped surface which indicates that the DODA cations cannot penetrate the large central cavity of the Keplerate. In contrast, the SAS of **{Mo_{132}}** for an 0.14 nm probe which is often referred to as the water-accessible (Connolly) surface, extends into the central cavity through each of the twenty non-circular (Mo-O) 9-ring openings (FIGURE 4B). The whole assembly of cluster and surfactant on this level of structural organization resembles a reversed micelle in which the hydrophilic cavity is completely filled by the large cluster anion. A molecular dynamics simulation of the SEC gives an idea of the packing of the surfactant molecules around the nano-sized cluster. The simulation in FIGURE 4C indicates that the cluster is completely shielded by the long hydrophobic octadecyl chains, which explains the good solubility of the SECs in organic solvents.

Analytical characterization of such large assemblies constitutes an enormous challenge. We used a host of different techniques to probe the structural integrity of the POM in the SEC. Raman, infrared and UV/vis spectroscopy are the methods of choice to prove the presence of the POM anion in the final surfactant-encapsulated assembly due to characteristic vibrational and elctronic transitions of the cluster anions. The surfactant shell can be investigated by ^{1}H-NMR spectroscopy, which reveals that the positively charged head groups of the surfactants point towards the negatively charged cluster surface. Due to the high electron density of the cluster anion, X-rays can be used to examine the particle core of the SEC. As a sensitive probe for electron-density differences in the scattering medium, the pair-distance distribution function (PDDF) obtained from small angle X-ray scattering (SAXS) provides information on the maximum dimension of the particle and its radius of gyration. In diluted systems where inter-particle interferences can be neglected, the PDDF profile is characteristic for the shape and internal structure of the particles. SAXS of SEC solutions confirms that single cluster anions are present in the SEC. Analysis of the scattering data is in agreement with the dimensions of the POM anions as determined by single crystal X-ray structure analysis. The SAXS measurements answer the important question whether or not the structural integrity of crystalline POMs is preserved in aqueous solution and in the SECs.

The reasons for spontaneously occurring SEC self-assembly are not well understood yet, since the accurate values of the contributing enthalpy and entropy terms are difficult to determine. Currently we assume that the process is mainly driven by an increase in Coulomb interactions: placing the cationic head-groups in close vicinity to the POM surface efficiently screens the electrostatic charge of the encapsulated anion. Hydrophobic interactions between the alkyl-chains of the close-packed surfactant shell may furthermore stabilize the SEC. Finally, the gain of hydration enthalpy upon release of counter anions (e.g. Cl^- or Br^-) into the aqueous phase, and the entropically favorable liberation of a huge number of ammonium cations and water molecules from the cluster surface may explain the driving force for SEC formation.

3. Thin Films of POMs

In spite of the numerous technical applications of thin films and coatings, only a few research activities so far have been focussed on methods to incorporate POMs as active functional components in extended two-dimensional assemblies. Thin films of POMs were successfully produced by electro-deposition.[27] However, the internal structure and the purity of the resulting films remain to be characterized in detail. Recently, is has been reported that certain POMs spontaneously adsorb on precious metal surfaces.[28] Although this is a simple and elegant approach, it is restricted to films of monolayer coverage and requires metal substrates. Well-defined composite films spontaneously form by adsorption of Keggin- and Dawson-type heteropolyoxotungstates underneath compressed Langmuir monolayers of cationic surfactants at the air-water interface. Subsequent repeated transfer of monolayers on solid substrates led to multi-lamellar thin films which have been characterized by X-ray diffraction and IR dichroism experiments. A practical limitation of this approach is that the hybrid POM-surfactant materials cannot be isolated in bulk substance.

We investigated two different approaches to integrate POMs into well-defined thin films and coatings. In the LbL approach, advantage is taken of the multiple negative charge of POMs polyanions which facilitates their adsorption at oppositely charged surfaces. We also discovered a novel route to produce well-defined thin films through our work on SECs. The hydrophobic nature of SECs permits to make Langmuir-Blodgett films. In addition, we found that SECs spontaneously assemble into highly ordered two-dimensional arrays.

3.1. Layer-by-Layer (LbL) Multilayer Thin Films

The layer-by-layer (LbL) method is based on sequential adsorption of oppositely charged species from dilute solutions and relies primarily on electrostatic attraction of the oppositely charged components. A scheme of this approach is depicted in FIGURE 5. The substrate is dipped into alternate solutions containing the charged species. The deposition process is self-terminating because the adsorption of polyelectrolytes within each deposition cycle causes a repeated charge reversal of the interface with concomitant electrostatic repulsion between the immobilized and the dissolved species. This method is generally suited for polyelectrolytes and provides control over film growth and film thickness at the nanometer-scale. It does not require specialized equipment, or substrates, and it is readily adapted for automated fabrication. The deposited films are mechanically robust and permeable for small molecules.

We adapted the LbL method to produce multi-layered thin films of POMs on quartz supports.[29] In the first step, the quartz substrate is charged positively by depositing a layer of poly(ethyleneimine) (PEI) on the cleaned substrate, followed by adsorption of poly(4-styrenesulfonate) (PSS) and poly(allylamine hydrochloride) (PAH). The (PEI/PSS/PAH) layer forms the base for subsequent POM deposition. Alternate adsorption of POM and PAH (as exemplified in FIGURE 5) produces thin films with a multilayer structure. Analysis by UV/vis-spectroscopy, ellipsometry, and quartz crystal resonators confirm that film growth is linear over several deposition cycles and highly reproducible. The films have a very uniform thickness as indicated by the appearance of interference or so-called Kiessig fringes in X-ray reflectance curves.

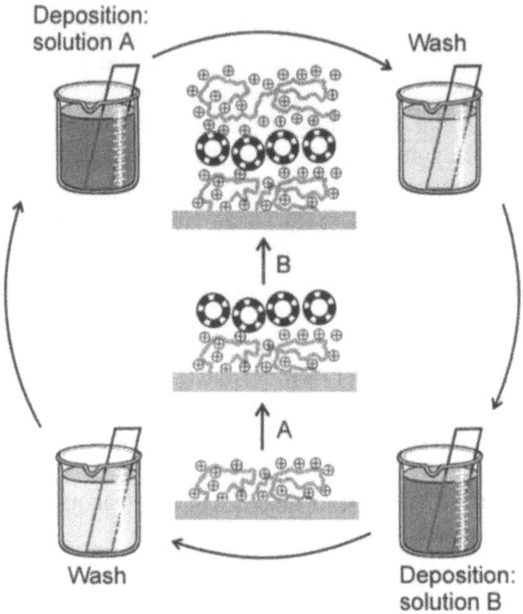

FIGURE 5. The Layer-by-Layer method relies primarily on electrostatic interactions of oppositely charged adsorbates. Multilayer film growth proceeds in a cyclic deposition process in which the substrate is immersed in dilute solutions of different, charged species interrupted by intermediate washing steps. The LbL method does not require specialized equipment, is readily adapted for automated production and can be applied to arbitrary surfaces. Since the deposition process is strictly sequential, combinations of different components in a single film are easily put into practice.

The robust and permeable nature of LbL films points to potential applications of POM-based multilayer composite films in electrochemistry, catalysis, and sensing, in particular with the nanoporous Keplerate type POMs. On-going investigations on LbL films that have been deposited on a quartz crystal resonator may lead into a practical approach to transduce mass changes of the coating as response to a molecular recognition event (e.g. shape-selective loading of the **{Mo$_{132}$}** nanocavities) into an electric signal with high sensitivity.

3.2. Langmuir-Blodgett (LB) films of SECs

The Langmuir-Blodgett (LB) technique was one of the first methods to fabricate thin films with long-range order and precise thickness control and played a key role in the development of molecular electronics.[30] The defined conditions of the air-water interface permit to explore the surface activity and provide molecular level control to construct layered materials using Langmuir-Blodgett (LB) film transfer. The general concept of Langmuir monolayers and LB transfer is depicted in FIGURE 6. A carefully weighed quantity of the surfactant is dissolved in a water-immiscible volatile solvent, and the solution is spread on the aqueous phase contained within a Langmuir trough. Upon evaporation of the solvent, a surfactant layer remains at the air-water interface, which is subsequently compressed with a moving barrier. During compression, the surface pressure, π, is measured and displayed as a function of the corresponding area, A, revealing details of the phase behavior of the monolayer as well as information about the size of the molecules. Molecules that form monolayers at the air-water interface usually consist of a hydrophilic part, e.g. a charged head group, and a hydrophobic part, e.g. long hydrocarbon chains. Such amphiphilic molecules are co-aligned with respect to each other at the air-water interface, with the hydrophilic part pointing towards the aqueous subphase. Langmuir monolayers can be transferred on solid substrates by dipping an appropriate substrate into the aqueous subphase at constant pressure. Repeated monolayer transfer often yields highly-ordered lamellar multilayers, but in contrast to the LbL method, LB transfer is limited to certain types of molecules and substrates, it requires specialized instrumentation, and the films are generally not at thermodynamic equilibrium.

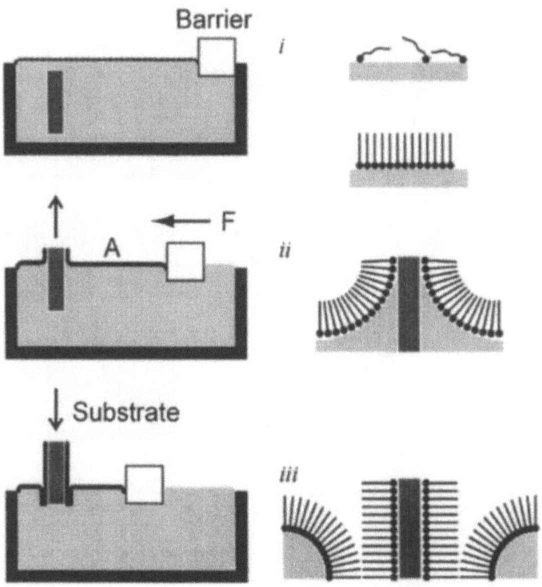

FIGURE 6. Langmuir-Blodgett transfer of monolayers on solid substrates: The water-insoluble material is spread at the air-water interface in a Langmuir trough of known area. Initially the molecules may be distributed statistically at the air-water-interface (*i*). By moving the barrier, the available area is successively decreased, leading to a compressed, highly-ordered monolayer. At a given surface pressure, the monolayer may be transferred on the solid support (*ii*). Repeated up- and down-stroke of the substrate results in lamellar multilayers (*iii*), the structure of which will depend on the nature of the hydrophobic compounds, the substrate surface characteristics and the composition of the aqueous subphase. In the example given above, a monolayer of a conventional surfactant comprised of a long-tail hydrophobic part and a hydrophilic head-group is transferred as a so-called Y-type film on the polar surface of a solid substrate (e.g. glass).

Due to the good water-solubility of most POMs, it is impossible to obtain POM Langmuir monolayers at the air-water interface. In contrast, we found that SECs can be directly spread at the air-water interface to yield a homogeneous SEC monolayer. Therefore, SECs are suitable for direct LB-processing: SECs are spread from chlorofrom solution on the water surface, the resulting SEC monolayer is compressed and the film is transferred at constant surface pressure onto the solid substrate. Analytical characterization of SEC Langmuir films by Brewster-angle microscopy, optical ellipsometry, and grazing angle X-ray diffraction [31] confirm monolayer coverage. From Langmuir isotherms, the surface area of a single $(DODA)_{20}(NH_4)$**{Mo$_{57}$V$_6$}** was determined to be 10.4 nm^2, which corresponds to an object with a diameter of 3.6 nm (FIGURE 7). The $DODA_{40}(NH_4)_2$**{Mo$_{132}$}** occupies an area of 15 nm^2, which corresponds to a diameter of 4.4 nm. Both values are in excellent agreement with the proposed structural model in which a single cluster anion resides within a close shell of surfactant molecules.

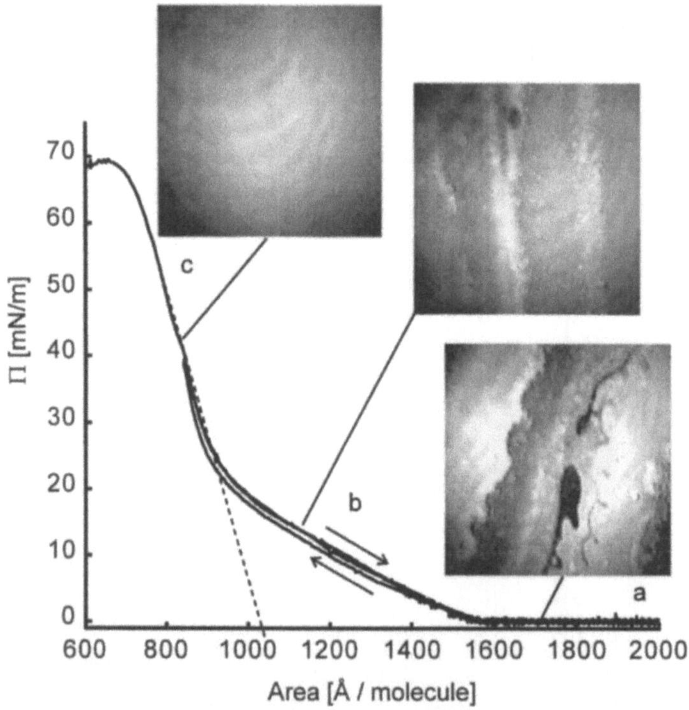

FIGURE 7. Surface-pressure (π-A) diagram of a $(DODA)_{20}(NH_4)$**{Mo$_{57}$V$_6$}** monolayer spread at the air-water interface. The inserted photographs stem from Brewster angle microscopic investigations of the macroscopic film structure at different surface pressures. (a) $\pi = 0$ mN; (b) $\pi = 20$ mN; (c) $y = 40$ mN. (Figure reproduced from Ref. 24, with permission).

LB transfer of SEC monolayers was achieved on several substrates, including silicon, quartz, and gold-sputtered glass slides. The substrate is immersed in the subphase before spreading and multilayers are formed by repeated LB transfer. The transfer ratio is close to unity in all cases. Investigation with optical ellipsometry and UV/vis-spectroscopy demonstrate that LB transfer is very reproducible and that film growth is essentially linear, that is, in each dipping cycle equivalent amounts of SECs are transferred on the substrate. X-ray reflectivity (XRR) of SEC LB multilayers on silicon substrates shows well-resolved Kiessig fringes, indicating a uniform film thickness. In addition, several Bragg reflections are discernible, which implies that these LB-films have an internal layered structure.

The static contact angle, γ, of $(DODA)_{20}(NH_4)\{\mathbf{Mo_{57}V_6}\}$ LB multilayers is 102° for water and 31° for hexadecane. Similarly, the water contact angle of $DODA_{40}(NH_4)_2\{\mathbf{Mo_{132}}\}$ LB multilayers amounts to 97°. These values clearly indicate the hydrophobic nature of the SEC-coated substrate surface and demonstrate how efficiently the DODA surfactant shell screens the underlying hydrophilic cluster. The absolute values of the contact angles suggest that the SEC alkyl chains are somewhat disordered. For comparison, a methyl-terminated surface typically has a contact angle of 110 – 115°.[32]

It is not yet clear how SECs are stabilized at the air-water interface because hydrophobic compounds typically aggregate and float as lenses on the water surface. Since SECs have finite water contact angles, partial wetting may cause a partial immersion of the spherical SEC into the aqueous subphase.[33] This could trap the SEC at the air-water interface and would prevent aggregation to droplets (see FIGURE 8).

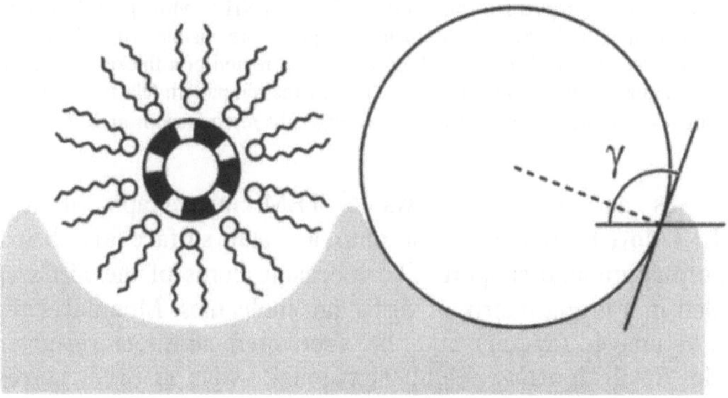

FIGURE 8. A simplified model of SECs floating at the air-water interface. Because SEC LB-films have a finite water contact angle, γ, it may be assumed that the SECs are partially wetted at the air-water interface, which would prevent aggregation of SECs to droplets. The scheme on the right suggests how the immersion depth relates to the contact angle γ.

TEM studies on SEC thin films deposited on solid substrates either by LB transfer or, alternatively, by simple evaporation of diluted SEC solutions, reveal that SECs have a strong tendency to self-assemble into extended, well-ordered two-dimensional arrays.

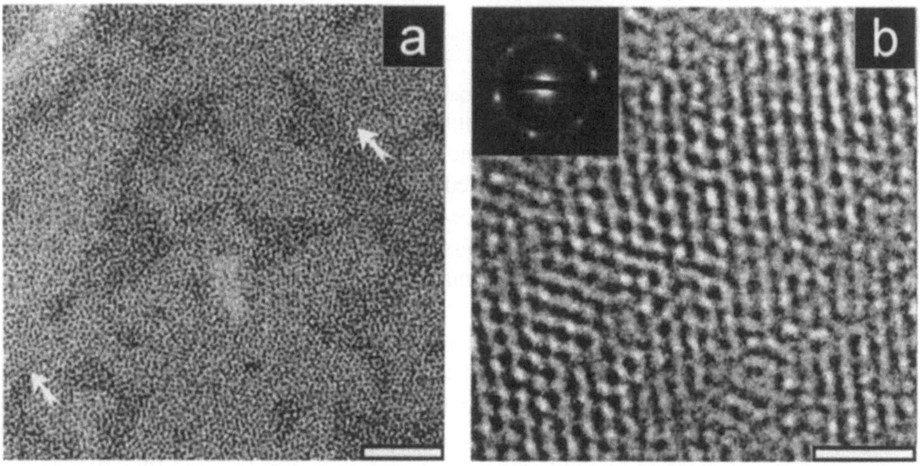

FIGURE 9. (a) TEM micrograph of a thin film of $DODA_{40}(NH_4)_2$**{Mo$_{132}$}** (scale bar 50 nm). Extended monolayers, interspersed with holes and doublelayer regions are visible. Arrows mark some regions of apparent hexagonal order. (b) TEM micrograph of an ordered region of a thicker film at high magnification (scale bar 20 nm). The insert shows a low-angle electron diffraction pattern (elastically filtered) recorded from a larger area (5 μm^2) containing the region shown. (Micrographs reproduced from Ref. 25, with permission)

FIGURE 9*a*, as an example, shows a TEM micrograph of a thin film of $DODA_{40}(NH_4)_2$**{Mo$_{132}$}** originally cast onto a water surface and lifted off with an amorphous porous graphite support. The inorganic cores of the SECs appear as dark spots embedded in a bright matrix of surfactant molecules. Monolayer regions, regions consisting of a bilayer (darker) and the uncovered substrate (brighter) are clearly distinguishable. Small domains exhibit hexagonal arrays of SECs corresponding to a two-dimensional close packing of spherical particles (arrows). Both the diameter of the dark objects (approx. 3 nm) as well as the average distance between them (approx. 4.5 nm) are consistent with the proposed SEC model. In thicker films, the order improves and becomes three-dimensional. An ordered region is shown in FIGURE 9*b*; the related electron diffraction pattern (insert) clearly reveals long-range order (spots rather than rings) and a threefold symmetry of the pattern of reflections corresponding to a spacing of 4.2 nm. The packing of the SECs may be tentatively described by a fcc lattice with a cubic unit cell axis of approximately 6 nm. The observed diffraction pattern then corresponds to the close packing of SECs in a (111) plane at normal orientation to the electron beam. It is noteworthy that upon heating thick SEC films, dewetting and terracing is observed, as in the case of block-copolymers.

4. Outlook on Future Investigations and Applications

Non-covalent interactions can be employed to control the supramolecular architecture of materials on several length scales. At the nanoscopic level, self-assembly of surfactants and POMs leads to discrete SECs possessing a core-shell structure with a chemically well-defined composition. Auto-assembly of SECs at the air-water interface or at solid supports results in two- or three-dimensional periodic arrays extending into macroscopic dimensions.

The huge number of commercially or synthetically available surfactants and POMs give rise to novel SEC structures with different functional properties. The influence of encapsulation on POM solubility, surface activity, and adhesion mentioned here grant a glimpse at the wealth of potential technical applications.

In future work, three fundamental questions will deserve particular attention. First, understanding the interactions and the driving forces for POM self-encapsulation will be important in designing custom surfactants for SECs with tailored properties. From a technological point of view it will be of interest to investigate the effect of encapsulation on the functional properties of the cluster core, for instance on electron transfer reactions. As an imaginative example, a catalytically active POM cluster may be encapsulated within a surfactant shell that – based upon size-exclusion or specific molecular recognition – serves to discriminate between different substrate molecules reaching the surface of the catalytically active site.

Second, investigations on the structure of SEC monolayers will help to define possible mechanisms for spreading this novel type of non-amphiphilic compounds at the air-water interface. Model studies using structurally uniform, monodisperse POM-based SECs may help to design SECs of technologically equally interesting inorganic compounds (e.g. quantum-confined clusters of semi-conductors or precious metals) that will be suitable for LB film transfer.

Finally, it will be important to address the question of how to self-assemble POMs into three-dimensional supramolecular architectures in a controlled and predictable manner. Our contributions to this topic currently include thick films of SECs and POM-based LbL films. Especially LbL processing provides an exceptionally simple and efficient method to implement POMs in a robust and permeable matrix, also in alliance with other functional components.

Despite of the progress made in this area of "nanotechnology", it should be pointed out that the implementation of POMs as active components in functional materials and molecular devices requires solutions to many unsolved practical problems, e.g. in terms of mechanical, thermal, and chemical stability. The increasing structural complexity of the investigated hybrid materials often requires major interdisciplinary and time-consuming efforts to apply the existing and to develop novel analytical methods. Further steps towards realisable technological applications of SECs and related materials will depend on the successful cooperation of scientists from many different disciplines. In this sense, the visionary claims of a nanotechnological paradigm may smooth the way for necessary fundamental research in directions that go beyond "classic" synthetic approaches in chemistry.

Acknowledgements
DGK gratefully acknowledges financial support by the Max-Planck-Society. DV thanks the Fonds der Chemischen Industrie for a Liebig fellowship. The Deutsche Forschungsgemeinschaft is acknowledged for financial support. Valuable discussions with Helmuth Möhwald and Achim Müller are gratefully acknowledged.

[1] (a) Timp, G.L. (1999) *Nanotechnology*, Springer, Berlin. (b) Ando, T.; Arakawa, Y.; Furuya, K. (1998), *Mesoscopic Physics and Electronics*, Springer, Berlin. (c) Mitura, S. (1999) *Nanotechnology in Materials Science*, Pergamon, Oxford.

[2] Lehn, J.-M. (1995) *Supramolecular Chemistry*, VCH Wiley, Weinheim.

[3] Stang, P.J.; Olenyuk, B. (1997) Self-assembly, symmetry, and molecular architecture: Coordination as the motif in the rational design of supramolecular metallacyclic polygons and polyhedra, *Acc. Chem. Res.* 30, 502–518.

[4] Claessens, C.G.; Stoddart, J.F. (1997), π-π interactions in self-assembly, *J. Phys Org. Chem.* **10**, 254–272.

[5] (a) Whitesides, G.M.; Simanek, E.E.; Mathias, J.P.; Seto, C.T.; Chin, D.N.; Mammen, M.; Gordon, D.M. (1995) Non-covalent Synthesis – Using Physical-Organic Chemistry to Make Aggregates, *Acc. Chem. Res.* **28**, 37–44. (b) Rebek, J. (1999) Reversible encapsulation and its consequences in solution, *Acc. Chem. Res.* **32**, 278–286.

[6] Valdes, C.; Spitz, U.P.; Toledo, L.M.; Kubik, S.W.; Rebek, J. (1995) Synthesis and self-assembly of pseudo-spherical homo- and heterodimeric capsules, *J. Am. Chem. Soc.* **117**, 12733–12745.

[7] Hasenknopf, B.; Lehn, J.-M.; Kneisel, B.O.; Baum, G.; Fenske, D. (1996) Self-assembly of a circular double helicate, *Angew. Chem. Int. Ed. Engl.***35**, 1838–1840.

[8] Watton, S.P.; Fuhrmann, P.; Pence, L.E.; Caneschi, A.; Cornia, A.; Abbati, G.L.; Lippard, S.J. (1998) A cyclic octadecairon(III) complex, the molecular 18-wheeler, *Angew. Chem. Int. Ed. Engl.* **36**, 2774–2776.

[9] Ashton, P.R.; Boyd, S.E.; Claessens, C.G.; Gillard, R.E.; Menzer, S.; Stoddart, J.F.; Tolley, M.S.; White, A.J.P.; Williams, D.J. (1997) A novel type of isomerism in [3]catenanes, *Chem. Eur. J.* **3**, 788–798.

[10] (a) Kozhevnikov, I. V. (1998) Catalysis by heteropoly acids and multicomponent polyoxometalates in liquid-phase reactions, *Chem. Rev.* **98**, 171–198. (b) Mizuno, N.; Misono, M. (1998) Heterogenous catalysis, *Chem. Rev.* **98**, 199–217.

[11] (a) Weinstock, I. A. (1998) Homogeneous-phase electron-transfer reactions of polyoxometalates, *Chem. Rev.* **98**, 113–170. (b) Sadakane, M.; Steckhan, E. (1998) Electrochemical properties of polyoxometalates as electrocatalysts, *Chem. Rev.* **98**, 219–237.

[12] Yamase, T. (1998) Photo- and electrochromism of polyoxometalates and related materials, *Chem. Rev.* **98**, 307–325.

[13] Rhule, J. T.; Hill, C. L.; Judd, D. A.; Schinazi, R. F. (1998) Polyoxometalates in medicine, *Chem. Rev.* **98**, 327–357.

[14] Katsoulis, D.E. (1998) A survey of applications of polyoxometalates, *Chem. Rev.* **98**, 359–387.

[15] (a) Müller, A.; Kögerler, P.; Kuhlmann, C. (1999) A variety of combinatorially linkable units as disposition: from a giant icosahedral Keplerate to multi-functional metal-oxide based network structures, *Chem. Commun.* **15**, 1347–1358. (b) Müller, A.; Krickemeyer, E.; Meyer, J.; Bögge, H.; Peters, F.; Plass, W.; Diemann, E.; Dillinger, S.; Nonnenbruch, F.; Randerath, M.; Menke, C. (1995) $[Mo_{154}(NO)_{14}O_{420}(OH)_{28}(H_2O)_{70}]^{(25\pm5)-}$ – A Water-Soluble Big Wheel With More Than 700 Atoms and a

relative molecular mass of about 24000. *Angew. Chem. Int. Ed. Engl.* **34**, 2122–2124. (c) Müller, A.; Krickemeyer, E.; Bögge, H.; Schmidtmann, M.; Beugholt, C.; Kögerler, P.; Lu, C.Z. (1998) Formation of a ring-shaped reduced "metal oxide" with the simple composition $[(MoO_3)_{176}(H_2O)_{80}H_{32}]$, *Angew. Chem. Int. Ed. Engl.* **37**, 1220–1223. (d) Müller, A.; Krickemeyer, E.; Bögge, H.; Schmidtmann, M.; Peters, F. (1998) Organizational forms of matter: An inorganic super fullerene and keplerate based on molybdenum oxide, *Angew. Chem. Int. Ed. Engl.* **37**, 3360–3363.

[16] Müller, A.; Krickemeyer, E.; Dillinger, S.; Bögge, H.; Plass, W.; Proust, A.; Dloczik, L.; Menke, C.; Meyer, J.; Rohlfing, R. (1994) New Perspectives in Polyoxometalate Chemistry by Isolation of Compounds Containing Very Large Moieties as Transferable Building Blocks: $(NMe_4)_5[As_2Mo_8V_4AsO_{40}] \bullet 3\ H_2O$, $(NH_4)_{21}[H_3Mo_{57}V_6(NO)_6O_{183}(H_2O)_{18}] \bullet 65\ H_2O$, $(NH_2Me_2)_{18}(NH_4)_6[Mo_{57}V_6(NO)_6O_{183}(H_2O)_{18}] \bullet 14\ H_2O$, and $(NH_4)_{12}[Mo_{36}(NO)_4O_{108}(H_2O)_{16}] \bullet 33\ H_2O$, *Z. anorg. allg. Chem.* **620**, 599–619.

[17] Keplerates with coordinated formate anions which possess a more hydrophilic inner surface have been described very recently. In this case, the positions of approximately 60 H_2O molecules residing within the Keplerate cavity could be determined directly by X-ray crystallography. Müller, A.; Fedin, V.P.; Kuhlmann, C.; Bögge, H.; Schmidtmann, M. (1999) A hydrogen-bonded cluster with 'onion-type' structure, encapsulated and induced by a spherical cluster shell: $[(H_2O)_n \subset (Mo^{VI}{}_{72}Mo^{V}{}_{60}O_{372})(HCO_2)_{30}(H_2O)_{72}]^{42-}$ *Chem. Commun.* 927–928.

[18] Simon, J.; André, J.; Skoulis, A. (1986) Molecular Materials I: Generalities, *Nouv. J. Chim.* **10**, 295–311.

[19] Decher, G. (1997) Fuzzy Nanoassemblies – Toward Layered Polymeric Multicomposites **277**, 1232–1237.

[20] Schmid, G. (1994) *Clusters and Colloids*, VCH, Weinheim.

[21] Gouzerh, P.; Proust, A. (1998) Main-Group Element, Organic, and Organometallic Derivatives of Polyoxometalates, *Chem. Rev.* **98**, 77–111

[22] Fendler, J.H. (1998) *Nanoparticles and Nanostructured Films*, VCH Wiley, Weinheim.

[23](a) Clemente-León, M.; Mingotaud, C.; Agricole, B.; Gómez-García, C.J.; Coronado, E.; Delhaes, P. (1997) Application of the Langmuir-Blodgett technique to polyoxometalates: Towards new magnetic films, *Angew. Chem. Int. Ed. Engl.* **36**, 1114–1116. (b) Janauer, C.G.; Dobley, A.; Guo, J.D.; Zavalij, P.; Whittingham, M.S. (1996) Novel tungsten, molybdenum, and vanadium oxides containing surfactant ions, *Chem. Mater.* **8**, 2096–2101. (c) Stein, A.; Fendorf, M.; Jarvie, T.P.; Müller, K.T.; Benesi, A.J.; Mallouk, T.E. (1995) Salt gel synthesis of porous transition-metal oxides, *Chem. Mater.* **7**, 304–313.

[24] Kurth, D.G.; Lehmann, P.; Volkmer, D.; Cölfen, H.; Koop, M.J; Müller, A.; Du Chesne, A. (2000) Surfactant-Encapsulated Clusters (SECs): $(DODA)_{20}(NH_4)[H_3Mo_{57}V_6(NO)_6O_{183}(H_2O)_{18}]$, a Case-Study, *Chem. Eur. J.*, in press.

[25] Volkmer, D.; Du Chesne, A.; Kurth, D.G.; Schnablegger, H.; Lehmann, P.; Koop, M.J.; Müller, A. (2000) Towards Nanodevices: Synthesis and Characterization of the Nanoporous Surfactant-Encapsulated Keplerate $(DODA)_{40}\ (NH_4)_2\ [(H_2O)_n \subset Mo_{132}O_{372}(CH_3COO)_{30}(H_2O)_{72}]$, *J. Am. Chem. Soc.*, in press.

[26] The molecular surface area of a single DODA cation as determined from the lamellar arrangement of DODA cations in the single crystal structure of the compound DODA · Br (monohydrate), CSD entry code CIYWOW20, is 56.7 Å^2. Okuyama, K.; Soboi, Y.; Iijima, N.; Hirabayashi, K.; Kunitake, T.; Kajiyama, T. (1988) Molecular and Crystal Structure of the Lipid-Model Amphiphile, Dioctadecylammonium Bromide Hydrate, *Bull. Chem. Soc. Jpn.* **61**, 1485–1490. Empirical values for the molecular surface area of a single DODA cation have been frequently determined from the Langmuir isotherms of compressed DODA monolayers spread at the air-water interface. The reported values range from 60–100 Å^2/ DODA molecule, depending on the chemical nature of counter anions within the aqueous subphase. See: (a) Marra, J. (1986) Effects of Counterion Specifity on the Interactions between Quaternary Ammonium Surfactants in Monolayers and Bilayers, *J. Phys. Chem.* **90**, 2145–2150. (b) Clemente-León, M.; Agricole, B.; Mingotaud, C.; Gómez-García, C.J.; Coronado, E.; Delhaes, P. (1997) Toward new organic/inorganic superlattices: Keggin polyoxometalates in Langmuir and Langmuir-Blodgett films, *Langmuir* **13**, 2340–2347.

[27] Ingersoll, D.; Kulesza, P. J.; Faulkner, L. F. (1994) Polyoxometalate-Based Layered Composite Films on Electrodes, *J. Electrochem. Soc.* **141**, 140–147.

[28] (a) Klemperer, W. G.; Wall, C. G. (1998) Polyoxoanion chemistry moves toward the future: From solids and solutions to surfaces, *Chem. Rev.* **98**, 297–306. (b) Kaba, M.S.; Song, I.K.; Duncan, D.C.; Hill, C.L.; Barteau, M.A. (1998) Molecular shapes, orientation, and packing of polyoxometalate arrays imaged by scanning tunneling microscopy, Inorganic Chemistry **37**, 398–406.

[29] Caruso, F.; Kurth, D.G.; Volkmer, D.; Koop, M.J.; Müller, A. (**1998**) Ultrathin molybdenum polyoxometalate-polyelectrolyte multilayer films, *Langmuir* **14**, 3462–3465.

[30] Kuhn, H.; Möbius, D. (1993) Monolayer Assemblies, in *Physical Methods of Chemistry Series, Vol. IX* B. Rossiter, W.; Baetzold, R. C. (eds.), John Wiley & Sons, 375–542.

[31] To be published.

[32] Bain, C. D.; Troughton, E. B.; Tao, Y.-T.; Evall, J.; Whitesides, G. M.; Nuzzo, R. G. (1989) Formation of Monolayer Films by the Spontaneous Assembly of Organic Thiols From Solution onto Gold, *J. Am. Chem. Soc.* **11**, 321–335.

[33] (a) Horvölgyi, Z.; Medveczky, G.; Zrinyi, M. (1993) On the structure formation of hydrophobed particles in the boundary layer of water and octane phases, *Colloid & Polymer Sci.* **271**, 396–403. (b) Horvölgyi, Z.; Nemeth, S.; Fendler, J. H. (1993) Spreading of hydrophobic silica beads at water-air interfaces, *Coll. Surf. A: Physicochem. Eng. Asp.* **71**, 327–335.

Polyoxometalate Chemistry: a Source for Unusual Spin Topologies

D. GATTESCHI, R. SESSOLI
Department of Chemistry, University of Florence, I-50144 Florence, Italy

A. MÜLLER, P. KÖGERLER
Department of Chemistry, University of Bielefeld, D-33501 Bielefeld, Germany

Abstract. Giant molybdenum oxide-based clusters in which pentagonal $\{(Mo)Mo_5\}$ units are linked by a large number of paramagnetic centers like 30 Fe^{III} or 20 VO^{2+} show novel and unusual types of spin topologies: here we present an Fe_{30} icosidodecahedron and a magnetic ring-shaped band built up by 10 $\{V^{IV}_3\}$ triangles sharing vertices.

Key words: molecular magnetism, clusters, polyoxometalates, topology

1. Introduction

Polyoxometalate chemistry serves as inexhaustible source for molecular models for different aspects, especially due to their versatile redox chemistry. These systems have attracted large interest in the last few years, and already several reviews have appeared on this subject [1-3]. Referring to molecular magnets, paramagnetic centers can be embedded in a structure-determining diamagnetic polyoxometalate (linker) framework which gives rise to properties which are intermediate between simple paramagnetism and bulk magnetism [4]. For example, in the $\{Mo_{57}M_6\}$-type cluster systems it is possible to place (or exchange) step-wise different (para-) magnetic centers M like $Fe^{II/III}$ and $V^{IV}O^{2+}$ in the respective linker positions, thus allowing some control over the cluster's magnetic properties or even the tuning of these [5].

Molecular clusters are currently investigated in order to study quantum size effects in magnets [6-9]. In fact some molecular clusters have been termed single-molecule magnets [10] in order to stress with a somewhat emphatic designation their unique properties. Polyoxometalates have been shown to give rise to a behavior reminiscent of that of magnetic multilayers [11] but there are difficulties in observing real bulk behavior in clusters containing $S = 1/2$ magnetic centers, because the spin carriers have an intrinsically quantum nature. Therefore very large numbers of metal ions must be assembled in order to observe complex magnetic behavior.

The recent generation of giant polyoxometalate cluster systems [12] has indeed allowed to incorporate this required high number of paramagnetic functions in a planned Archimedean type of reaction route. This allows to generate a variety of discrete nanomagnets, e.g. a molecular cluster with the highest number of paramagnetic centers, reaching an unprecedent $S = 75$ state at room temperature.

M.T. Pope and A. Müller (eds.), Polyoxometalate Chemistry, 319–328.

2. Molecular Magnets with Unusual Spin Topology Based on Linking Pentagonal Units with Paramagnetic Centers

Of special interest in respect to molecular magnets are highly symmetrical, spherically shaped molecular systems which can accommodate paramagnetic centers on their surfaces. The principal route to nanostructured spherical systems with icosahedral symmetry (C_5 C_3 C_2) is based on the linking of pentagonal building units *via* spacers [12,13]. In the present case of polyoxomolybdates, an archetypical unit is the {(Mo)Mo_5}-type group structurally abundant in many giant clusters, and consisting of a central pentagonal MoO_7 bipyramid sharing its equatorial edges with five additional MoO_6 octahedra (Fig. 1). These pentagonal units are subsequently linked to form structures in which the C_5 symmetry of these pentagons is retained.

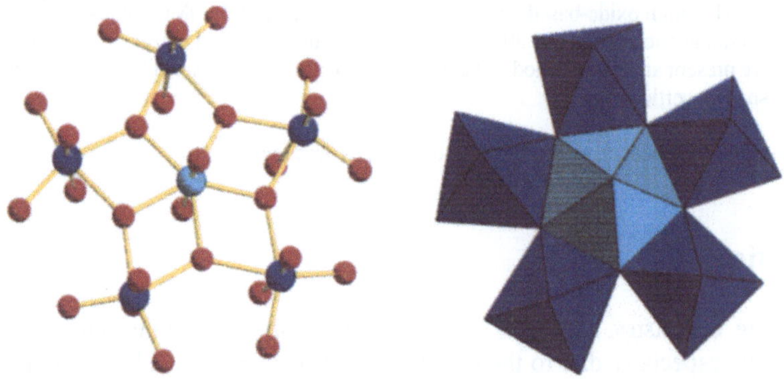

Figure 1. Structure of the $[(Mo)Mo_5O_{21}]^{6-}$ building group in ball-and-stick (left) and polyhedral (right) representation. The central pentagonal MoO_7 bipyramid (bright blue) shares edges with the five equatorial MoO_6 octahedra (blue).

In principle, the clusters referred to here can be divided – from a functional point of view – into (1) a basically diamagnetic polyoxomolybdate framework (mostly based on {(Mo)Mo_5} groups) and (2) a framework of paramagnetic centers – such as V^{IV}, Fe^{II} or Fe^{III} – with unusual spin topology connecting the pentagonal polyoxomolybdate fragments.

2.1. A Keplerate with 30 High-Spin-Iron(III) Centers

The reaction of polyoxomolybdate solutions of appropriate pH value in which {(Mo)Mo_5} groups exist as virtual units and Fe^{III} ions in aqueous media results in the formation of a water-soluble icosahedral {$Mo_{72}Fe_{30}$}-type cluster. This type of cluster can also be obtained by replacing binuclear $[Mo^{V}_2O_4]^{2+}$ linkers of a related cluster by Fe^{III}, a reaction which results in $[Mo_{72}Fe^{III}_{30}O_{252}(CH_3COO)_{12}\{Mo_2O_7(H_2O)\}_2$-- $\{H_2Mo_2O_8(H_2O)\}(H_2O)_{91}]$ in high yields (Fig. 2) [14]. The 30 Fe^{III} centers are bound to two oxygen atoms of two MoO_6 octahedra of two {(Mo)Mo_5} groups, resulting in a (slightly distorted) octahedral $[Fe^{III}O_{eq,4}(OH_2)_{ax,2}]$ unit. While the central Mo positions

Figure 2. Polyhedral representation of the {$Mo_{72}Fe_{30}$}-type Keplerate (upper part) with the inscribed icosahedron (defined by the 12 central positions of the {(Mo)Mo_5} groups, lower left) and the inscribed icosidodecahedron (formed by the 30 Fe centers, lower right). The diameter of the cluster is ca. 25 Å.

of the pentagonal {(Mo)Mo_5} groups span an icosahedron, the 30 Fe centers ($S = 5/2$) span an icosidodecahedron – one of the 13 Archimedean solids having 20 triangles and 12 pentagons. Due to the high S values of the magnetic centers the {Fe_{30}} system can be treated with the classical Heisenberg model [15]. This also allows to explain the Curie-Weiss form of the susceptibility which has even been observed in the widespread temperature range of 1.8 to 300 K with a Weiss temperature of Θ = ca. –20 K. At 300 K the measured χT value of 131 emu K mol^{-1} almost reaches the theoretical spin-only value of 131.25 emu K mol^{-1} for 30 uncorrelated $S = 5/2$ centers.

In a solid state reaction, the {$Mo_{72}Fe_{30}$} Keplerate can be furthermore condensed to form layer structures in which every {$Mo_{72}Fe_{30}$} entity is linked to four nearest neighboring entities *via* linear Fe-O-Fe bridges [16]. Here, the χT value at 300 K decreases to 113 emu K mol^{-1} which compares to 113.75 emu K mol^{-1} for 26 $S = 5/2$ centers. This is due to the fact that the iron centers of the Fe-O-Fe groups are strongly coupled and obviously do not significantly contribute to the overall magnetic moment (Fig. 3) [17].

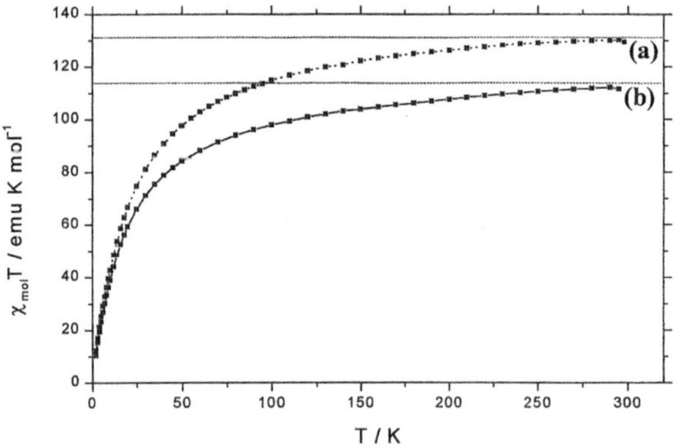

Figure 3. $\chi_{mol}T$ vs. T graphs of the (discrete) $\{Mo_{72}Fe_{30}\}$ Keplerate (a) and the corresponding layer compound $[Mo_{72}Fe_{30}O_{252}(CH_3COO)_{12}\{Mo_2O_7(H_2O)\}_2\{H_2Mo_2O_8(H_2O)\}(H_2O)_{91}]$ · ca. 150 H_2O (b). The grey horizontal lines indicate the spin-only limits for 30 (upper line) and 26 (lower line) uncorrelated S = 5/2 centers.

2.2. A $\{Mo_{75}V_{20}\}$-type Cluster Containing 20 Vanadium(IV) Centers

$\{(Mo)Mo_5\}$ pentagons can also be used in presence of other building blocks to construct derivatives of the afore-mentioned $\{Mo_{72}Fe_{30}\}$-type cluster which has 12 of these pentagonal $\{(Mo)Mo_5\}$ units: this can for instance lead to the cluster anion $[\{(Mo)Mo_5O_{21}(H_2O)_3\}_{10}\{NaSO_4\}_{10}\{V^{IV}O(H_2O)\}_{20}\{Mo^{VI}O_3(H_2O)\}_{10}\{Mo^{VI}O_2(H_2O)_2\}_5]^{20-}$ (≡ $\{Mo_{75}V_{20}\}$) [18] which can e.g. be described with reference to one of the Archimedean solids, namely the icosidodecahedron with 30 corners and 20 faces (Fig. 4), which is formed by 12 pentagonal and 20 triangular faces (ten $\{V_3\}$ (see below) and ten $\{VMo_2\}$ triangles) built up by 20 V^{IV} and 10 Mo^{VI} centers. While ten of the twelve pentagonal faces of the icosidodecahedron are each formed by one Mo^{VI} and four V^{IV} centers and are capped by $\{(Mo)Mo_5\}$ pentagons, the two remaining pentagonal faces both spanned by five Mo^{VI} centers (top and bottom, lying perpendicular to the principal C_5 axis of the cluster) are capped by one (under-occupied) array built up by five MoO_6 octahedra. In contrast to the $\{(Mo)Mo_5\}$ units the central pentagonal bipyramidally coordinated Mo^{VI} center is missing in these $\{Mo\}_5$ arrays. Interestingly the 20 V^{IV} centers build up a ring of 10 linked triangles (V-V: 6.27-6.46 Å). These vanadium centers are coordinated to four bridging oxygen atoms, while the 20 V^{IV} positions form two groups: (1) 10 equatorial V^{IV} centers showing outward-projecting H_2O ligands and *trans*-terminal oxygen, and (2) the 10 remaining V^{IV} centers exhibiting outward-projecting terminal oxygen and *trans*-H_2O ligands. This results in a novel equatorial paramagnetic ring-shaped band with very strong antiferromagnetic exchange interactions. In analogy to the $\{Mo_{57}M_6\}$ systems [5], we believed that the 20 $V^{IV}O^{2+}$ groups should be replaceable by Fe^{III} centers.

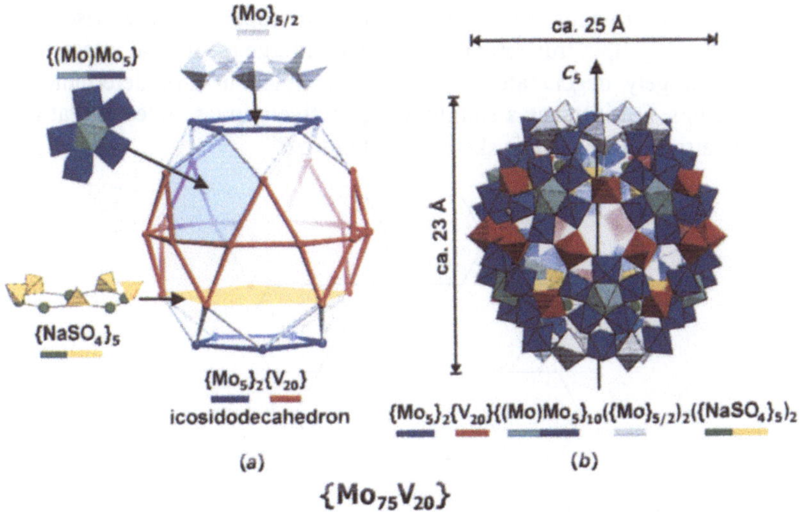

Figure 4. Structure of the $\{Mo_{75}V_{20}\}$-type cluster $Na_{20}[Mo_{75}V_{20}O_{270}(H_2O)_{70}Na_{10}(SO_4)_{10}] \cdot$ ca. 150 H_2O. (a) $\{V_{20}Mo_{10}\}$ framework structure of the $\{Mo_{75}V_{20}\}$-type cluster with 30 corners and 32 faces (20 triangles and 12 pentagons) spanned by 20 V^{IV} (red) and 10 Mo^{VI} (blue) centers emphasizing the ring-shaped band formed by 10 V^{IV}_3 triangles (red). Additional details: (1) One of the $\{(Mo)Mo_5\}$ capping 10 pentagonal faces, (2) pentagonal under-occupied array built up by five of the MoO_6 octahedra capping the other two pentagonal faces (grey), and (3) one of the two encapsulated $\{NaSO_4\}_5$-type rings (Na atoms: green, SO_4 tetrahedra: yellow). (b) Polyhedral representation of the (approximately) spherical structure of the complete $\{Mo_{75}V_{20}\}$-type cluster.

From the magnetic point of view the $\{Mo_{75}V_{20}\}$ cluster is interesting because of the ring of ten V^{IV} (d^1) triangles sharing two vertices with two neighboring triangles, as shown in Figure 5.

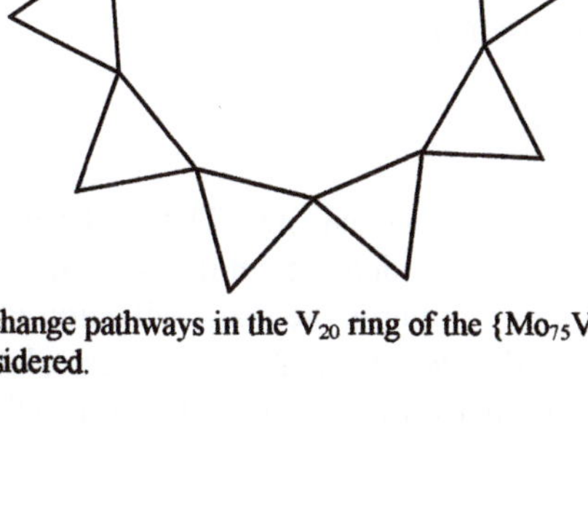

Figure 5. Topology of the exchange pathways in the V_{20} ring of the $\{Mo_{75}V_{20}\}$ cluster. Only nearest neighbor interactions are considered.

In the case of antiferromagnetic coupling this spin topology gives rise to strong frustration effects [19]. Spin frustration is currently of large theoretical interest, because it gives rise to largely degenerate ground states, which in turn determine unusual magnetic properties. For instance a spin topology actively investigated is that of the so called kagomé lattice [20], which is sketched in Figure 6.

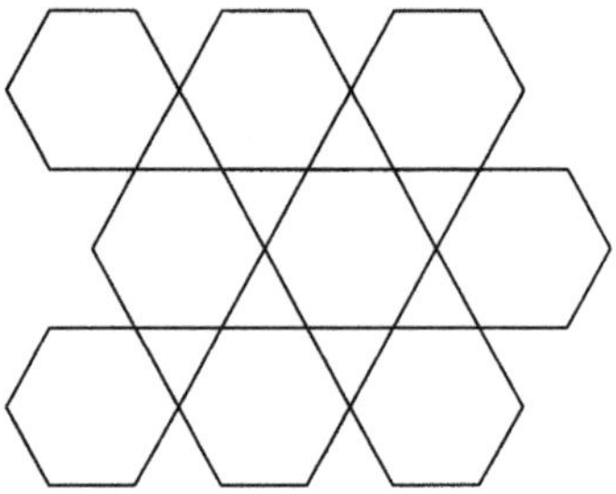

Figure 6. Scheme of a kagomé lattice.

In fact degenerate states are always very sensitive to small perturbations, and a large variability of properties can be expected. A typical and clarifying example is the extensive physics of Jahn-Teller systems and of mixed valence compounds.
In molecular magnetism, rings have been largely investigated in order to extrapolate the thermodynamic properties of infinite chains, and/or of larger rings. In this sense the V_{20} ring in $\{Mo_{75}V_{20}\}$ is a model for a chain as depicted in Figure 7, also indicating possible next-nearest neighbor interactions to be discussed below.

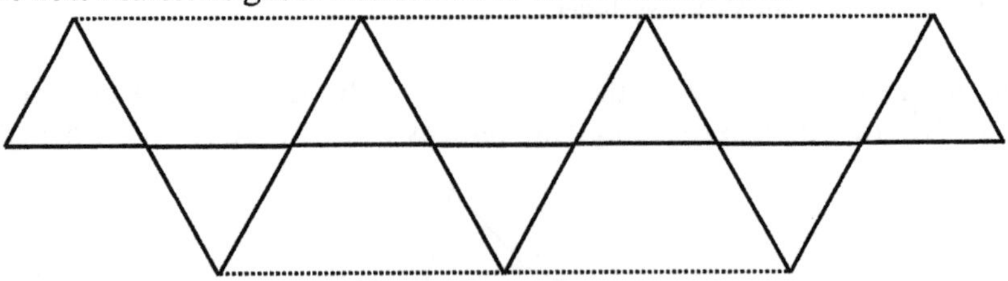

Figure 7. Topology of the exchange interactions in a chain equivalent to the V_{20} ring.

The most straightforward approach to calculate special thermodynamic properties of $\{Mo_{75}V_{20}\}$, and in particular of the magnetic susceptibility, is that of calculating the energy levels of the spin hamiltonian

$$H = \Sigma_{i \neq j} J_{ij}\, \mathbf{S_i} \cdot \mathbf{S_j} \quad (1)$$

where the sum is over all the spins of the ring. The total number of states is however very high, $(2S + 1)^{20} = 1\ 048\ 576$, which – even using symmetry arguments in order to reduce the size of the matrices – requires the diagonalization of matrices still too big to be handled. Therefore we decided to make some calculations on smaller rings, extrapolating the magnetic properties of the relevant ten membered ring. The basis for this is the rapid convergence of the calculated magnetic susceptibility in antiferromagnetic rings independent of the number of spins in the ring.

The procedure therefore implies calculating χ for rings of increasing size. If the values are considered per vanadium ion it is seen that at high temperatures the calculated values rapidly converge and become essentially identical to each other. Therefore it may be reasonably assumed that the values remain approximately constant for the larger rings. Naturally the treatment can be used down to the lowest temperature where the calculated values for different rings converge.

Sample calculations were performed with rings of 3, 4, 5, 6, 7, 8, and 9 triangles. For the smaller rings a large range of coupling constants was investigated, while for the larger rings only a limited set of values was explored. The results are best described by referring to a ring of six triangles containing 12 spins, as sketched in Figure 8.

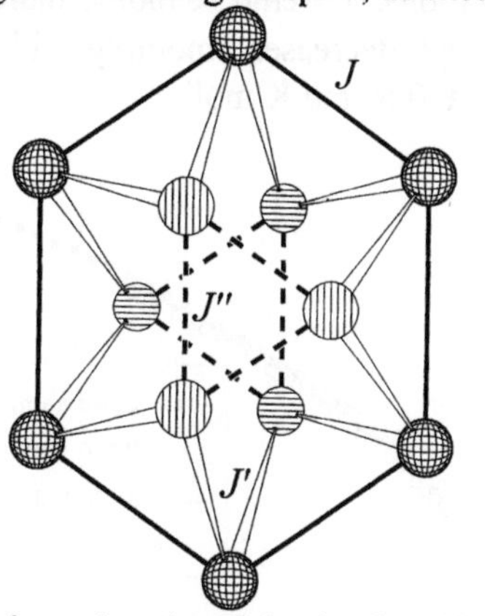

Figure 8. Sketch of the exchange pathways in a ring of six triangles with three vertices pointing above and three below the main plane containing the six spins, showing the interaction pathways J, J', and J''.

The results of sample calculations with $J = 200\ cm^{-1}$ and J' variable from 0 to 150 cm^{-1} are shown in Figure 9.

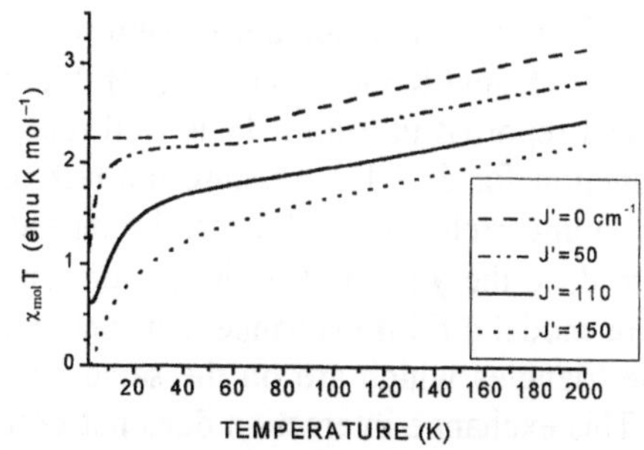

Figure 9. Calculated χT values for a cluster comprising a ring of six triangles (12 spins) and a different ratio of J'/J. For J'/J = ca. 0.55 the lowest singlet and triplet states are almost degenerate and χT tends to the value expected for two non-correlated spins 1/2. The low temperature value of six non-correlated spins1/2 is instead observed for $J' = 0$.

If $J' = 0$, the limit value of χT at low temperature is 2.25 emu K mol^{-1}, in agreement with six uncoupled electrons. This value is easily justified considering that if $J' = 0$ the centers on the apexes of the triangles are uncorrelated to the other spins. The ring of triangles therefore behaves like a six membered ring, plus six uncorrelated spins. If the J' constant is allowed to differ from zero, χT goes to zero at increasingly higher temperatures corresponding to the fact that the ground state becomes a singlet, well separated from excited states except for a narrow range around $J' = 110$ cm^{-1}, where the singlet and triplet states are almost degenerate as a consequence of spin frustration and χT tends to the value expected for two uncorrelated spins $S = 1/2$.
The temperature dependence of χT for $\{Mo_{75}V_{20}\}$ is shown in Figure 10. At room temperature the value of 4.8 emu K mol^{-1} is much smaller than expected for 20 uncoupled oxovanadium(IV) ions, 7.5 emu K mol^{-1}, indicating that there is a sizeable antiferromagnetic coupling. χT decreases smoothly with decreasing temperature and shows a hint of a plateau at ca. 0.8 emu K mol^{-1}.

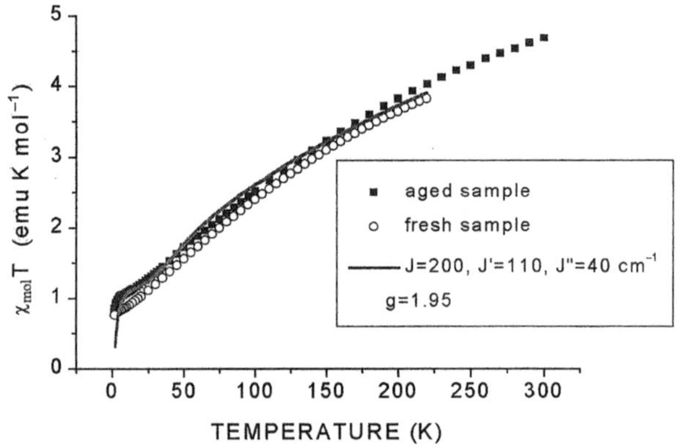

Figure 10. Temperature dependence of χ_{mol}T of fresh and aged samples of the $\{Mo_{75}V_{20}\}$ compound. The solid line corresponds to the calculated value for a ring of six triangles and rescaled to 20 spins (see text).

The experimental values of χT at low temperature seem to level at the value of two uncorrelated $S = 1/2$ (χT = 0.75 emu K mol^{-1} for $g = 2.0$). Moreover the magnetization curve recorded at 2.4 K and reported in Figure 11 is well reproduced by 2.1 times the value of the Brillouin function for $S = 1/2$. Despite the fact that the behavior at low temperature, typical of two uncorrelated $S = 1/2$, can be reproduced when $J'/J = 0.55$, such a model cannot reproduce the χT variation in a wide range of temperatures. We therefore introduced in our model a third exchange constant, connecting two vanadium ions of the apexes of the triangles which are on the same hemisphere around the six membered ring (Fig. 8). This exchange interaction does not destroy the spin frustration as the apexes of the triangles form rings of odd numbers of spins. It is interesting to notice that this particular behavior is different for rings of 4n triangles. In fact for these the polygons connected through J'' have an even number of sites, and the ground state within the polygons is $S = 0$, and therefore no plateaus are observed in this case. This alternating behavior is typical of spin frustration.

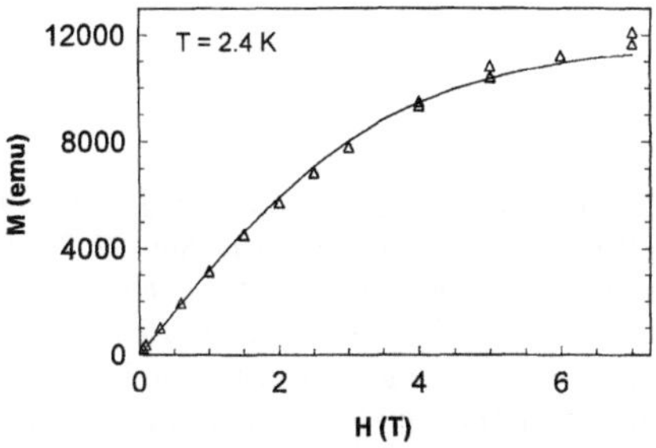

Figure 11. Field dependence of the molar magnetization (per cluster) at 2.4 K. The line corresponds to the calculated value for 2.1 uncorrelated spins $S = 1/2$ using the Brillouin function.

The introduction of a next-nearest neighbor exchange interaction is justified by previous observations that molybdate bridges are extremely efficient in transmitting the magnetic interaction between oxovanadium(IV) ions, even when they are separated by long distances [5].

The fit of the experimental data was perfomed by using the calculated curves for rings of six triangles and rescaling the χT values from 12 to 20 vanadium centers. The results are shown in Figure 10. The fit can be considered as satisfactory. It must be stressed here that the low temperature value in the case of frustration effects is expected to be 2 times the χT value for $S = 1/2$ for the model cluster of 12 spins and the real cluster of 20 spins. The rescaling procedure of the calculated χT values, which is therefore no longer valid for the low temperature limit, overestimates the low temperature χT as observed in Figure 10. The largest value for a coupling constant is needed for J, 200 cm^{-1}, and it correlates well with that previously reported for a similar compound [5]. Also the value of J', 110 cm^{-1}, compares well with that previously reported, while J'', 40 cm^{-1}, is significantly larger than those previously observed but its value should be considered only as indicative, as possibly affected by the rescaling error mentioned above.

Minor differences in the low temperature behavior due to the aging of the samples can be justified with minor variations of the J parameters.

3. Conclusion

These results confirm that large molybdenum oxide-based clusters may embed magnetic fragments which are particularly exciting, because they can present novel spin topologies and magnetic behaviors. Further it is confirmed that the molybdate groups are far from being inert from the magnetic point of view and they may cause strong antiferromagnetic coupling between metal ions separated by more than 0.6 nm, of course depending on the types of magnetic centers. For example, the coupling is much stronger for $V^{IV}O^{2+}$- than for Fe^{III}-containing systems.

References

1. D. Gatteschi, L. Pardi, A. L. Barra, and A. Müller: *Mol. Eng.* **3**, 157 (1993).
2. A. Müller, F. Peters, M. T. Pope, and D. Gatteschi: *Chem. Rev.* **98**, 239 (1998).
3. E. Coronado and C. J. Gomez-Garcia: *Chem. Rev.* **98**, 273 (1998).
4. D. Gatteschi, A. Caneschi, L. Pardi, and R. Sessoli: *Science* **265**, 1054 (1994).
5. A. Müller, W. Plass, E. Krickemeyer, R. Sessoli, D. Gatteschi, J. Meyer, H. Bögge, M. Kröckel, and A. X. Trautwein: *Inorg. Chim. Acta* **271**, 9 (1998); D. Gatteschi, R. Sessoli, W. Plass, A. Müller, E. Krickemeyer, J. Meyer, D. Sölter, and P. Adler: *Inorg. Chem.* **35**, 1926 (1996).
6. R. Sessoli, D. Gatteschi, A. Caneschi, and M. A. Novak: *Nature (London)* **365**, 141 (1993).
7. J. R. Friedman, M. P. Sarachik, J. Tejada, and R. Ziolo: *Phys. Rev. Lett.* **76**, 3830 (1996).
8. L. Thomas, F. Lionti, R. Ballou, D. Gatteschi, R. Sessoli, and B. Barbara: *Nature (London)* **383**, 145 (1996).
9. W. Wernsdorfer and R. Sessoli: *Science* **284**, 133 (1999).
10. G. Aromi, S. M. J. Aubin, M. A. Bolcar, G. Christou, H. J. Eppley, K. Folting, D. N. Hendrickson, J. C. Huffman, R. C. Squire, H. L. Tsai, S. Wang, and M. W. Wemple: *Polyhedron* **17**, 3005 (1998).
11. D. Gatteschi, L. Pardi, A. L. Barra, A. Müller, and J. Döring: *Nature (London)* **354**, 463 (1991); A. L. Barra, D. Gatteschi, L. Pardi, A. Müller, and J. Döring: *J. Am. Chem. Soc.* **114**, 8509 (1992).
12. A. Müller, P. Kögerler, and C. Kuhlmann: *Chem. Commun.* 1347 (1999).
13. A. Müller, E. Krickemeyer, H. Bögge, M. Schmidtmann, and F. Peters: *Angew. Chem., Int. Ed.* **37**, 3360 (1998).
14. A. Müller, S. Sarkar, Q. S. N. Shah, H. Bögge, M. Schmidtmann, Sh. Sarkar, P. Kögerler, B. Hauptfleisch, A. Trautwein, and V. Schünemann: *Angew. Chem., Int. Ed.* **38**, 3238 (1999).
15. A. Müller, M. Luban, C. Schröder, P. Kögerler, M. Axenovich, S. Budko, and P. C. Canfield: in preparation.
16. A. Müller, E. Krickemeyer, S. K. Das, P. Kögerler, S. Sakar, H. Bögge, M. Schmidtmann, and Sh. Sarkar: *Angew. Chem., Int. Ed.* **39**, 1612 (2000).
17. K. S. Murray: *Coord. Chem. Rev.* **12**, 1 (1974); D. M. Kurtz: *Chem. Rev.* **90**, 585 (1990); H. Weihe and H. U. Güdel: *J. Am. Chem. Soc.* **119**, 6539 (1997).
18. A. Müller, M. Koop, H. Bögge, M. Schmidtmann, F. Peters, and P. Kögerler: *Chem. Commun.* 1885 (1999).
19. J. Vannimenous and G. Toulouse: *J. Phys. C* **10**, 537 (1977).
20. K. Awaga and N. Wada in: *Magnetism: a Supramolecular Function*; ed. O. Kahn, Kluwer, Dordrecht (1996), Nato ASI Series C Vol. 484.

Heteropolyanions: Molecular Building Blocks for Ultrathin Oxide Films

JASON D. POWELL, ANDREW A. GEWIRTH,* AND WALTER G. KLEMPERER*
Frederick Seitz Materials Research Laboratory and Department of Chemistry, University of Illinois at Urbana-Champaign, Urbana, Illinois 61801

Abstract. α-Phosphovanadotungstate monolayers are formed on Ag(111) by reaction of the surface with α-$PVW_{11}O_{40}^{4-}$ in acidic aqueous solution. The ordered domains formed on smooth Ag(111) terraces are extremely large, on the average over 1,000 nm^2 in area. Comparison of scanning tunneling microscopy (STM) images of the Ag(111) surface during immersion in solutions containing α-$PVW_{11}O_{40}^{4-}$ and α-$PVW_{11}O_{40}^{5-}$ show that the former, oxidized form is required for the formation of ordered domains. A combination of UV-visible and ICP-atomic absorption spectroscopy confirmed that α-$PVW_{11}O_{40}^{4-}$ is an effective etchant for the Ag(111) surface.

Key Words: Polyoxoanion, heteropolyanions, polyoxotungstate, Keggin ion, self-assembled monolayer, large domain size, long-range order, etching, corrosion, scanning tunneling microscopy (STM), and scanning probe microscopy.

1. Introduction

Self-assembled monolayers (SAMs), ordered molecular monolayers formed by reaction of a surface with a solution of the appropriate molecular precursor, provide an attractive opportunity for studying structurally well-defined, functional surfaces[1]. Highly ordered, adherent monolayers of thiolates have been prepared in this fashion on gold surfaces[1], and analogous inorganic monolayers of heteropolyanions have been prepared in the same fashion on silver and gold surfaces[2-4]. In both systems, however, ordering is restricted to domains on the order of 200 nm^2 in area[2-5]. We have therefore examined a variety of different heteropolyanion systems in an empirical attempt to obtain significantly longer-range order. We report here successful reaction of the α-phosphovanadotungstate α-$PVW_{11}O_{40}^{4-}$ ion with Ag(111), which reproducibly results in the formation of ordered monolayers with domain sizes averaging above 1,000 nm^2.

2. Experimental Section

Potassium salts of α-$PVW_{11}O_{40}^{4-}$[6] and α-$PVW_{11}O_{40}^{5-}$[7] were prepared following literature procedures and were characterized by IR spectroscopy[6,8], elemental analysis, and cyclic voltammetry[7,9,10]. Silver substrates were prepared by evaporating a 210-nm film onto mica in a vacuum bell jar system (Norton Company). Scanning tunneling

M.T. Pope and A. Müller (eds.), Polyoxometalate Chemistry, 329–334.

microscopy (STM) images were measured in constant current mode (E_{tip} = 200 mV, i_{tip} = 2.0 nA) using a NanoScope IIIe electrochemical STM (Digital Instruments, Santa Barbara, CA). All images were measured in situ on Ag(111) immersed in aqueous solutions 0.5 mM in heteropolyanion and 0.1 M in sulfuric acid (J. T. Baker Ultrex II ultra-pure). Each image was "flattened" by subtracting an average plane from the raw data. Bright features of the image represent higher features on the surface; the grayscale was adjusted to clarify topographic features. UV-visible absorption spectra were obtained using a Hewlett-Packard HP 8452A diode-array spectrophotometer. Silver ion concentrations were measured using ICP-atomic absorption data obtained by the Microanalysis Laboratory at the University of Illinois at Urbana-Champaign.

3. Results

In situ STM images of a Ag(111) surface exposed to 0.5 mM α-$K_4PVW_{11}O_{40}$ solution in 0.1 M H_2SO_4 are shown in Figs 1 and 2. After exposure to solution for approximately 15 minutes, STM images displayed bright, textured regions and dark, relatively featureless regions as shown in Fig. 1, where the height of the bright regions above the dark regions is 0.8 ± 0.2 nm. This distance is consistent with α-$PVW_{11}O_{40}^{4-/5-}$ monolayers on Ag(111), since the heteropolyanion has a diameter of ca. 1 nm, but inconsistent with Ag(111) terraces, since Ag(111) steps are only 0.24 nm high. The area of the textured region increased with time until the entire surface was covered with textured domains after approximately 30 minutes. A region in the center of such a domain is shown in Fig. 2. Actual domain sizes were larger than the 900 nm^2 area shown, but the detailed pattern is less visible in larger images due to limited instrumental resolution. These large textured domains remained visible on the Ag(111) surface during several hours of exposure to the

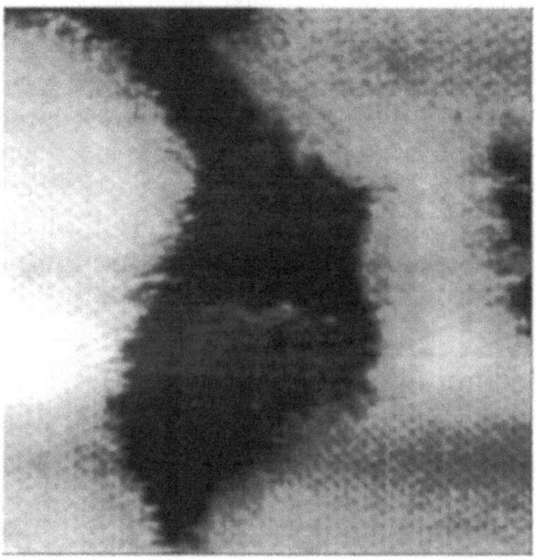

Fig. 1. 60 nm × 60 nm in situ STM image of Ag(111) surface immersed in α-$PVW_{11}O_{40}^{4-}$ solution for 15 minutes.

Fig. 2. 30 nm × 30 nm in situ STM image of a Ag(111) surface after exposure to α-$PVW_{11}O_{40}^{4-}$ solution for 30 minutes.

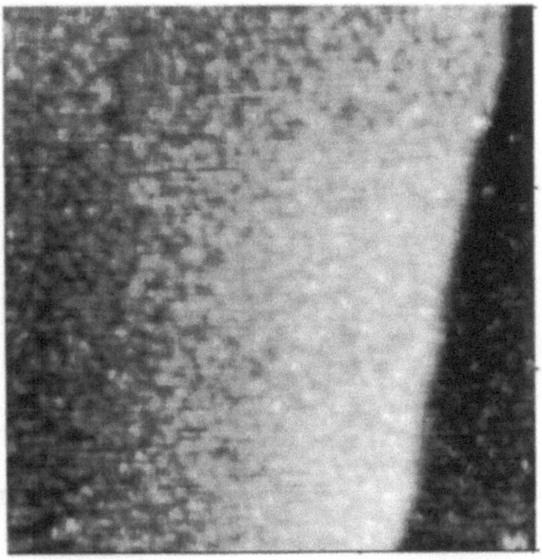

Fig. 3. 60 nm × 60 nm in situ STM image of a Ag(111) surface after exposure to α-$PVW_{11}O_{40}^{5-}$ solution for 30 minutes.

α-$PVW_{11}O_{40}^{4-}$ solution. Large Ag(111) terraces appeared to be a prerequisite for the formation of large domains in that Ag substrates with small terraces led to smaller textured domains. As shown in Figs 1 and 2, the textured pattern is hexagonal and periodic with a 1.34 ± 0.06 nm repeat distance. The angle between the hexagonal unit cell vectors and those of the Ag(111) surface, measured before introduction of the α-$PVW_{11}O_{40}^{4-}$, was 14 ± 5°.

In an attempt to determine the effect of counter-cation on the observed surface structure, experiments similar to those just described for α-$K_4PVW_{11}O_{40}$ were performed using solutions of the heteropoly acid α-$H_4PVW_{11}O_{40}$ and the guanidinium salt α-$(C(NH_2)_3)_4PVW_{11}O_{40}$. These experiments yielded STM images indistinguishable from those using the potassium salt.

An in situ STM image of a Ag(111) surface in an aqueous solution 0.5 mM in α-$K_5PVW_{11}O_{40}$ and 0.1 M in H_2SO_4 is shown in Fig. 3. STM images displayed the irregularly textured pattern shown in Fig. 3 after less than 10 minutes of exposure to α-$PVW_{11}O_{40}^{5-}$ solution. Unlike the textured pattern shown in Figs 1 and 2, this pattern is not periodic, and nearest-neighbor separations between the bright features observed ranged between 1.25 and 1.83 nm. The height of the step edge visible in the right-hand portion of the image is 0.3 ± 0.1 nm, which is consistent with the 0.24 nm height of a Ag(111) step edge.

A progressive color change from yellow through brown to violet was observed for α-$PVW_{11}O_{40}^{4-}$ reaction solutions after exposure to the Ag surface. This color change was attributed to conversion of α-$PVW_{11}O_{40}^{4-}$ to α-$PVW_{11}O_{40}^{5-}$ by studying the reaction of α-$PVW_{11}O_{40}^{4-}$ with a Ag wire substrate using UV-visible spectroscopy. Heteropolyanion

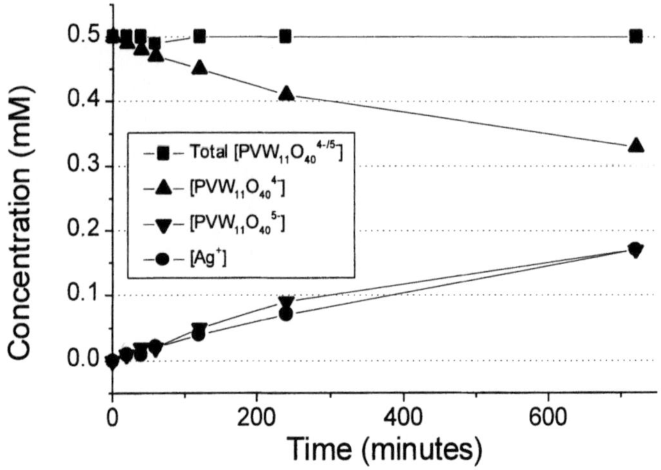

Fig. 4. Plot of solution concentrations as a function of time for the reaction of Ag wire with a 0.5 mM α-$K_4PVW_{11}O_{40}$ in 0.1 M aqueous H_2SO_4 solution.

concentrations were determined by monitoring the absorption at 410 nm (ε=1500 $M^{-1}cm^{-1}$[7]) and at 520 nm (ε=750 $M^{-1}cm^{-1}$[8]) for α-$PVW_{11}O_{40}^{4-}$ and α-$PVW_{11}O_{40}^{5-}$, respectively. The concentration of Ag^+ in solution was also monitored as a function of time using ICP-atomic absorption spectroscopy. These results are shown in Fig. 4, where α-$PVW_{11}O_{40}^{4-}$ concentrations are indicated by triangles pointing upwards, α-$PVW_{11}O_{40}^{5-}$ concentrations by triangles pointing downwards, and Ag^+ concentrations by circles. The sum of α-$PVW_{11}O_{40}^{4-}$ and α-$PVW_{11}O_{40}^{5-}$ concentrations is indicated by squares.

4. Discussion

According to the STM results described above, acidic aqueous solutions of the α-phosphovanadotungstate α-$PVW_{11}O_{40}^{4-}$ ion react with Ag(111) surfaces to form ordered hexagonal monolayers with a translational repeat distance of 1.34 ± 0.06 nm. This distance matches the √21 (1.324 nm) interatomic separations on the Ag(111) surface, suggesting a commensurate overlayer and the (√21×√21)R10.9° superlattice is shown in Fig. 5. The Ag(111) hexagonal superlattice with the next longest spacing[11], 5×5, has a spacing of 1.445 nm while that with the next shortest spacing[11], (√19×√19)R23.41°, has a spacing of 1.260 nm. The (√21×√21)R10.9° superlattice implies a 10.9° angle between the superlattice unit cell vectors and the Ag(111) unit cell vectors, consistent with the 14 ± 5° observed angle. This angle is inconsistent with both the (√19×√19)R23.41° superlattice (23.41°) and the 5×5 superlattice (0°).

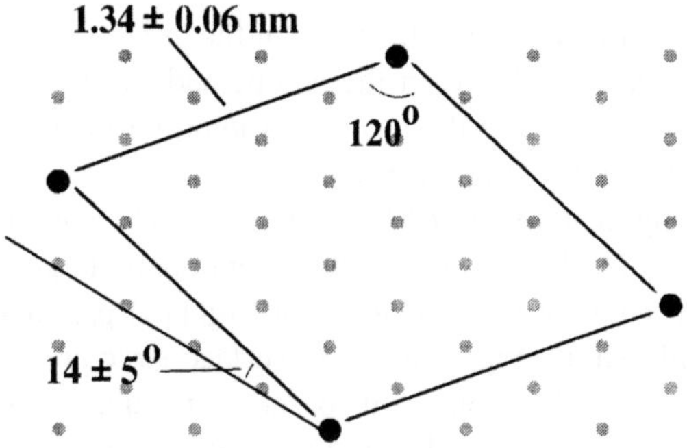

Fig. 5. Superlattice constructed using dimensions measured for the monolayer formed on Ag(111) upon exposure to α-$PVW_{11}O_{40}^{4-}$ solution. The gray spots show the spacing of the Ag(111) lattice relative to the overlayer superlattice.

A hexagonal molecular model for the proposed (√21×√21)R10.9° overlayer is proposed in Fig. 6a, where α-$PVW_{11}O_{40}$ anions are oriented such that a C_3 axis of the idealized T_d $M_{12}O_{40}$ (M = V, W) cage is perpendicular to the Ag(111) surface. In this model, the heteropolyanions are rotated about this C_3 axis to the position that maximizes interionic oxygen-oxygen separations. When the anions are oriented as in Fig. 6a, three terminal oxygen atoms in each anion are in contact with the Ag(111) surface. The spacing between these three atoms (0.508 nm) closely matches the √3 (0.500 nm)

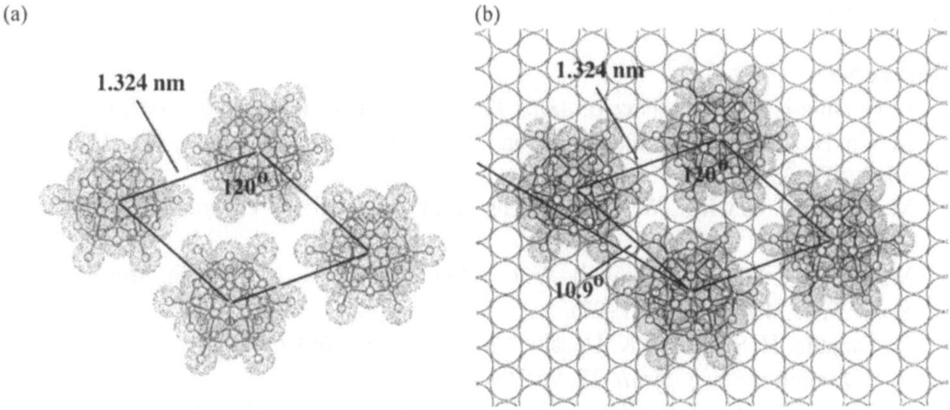

Fig. 6. (a) Proposed model for the monolayer formed by α-$PVW_{11}O_{40}^{4-}$ on Ag(111) in 0.1 M H_2SO_4. Shading around the molecules represents the van der Waals surface of the individual atoms in the molecule (r_O = 1.4 Å). (b) Proposed model showing one possible mode of interaction between anions and the Ag(111) surface, through terminal oxygen atoms binding in threefold hollow sites on the surface. Silver atoms are large open circles.

distance on the Ag(111) surface. This suggests that the ($\sqrt{21}\times\sqrt{21}$)R10.9° unit cell might arise from a specific interaction between the anions and the Ag(111) surface. Fig. 6b shows one possible structure, where the heteropolyanion terminal oxygen atoms occupy threefold hollow sites on the Ag(111) surface.

The most striking feature of α-$PVW_{11}O_{40}$-Ag(111) monolayers prepared using acidic aqueous solutions of α-$PVW_{11}O_{40}^{4-}$ is the exceptionally large size of their domains (see Fig. 2). This is in contrast to monolayers on Ag(111) prepared from α-$SiW_{12}O_{40}^{4-}$ [2-4], which display only short-range order, and surface structures on Ag(111) prepared from α-$PVW_{11}O_{40}^{5-}$ (see Fig. 3), which fail to show any translational symmetry. The α-$PVW_{11}O_{40}^{4-}$ anion differs from the α-$PVW_{11}O_{40}^{5-}$ and α-$SiW_{12}O_{40}^{4-}$ anions in that it is a sufficiently strong oxidizing agent to oxidize the Ag substrate, and data shown in Fig. 4 clearly demonstrate that α-$PVW_{11}O_{40}^{4-}$ reacts stoichiometrically according to Eq. 1 with

$$\alpha\text{-}PVW_{11}O_{40}{}^{4-}{}_{(aq)} + Ag_{(s)} \rightarrow \alpha\text{-}PVW_{11}O_{40}{}^{5-}{}_{(aq)} + Ag^{+}{}_{(aq)} \qquad (1)$$

Ag to form α-$PVW_{11}O_{40}^{5-}$ and Ag^+, as expected on the basis of the reduction potentials involved (Ag^+/Ag = +0.56 vs. SCE, and α-$PVW_{11}O_{40}^{4-}/\alpha$-$PVW_{11}O_{40}^{5-}$ = +0.60 vs. SCE)[8]. We are currently investigating the possibility that this redox activity might be responsible for the formation of unusually large domains formed when Ag(111) is immersed in acidic aqueous α-$PVW_{11}O_{40}^{4-}$ solution.

5. Acknowledgements

Support from Department of Energy Grant DE-FG02-91ER45349 through the Frederick Seitz Materials Research Laboratory at the University of Illinois is gratefully acknowledged.

References

1. L. H. Dubois and R. G. Nuzzo: *Annu. Rev. Phys. Chem.* **43**, 437-463 (1992).
2. M. Ge, B. Zhong, W. G. Klemperer and A. A. Gewirth: *J. Am. Chem. Soc.* **118**, 5812-5813 (1996).
3. M. Ge, A. A. Gewirth, W. G. Klemperer and C. G. Wall: *Pure Appl. Chem.* **69**, 2175-2178 (1997).
4. M. Ge, B. K. Niece, C. G. Wall, W. G. Klemperer and A. A. Gewirth: *Mat. Res. Soc. Symp. Proc.* **451**, 99-108 (1997).
5. R. Yamada, H. Sakai and K. Uosaki: *Chem. Lett.*, 667-668 (1999).
6. P. J. Domaille: *Inorg. Synth.* **27**, 96-104 (1990).
7. D. P. Smith, H. So, J. Bender and M. T. Pope: *Inorg. Chem.* **12**, 685-688 (1973).
8. D. P. Smith and M. T. Pope: *Inorg. Chem.* **12**, 331-336 (1973).
9. N. A. Polotebnova, G. M. Shinik and N. A. Dunaevskaya: *Russ. J. Inorg. Chem.* **18**, 417-418 (1973).
10. P. Courtin: *Rev. Chim. Min.* **8**, 75-85 (1971).
11. D. G. Frank, in *The Handbook of Surface Imaging and Visualization*; edited by A. T. Hubbard; CRC Press, Inc., Boca Raton (1995) 289-354.

Selective Oxidation of Hydrocarbons with Hydrogen Peroxide Catalyzed by Iron-substituted Silicotungstates

NORITAKA MIZUNO
Department of Applied Chemistry, Graduate School of Engineering, The University of Tokyo, Hongo, Bunkyo-ku, Tokyo 113-8656, Japan

(Received: 4 October 1999)

Abstract. The Keggin-type *di*-iron-substituted silicotungstate, γ-$SiW_{10}\{Fe(OH_2)\}_2O_{38}^{6-}$, can catalyze the selective oxidation of various alkanes including methane with hydrogen peroxide. The tetrabutylammonium salt catalytically oxidized cyclohexane, *n*-hexane, *n*-pentane, and adamantane in acetonitrile. Even lower alkanes such as methane, ethane, propane, and *n*-butane were catalytically oxidized. It is remarkable that the efficiency of hydrogen peroxide utilization to oxygenated products reached up to ca. 100% for the oxidation of cyclohexane and adamantane. The efficiency and activity for the utilization of hydrogen peroxide greatly depended on the iron centers and *di*-iron-substituted γ-$SiW_{10}\{Fe(OH_2)\}_2O_{38}^{6-}$ showed the highest efficiency of hydrogen peroxide utilization and conversion. Such a structure dependency of the catalysis is significant and the remarkable catalytic performance of *di*-iron-substituted polyoxometalate may be related to the catalysis by methane monooxygenase. It was also demonstrated that the water-soluble potassium salt catalytically oxidized the lower alkanes with hydrogen peroxide in water.
Key words: Iron-substituted silicotungstates, oxidation, alkanes and alkenes, hydrogen peroxide

1. Introduction

Catalytic oxidation is used for the conversion of petroleum-derived hydrocarbons to commodities as well as in the manufacture of fine chemicals. In the bulk chemical industry, classical processes that are environmentally unacceptable have been largely supplanted by cleaner, catalytic alternatives. Yet, stoichiometric (non-catalytic) oxidation is still widely used, and large amounts of byproducts (particularly salts) are formed in the fine chemicals industry. These oxidation processes require new catalytic, low-salt technologies [1-3].

Among hydrocarbons, the oxidation of alkanes and alkenes has attracted much attention. With respect to the oxidants, fine chemicals production allows the choice of various oxygen donors such as peroxides. Among them, hydrogen peroxide is a preferable oxidant because of the simplicity of handling, the environmentally friendly nature of coproduct (water), the high oxygen atom efficiency, and the versatility [1-3].

The *di*nuclear iron-oxo centers of hemerythrin [4], ribonucleotide

M.T. Pope and A. Müller (eds.), Polyoxometalate Chemistry, 335–345.

reductase [5], and the purple acid phosphatases [6] catalyze not only the oxidation of methane to methanol but also the oxygen atom transfer from O_2 into alkanes, alkenes, ethers, and so on [7,8]. While a number of structural models with *di*-iron center have been reported, functional models with *di*-iron center for oxidations are fewer in numbers [9-14] mainly because of the instability of organic ligands towards oxygen donors.

Various elements can be introduced into polyoxometalates and the countercations and the numbers and structures can also be controlled. The additional attractive aspects of polyoxometalates in catalysis are their inherent stability towards oxygen donors such as molecular oxygen and hydrogen peroxide. For example, high stability towards hydrogen peroxide has been reported for manganese- or iron-substituted polyoxometalates [15-22]. Therefore, polyoxometalates are useful catalysts for liquid-phase oxidations of various organic substrates with hydrogen peroxide.

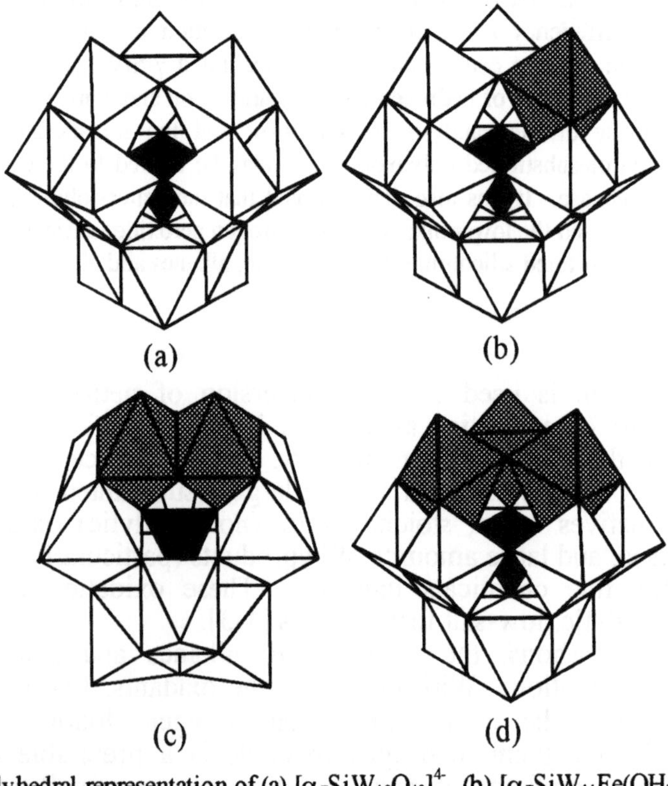

Fig. 1. Polyhedral representation of (a) $[\alpha\text{-SiW}_{12}O_{40}]^{4-}$, (b) $[\alpha\text{-SiW}_{11}Fe(OH_2)O_{39}]^{5-}$, (c) $[\gamma\text{-SiW}_{10}\{Fe(OH_2)\}_2O_{38}]^{6-}$, and (d) $[\alpha\text{-SiW}_9\{Fe(OH_2)\}_3O_{37}]^{7-}$ Keggin-type polyoxometalates. Iron atoms are represented by hatched octahedra. WO_6 octahedra occupy the white octahedra and an SiO_4 group is shown as the internal black tetrahedron.

There are several examples of iron-containing polyoxometalates, *mono* and *tri*iron-substituted Keggin-type silicotungstates [19,20], Fe_2Ni-substituted Keggin-type phosphotungstate [21], and *tetra*iron-containing Dawson- and Keggin-derived sandwich compounds [16,17]. However, *di*iron-substituted silicotungstate, $\gamma\text{-}SiW_{10}\{Fe(OH_2)\}_2O_{38}^{6-}$, has never been reported. Recently, we have reported the synthesis of $\gamma\text{-}SiW_{10}\{Fe(OH_2)\}_2O_{38}^{6-}$ (**I**) [22,23].

Recently, we have also reported (1) that the *tetra-n*-butylammonium salt of *di*iron-substituted silicotungstate (TBA-**I**) can catalyze selective oxidation of alkanes and alkenes with highly efficient utilization of hydrogen peroxide [23-28], (2) that even methane is catalytically oxidized by **I** with hydrogen peroxide both in acetonitrile and water [26-28], and (3) that structures of iron centers (Fig. 1) remarkably influence the catalytic activities [24], TBA-**I** being specifically the most active among *non*, *mono*, *di*, and *tri*iron-substituted silicotungstates [23-28]. Here, the work on the oxidation catalysis by iron-substituted polyoxometalates is reviewed.

2. Oxidation of alkanes in organic solvents [23,25,27,28]

The oxidation of various alkanes with hydrogen peroxide was carried out in the presence of TBA-**I** in acetonitrile at 305 K. The results are shown in Table 1. The main products were the corresponding alcohols, ketones, aldehydes, and acids. The turnover numbers (estimated by moles of oxidizing equivalent in all products per mole of catalyst) for the oxidation of methane, ethane, propane, *n*-butane, *n*-pentane, *n*-hexane, cyclohexane, and adamantane were 25, 64, 42, 36, 19, 33, 53, and 57, respectively. This shows that the reactions are catalytic. To our knowledge, such high turnover number of 25 has never been reported for oxidation of methane with hydrogen peroxide by *di*-iron containing model complexes having organic ligands.

It is remarkable that the efficiencies of hydrogen peroxide utilization to oxidized products for the oxidation of cyclohexane and adamantane were almost 100%. Such high efficiency in the oxidation of cyclohexane has never been reported: For example, the efficiency was higher than those in the cyclohexane oxidations with hydrogen peroxide on the catalysts (efficiency in the parentheses), $Fe(PA)_2$ (PA, picolinic acid; 85%) [11], $FeCl_3$/py/Ph_2S/picolinic acid system (79%) [29], Mn(TDCPP)Cl (H_2TDCPP, 5,10,15,20-tetrakis-2',6'-dichlorophenyl porphyrin; 52%) [30], TS-1 (32%) [31], $[PW_9O_{37}\{Fe_{3-x}Ni_x(OAc)_3\}]^{(9+x)-}$ (x = predominantly 1) (14%) [21], $[Fe_2O(bipy)_4(OH_2)_2][ClO_4]_4$ (8%) [32], or $Fe_3O(OAc)_6(H_2O)_3$ (4%) [33].

The dependency of the efficiency on reaction temperatures and concentrations of hydrogen peroxide was investigated. The efficiency was slightly decreased to 95% by increasing reaction temperature from 305 K to 356 K and still high. The efficiency was kept almost 100% below the amount of H_2O_2 of 1 mmol and decreased to 48 % at the amount of H_2O_2 of 2 mmol.

TABLE 1

Oxidation of various alkanes with hydrogen peroxide catalyzed by TBA-**I** in acetonitrile at 305 K

Substrate	Conversion[a] / %	Product	Selectivity / %	Efficiency[b] / %
Methane[c]	0.4	Methanol	73	20
		Carbon dioxide	27	
Ethane[c]	1.2	Acetaldehyde	57	52
		Acetic acid	40	
		Carbon dioxide	3	
Propane[c]	4.0	Acetone	86	33
		Acetaldehyde	12	
		Carbon dioxide	2	
n-Butane[c]	11.0	Methylethylketone	82	28
		Acetaldehyde	17	
		Carbon dioxide	1	
n-Pentane[d]	9.0	Pentanols[e]	33	74
		Pentanones[e]	67	
n-Hexane[d]	15.0	Hexanols[f]	25	83
		Hexanones[f]	75	
Cyclohexane[d]	25.0[g]	Cyclohexanol	32[g]	99
		Cyclohexanone	68[g]	
Adamantane[h]	42.0	1-Adamantanol	20	99
		2-Adamantanol	71	
		2-Adamantanone	9	

Catalyst, 8 μmol; H_2O_2, 1 mmol; reaction time, 96 h; acetonitrile, 6 mL.

[a] (Mol of products/mol of substrate used) x 100.

[b] ([alcohol]+2[ketone]+2[aldehyde]+3[acid]+4[CO_2]) / $[H_2O_2]_c$ × 100 (%), where $[H_2O_2]_c$ is the concentration of H_2O_2 consumed.

[c] The pressures of methane, ethane, propane, and *n*-butane were 50, 28, 6, and 2 atm, respectively.

[d] Substrate, 1 mmol.

[e] 1-ol : 2-ol : 3-ol = 5 : 27 : 68; 1-one : 2-one : 3-one = trace: 35 : 65.

[f] 1-ol : 2-ol : 3-ol = 11 : 35 : 54 ; 1-one : 2-one : 3-one = trace : 47 : 53.

[g] Conversion to dicyclohexyl was < 1%.

[h] Solvent, acetonitrile/benzene = 6 mL / 3 mL.

The catalytic activity and selectivity of $\gamma\text{-SiW}_{10}\{Fe(OH_2)\}_2O_{38}^{6-}$ were solvent-dependent: The oxidation hardly proceeded in dimethylformamide and methanol, while the oxidation of cyclohexane proceeded in acetonitrile and 1,2-dichloroethane. The highest conversion was obtained for acetonitrile.

3. Oxidation of alkanes in water [26-28]

The oxidation of lower alkanes with hydrogen peroxide in water by water-soluble potassium salt of $\gamma\text{-SiW}_{10}\{Fe(OH_2)\}_2O_{38}^{6-}$ was carried out since the use of water as a solvent in homogeneous transition-metal catalysis has been of growing importance because of its environmentally friendly nature and catalytic oxidation of alkanes in water is of great interest on the standpoint of enzymatic catalysis. The data are shown in Table 2. In the case of oxidation of methane, hydrogen peroxide was completely consumed and the amounts of respective products little changed after 24 h. The resulting hydrogen peroxide decomposed into molecular oxygen, of which the amount was also quantitatively confirmed. The main products were methylformate (yield, 10.3 μmol)

TABLE 2

Oxidation of lower alkanes with hydrogen peroxide catalyzed by potassium salt of **I** in water at 353 K

Substrate	Conversion[a] / %	Product	Selectivity / %	Efficiency[b] / %
Methane[c,d]	0.1	Methylformate	54	5
		Methanol	1	
		Formic acid	1	
		Carbon dioxide	44	
Ethane[c,e]	0.3	Acetic acid	77	16
		Acetaldehyde	10	
		Carbon dioxide	13	
Propane[c,e]	1.1	Acetone	82	12
		Propionic acid	11	
		Carbon dioxide	7	
n-Butane[c,e]	0.6	Methylethylketone	52	6
		Carbon dioxide	48	

Catalyst, 5 μmol; reaction time, 24 h; water, 1.8 or 1.9 mL.

[a,b] See Table 1.

[c] The pressures of methane, ethane, propane, and *n*-butane were 50, 28, 6, and 2 atm, respectively.

[d] H_2O_2, 2.4 mmol.

[e] H_2O_2, 1.0 mmol.

and carbon dioxide (17.8 μmol). The total turnover number reached to 23 (8.4 to selective oxidation products) after 48 h, showing that **I** is catalytically active. Such catalytic oxidation of methane by *di*-iron containing model complex catalysts under environmentally friendly conditions (i.e., with hydrogen peroxide in water) has never been reported. The amounts of products significantly increased with the increase in the reaction temperature from 303 K to 353 K, and decreased by further increase in the reaction temperature to 403 K, where the decomposition of hydrogen peroxide to form dioxygen and water was dominant at 403 K. Therefore, reaction was carried out at 353 K. The other lower alkanes were also catalytically oxidized to the corresponding ketones, aldehydes, and acids. Solubilities of methane (50 atm) in acetonitrile and water were 510 ± 50 and 70 ± 10 mmol/L, respectively. The lower conversions in water than those in acetonitrile probably results from the lower solubility.

4. Oxidation of alkenes [25]

The oxidation of cyclooctene with hydrogen peroxide was carried out in the presence of TBA–I for 24 h at 305 K. Cyclooctene oxide is the major product and no induction period was observed for the formation. The selectivity to cyclooctene oxide was 96% after 24 h. Only small amounts of 2-cycloocten-1-ol and 2-cycloocten-1-one were observed. Neither acids nor carbon oxides were observed. After 24 h, hydrogen peroxide was completely consumed and the reaction stopped. It is remarkable that the efficiency of hydrogen peroxide utilization to products was almost 100%.

The efficiency on TBA-**I** was higher than those reported for iron complexes in oxidations of cyclohexene and cyclooctene with hydrogen peroxide; in cyclohexene oxidation, $Fe(PA)_2$ (PA = picolinic acid) (efficiency, 59 %) [11], $Fe(cyclam)(CF_3SO_3)_2$ (cyclam = 1,4,8,11-tetraazacycloteetradecane) (42 %) [34], $FeCl_3$ (11 %) [35], and $Fe_3O(OAc)_6(H_2O)_3$ (4.2 %) (33); in cyclooctene oxidation, $Fe_2O(L)_4(H_2O)_2(ClO_4)_4$ (L = (-)4,5 pinene bipyridine) (36 %) [36] and $[Fe^{III}_4(H_2O)_2(PW_9O_{34})_2]^{6-}$ (0.32 %) [17]. Gif system shows high efficiency and conversion for alkane oxidations, but inactive for alkene epoxidations [11,34,37]. To date, almost 100% efficiency in the epoxidation of cyclooctene with hydrogen peroxide has also been achieved for Mn(TDCPP)Cl/Im system (H_2TDCPP = 5,10,15,20-tetrakis-2',4',6'-terimethylphenyl)porphirin) [30], TS-1, and Ti-beta [31,38]. These catalysts show high efficiency of hydrogen peroxide utilization for alkene oxidations, but lower efficiency for alkane oxidations. In contrast with the iron, manganese, or titanium complexes as described above, γ-$SiW_{10}\{Fe(OH_2)\}_2O_{38}^{6-}$ showed high efficiency of hydrogen peroxide utilization for both oxidations of alkanes and alkenes.

Results of oxidation reactions of various alkenes catalyzed by TBA-**I** are summarized in Table 3. Not only cyclooctene but also 2-octene and cyclohexene were catalytically oxidized with ≥ 92% efficiency of hydrogen peroxide utilization. The conversion and efficiency decreased for the oxidation of 1-octene with the electron deficient double bond. The epoxidation and oxidative cleavage was observed for styrene; the oxidation of styrene with hydrogen peroxide gave styrene oxide and benzaldehyde with % selectivities of 76:24, respectively. The efficiency for oxidation of cyclooctene and cyclohexene decreased to 27 - 66% with increases in the molar cycloalkenes to hydrogen peroxide ratios from 0.2 to 0.5 - 1.5 while 12-tungstophosphoric acid combined with cetylpyridinium chloride shows ca. 60% efficiency for epoxidation of

TABLE 3

Oxidation of Alkenes Catalyzed by TBA-**I** at 305 K

Substrate	Conversion[a] / %	Product	Selectivity / %	Efficiency[b] / %
Cyclooctene	19	Cyclooctene oxide	96	99
		2-Cycloocten-1-ol	2	
		2-Cycloocten-1-one	2	
2-Octene	17	2,3-Octene oxide	89	92
		2-Octen-4-ol	6	
		2-Octen-4-one	5	
1-Octene	10	1,2-Octene oxide	85	53
		1-Octen-3-ol	8	
		1-Octen-3-one	7	
Cyclohexene	16	Cyclohexene oxide	48	99
		2-Cyclohexen-1-ol	26	
		2-Cyclohexen-1-one	26	
Styrene	9	Styrene oxide	76	40
		Benzaldehyde	24	
trans-Stilbene[c]	4	*trans*-Stilbene oxide	66	27
		Benzaldehyde	34	

Catalyst, 8 μmol; substrate, 1000 μmol; H_2O_2, 200 μmol; acetonitrile, 6 mL; reaction time, 24 - 96 h.

[a,b] See Table 1.

[c] Acetonitrile, 9 mL.

cyclooctene with the molar cyclooctene to hydrogen peroxide ratio of 1.5 [39,40].

5. Structure dependence of intrinsic catalytic activity [23,24,27]
Figure 2 shows the effect of structures of iron sites on intrinsic catalytic activities for oxidation of not only cyclohexane but also cyclohexene and *trans*-stilbene. To compare the intrinsic catalytic activities, the rates were calculated below the conversion of 1%, where the conversions linearly increased with time. The catalytic activity of TBA–I for oxidation of cyclohexane, cyclohexene, and *trans*-stilbene was 10^1 - 10^2 times higher than those of *non*-, *mono*- and *tri*-iron-substituted silicotungstates, showing that the *di*-iron site in γ-$SiW_{10}\{Fe(OH_2)\}_2O_{38}{}^{6-}$ is an effective center for oxidation of hydrocarbons. Such a structure dependency of

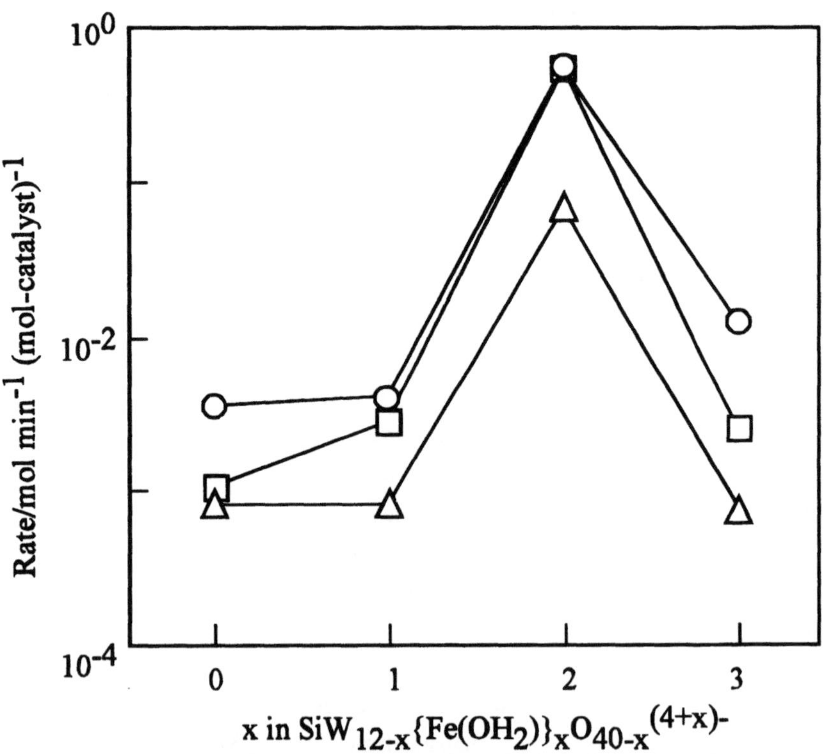

Fig. 2. Effect of Fe^{3+}-substitution for W^{6+} in $SiW_{12}O_{40}{}^{4-}$ on catalytic activity for oxidation at 305 K. □, cyclohexane; ○, cyclohexene; △, *trans*-stilbene.
Reaction conditions for cyclohexane and cyclohexene, see Table 1. Reaction conditions for *trans*-stilbene; catalyst, 1.3 - 1.6 μmol; substrate, 0.11 mmol; H_2O_2, 0.22 mmol; acetonitrile, 3 mL. Rates evaluated below 1% conversion.

the catalysis is noticeable and the catalytic performance of *di*-iron-containing polyoxometalate is possibly related to the catalysis by methane monooxygenase.

The rate of 5.4 x 10^{-1} molmin^{-1}(mol-catalyst)$^{-1}$ for the oxidation of cyclohexane is higher than those (1 x 10^{-2} - 4 x 10^{-1} molmin^{-1}(mol-catalyst)$^{-1}$ without additives) reported for various *mono*- [12,41,42], *di*- [12,42,43], and *tri*- [33,44] iron complexes and lower than that for $[Fe_2O(bpy)_4(OH_2)_2]ClO_4$, which was unfortunately much deactivated by hydrogen peroxide [32].

Acknowledgments

I gratefully acknowledge a number of students and collaborators for their experiments and discussions. This work was supported in part by a Grant-in-Aid for Scientific Research from the Ministry of Education, Science, Sports and Culture of Japan.

References

1. R. A. Sheldon: *Top. Current Chem.* **164**, 23 (1993).
2. A. Sobkowiak, H. Tung, and D. T. Sawyer, *Prog. Inorg. Chem.* **40**, 291 (1992).
3. C. L. Hill and C. M. Prosser-McCartha: *Coord. Chem. Rev.* **143**, 407 (1995).
4. D. M. Kurtz, Jr., D. F. Shriver, and I. M. Koltz: *Coord. Chem. Rev.* **24**, 145 (1977).
5. T. Joelson, U. Uhlin, H. Eklund, B.-M. Sjöberg, S. Hahne, and M. Karlsson: *J. Biol. Chem.* **259**, 9076 (1984); B.-M. Sjöberg, T. M. Loehr, and J. Sanders-Loehr: *Biochemistry* **21**, 96 (1982); L. Petersson, A. Gräslund, A. Ehrenberg, B.-M. Sjöberg, and P. Reichard: *J. Biol. Chem.* **255**, 6706 (1980); C. L. Atkin, L. Thelander, P. Reichard, and G. Lang: *J. Biol. Chem.* **248**, 7464 (1973); P. Nordlund, B.-M. Sjöberg, and H. Eklund: *Nature* **345**, 593 (1990); R. Davydov, S. Kuprin, A. Gräslund, and A. Ehrenberg: J. Am. Chem. Soc. **116**, 11120 (1994).
6. J. W. Pyrz, J. T. Sage, P. G. Debrunner, and L. Que, Jr.: *J. Biol. Chem.* **261**, 11015 (1986); B. A. Averill, J. C. Davis, S. Burman, T. Zirino, J. Sanders-Loehr, T. M. Loehr, J. T. Sage, and P. G. Debrunner: *J. Am. Chem. Soc.* **109**, 3760 (1987); B. C. Antanaitis, T. Strekas, and P. Aisen: *J. Biol. Chem.* **257**, 3766 (1982); R. B. Lauffer, B. C. Antanaitis, P. Aisen, and L. Que, Jr.: *J. Biol. Chem.* **258**, 14212 (1983); D. F. Hunt, J. R. Yates, III, J. Shabanowitz, N.-Z. Zhu, T. Zirino, B. A. Averill, S. T. Daurat-Larroque, J. G. Shewale, R. M. Roberts, and K. Brew: *Biochem. Biophys. Res. Commun.* **144**, 1154 (1987); J. B. Vincent, M. W. Crowder, and B. A. Averill: *Biochemistry* **30**, 3025 (1991).
7. J. Colby and H. Dalton: *Biochem. J.* **157**, 495 (1976); J. Colby, D. I. Stirling, and H. Dalton: *Biochem. J.* **165**, 395 (1977); J. Colby and H. Dalton: *Biochem. J.* **171**, 461 (1978); J. Green and H. Dalton: *Biochem. J.* **236**, 155 (1986); J. Green and H. Dalton: *J. Biol. Chem.* **263**, 17561 (1988); D. D. S. Smith and H. Dalton: *Eur. J. Biochem.* **182**, 667 (1989).
8. B. G. Fox and J. D. Lipscomb: *Biochem. Biophys. Res. Commum.* **154**, 165 (1988);

B. G. Fox, W. A. Froland, J. E. Dege, and J. D. Lipscomb: *J. Biol. Chem.* **264**, 10023 (1989).

9. D. H. R. Barton, J. Boivin, M. Gastiger, J. Morzycki, R. S. Hay-Motherwell, W. B. Motherwell, N. Ozbalik, and K. M. Schwartzentruber: *J. Chem. Soc., Perkin Trans. 1* 947 (1986).
10. N. Kitajima, M. Ito, H. Fukui, and Y. Moro-oka: *J. Chem. Soc., Chem. Commun.* 102 (1991) .
11. C. Sheu, S. A. Richert, P. Cofré, B. Ross, Jr., A. Sobkowiak, D. T. Sawyer, and J. R. Kanofsky: *J. Am. Chem. Soc.* **112**, 1936 (1990).
12. R. H. Fish, M. S. Konings, K. J. Oberhausen, R. H. Fong, W. M. Yu, G. Christou, J. B. Vincent, D. K. Coggin, and R. M. Buchanan: *Inorg. Chem.* **30**, 3002 (1991).
13. J. B. Vincent, J. C. Huffman, G. Christou, Q. Li, M. A. Nanny, D. N. Hendrickson, R. H. Fong, and R. H. Fish: *J. Am. Chem. Soc.* **110**, 6898 (1988).
14. R. A. Leising, J. Kim, M. A. Pérez, and L. Que, Jr.: *J. Am. Chem. Soc.* **115**, 9524 (1993).
15. M. Bösing, A. Nöh, I. Loose, and B. Krebs: *J. Am. Chem. Soc.* **120**, 7252 (1998).
16. X. Zhang, Q. Chen, D. C. Duncan, C. F. Campana, and C. L. Hill: *Inorg. Chem.* **36**, 4208 (1997).
17. X. Zhang, Q. Chen, D. C. Duncan, R. J. Lachicotte, and C. L. Hill: *Inorg. Chem.* **36**, 4381 (1997).
18. R. Neumann and M. Gara: *J. Am. Chem. Soc.* **116**, 5509 (1994).
19. F. Zonnevijlle, C. M. Tourné, and G. F. Tourné: *Inorg. Chem.* **21**, 2751 (1982).
20. J. Liu, F. Ortéga, P. Sethuraman, D. E. Katsoulis, C. E. Costello, and M. T. Pope: *J. Chem. Soc., Dalton Trans.* 1901 (1992).
21. N. Mizuno, T. Hirose, M. Tateishi, and M. Iwamoto: *J. Mol. Catal.* **88**, L125 (1994).
22. C. Nozaki, I. Kiyoto, Y. Minai, M. Misono, and N. Mizuno: *Inorg. Chem.* in press.
23. N. Mizuno, C. Nozaki, I. Kiyoto, and M. Misono: *J. Am. Chem. Soc.* **120**, 9267 (1998).
24. N. Mizuno, I. Kiyoto, C. Nozaki, and M. Misono: *J. Catal.* **181**, 171 (1999).
25. N. Mizuno, C. Nozaki, I. Kiyoto, and M. Misono: *J. Catal.* **182**, 285 (1999).
26. N. Mizuno, Y. Seki, Y. Nishiyama, I. Kiyoto, and M. Misono: *J. Catal.* **184**, 550 (1999).
27. N. Mizuno, M. Misono, Y. Nishiyama, Y. Seki, and I. Kiyoto: *Res. Chem. Interm.* in press.
28. N. Mizuno, Y. Nishiyama, I. Kiyoto, and M. Misono: *Stud. Surf. Sci. Catal.* in press.
29. D. H. R. Barton, B. Hu, D. K. Taylor, and R. V. Rojas Wahl: *Tetrahedron Lett.* **37**, 1133 (1996).
30. P. Battioni, J. P. Renaud, J. F. Bartoli, M. R.-Artiles, M. Fort, and D. Mansuy: *J. Am. Chem. Soc.* **110**, 8462 (1988).
31. M. G. Clerici: *Appl. Catal.* **68**, 249 (1991).
32. S. Ménage, J. M. Vincent, C. Lambeaux, and M. Fontecave: *J. Chem. Soc., Dalton Trans.* 2081 (1994).

33. W. Nam and J. S. Valentine: *New J. Chem.* **13**, 677 (1989).
34. W. Nam, R. Ho, and J. S. Valentine: *J. Am. Chem. Soc.* **113**, 7052 (1991).
35. H. Sugimoto and D. T. Sawyer: *J. Org. Chem.* **50**, 1784 (1985).
36. C. D.-Toia, S. Ménage, C. Lambeaux, and M. Fontecave: *Tetrahedron Lett.* **38**, 3727 (1997).
37. A. E. Shilov and G. B. Shul'pin: *Chem. Rev.* **97**, 2879 (1997).
38. A. Corma, P. Estere, A. Martínez, and S. Valencia: *J. Catal.* **152**, 18 (1995).
39. Y. Ishii, K. Yamawaki, T. Ura, H. Yamada, T. Yoshida, and M. Ogawa: *J. Org. Chem.* **53**, 3587 (1988).
40. Y. Ishii and M. Ogawa: *Heteroatom Chem. (Tokyo)* **3**, 121 (1990).
41. C. Kim, K. Chen, J. Kim, and L. Que, Jr.: *J. Am. Chem. Soc.* **119**, 5964 (1997).
42. H.-C. Tung, C. Kang, and D. T. Sawyer: *J. Am. Chem. Soc.* **114**, 3445 (1992).
43. S. Ito, T. Okuno, H. Matsushima, T. Tokii, and Y. Nishida: *J. Chem. Soc., Dalton Trans.* 4479 (1996).
44. G. Balavoine, D. H. R. Barton, J. Boivin, and A. Gref: *Tetrahedron Lett.* **31**, 659 (1990).

Aerobic Oxidations Catalyzed by Polyoxometalates

A. M. KHENKIN, R. BEN-DANIEL, A. ROSENBERGER, I. VIGDERGAUZ AND R. NEUMANN*
Casali Institute of Applied Chemistry, The Hebrew University, Jerusalem, Israel 91904

(Received: 4 October 1999)

Abstract. Two major reaction modes have been perceived for the catalytic activity of polyoxometalates in oxidation reactions. In one case, the catalytic cycle has been described by the division of the reaction into two stages. First, the substrate is oxidized by consecutive electron and proton transfer by the polyoxometalate in the oxidized form to yield the product and the reduced polyoxometalate catalyst. The reduced polyoxometalate catalyst is then reoxidized, *importantly by molecular oxygen* to form water, in the second and possibly separate stage completing the catalytic cycle. The polyoxometalates often most effective in this reaction are the phosphovanadomolybdates of the Keggin structure, $H_{3+x}PV_xMo_{12-x}O_{40}$ (x = 1 - 6, especially x = 2). Now, new investigations of the reactivity of $PV_2Mo_{10}O_{40}^{5-}$ with aldehydes and quinones enables the differentiation between the reactivity of the five inseparable isomers of α-$H_5PV_2Mo_{10}O_{40}$ using ^{31}P NMR and ESR spectroscopy, and UV-vis absorption-time profiles. The 1,11 isomer with vanadium in distal positions is the most abundant, although the isomers with vanadium in vicinal positions appeared to be the most kinetically viable. For example, alkane/aldehyde/O_2 oxidizing systems were found to be quite effective and selective for oxidation of alkanes to ketones. Further studies of $PV_2Mo_{10}O_{40}^{5-}$ - quinone interactions has shown the formation of semiquinone intermediates. The later are active in the dehydrogenation of benzylic and allylic alcohols to aldehydes and can be used as models for the reactivity of $PV_2Mo_{10}O_{40}^{5-}$ on carbon supports . The second reaction type views the oxidation catalyzed by the polyoxometalate as an interaction with a primary oxidant. This interaction yields an activated catalyst intermediate *eg* a peroxo, hydroperoxo or oxo species which can be used to oxidize the organic substrate. In this mode, one can consider reaction at a transition metal substituted position within the polyoxometalate. Here the polyoxometalate acts as an "inorganic ligand" for transition metals such as cobalt, manganese, ruthenium, etc. In mechanistic scenarios for such reactions, the catalytically active site is a tetragonally (pyrimidal) oxo coordinated transition metal while the polyoxometalate as a whole functions as a ligand with a strong capacity for accepting electrons. In this last group of oxidation reactions the actual reaction mechanism certainly varies as a function of the transition metal and oxidant, but can be conceived to take place via a general intermediate "transition metal - oxidant" species. The ruthenium substituted "sandwich" type polyoxometalate, $[WZnRu^{III}_2(ZnW_9O_{34})_2]^{11-}$, has been shown to activate molecular oxygen in a dioxygenase type mechanism, and selectively catalyze thereby a) the hydroxylation of alkanes at the tertiary carbon position and b) the stereoselective epoxidation of alkenes. For comparison, catalytic oxidation of a novel ruthenium substituted polyoxometalate, $[Ru^{II}(H_2O)W_{17}O_{55}F_6NaH_2]^{9-}$, in similar reactions appears to occur by a metal catalyzed autooxidation.

Key words: Molecular oxygen, Redox, Quinone, Electron Transfer, Dioxygenase

* To whom correspondance should be addressed
Present address: Department of Organic Chemistry, Weizmann Institute of Science, Rehovot, Israel 76100

M.T. Pope and A. Müller (eds.), Polyoxometalate Chemistry, 347–362.

Introduction

The diversity of research in polyoxometalates, as born out in this symposium, is significant and includes their application in fields dealing with synthetic strategy, structure, and applications as materials in both solution and the solid state. Relevant to this presentation are only the uses of polyoxometalates as catalysts. The use of heteropoly acids in acid catalysis and was based on the realization that they are very strong Bronsted acids. Especially, $H_3PW_{12}O_{40}$ and $Cs_{2.5}H_{0.5}PW_{12}O_{40}$ were effectively used in many typical acid catalyzed reactions and it is probable that research will expand to cases where "superacidity" may be required.[1]

Here we are concerned with the use polyoxometalates as oxidation catalysts, particularly using molecular oxygen as oxidant. Various aspects of these research efforts in oxidation catalysis have been reviewed. First, Matveev and his coworkers[2] described some of their own early research carried out during the late 1970's to the early 1980's. Tsigdinos also reviewed the earlier research.[3] In 1987 Misono described some of the earlier work in heterogeneous catalysis.[4] Then he[5] and others[6] expanded on the earlier work. Ishii[7] has compiled his work on the use of hydrogen peroxide and Kozhevnikov discussed the use of heteropolyanions in fine chemical synthesis.[8] Hill and ourselves have also reviewed oxidation catalyzed by polyoxometalates.[9] Polyoxometalate (POM) catalyzed oxidations may be categorized by the principle mode of the catalytic reaction. There are two major reaction mechanisms. In the first case, the catalytic cycle can be best described by the division of the reaction into two stages, eqn. 1. Initially, the substrate is oxidized by the polyoxometalate in the oxidized form to yield the product and the reduced polyoxometalate catalyst. The reduced polyoxometalate catalyst is then reoxidized, often by oxygen to form water, in the second stage completing the catalytic cycle.

$$\begin{aligned} &\text{Substrate} + \text{POM}_{ox} \longrightarrow \text{Product} + \text{POM}_{red} + 2\text{H}^+ \\ &\text{POM}_{red} + \text{O}_2 + 2\text{H}^+ \longrightarrow \text{POM}_{ox} + \text{H}_2\text{O} \end{aligned} \qquad (1)$$

In the liquid phase, the oxidation of the substrate is often a dehydrogenation (electron and proton transfer from the substrate to the catalyst) and regeneration of the catalyst implies electron donation of oxygen to the catalyst with co-formation of water or insertion.

The second reaction type views the oxidation catalyzed by the polyoxometalate as an interaction with a primary oxidant. This interaction yields an activated catalyst intermediate *eg* a peroxo, hydroperoxo or high valent oxo species which can be used to oxidize the organic substrate, eqn. 2.

$$\begin{aligned} &\text{Oxidant} + \text{POM} \longrightarrow [\text{POM–Ox}]_{activated} \\ &[\text{POM–Ox}]_{activated} + \text{Substrate} \longrightarrow \text{POM} + \text{Product} \end{aligned} \qquad (2)$$

The polyoxometalates function as "inorganic ligands" for transition metals such as manganese, iron or ruthenium; the catalytically active site is at the substituted transition metal center while the polyoxometalate functions as a ligand with a strong capacity for accepting electrons. In this last group of oxidation reactions the actual reaction mechanism certainly varies as a function of the transition metal and oxidant, but can be considered as taking place via a general intermediate "transition metal - oxidant" species.

The polyoxometalate reduction/substrate oxidation reaction mode; the $[PV_2Mo_{10}O_{40}]^{5-}$ catalyst and oxygen as primary oxidant. Phosphovanadomolybdates of the Keggin structure, $H_{3+x}PV_xMo_{12-x}O_{40}$ (x = 1 - 6, especially x = 2, see figure 1a below) dominate reports on polyoxometalates in liquid phase oxidation catalysis in reactions occurring according to the set of reactions as presented in eqn. 1. Original use of the polyoxometalate as a co-catalyst in the Pd(II) catalyzed oxidative hydration of alkenes (Wacker process),[10] indicated that the reoxidation with molecular oxygen takes place only in the coordination sphere of $H_5PV_2Mo_{10}O_{40}$; free VO^{2+} is not reoxidized.[11] These findings have now led to the preferred use of $H_5PMo_{10}V_2O_{40}$ in many aerobic oxidation reactions, notably the oxidative dehydrogenation of cyclic dienes such as dihydroanthracene and α-terpinene with $H_5PMo_{10}V_2O_{40}$,[12] the oxidative dehydrogenation of alcohols and amines to aldehydes and imines,[13] and the oxidation of phenols.[14] The specific mechanism of the generalized reaction scheme described in eqn. 1 is a matter of very significant interest. In non-aqueous media using the α-terpinene to *p*-cymene oxidative dehydrogenation as the model reaction[15] we have found that the reaction mechanism involves two steps. First, the fast formation of a catalyst-substrate complex based on electron transfer. After the completion of the catalyst reduction and substrate dehydrogenation, the catalyst reoxidation was by a four electron reduction of molecular oxygen to water with two equivalents of $H_5PMo_{10}V_2O_{40}$ catalyst through a μ-peroxo intermediate[15] or an outer sphere molecular oxygen reduction.[16]

The fast catalyst - substrate interaction by electron transfer can now be exploited in new ways. First, the reaction between an aldehyde and $H_5PMo_{10}V_2O_{40}$ leads to eventual formation of a acyl radical and the reduced $H_6PV^VV^{IV}Mo_{10}O_{40}$. By using ^{31}P NMR and ESR spectroscopy, and UV-vis time *versus* absorption profiles one can differentiate between the five isomers of α-$H_5PMo_{10}V_2O_{40}$. Interestingly, our analysis shows that the various isomers are not found in their "theoretical statistical" abundance but rather, distal vanadium substitution is preferred over vicinal vanadium substitution. However, the vicinal isomers appear the most active in the electron/proton transfer. The acyl radical formed can be used effectively in autooxidative formation of ketones from alkanes via acylperoxo intermediates.[17] Second, reactions of neutral (as opposed to acidic) $PMo_{10}V_2O_{40}^{5-}$ with quinones lead to formation of polyoxometalate-semiquinone complexes or intermediates. Such intermediates are active for electron transfer from benzylic and allylic alcohols leading the formation of aldehydes by oxidative dehydrogenation or electron transfer from alkylaromatics leading to formation of dimers. The polyoxometalate - quinone systems are working homogeneous models for aerobic heterogeneous oxidation with $H_5PMo_{10}V_2O_{40}$ supported on carbon.

The polyoxometalate oxidation/substrate oxidation reaction mode - low valent transition metal substituted polyoxometalates. Since all polyoxometalates have high valent addenda W, Mo and V atoms, polyoxometalates may have catalytic properties in the activation of hydrogen peroxide, alkyl hydroperoxides and to a lesser degree peracids common to these classes of compounds. Here we are interested in oxidation by the oxidation/substrate oxidation reaction mode, eqn. 2, with low valent transition metal substituted polyoxometalates. If the transition metal is the heteroatom and not available for coordination, compounds such as $Co^{III}W_{12}O_{40}^{5-}$ and $Ce^{IV}Mo_{12}O_{42}^{8-}$ with high oxidation potential allowed non-catalytic outer sphere oxidation of *p*-xylene and *p*-methoxytoluene to the corresponding aldehydes.[18] If, however, the substituted transition metal is in a terminal position, coordination of oxidants is possible (usually in place of a labile aquo ligand. This led to quite a few and varied catalytic systems for alkene and alkane oxidation with Keggin, Wells-Dawson and "sandwich" type transition metal substituted polyoxometalates containing, Mn, Fe, Co, Cu, Rh and Ru using iodosobenzene,[19] N-oxides,[20] periodate,[21] ozone,[22] *t*-butylhydroperoxide[23] and hydrogen peroxide[24] as oxidants. In the presence of molecular oxygen as primary oxidant and in the liquid phase reactions are often dominated by autooxidation pathways which in certain important cases can be utilized for preparation of basic chemicals of industrial importance. In fine chemical synthesis these autooxidation pathways often lead to non-selective reactions. Use of polyoxometalates as autooxidation catalysts have been reported. For example, mono and trisubstituted Keggin compounds of the formula $M_3(H_2O)_3PW_9O_{37}^{q-}$ were used in the autooxidation of alkanes.[25]

An important but very elusive goal in the chemistry of ground state molecular oxygen is its activation in the absence of reducing agents and autooxidation pathways. We have recently used the ruthenium substituted "sandwich" compound, $\{[(WZnRu_2(H_2O)(OH)][(ZnW_9O_{19})_2]\}^{11-}$ [26] to activate molecular oxygen by a dioxygenase reaction mode,[27] eqn. 3.

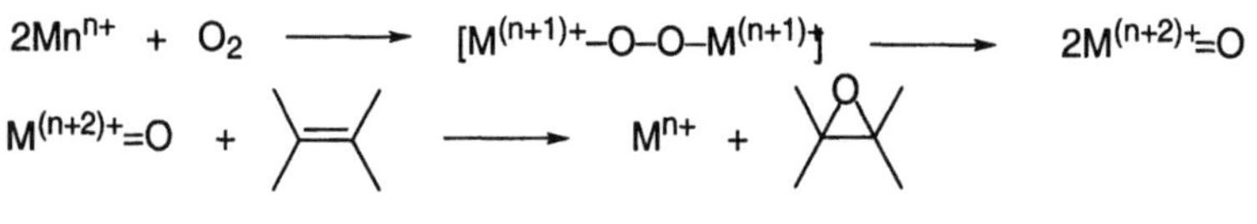

(3)

In this system, we have discounted an autooxidation pathway and provided evidence for a dioxygenase mechanism by (a) showing a substrate (adamantane): dioxygen reaction stoichiometry of 2:1, (b) selective formation of the *trans*-oxide from *trans*-cyclooctene and 1-adamantanol from adamantane (c) the fact that free radical scavengers do not inhibit catalytic hydroxylation of adamantane, (d) the kinetic isotope effect observed in 1,3-adamantane-d_2 hydroxylation, (e) the lack of reactivity of alkylaromatics such as cumene, (f) the absence of $H_2{}^{18}O$ formation when using $^{18}O_2$ as oxidant and g) the isolation of a ruthenium oxo or peroxo species, its identification by IR and UV-vis and its use in the stoichiometric epoxidation of alkenes. The conclusions have been supported by kinetic analyses. This catalyst promises to be a model for development of new dioxygenase type reactions, useful for example for the aerobic epoxidation of alkenes, and catalysts with increased reactivity.

Although $\{[(WZnRu_2(H_2O)(OH)][(ZnW_9O_{19})_2]\}^{11-}$ showed definite dioxygenase type activity in the oxidation of active alkanes and alkenes, the turnover frequency is usually low, especially for difficult to oxidize terminal olefins such as propylene. One approach examined to increase catalytic activity was to partially substitute fluorine into the polyoxometalate framework. The logic behind this approach was to increase the electropositivity of the ruthenium center and thereby the activity of the desired ruthenium-oxo intermediate. In this context, we prepared a novel compound, the ruthenium substituted polyfluorooxometalate, $[Ru^{II}(H_2O)W_{17}O_{55}F_6NaH_2]^{9-}$, which has a quasi Wells-Dawson structure. This new polyoxometalate shows improved activity at typically 60- 70% selectivity towards aerobic epoxidation of terminal linear alkenes. Unfortunately, this polyfluorooxometalate does not appear to activate molecular oxygen by a dioxygenase mechanism, but rather appears to be an initiator of autooxidation.

Results and Discussion

The reaction of α-$H_5PV_2Mo_{10}O_{40}$ with aldehydes. The $H_5PV_2Mo_{10}O_{40}$ polyoxometalate can react by consecutive electron and proton transfer to yield radical species.[15] Such a reaction with an aldehyde will yield an acyl radical as a reactive intermediate. In the presence of molecular oxygen, this reaction will initiate the formation of acyl peroxo intermediates and the oxygenation of alkanes to ketones.[17] There is therefore an interest in further studying the polyoxometalate - aldehyde reaction. The ^{31}P

NMR spectrum of $H_5PV_2Mo_{10}O_{40}$ in acetone-d_6, figure 1a, shows five peaks at -3.96, -3.42, -3.37, -3.31 and -3.22 ppm attributable to the five predicated isomers of α-$H_5PV_2Mo_{10}O_{40}$, figure 2 (the additional sixth peak at -4.38 ppm is from $H_4PVMo_{11}O_{40}$). In the past, it had been noted[28] that a statitistical peak ratio of 2 : 2 : 2 : 4: 1 for the 1,2; 1,4; 1,5; 1,6 and 1,11 isomers, respectively, was to be expected, but instead a ratio of 2 : 2 : 4 : 1 : 6 was observed. Especially noteworthy is the intense peak at a highest field, -3.96 ppm, compared to the other four less intense peaks at the lower field. A partially reduced $H_5PV_2Mo_{10}O_{40}$ polyoxometalate (light green solution) was obtained after one hour by addition of excess isobutyraldehyde under nitrogen. The new ^{31}P NMR spectrum showed the same five peaks (slightly shifted due to paramagnetism) but at different intensities, figure 1b. The lower field peaks were reduced in intensity while the higher field peak remained unchanged relative to the peak from $H_4PVMo_{11}O_{40}$ used as an internal standard. After further reduction (5 hours), the low field peaks almost entirely disappeared and after 2 days the high field peak also vanished.

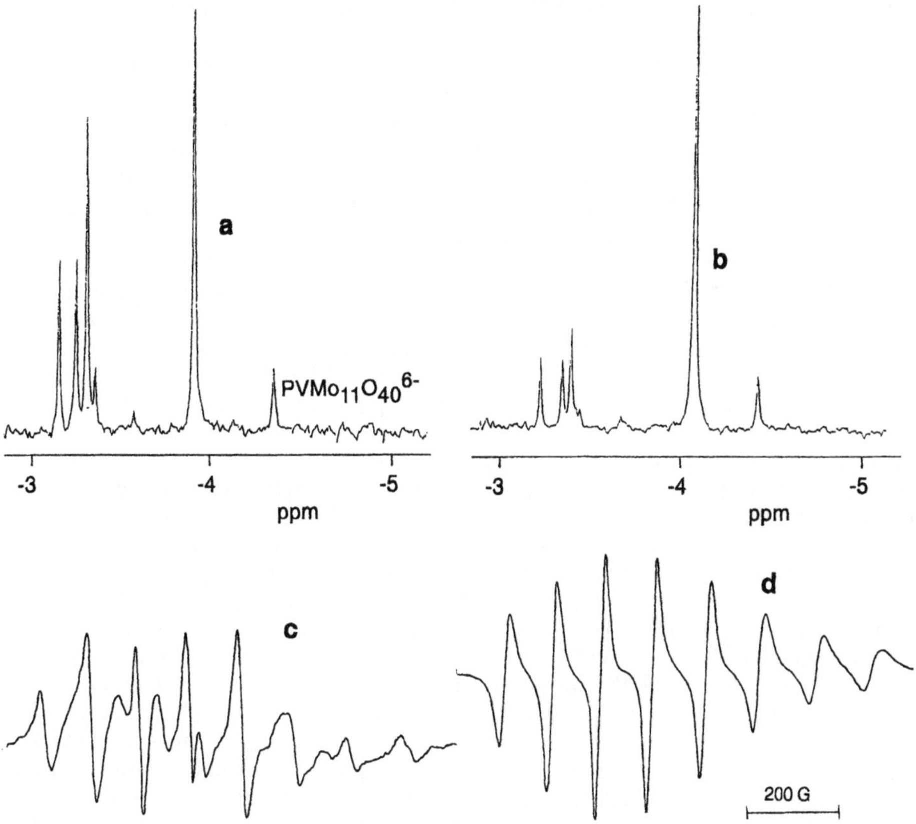

Figure 1: The ^{31}P NMR and ESR spectra of α-$H_5PV_2Mo_{10}O_{40}$.

(a) ^{31}P NMR spectrum 10 mM $H_5PV_2Mo_{10}O_{40}$ in acetone d_6 prior to addition of isobutyraldehyde. (b) ^{31}P NMR spectrum one hour after addition of 5 equivalents of isobutyraldehyde under nitrogen. (c) ESR spectrum - 10 mM $H_5PV_2Mo_{10}O_{40}$ solution in acetone one hour after addition of 5 equivalents isobutyraldehyde under nitrogen at ambient temperature. (d) as in (c) 18 hours after addition of isobutyraldehyde.

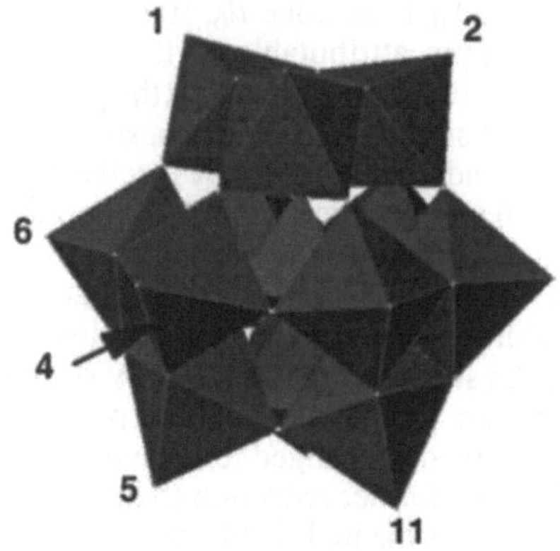

Figure 2: A representation of the α-$H_5PV_2Mo_{10}O_{40}$ and the five isomers.

Following the ^{31}P NMR experiment, a similar experiment using ESR spectroscopy was carried out. Fully oxidized $H_5PV_2Mo_{10}O_{40}$ is of course ESR silent, but addition of 5 equivalents of isobutyraldehyde under nitrogen after one hour yielded a 15 line spectra, figure 1c, and continued reduction for a further 13 hours yielded an 8-line spectrum, figure 1d. We believe that both the ^{31}P NMR and ESR experiments are best explained by rejecting the notion that α-$H_5PV_2Mo_{10}O_{40}$ exists as a statistical distribution of isomers. Rather, we postulate that distal isomers, especially the 1,11 isomer, are expected to be most stable as these isomer(s) have an improved distribution of charge (V^{IV} or V^{V} *verrsus* Mo^{VI}) in the polyoxometalate. This hypothesis is supported by the well known fact that, increasing the number of vanadium atoms in $H_{3+x}PV_xMo_{12-x}O_{40}$ destabilizes the Keggin structure, due mainly to the increase in negative charge. Indeed, in the ^{31}P NMR spectra the peak representing the 1,11 α-$PV_2Mo_{10}O_{40}^{5-}$ isomer is anticipated to be positioned at a higher field compared to the other α-$PV_2Mo_{10}O_{40}^{5-}$ isomers, since the order in the ^{31}P NMR absorption (high field to low field) is α-$PMo_{12}O_{40}^{3-}$ > α-$PVMo_{11}O_{40}^{4-}$ > α-$PV_2Mo_{10}O_{40}^{5-}$ > α-$PV_3Mo_9O_{40}^{6-}$. Since it has already shown in the past[15] with support from other computational and experimental information[29] that redox interactions between α-$PV_2Mo_{10}O_{40}^{5-}$ and organic substrates are to be preferred at vicinal vanadium positions, one would indeed initially expect the broadening and disappearance in the ^{31}P NMR spectrum of the peaks attributable to vicinal isomers at the lower fields due to formation of the paramagnetic species, while the higher field peak attributable to the 1,11 isomer would remain unchanged. The complementary ESR experiment shows that upon reaction of the isobutyraldehyde with the vicinal isomers of α-$H_5PV_2Mo_{10}O_{40}$ a mono-reduced polyoxometalate was formed. This leads to the observation of a 15 line spectrum due to additional splitting of the 8 lines at the V^{IV} center ($I = 7/2$ for ^{51}V) by a nearest neighbor vanadium atom, especially in the 1, 2 and 1, 6 configurations. As the reduction of the polyoxometalate mixture continues, the 1, 2 and 1, 6 isomers become doubly reduced and because of antiferromagnetic coupling these isomers become ESR silent. The slower reduction of the more distal isomeric forms, especially the 1,11 isomer leads to a simple 8 line spectrum with no nearest neighbor interaction.

The ^{31}P NMR and ESR experiments described above clearly attested to different reactivity of isobutyraldehyde with different isomers of α-$H_5PV_2Mo_{10}O_{40}$. Further kinetic profiles of the reduction of α-$H_5PV_2Mo_{10}O_{40}$ by isobutyraldehyde were sought out to quantify the relative reactivities of the isomers. By monitoring with UV-vis spectroscopy at 750 nm, the reduction of α-$H_5PV_2Mo_{10}O_{40}$ a complex behavior was detected, figure 3. The reduction of α-$H_5PV_2Mo_{10}O_{40}$, measured by heteropoly blue formation, occurs in three steps. First, over a period of 4 hours there was a relatively fast increase in optical density from 0 to ~0.5, *ie* reduction, with an approximate first order behavior. A second regime showed slower reduction (~0.5 to ~1.0) again with an approximate first order behavior occurs over the following 20 hours. Finally there is a even slower reduction to an optical density of 1.2 over the following 3 days. In the re-oxidation reaction under oxygen of the fully reduced polyoxometalate again three parallel regimes were observed over a 400 second time period.

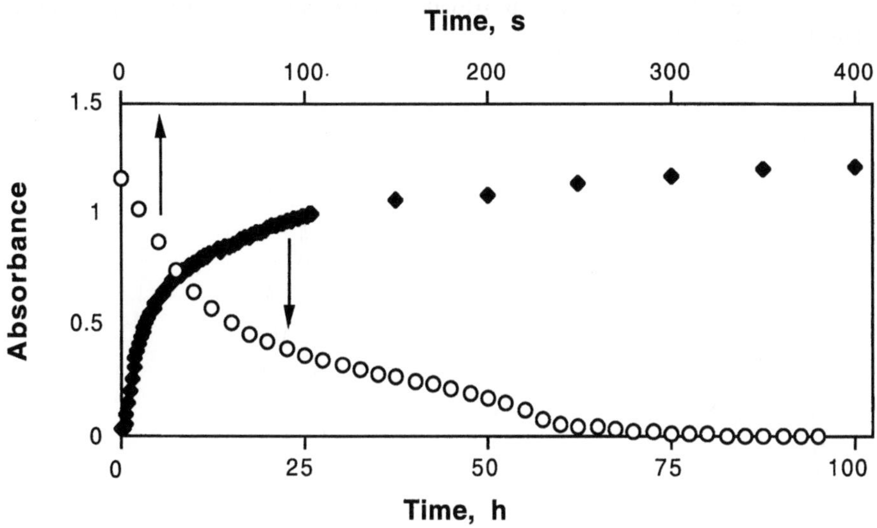

Figure 3: Absorption - time curves of α-$H_5PV_2Mo_{10}O_{40}$ reduction by isobutyraldehyde and reoxidation by molecular oxygen.

Absorption of α-$H_5PV_2Mo_{10}O_{40}$ was measured at 750 nm at 26 °C, ♦ - reduction of 1 mM a-$H_5PV_2Mo_{10}O_{40}$ in acetonitrile upon addition of 25 equivalents isobutyraldehyde. O - Re-oxidation of reduced α-$H_5PV_2Mo_{10}O_{40}$ under 1 atm dioxygen.

Since the isomeric mixture of α-$H_5PV_2Mo_{10}O_{40}$ is only approximately defined, it would be overly bold to exactly define which reaction is occurring in each segment of the kinetic profile. Nonetheless to a reasonable degree of confidence, the first segment is mostly clearly connected with mono reduction of the more reactive vicinal isomers. The intermediate and third segments are attributed to a combination of di-reduction of the vicinal and the mono-reduction of the more distal isomers. Additionally, using the method of initial rates the reaction of isobutyraldehyde with the most reactive isomers of α-$H_5PV_2Mo_{10}O_4$ may be studied under an inert nitrogen atmosphere at different temperatures. The initial slope is a related to the reaction of isobutyraldehyde with the most reactive isomer(s) of α-$H_5PV_2Mo_{10}O_{40}$. An Arrhenius plot of the initial rate constants as a function of temperature yielded a surprisingly high activation energy of 22 Kcal/mol ($\Delta H^{\ddagger}_{25°C}$ = 21.4 Kcal/mol, $\Delta S^{\ddagger}_{25°C}$ = -24 e.u.).

A further understanding of the $PV_2Mo_{10}O_{40}^{5-}$ – isobutyraldehyde reaction was attained using ^{51}V NMR spectroscopy, figure 4. The ^{51}V NMR spectrum of

$H_5PV_2Mo_{10}O_{40}$ in acetone-d_6 was not well resolved as opposed to the ^{31}P NMR spectrum with only two distinct peaks at -525.4 and -534.6 ppm. Upon addition of 5 equivalents of isobutyraldehyde under N_2 a new spectrum was measured after one hour, figure 4, right. At least one additional peak was observed down-field at -521.1 ppm. The small magnitude of the down-field shift indicates that this peak is attributable to a complex between a vanadium atom in α-$H_5PV_2Mo_{10}O_{40}$ and isobutyraldehyde. Such complexes between aldehydes and polyoxometalates have already been described and investigated in the past. It has been shown that dioxymethylene moieties are formed on the polyoxometalate, leading to formulation of metal acetal or hemiacetal compound, MO–OHCR.[30] A similar intermediate is postulated in this case. The formation of this intermediate is the key to the aerobic autoxidation of alkanes, which occurs by a separate and well-known catalytic cycle.

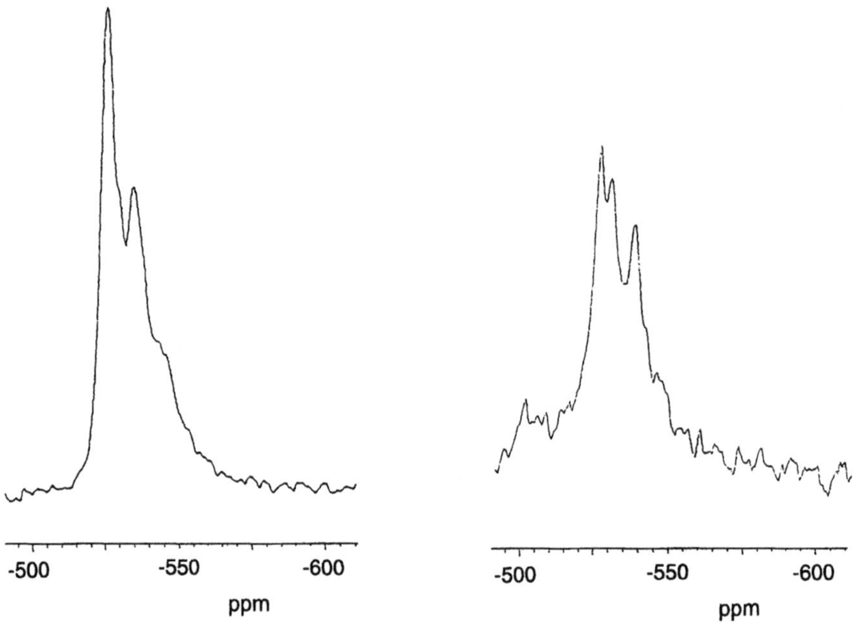

Figure 4: ^{51}V NMR spectra of α-$H_5PV_2Mo_{10}O_{40}$ in acetone-d_6.

(left) spectrum 10 mM α-$H_5PV_2Mo_{10}O_{40}$ in acetone. (right) spectrum after addition of 5 equivalents of isobutyraldehyde under N_2.

Quinones as models of surface active species in carbaon and the reaction of quinones with $PV_2Mo_{10}O_{40}{}^{5-}$. Although quite a few catalytic redox applications of the acidic $H_5PV_2Mo_{10}O_{40}$ polyoxometalate have been described, the high acidity of this compound is sometimes undesirable and leads to much reduced yields. For example, in the oxidative dehydrogenation of benzyl alcohol, use of $H_5PV_2Mo_{10}O_{40}$ leads to formation of dibenzylether as major product instead of benzaldehyde, eqn. 4

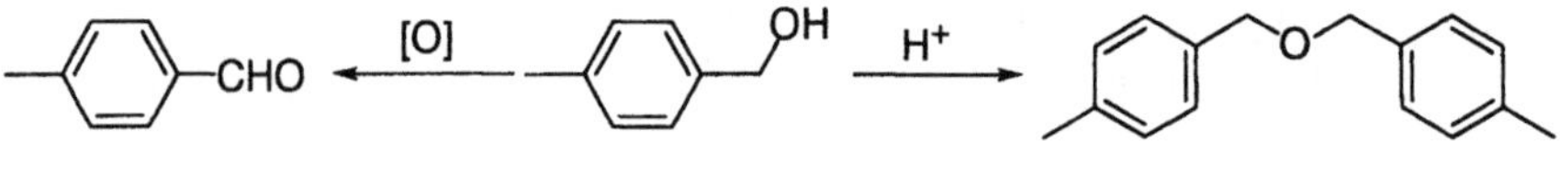

(4)

The use of neutral salts such as [$^{4}Q_5PV_2Mo_{10}O_{40}$ ($^{4}Q \equiv (n\text{-}C_4H_9)_4N^+$) and $Na_5PV_2Mo_{10}O_{40}$ as catalysts in monophasic and water/organic biphasic system were inactive. Silica and alumina supports were oxidatively inactive showing only some mild acid catalysis. Only $Na_5PV_2Mo_{10}O_{40}$/carbon was active. Strangely it has caught our attention that also in all the literature reports,[13,31] active carbon was used as the support. This observation leads to two possible conclusions (a) that either the presence of protons was vital to sustain a catalytic cycle or (b) that for non-protic, neutral $PV_2Mo_{10}O_{40}^{5-}$, active carbon was not an inert matrix but played an integral part in the catalytic cycle. As it is known that in the presence of oxidizing agents carbon surfaces are rich with oxygen containing moieties,[32] we set out to investigate the possibility that quinones are carbon surface active species for oxydehydrogenation of alcohols to aldehydes. These quinones, formed in heterogeneous systems by the surface reaction of $PV_2Mo_{10}O_{40}^{5-}$ on carbon, were thus studied as co-catalysts in the presence of $PV_2Mo_{10}O_{40}^{5-}$ in homogeneous (liquid phase) reaction media according to the following scheme 1:

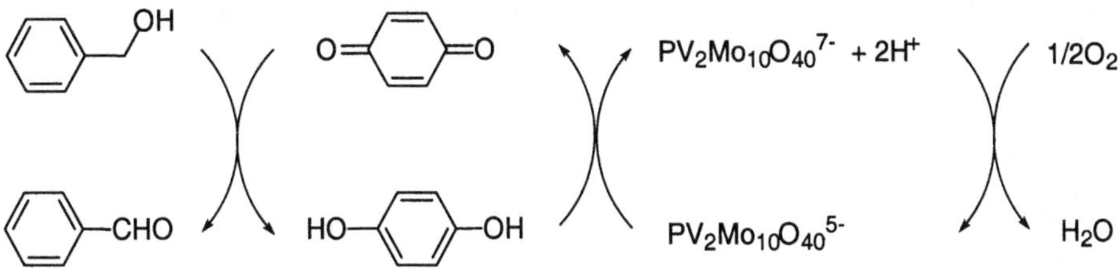

Scheme 1

Similar quinone-$H_5PV_2Mo_{10}O_{40}$ systems have been used in the palladium catalyzed oxidation of alkenes and conjugated dienes.[33] There, however, the primary functions of the quinone were to act as a coordinating and activating ligand to the palladium(II) center.

Table 1. Oxidation of benzylic and allylic alcohols catalyzed by $PV_2Mo_{10}O_{40}^{5-}$–*p*-chloranil.

Substrate	Product	Chemical yield, mol%[a] Method A	Method B
4-methoxybenzyl alcohol	4-methoxybenzaldehyde	85	75
4-methylbenzyl alcohol	4-methylbenzaldehyde	82	63
4-chlorobenzyl alcohol	4-chlorobenzaldehyde	77	59
benzyl alcohol	benzaldehyde	81	62
4-nitrobenzyl alcohol	4-nitrobenzaldehyde	65	52
cinnamyl alcohol	cinnamyl aldehyde	85	80
1-phenylethanol	acetophenone	87	80
benzhydrol	benzophenone	94	84
cis-2-hexene-1-ol	*cis*-2-hexenal	35	24
trans-2-hexene-1-ol	*trans*-2-hexenal	33	20
1-octene-3-ol	1-octene-3-one	76	34
2-cyclohexene-1-ol	2-cyclohexene-1-one	88	37

Reaction conditions: Method A - 1 mmol substrate, 0.05 mmol *p*-chloranil, 1 mL decalin, 0.015 mmol $Na_5PV_2Mo_{10}O_{40}$, 1 mL water, 1 atm O_2, 90 ± 2 °C, 18 h. Method B - 1 mmol substrate, 0.05 mmol *p*-chloranil, 0.015 mmol $^{4}Q_5PV_2Mo_{10}O_{40}$, 1 mL acetonitrile, 1 atm O_2, 110 ± 2 °C, 18 h. (a) the chemical yield was determined by GC using reference standards and calibrated mixtures.

The use of quinones as co-catalyst was tested (**A**) in a biphasic liquid-liquid system where $Na_5PV_2Mo_{10}O_{40}$ was dissolved in water and reacted with the substrate and quinone dissolved in an immiscible organic phase. In a second mode (**B**), the polyoxometalate, $^4Q_5PV_2Mo_{10}O_{40}$, quinone and substrate were all dissolved in a single solvent. In Table 1 are presented the results for the oxydehydrogenation of a variety of benzylic and allylic alcohols using 2,3,5,6-tetrachloro-1,4-benzoquinone (*p*-chloranil) as co-catalyst. Oxydehydrogenation reaction was possible using both methods. The two phase method is preferable in terms of yield and recovery of the polyoxometalate. As expected electron donating substituents in benzylic alcohols increased the yield and electron withdrawing groups decrease the yield, although the effect is rather small. Allylic alcohols are significantly less reactive than benzylic alcohols except for the conjugated cinnamyl alcohol. The selectivity of the reactions is high, >95%. The only by-product formed is not the carboxylic acid formed by over oxidation of the aldehyde, but rather a ring coupled product (especially for 4-methoxybenzyl alcohol). The reactivity of a series of quinones was also compared using the biphasic system and the water soluble $Na_5PV_2Mo_{10}O_{40}$, figure 5. The reactivity and yield in the benzyl alcohol oxidative dehydrogenation is a function of the oxidation potential of the quinone.[34]

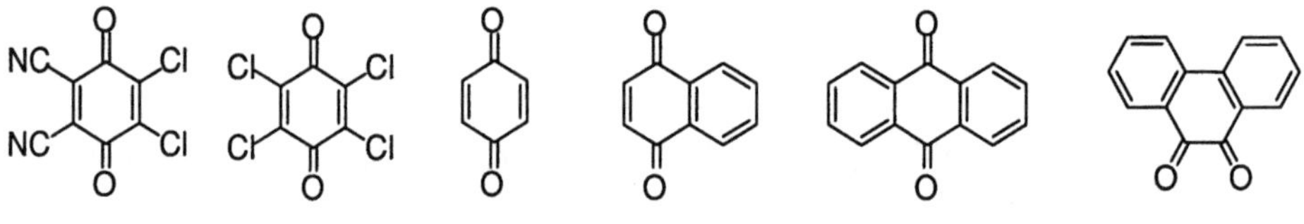

84 81 69 36 14 0

Figure 5: Yield in the oxidation of benzyl alcohol catalyzed by $Na_5PV_2Mo_{10}O_{40}$ and various quinones.

Reaction conditions: 1 mmol benzyl alcohol, 0.05 mmol quinone, 1 mL decalin, 0.015 mmol $Na_5PV_2Mo_{10}O_{40}$, 1 mL water, 1 atm O_2, 90 ± 2 °C, 18 h. (a) the chemical yield was determined by GC using reference standards and calibrated mixtures.

According to scheme 1 an empirical rate equation for the entire benzylic alcohol oxidative dehydrogenation reaction scheme, taking into account all the components may be written as follows, eqn. 5.

$$-d[MBA]/dt = k_{obs}[MBA]^a[TCBQ]^b[POM]^cP_{O_2}{}^d \quad (5)$$

MBA = 4-methylbenzyl alcohol, TCBQ = p-chloranil, POM = $^4Q_5PV_2Mo_{10}O_{40}$.

The kinetic studies for 4-methylbenzyl alcohol oxydehydrogenation showed that (a)the reaction appears to be zero order in 4-methylbenzyl alcohol (a = 0), (b) the observed activation energy, E_a, was 13.8 Kcal/mol (lnA = 14.74, $\Delta H^{\ddagger}{}_{25}$ = 13.2 Kcal/mol, $\Delta S^{\ddagger}{}_{25}$ = -31.1 cal/mol K), (c) the reaction rate was independent of the oxygen pressure, (d) the reaction order in TCBQ was 0.6 and (e) the reaction order in POM was 0.5. leading to the following empirical rate law, eqn. 6

$$-d[MBA]/dt = k_{obs}[TCBQ]^{0.7}[POM]^{0.5}P_{O_2}{}^0 \text{ where } k_{obs} = 2.5 \times 10^6(\exp)^{-13.2/RT} \quad (6)$$

According to scheme 1, the entire catalytic oxidative dehydrogenation reaction cycle may be divided into three separate reactions. If the rate of the reaction is measured as

-d[MBA]/dt and assuming that the rate determining step is the bimolecular elementary reaction between MBA and TCBQ then the reaction should theoretically be first order both in MBA and TCBQ. However, under the general reaction conditions used MBA:$TCBQ_0 \geq 10$, TCBQ is in a limited concentration *ie* catalytic. Therefore, the reaction instead appears to be zero order, as observed, in MBA and should be pseudo first order in TCBQ. The fact, however, that the observed reaction order in TCBQ was only partial (0.7) and that the rate was also dependent on and of partial order in POM clearly argued for a more complicated kinetic scheme although the fact that the reaction was zero order in molecular oxygen indicates that the reoxidation step is not rate determining. Separate kinetic investigations of each reaction step according to scheme 1 showed that the reoxidation of the hydroquinone by the oxidized polyoxometalate and the reoxidation of the polyoxometalate by molecular oxygen were much faster than the rate for the entire reaction while the dehydrogenation of MBA by TCBQ was significantly slower. These findings together with the non-integer reaction orders, leads to the postulation that an interaction between TCBQ and POM could lead to new "TCBQ-POM" species or complexes of catalytic significance. Indeed, there is much evidence from the literature that oquinone compounds may insert[35] and carbonyl moieties in general may interact[17,30] with M–O bonds of polyoxometalates

Evidence for such a "TCBQ-POM" species was gained using ESR spectroscopy. The mixing of 1 equivalent of TCHQ (a potential two electron reductant) with 1.0 equivalent ${}^{4}Q_5PV^{V}{}_2Mo_{10}O_{40}$ yield an 8-line ESR spectrum, figure 6 top, as discussed above. Now, addition of five equivalents of TCBQ (ratio TCBQ:POM = 6), clearly shows the previously observed 8-line spectrum (g = 1.98) with an overlapping singlet at g = 2.00, figure 7 bottom. This singlet is easily attributed to the formation of the 2,3,5,6-tetrachloro-1,4-benzosemiquinone[36] SQ by electron transfer from the doubly reduced POM to TCBQ, eqn. 7.

$$POM^{2-} + Q \underset{}{\overset{K_1}{\rightleftharpoons}} POM^{1\cdot-}\text{–}SQ \underset{}{\overset{K_2}{\rightleftharpoons}} POM^{1\cdot} + SQ \qquad (7)$$

It is not entirely clear if SQ exists in solution as is or as a complex, SQ-POM. We were not successful in isolating any such intermediate. Further support for a SQ or SQ-POM species was obtained using cyclic voltammetry (CV). Thus, the CV of TCBQ showed two reversible wave with oxidation potentials of -0.12 V and 0.6 V using a silver electrode whereas a 1:1 mixture of TCBQ and POM showed a new reversible wave at 0.8 V. This indicates the formation of a strong oxidant for the TCBQ-POM system compared to those of TCBQ and POM (0.68 V) alone. Additionally, UV-vis measurements showed no significant decrease in the optical density at 750 nm, ie oxidation of POM_{red} upon addition of the quinone oxidant. This clearly indicates that K_2 (eqn. 7) is low as it would be expected that POM^{1-} would have a ~50% lower optical density relative to POM^{2-} or $POM^{1\cdot-}$–SQ as optical densities at 750 nm are proportional to the degree of reduction. The existence of a SQ-POM intermediate complex is reasonable.

The possible existence of SQ-POM intermediate species under catalytically relevant conditions requires one to check if such species are in fact catalytically relevant oxidants. In order to differentiate between the oxidative activity of TCBQ and SQ-POM, conditions should be preferably found where one is active and the other is not or *vice versa*. For the oxidation of methyl aromatics to aldehydes or ketones, it was found that chloranil alone was inactive, however in the presence of ${}^{4}Q_5PV_2Mo_{10}O_{40}$ there was a significant reaction, scheme 2 Thus, the reaction of 1 mmol 4-methoxytoluene with catalytic amounts of *p*-chloranil (0.1 mmol) and polyoxometalate (0.01 mmol) in 8 mL benzonitrile at 140 °C and 1 atm O_2 led to a 16 % conversion of 4-methoxytoluene to 75% 4-methoxybenzaldehyde and 25% ring coupled by-product.

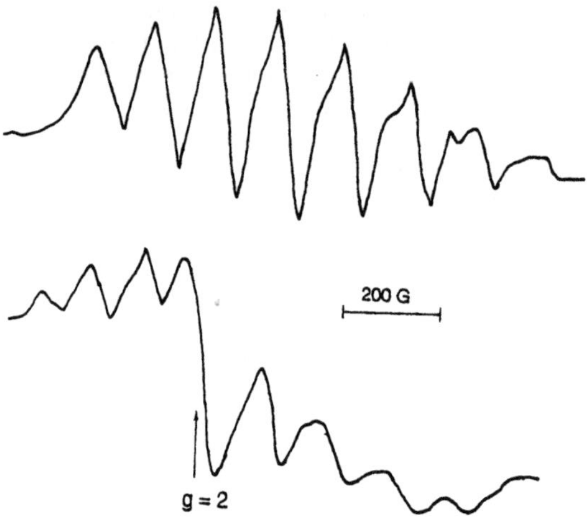

Figure 6: ESR spectra at room temperature of reduced $H_2{}^4Q_5PV_2Mo_{10}O_{40}$ in the presence of *p*-chloranil.

top - spectrum of 2 mM polyoxometalate after addition of 0.5 equivalents 2,3,5,6-tetrachloro-1,4-dihydroxybenzene. bottom - spectrum of 2 mM polyoxometalate after addition of 1.0 equivalents 2,3,5,6-tetrachloro-1,4-dihydroxybenzene and then followed by addition of 5.0 equivalents *p*-chloranil.

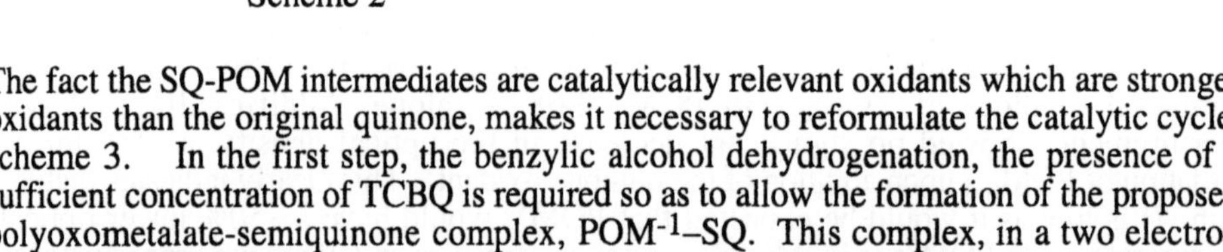

Scheme 2

The fact the SQ-POM intermediates are catalytically relevant oxidants which are stronger oxidants than the original quinone, makes it necessary to reformulate the catalytic cycle, scheme 3. In the first step, the benzylic alcohol dehydrogenation, the presence of a sufficient concentration of TCBQ is required so as to allow the formation of the proposed polyoxometalate-semiquinone complex, POM^{-1}–SQ. This complex, in a two electron oxidation, dehydrogenates the benzylic alcohol to yield the benzaldehyde and TCHQ in the rate determining step. Determination of this step as rate limiting is also supported by the results of a competitive reaction between benzyl alcohol and benzyl alcohol-d_7 where a k_H/k_D ratio of 4.3 was observed. In a second relatively faster reaction, the oxidized polyoxometalate, POM^0 is aerobically regenerated. The oxidized polyoxometalate then can regenerate the quinone, allowing again for formation of the POM^{-1}–SQ oxidant. The fact that the reduced polyoxometalate may (a) react with quinone to produce semiquinone and (b) be aerobically oxidized accounts for the non-elementary rate equation obtained for this reaction.

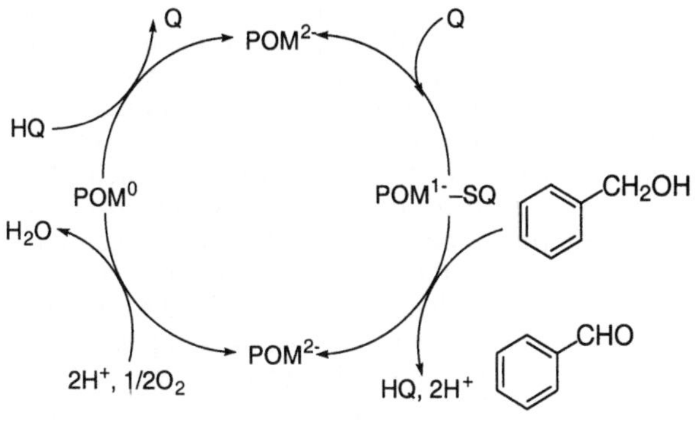

Scheme 3

Aerobic epoxidation of alkenes catalyzed by a new ruthenium substituted polyfluorooxometalate, $[Ru^{II}(H_2O)W_{17}O_{55}F_6NaH_2]^{9-}$. As noted in the introduction we have previously found that the ruthenium substituted "sandwich" type polyoxometalate shows dioxygenase type activity in aerobic alkane hydroxylation and alkene epoxidation. Unfortunately in liquid phase reactions, the catalytic activity remained low, and a more active catalysts was sought. Here, we describe the synthesis, characterization and catalytic activity of a new ruthenium containing polyoxometalate with partial fluoride substitution. In order to prepare the ruthenium substituted polyfluorooxometalate, PFOM, the previously known zinc substituted, $[Zn(H_2O)H_2F_6NaW_{17}O_{55}]^{9-}$ PFOM was used as starting material.[36] Accordingly, a lacunary PFOM was prepared *in situ* by reacting sodium tungstate with 49 % HF until the pH reaches 4.5. The white precipitate, of unknown composition, formed was removed by filtration. Zinc acetate was added to the filtrate containing the lacunary PFOM, yielding the desired zinc substituted PFOM, $[Zn(H_2O)H_2F_6NaW_{17}O_{55}]^{9-}$. The zinc atom was exchanged using $Ru(DMSO)_4Cl_2$ as the source of ruthenium. After 10 hours, a yellow-brown precipitate tentatively formulated as the mixed addenda PFOM, $[H_2F_6NaRu^{II}W_{17}O_{56}]^{9-}$ was obtained and recrystallized. Characterization (see below) showed that indeed the quasi Wells-Dawson compound, $[H_2F_6NaRu^{II}(H_2O)W_{17}O_{56}]^{9-}$ was obtained, figure 7.

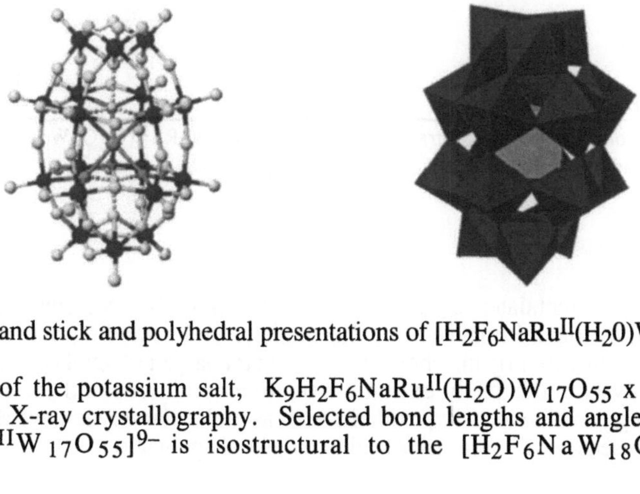

Figure 7: Ball and stick and polyhedral presentations of $[H_2F_6NaRu^{II}(H_20)W_{17}O_{55}]^{9-}$

The structure of the potassium salt, $K_9H_2F_6NaRu^{II}(H_2O)W_{17}O_{55}$ x $15H_2O$ was determined by X-ray crystallography. Selected bond lengths and angles, show that $[H_2F_6NaRu^{II}W_{17}O_{55}]^{9-}$ is isostructural to the $[H_2F_6NaW_{18}O_{56}]^{7-}$ and

$[H_2F_6NaV^VW_{17}O_{56}]^{8-}$ anions reported previously.[38] There is central core consisting of a sodium atom surrounded by six crystallographically identical fluorine atoms (distance 2.17 -2.23Å) in a triganol prism coordination. It was not possible to resolve the position of the one ruthenium atom among the remaining 17 tungsten atoms, nor could one determine if ruthenium was exclusively in a belt or capped position . The 12 tungsten atoms in the belt positions have a distorted octahedral (tetragonal) coordination, with W(V) — O bond lengths of ~1.9Å for the bridging oxygen positions, W(V) — O bond lengths of ~1.7Å for the terminal oxygen positions and longer W(V)—F bonds, 2.2-2.3Å. The six tungsten atoms in the capped position have a very similar tetragonal coordination with W(V) — O bond lengths of ~1.9Å for the bridging oxygen positions, W(V) — O bond lengths of ~1.7Å for the terminal oxygen positions and longer W(V)—O74 bonds, ~2.2Å. The position of the hydrogen atoms was not resolved in the x-ray structure, but is surmised to be between the O74 oxygen and F1, F2 and F3 forming F_3HO tetrahedra as observed for the vanadium substituted analog.

The ^{19}F NMR spectrum (CF_3COOH as reference) of $K_9H_2F_6NaRu^{II}W_{17}O_{55}$ in D_2O, and the ^1H NMR and ^{19}F NMR of $^4Q_9H_2F_6NaRu^{II}(H_2O)W_{17}O_{55}$ in acetonitrile were almost identical to the vanadium substituted compound previously characterized and leads to the conclusion that the ruthenium atom is both in the capped and belt position. Importantly, the formulation of the ruthenium substituted PFOM as a ruthenium (II) species was strongly supported by (a) measurement of the magnetic susceptibility which showed the compound to be diamagnetic. The compound is ESR silent. Calorimetric and gravimetric measurements show that the compound is stable up to 350 °C.

The catalytic activity for alkene epoxidation under 1 atm molecular oxygen is summarized in Table 2. One may observe the $[H_2F_6NaRu^{II}W_{17}O_{55}]^{9-}$ has significant catalytic capacity with turnover frequencies of 3-10 per hour. Unfortunately, however, selectivity was limited. There were significant quantities of autooxidation products formed indicated that $[H_2F_6NaRu^{II}W_{17}O_{55}]^{9-}$ does not catalyze aerobic epoxidation by a dioxygenase mechanism but rather by a autooxidation mechanism.

Table 2: Aerobic alkene epoxidation catalyzed by $[H_2F_6NaRu^{II}W_{17}O_{55}]^{9-}$.

Substrate	Turnover number	Selectivity, mol %
1-octene	90	68
trans-2-octene	110	70
cyclooctene	178	88
cyclohexene	103	10
cis-2-hexenol	237	43
trans-2-hexenol	168	21
cis-3-hexenol	171	46
trans-3-hexenol	106	45

Reaction conditions: 1 mmol substrate, 1 µmol $^4Q_9H_2F_6NaRu^{II}(H_2O)W_{17}O_{55}$, in 1 mL 1,2-dichloroethane under 1 atm O_2 at 80 °C for 24 h. Analysis by GC.

Conclusion

The use of polyoxometalates in oxidation chemistry has advanced considerably over the past decade. We have improved our mechanistic understanding and developed new synthetic procedures for organic chemistry. Given the large potential of polyoxometalates as stable catalysts under fairly extreme oxidative conditions and the possibilities of activating benign and inexpensive oxidants such as molecular oxygen, we are hopeful that the academic research will eventually lead to industrial application

References

1. M. Misono and T. Okuhara: *Chemtech*, 23 (1993).
2. K. I. Matveev and I. V. Kozhevnikov: *Kinet. Catal.* **21**, 855 (1980); I. V. Kozhevnikov and K. I. Matveev: *Russ. Chem. Rev.* **51**, 1075 (1982); I. V. Kozhevnikov and K. I. Matveev: *Appl. Catal.* **5**, 135 (1983).
3. G. A. Tsigdinos: *Top. Curr. Chem.* **26**, 1 (1978).
4. M. Misono: *Catal. Rev.-Sci. Eng.* **29**, 269 (1987).
5. N. Mizuno and M. Misono: *J. Mol. Catal.* **86**, 319 (1994); T. Okuhara, N. Mizuno and M. Misono: *Adv. Catal.* **41**, 113 (1996).
6. R. J. J. Jansen, H. M. Vanveldhuizen, M. A. Schwegler and H. van Bekkum: *Rec. Trav. Chim. Pays Bas* **113**, 115 (1994); Y. Ono in *Perspectives in Catalysis*, J. M. Thomas and K. Zamaraev eds., Blackwell, Oxford, 431 (1992).
7. Y. Ishii and M. Ogawa in *Reviews on Heteroatom Chemistry*, S. Oae ed., MYU, Tokyo, (1990).
8. I. V. Kozhevnikov: *Catal. Rev.-Sci. Eng.* **37**, 311 (1995); I. V. Kozhevnikov: *Russ. Chem. Rev.* **62**, 473 (1993).
9. C. L. Hill, D. C. Duncan and M. K. Harrup: *Comments Inorg. Chem.* **14**, 367 (1993); C. L. Hill and C. M. Prosser-McCartha: *Coord. Chem. Rev.* **143**, 407 (1995); R. Neumann: *Prog. Inorg. Chem.* **47**, 317 (1998).
10. K. I. Matveev: *Kinet. Catal.* **18**, 716 (1977); K. I. Matveev and I. V. Kozhevnikov: *Kinet. Catal.* **21**, 855 (1980); I. V. Kozhevnikov, V. E. Tarabanko, K. I. Matveev and V. D. Vardanyan: *React. Kinet. Catal. Lett.* **7**, 297 (1977); K. Urabe, F. Kimura and Y. Izumi: *VII Int. Cong. Catal.* C14 (1980); J. R. Grate, D. R. Mamm and S. Mohajan: *Mol. Eng.* **3**, 205 (1993); J. R. Grate, D. R. Mamm and S. Mohajan in *Polyoxometalates: From Platonic Solids to Anti-Retroviral Activity:* M.T. Pope; A. Müller eds. Kluwer: The Netherlands, 27 (1994). G. Centi, S. Perathoner and G. Stella: *Stud. Surf. Sci. Catal.* **88**, 393 (1994).
11. K. I. Matveev, E. G. Zhizhina, M. B. Shitova and L. I. Kuznetsova: *Kinet. Catal.* **18**, 380 (1977); L. I. Kutsenova and K. I. Matveev: *React. Kinet. Catal. Lett.* **3**, 305 (1975); V. M. Berdnikov, L. I. Kutsenova, K. I. Matveev, N. P. Kirik and E. N Yurchenko: *Sov. J. Coord. Chem.* **5**, 39 (1979).
12. R. Neumann and M. Lissel: *J. Org. Chem.* **54**, 4607 (1989).
13. R. Neumann and M. Levin: *J. Org. Chem.* **56**, 5707 (1991).
14. M. Lissel, H. Jansen van de Wal and R. Neumann: *Tetrahedron Lett.* **33**, 1795 (1992); O. A. Kholdeeva, A. V. Golovin, R. A. Maksimovskaya and I. V. Kozhevnikov: *J. Mol. Catal.* **75**, 235 (1992).
15. R. Neumann and M. Levin, *J. Am. Chem. Soc.* **114**, 7278 (1992).
16. C. L. Hill and D. C. Duncan: *J. Am. Chem. Soc.* **119**, 243 (1997).
17. A. M. Khenkin, A. Rosenberger and R. Neumann: *J. Catal.* **182**, 82 (1999).
18. L. Eberson: *J. Am. Chem. Soc.* **105**,3192 (1983); L. Eberson and L.-G. Wistrand: *Acta Chem. Scand.* **B34**, 349 (1980); L. Eberson and L. Jönsson: *Acta Chem. Scand.* **B40**, 79 (1986); U. Bietti, E. Baciocchi and J. B. F. N. Engberts, *Chem. Commun.* 1307 (1996).
19. C. L. Hill and R. B. Brown: *J. Am. Chem. Soc.* **108**, 536 (1986); D. Mansuy, J.-F. Bartoli, P. Battioni, D. K. Lyon and R. G. Finke: *J. Am. Chem. Soc.* **113**, 7222 (1991).
20. X. Zhang, K. Sasaki and C. L. Hill: *J. Am. Chem. Soc.* **118**, 4809 (1996).
21. R. Neumann and C. Abu-Gnim: *J. Chem. Soc., Chem. Commun.* 1324 (1989); R. Neumann and C. Abu-Gnim: *J. Am. Chem. Soc.* **112**, 6025 (1990); E. Steckhan and C. Kandzia: *Synlett* 139 (1992); M. Bressan, A. Morvillo and G. Romanello: *J. Mol. Catal.* **77**,283 (1992).

22. R. Neumann and A. M. Khenkin: *Chem. Commun.* 1967 (1998).
23. M. Faraj and C. L. Hill: *J. Chem. Soc., Chem. Commun.* 1487 (1987); R. Neumann and A. M. Khenkin: *Inorg. Chem.* **34**, 5753 (1995).
24 A. M. Khenkin and C. L. Hill: *Mendeleev Commun.* 140 (1993);, R. Neumann and M. Gara: *J. Am. Chem. Soc.* **116**, 5509 (1994); R. Neumann and M. Gara: *J. Am. Chem. Soc.* **117**, 5066 (1995); R. Neumann and A. M. Khenkin: *J. Mol. Catal.* **114**, 169 (1996). R. Neumann and D. Juwiler, *Tetrahedron* **47**, 8781, (1996); R. Neumann, A. M. Khenkin, D. Juwiler, H.Miller and M. Gara: *J. Mol. Catal.* **117**, 169 (1997); N. Mizuno, C. Nozaki, I. Kiyoto and M. Misono: *J. Am. Chem. Soc.* **120**, 9267 (1998).
25. D. Qin, G. Wang and Y. Wu: *Stud. Surf. Sci. Catal.* **82**, 603 (1994); J. E. Lyons, P. E. Ellis and V. A. Durante: *Stud. Surf. Sci. Catal.* **67**, 99 (1991); N. Mizuno, T. Hirose, M. Tateishi and M. Iwamoto: *J. Mol. Catal.* **88**, L125 (1994); N. Mizuno, M. Tateishi, T. Hirose and M. Iwamoto: *Chem. Lett.* 2137 (1993); R. Neumann and M. Dahan: *J. Chem. Soc., Chem. Commun.* 171 (1995).
26. R. Neumann, A. M. Khenkin and M. Dahan: *Angew. Chem., Int. Eng. Ed.*, **34**, 1287 (1995).
27. R. Neumann and M. Dahan: *Nature* **388**, 353 (1997); R. Neumann amd M. Dahan: *J. Am. Chem. Soc.* **120**, 11969 (1998).
28. M. T. Pope,S. E. O'Donnell and R. A. Prados: *J. Chem. Soc., Chem. Commun.* 22 (1975); S. E. O'Donnell and M. T. Pope: *J. Chem. Soc., Dalton Trans.* 2290 (1976); L. Pettersson, I. Andersson, A. Selling and J. H. Grate: *Inorg. Chem.* **33**, 982 (1994); M. T. Pope, S. E. O'Donnell and R. A. Prados: *Adv. Chem.* **150**, 85 (1976).
29. E. N. Yurchenko, H. Miessner and A. Trunschke: *J. Struct. Chem. (Engl. Trans.)* **30**, 22 (1989); W. Klemperer and G. Shum: *J. Am. Chem. Soc.* **100**, 4891 (1978).
30. G. A. Popova, A. A. Budneva and T. V. Andrushkevich: *React. Kinet. Catal. Lett.* **61**, 353 (1997); Y. Konishi, K. Sakata, M. Misono and Y. Yoneda: *J. Catal.* **77**, 169 (1982).
31. R. Neumann and I. Dror: *Appl. Catal. A: General* **172**, 67 (1998); K. Nakayama, M. Hamamoto, Y. Nishiyama and Y. Ishii: *Chem. Lett.* 1699 (1993); M. Hamamoto, K. Nakayama, Y. Nishiyama and Y. Ishii: *J. Org. Chem.* **58**, 6421 (1993); R. J. J. Jansen, H. M. van Veldhuizen and H. van Bekkum: *J. Mol. Catal.* **107**, 241 (1996); R. D. Gall, C. L. Hill and J. E. Walker: *J. Catal.* **159**, 473 (1996); R. D. Gall, C. L. Hill and J. E. Walker: *Chem. Mater.* **8**, 2523 (1996); R. D. Gall and C. L. Hill: *J. Mol. Catal.* **114**, 103 (1996).
32. D. W. van Krevelen, *Coal: Typology, Chemistry, Physics, Constitution*, Elsevier, Amsterdam, 1961.
33. H. Grennberg, K. Bergstad and J.-E. Bäckvall: *J. Mol. Catal.* **113**,355 (1996); T. Yokota, S. Fujibayashi, Y. Nishiyama, S. Sakaguchi and Y. Ishii: *J. Mol. Catal.* **114**, 113 (1996); K. Bergstad, H. Grennberg and J.-E. Bäckvall: *Organometallics* **17**, 45 (1998).
34. D. Walker and J. D. Heibert: *Chem. Rev.* **67**, 153 (1967).
35. S. Liu, S. N. Shaikh and J. Zubieta: *Inorg. Chem.* **26**, 4303 (1987); S. Liu, S. N. Shaikh and J. Zubieta: *Inorg. Chem.* **27**, 3064 (1988).
36. T. L. Jorris, Ph.D. Thesis, Georgetown University, 1987.
37. F. Chauveau, P. Doppelt and J. Lefebvre: *Inorg. Chem.* **19**, 2803 (1980); T. L. Jorris, M. Kozik and L. C. W. Baker: *Inorg. Chem.* **29**, 4584, (1990); S. H. Wasfi, C. E. Costello, A. L. Rheingold, B. S. Haggerty, *Inorg. Chem.* **30**,1788 (1991); A. M. Khenkin and R. Neumann, submitted for publication.

Polyoxoanions in Catalysis: From Record Catalytic Lifetime Nanocluster Catalysis to Record Catalytic Lifetime Catechol Dioxygenase Catalysis

RICHARD G. FINKE
Department of Chemistry, Colorado State University, Ft. Collins, CO 80523, U.S.A.

(Received: 14 December 1999)

Abstract: Following an introductory overview of the 8 newer classes of polyoxoanion-based catalysts in the last 20 years, as well as the 3 subclasses of polyoxoanion-based catalysts investigated by the Finke Group, highlights are presented of the research which has led to two of the 8 newer classes of polyoxoanion-based catalysts: polyoxonion-stabilized transition metal nanocluster "soluble heterogeneous catalysts" and polyoxoanion-based precatalysts for catechol dioxygenase activity. In both cases, the resultant catalysts show record catalytic lifetime as well as high catalytic activity. An introduction to dioxygenase catalysis, including some of the key goals in the area, is also provided. The highlights of a recently reported, polyoxoanion•Fe^{II}-based stiochiometric catechol extradiol dioxygenase are then presented. Key findings from polyoxoanion-derived catechol dioxygenase catalysts are presented next, catalysts which exhibit record catalytic activity as well as record catalytic lifetime. A summary, emphasizing the 4 key components of research which have led to the polyoxoanion-stabilized nanocluster and polyoxoanion-derived dioxygenase catalysts, is also provided.

Key Words: polyoxoanions in catalysis; polyoxoanion-supported catalysts and catalyst precursors; homogeneous and heterogeneous catalysis; nanocluster "soluble heterogeneous catalysts"; hydrogenation catalysis; oxygenation catalysis; catechol dioxygenase catalysis; kinetic and mechanistic studies; autocatalytic catalyst evolution mechanisms; approaches to developing new polyoxoanion-based catalysts.

1. Introduction

Polyoxoanions [1,2,3] continue to attract considerable interest as catalysts or catalyst precursors [4]. The reasons for such high interest in polyoxoanions in catalysis include: (i) that they are a robust, all-inorganic ligand type; (ii) that they are an oxidation resistant, if not inert, ligand type; (iii) that a wide range of possible complexes and ligand types are available; (iv) that they are discrete, modifiable and fully characterizable at the molecular level (e.g., by X-ray crystallography, FAB-mass spectroscopy, multinuclear NMR, and so on), and (v) that polyoxoanions have properties varying all the way from strong acids (e.g., $(H_3O)_4SiW_{12}O_{40}$ and $(H_3O)_6P_2W_{18}O_{62}$), and their weakly coordinating conjugate bases ($SiW_{12}O_{40}^{4-}$ and $P_2W_{18}O_{62}^{6-}$), to *strong bases* (e.g., $(BuN_4)_7SiW_9Nb_3O_{40}$ and $(BuN_4)_9P_2W_{15}Nb_3O_{62}$) and their strongly coordinating polyoxoanion components (e.g., $SiW_9Nb_3O_{40}^{7-}$ and $P_2W_{15}Nb_3O_{62}^{9-}$). One additional, very important reason for the high interest in polyoxoanions in catalysis is (vi) that only a small fraction of the conceivable, custom designed polyoxoanions have been synthesized, characterized and then exploited in catalysis. For this reason, the applications of polyoxoanions in catalysis should continue to attract interest for some time to come.

A corollary to point (vi) above, however, is that the new, truly novel applications of polyoxoanions require considerable effort in the synthesis and

M.T. Pope and A. Müller (eds.), Polyoxometalate Chemistry, 363–390.

characterization of the requisite, new types of polyoxoanions, a conclusion which has been apparent for some time now (see the concluding statement on p. 278 elsewhere[5]). Fortifying this point is the realization that of ca. 600 papers and 120 patents reviewed on the applications of polyoxoanions, ca. 80-85% of those reports are in catalysis and the overwhelming majority of these reports examine (only) *the classical Keggin anion* [6]—that is, the majority of the applications to date of polyoxoanions in catalysis involve commercially available, off-the-shelf, Keggin anions (e.g., $PW_{12}O_{40}^{3-}$ or $SiW_{12}O_{40}^{4-}$). Those studies are, then, limited to polyoxoanions that are strongly acidic with weakly coordinating conjugate acid polyoxoanions.

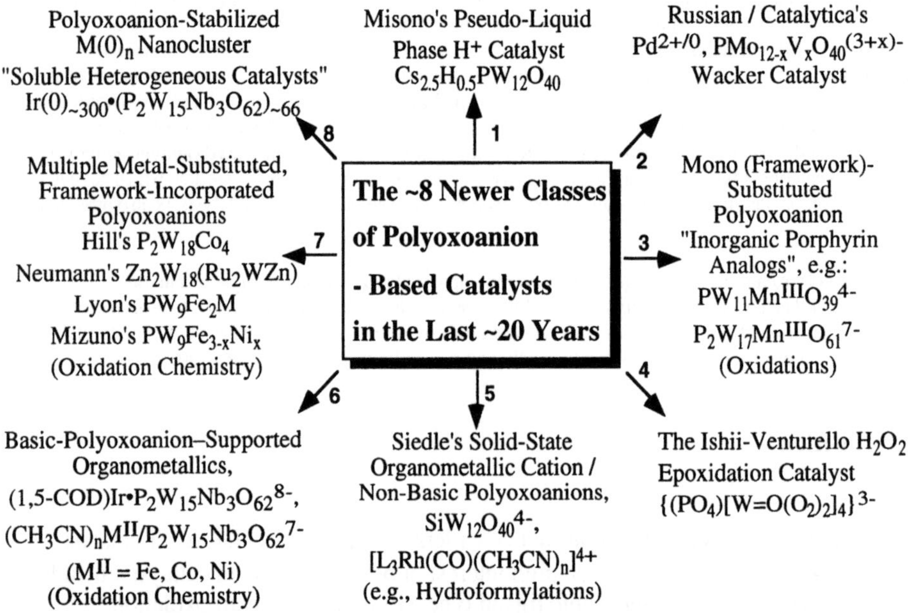

Fig. 1. The eight newer classes of polyoxoanion-based catalysis in the last ca. 20 years[7,8], virtually all of which are easily identifiable by the fact that they involve *new synthetic work* and are patented, are, arguably and in our opinion: (1) Misono's pseudo-liquid phase catalysts [9], (2) Russian Wacker-type $Pd^{2+/0}$ and V-containing polyoxoanion catalysts (as further developed and patented by Catalytica; see p. 281 elsewhere [2]); (3) Baker-type [10a] polyoxoanion single-metal-substituted, *framework-incorporated* catalysts, such as the $PW_{11}MnO_{39}^{4-}$ or $P_2W_{17}MnO_{61}^{7-}$ "inorganic porphyrin [10b,11] analog catalysts" developed by Pope, Hill, Lyons, Neumann and ourselves [11a]; (4) the Ishii-Venturello $[(PO_4)(W{=}O(O_2)_2]^{3-}$ epoxidation catalyst employing H_2O_2 and investigated mechanistically by Hill and co-workers [12]; (5) Siedle's new solid-state catalysts composed of organometallic cations-plus-polyoxoanions [13]; (6) polyoxoanion-*supported* catalysts [14]; (7) a rapidly developing area of *multiple-metal substituted*, framework-incorporated catalysts, such as Lyons' putative "PW_9Fe_2M" (M = Ni, Fe, Mn, Co, Zn) complexes [15] or Mizuno's recent work with the $PW_9Fe_{3-x}Ni_x$ complexes [16], Hill's derivatives [17] of the $P_2W_{18}M_4$ or $P_4W_{30}M_4$ type systems [18], or Neumann's work with the interesting $P_2W_{18}M_4$ analogs such as [19] $Zn_2W_{18}(Mn_2ZnW)$ or $Zn_2W_{18}(Ru_2ZnW)$[20]; and (8) polyoxoanion-stabilized nanocluster "soluble heterogeneous catalysts" [21,22,23].[1]

---

1. Although polyoxoanion purists might be tempted to exclude these nanocluster catalysts from the list, since the catalysis is not at the polyoxoanion itself, this would be an error: the polyoxoanion is *the* integral and

In the above context, it is useful to categorize the *new classes* of polyoxoanions in catalysis over the last ca. 20 years, and this is done in Figure 1 [7,8]. Figure 1 was first constructed for use in a lecture given at the 1998 Dallas ACS Meeting [7]. After the talk, several people remarked "you should publish that", and Figure 1 has since appeared elsewhere [8]. It is reproduced here for the convenience of the reader and as a lead-in to the research to be discussed herein. Note that it is *not* the purpose of this figure to claim priority for any specific work, nor is it claimed that Figure 1 is *the* absolutely complete, or unique, way to classify the newer classes of polyoxoanions in catalysis in the last ~20 years. This figure does, however, do a reasonable job of making apparent the major new classes of polyoxoanions in catalysis. It is also useful in helping guide future research—the original, primary goal behind the construction of Figure 1.

A main conclusion made apparent by Figure 1 is that both homogeneous and heterogeneous catalysis with polyoxoanions *is still a young, wide-open field for those willing to do new synthetic polyoxoanion chemistry.* Note also that a summary of the presently most commercially significant classes of polyoxoanions in catalysis would be a rather different figure, one which would emphasize classical polyoxoanions serving as strong-acid catalysts [24], and a figure which would include only a few of the newer subclasses of polyoxoanions in catalysis shown in Figure 1.[2] There are also emerging examples that time will likely show need to be added to Figure 1, for example Augustine's use of polyoxoanions to anchor cationic organometallic precatalysts to solid supports [25].

---

crucial part of the higher stability and isolability—and the unprecedented nanocluster catalysis in solution with limited metal-particle agglomeration—exhibited by these novel "soluble heterogeneous catalysts"; see elsewhere [21a,b]. In addition, the range of *possible* types of catalytic reactions by these "soluble heterogeneous catalysts" is large compared to those of any other single class of polyoxoanion-based catalysts since they potentially include many of the wide range of reactions exhibited by heterogeneous, metal-particle catalysts.

2. The *most significant reactions* among the classes shown in Figure 1, *and at present,* are probably the Misono strong acid catalysis and Ishii-Venturello H_2O_2 epoxidation chemistry. One should also note that the oldest, and the (at least presently) commercially most important areas of polyoxoanions in catalysis, employ off-the-shelf, classical polyoxoanions such as $PW_{12}O_{40}^{3-}$ or $SiW_{12}O_{40}^{4-}$ for strong acid or solid-state oxidation catalysis [24]. In the strong acid catalysis area, the SiO_2- heteropoly acid inclusion complex catalysts should perhaps be included in Figure 1 (see p.184 elsewhere [3]).

Other catalytic chemistries are not included in Figure 1 because the polyoxoanions employed are not novel, although in some cases the *use* of known polyoxoanions is novel and, hence, noteworthy for that, different reason. Examples include the patented, Weinstock-Hill wood pulp delignification reactions with V-substituted polyoxometalates (see p. 186 elsewhere [3]), or Augustine's use of polyoxoanions as anchors for cationic, organometallic precatalysts [25]. The latter catalysts are novel in that: (i) they show little leaching; (ii) their activity is not diminished vs. the activity of the precatalyst in solution—and, indeed, the activity is sometimes higher following support of the cationic organometallics, an unusual but desirable situation; and (iii) their % ee in enantioselective hydrogenation reactions is as high if not higher than that of the precatalyst in solution, also an uncommon result [25]. The evolving Augustine application of polyoxoanions is, then, a use of (to date) classical polyoxoanions that one should probably add even now to the solid-state, #5 classification in Figure 1.

2. Subclasses of Polyoxoanions Investigated by the Finke Research Group

Figure 2 shows the three subclasses of polyoxoanion-transition metal complexes investigated by the PI's research group. Our efforts in polyoxoanion framework-incorporated, "inorganic porphyrin" catalysts are detailed elsewhere [11]. One important conclusion from that work is that the $P_2W_{17}Mn^{III}O_{61}{}^{7-}$ complexes do, indeed, behave as "inorganic porphyrins" in terms of their proven selectivity, activity and other reactivity (e.g., regioselectivity in limonene oxygenations or *cis*-, *trans*- product stereoselectivity in *cis*-stilbene oxygenations).

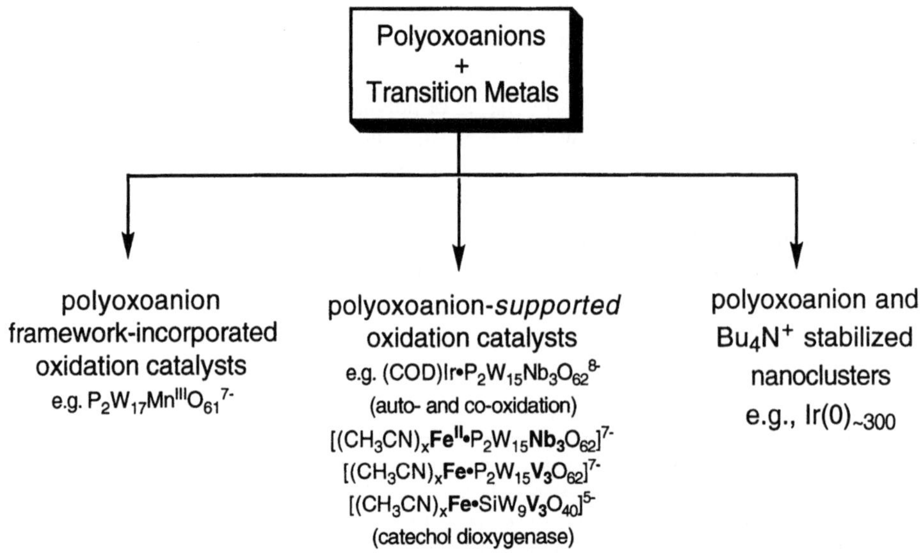

Fig. 2. The three subclasses of polyoxoanion-transition metal complexes in catalysis investigated by the Finke research group.

Polyoxoanion-supported catalysts are a subclass of polyoxoanions in catalysis that have been developed in the PI's labs over the last ca. 15 years. Summarized in a previous volume in this series along with lead references [5] is our use of the "molecular cookie cutter" approach, namely using basic, custom-made polyoxoanions such as $P_2W_{15}Nb_3O_{62}{}^{9-}$ to obtain a molecular-level analog of solid-oxide-supported heterogeneous catalysts, Figure 3.

Polyoxoanion-*Supported* Catalysts and Precatalysts $[(1,5\text{-COD})M\bullet P_2W_{15}Nb_3O_{62}^{8-}]$ (e.g., M = Ir^I, Rh^I)

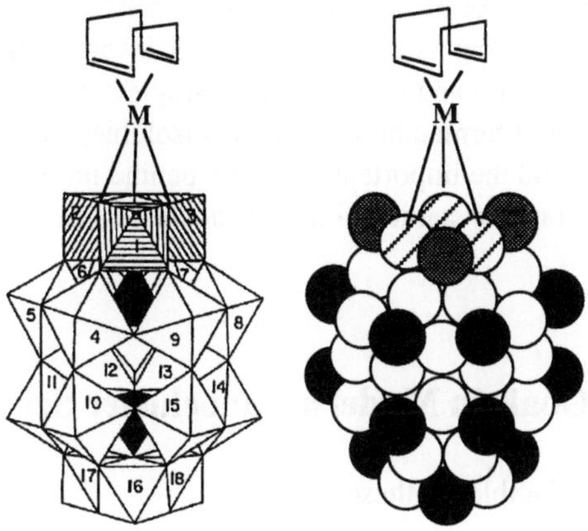

Synthesis: 9 Steps, *Inorg. Syn.* **1997**, *31*, 167.

Improved Synthesis: *Inorg. Chem.* **1996**, *34*, 7905.
[2 Steps, 4 days (30%) shorter, 59% higher yield (116g vs. previous 73g)].

Characterization: *Inorg. Chem.* **1990**, *29*, 1785 (comm. ed.)
Inorg. Chem. **1995**, *34*, 1413 (full paper)

Analysis (all elements: 100.18% total); ^{31}P, ^{17}O, ^{183}W, ^{13}C, ^{1}H NMR; MW (FAB-MS; solution MW); X-ray diffraction ($P_2W_{15}Nb_3O_{62}^{9-}$).

Key Prior Polyoxoanion Literature: W. Klemperer, V. Day and co-workers.

Fig. 3. The molecular "cookie-cutter" approach to oxide-supported catalysts exemplified by the prototype $P_2W_{15}Nb_3O_{62}^{9-}$–supported organometallics $M(1,5\text{-COD})^+$ (M = Ir^I, Rh^I).

3. Polyoxoanion-Stabilized Nanocluster "Soluble Heterogeneous Catalysts"

Transition-metal nanoclusters and their catalytic properties [21b] are one very active subarea of the currently important area of nano-scale chemistry . Some of the key goals in transition-metal nanoclusters are outlined in Figure 4.

Summarized elsewhere are other important background points for understanding this area, items such as [21a]: the key differences between modern transition-metal nanoclusters and classical colloids; the definition of near-monodisperse nanoclusters (those of ≤±15% size dispersion); the definition of higher stability, so-called "magic number" size nanoclusters; Schwartz's modern definition of heterogeneous (multi-active site) and homogeneous (single active site) catalysts (so that zeolites, for example, are best termed homogeneous-insoluble, rather than heterogeneous catalysts); and the important literature behind inorganic (charge or electrostatic) and organic (steric) stabilization of nanoclusters and colloids.

Key Research Goals in Modern Nanocluster Science Include:

- Rational, reproducible syntheses
- Syntheses yielding near monodisperse nanoclusters (± ≤15% size range)
- Isolable, redissolvable, well-characterized nanoclusters
- Well-defined composition
- Shape as well as size control
- Reproducible (± ≤15%) catalytic activity
- High catalytic activity in solution, with long lifetimes
- High—"single-site"—catalytic selectivity
- Mechanistic information (e.g., their mechanism of formation, and for their catalytic reactions)
- Understanding of why "magic number" (= full shell) nanoclusters form (M_{13}, M_{55}, M_{147}, M_{309}, M_{561}, M_{923}. . .)
- Hetero bi-, tri- and higher-metallic nanoclusters of known composition and structure
- 2-D, 3-D and other nanocluster assemblies (wires, nanoclusters plus DNA, nanocluster-based devices, etc.)

Fig. 4. Some of the key research goals in modern nanocluster science.

20±3 Å $Ir(0)_{\sim 190\text{-}450}$ (i.e., $Ir(0)_{\sim 300}$) Nanoclusters: Their Synthesis and Characterization

***Average* Stoichiometry of Formation:**

$$300\ (1,5\text{-COD})Ir\bullet P_2W_{15}Nb_3O_{62}^{8-} + 750\ H_2 \longrightarrow 300$$

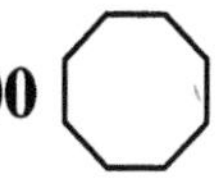

$$+$$

$$+\ Ir(0)_{\sim 300} + 300\ [P_2W_{15}Nb_3O_{62}^{9-}] + 300\ H^+$$

$$\downarrow$$

$$150\ \{H_2O + [(P_2W_{15}Nb_3O_{61})_2\text{—}O]^{16-}\}$$

Characterization:

- Elemental Analysis (on isolated material); *average* composition:

$$Ir(0)_{\sim 300}[(P_2W_{15}Nb_3O_{61})_2\text{—}O]_{33}(Bu_4N)_{\sim 300}Na_{\sim 228}$$

TEM; HR-TEM; Electron Diffraction [ccp Ir(0)]; Electrophoresis (proves they are *anions*); IR; UV-Vis; STM

Fig. 5. The stoichiometry of the synthesis, and the key characterizational methods used, for the polyoxoanion- and Bu_4N^+- stabilized $Ir(0)_{\sim 300}$ nanoclusters.

One of the novel findings from placing polyoxoanion-supported (1,5-COD)Ir• $P_2W_{15}Nb_3O_{62}^{8-}$ under H_2 was our discovery of *polyoxoanion-stabilized nanoclusters* [21a; 23]. Acttually, the nenoclusters were discovered only as the results of five years of painstaking mechanistic investigations during which the classic mechanistic problem of "is it homogeneous or heterogeneous catalysis?" was solved in a more general way [22]. That work involved disproving the prior myth that nanocluster, "heterogeneous" catalysts could not have a reproducible catalytic reactivity of $\leq\pm15\%$, that is, a reproducibility as good as that of many small molecule, monometallic homogeneous catalysts [22]. Figure 5 shows the stoichiometry of synthesis under H_2 of the 20±3 Å $Ir(0)_{\sim 190}$ to $Ir(0)_{\sim 450}$ nanoclusters, plus the key characterization methods employed [22]. Note that, as discussed in greater detail in nearly all of our nanocluster publications [e.g., 21a, 22, 23], the convenient notation of $Ir(0)_{\sim 300}$ is used as a *shorthand representation* of the true, 20±3 Å and thus still rather narrow ("near monodisperse" [21a]) ~15% size dispersion of $Ir(0)_{\sim 190}$ to $Ir(0)_{\sim 450}$ nanoclusters that are actually present.

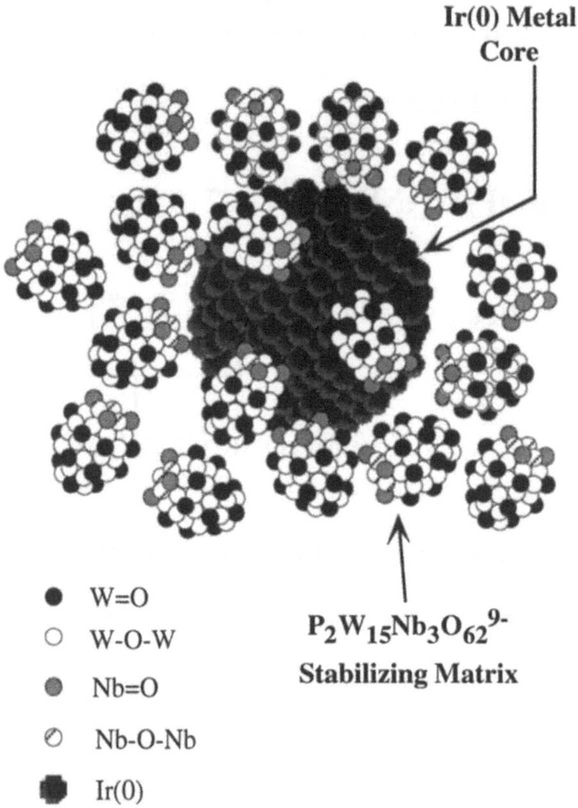

Fig. 6. Idealized, roughly-to-scale representation of a $P_2W_{15}Nb_3O_{62}^{9-}$ polyoxoanion and Bu_4N^+ stabilized 20 ± 3 Å $Ir(0)_{\sim300}$ nanocluster, $[Ir(0)_{\sim300}(P_4W_{30}Nb_6O_{123}^{16-})_{\sim33}](Bu_4N)_{\sim300}Na_{\sim228}$. For the sake of clarity, only 17 of the polyoxoanions are shown, and the polyoxoanion is shown in its monomeric, $P_2W_{15}Nb_3O_{62}^{9-}$ form (and not as its Nb-O-Nb bridged, anhydride, $P_4W_{30}Nb_6O_{123}^{16-}$ form). The ~330 Bu_4N^+ and ~228 Na^+ cations have also been omitted, again for the sake of clarity.

Figure 6 shows a representation of one 20 Å $Ir(0)_{\sim300}$ nanocluster and a portion of its stabilizing polyoxoanions [22]. This picture, plus a look back at the stoichiometry in Figure 5, reveals that a remarkable self-assembly reaction of >300 steps must (as discussed elsewhere [26]) be involved in the formation of these reproducibly formed [22, 23], near-monodisperse nanoclusters. The key details of the mechanism of formation of these nanoclusters, as well as the remarkable finding that the >300 step mechanism can be described by *only two, pseudo-elementary kinetic steps* [26] (one of which is repeated ca. 300 times), are available elsewhere for the interested reader [26]. One important result of the reproducible formation of these nanoclusters is that the resultant catalyst exhibits a ≤±15% catalytic reproducibility which, previously, had been reserved for small molecule, organometallic catalysts [22].

Why Are $P_2W_{15}Nb_3O_{62}^{9-}$ Polyoxoanion-Stabilized Nanoclusters So Stable?

A *Combined*, High Charge ("Inorganic"; polyoxoanion) *Electostatic* and *Steric* ("Organic"; $(Bu_4N)_9^{9+}$) Colloid-Type Stabilization

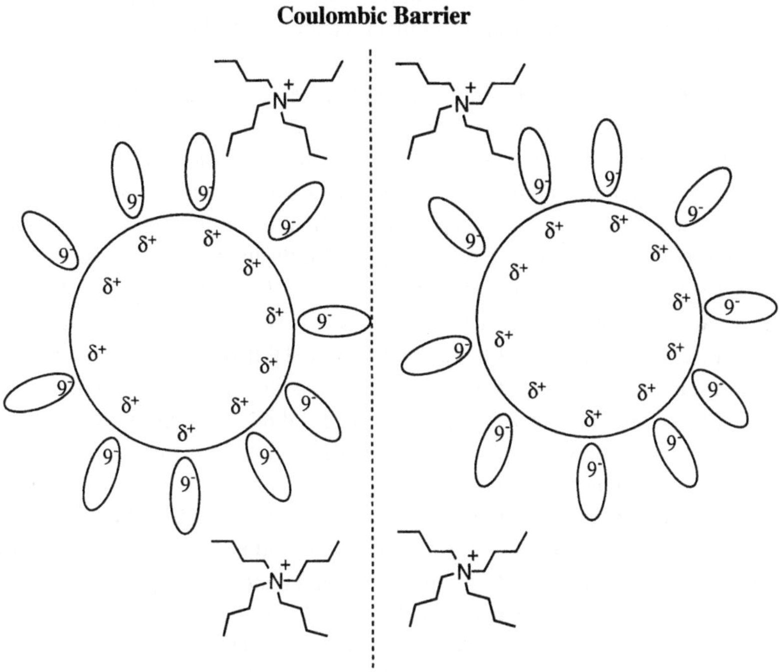

(1) C. S. Hirtzel, R. Rajogopalan: "Colloidal Phenomena: Advanced Topics" Noyes: New Jersey (1985), pp 27-39, 73-87.
(2) R. J. Hunter: "Foundations of Colloid Science" Oxford University: New York (1987), vol 1, pp 316-492.

Fig. 7. Rationalization of the high stability of polyoxoanion-stabilized nanoclusters in terms of well established colloidal electrostatic and steric stabilization theory. Unusual, however, is the *combined, intrinsic* high polyoxoanion charge and associated high tetrabutyl ammonium and polyoxoanion steric stabilization components.

Of special interest back in Figure 6 is the basic, highly charged, $P_2W_{15}Nb_3O_{62}^{9-}$ polyoxoanion, which is actually present primarily as its Nb-O-Nb bridged, anhydride form, $[P_2W_{15}Nb_3O_{61}–O–P_2W_{15}Nb_3O_{61}]^{16-}$ [22,23]. Figure 7 shows, in a schematic way, why the high overall charge of the polyoxoanion, in combination with its associated Bu_4N^+ (and some Na^+ [22]) counter cations, provide a novel, high level of stabilization due to a little precedented "combined high charge plus significant steric bulk present intrinsically within the *poly*(oxo)anion and *poly*-Bu_4N^+ cation components of $(Bu_4N^+)_9(P_2W_{15}Nb_3O_{62}^{9-})$", as we first noted elsewhere [23]. One way to look at the polyoxoanion stabilization, in comparison to the classic colloidal stabilizer Cl^-, is that the polyoxoanion is a much larger, ca. 12 x 15Å, "nine-minus Cl^-", one that intrinsically comes with the associated steric ("organic") stabilization provided by the multiple Bu_4N^+ (and Na^+ [22]) counter cations. It is clear that the polyoxoanion component is the key to the unusual combination of high stability, yet also high catalytic activity (and, therefore, relatively high accessibility to the catalytic surface) present in these novel composition-of-matter, $Ir(0)_{\sim 300}$•polyoxoanion nanoclusters.

We have gone on to make polyoxoanion-stabilized Rh(0) nanoclusters [27] and the syntheses of a variety of other transition metal nanoclusters (Pd, Pt, others) are also in progress. Of interest is that the polyoxoanion-stabilized Rh(0) nanoclusters currently hold the record for catalytic activity, stability and thus record catalytic lifetime, 190,000 total turnovers (TTOs) of cyclohexene hydrogenation, in comparison to all known, transition-metals nanocluster catalysts [28]—a claim that is fortified by our comprehensive, 197 reference-containing review on modern transition-metal nanoclusters in catalysis [21b]. Of further interest is that, under identical conditions, Wilkinson's hydrogenation catalyst, $RhCl(PPh_3)_3$, exhibits an order of magnitude fewer TTOs (19,000), while a classical, low surface area, 5%Rh(0) /Al_2O_3 heterogeneous catalyst deactivates after less than a factor of 2 more TTOs (350,000). A significant observation is that the insolubility of the polyoxoanion-stabilized Rh(0) nanocluster catalyst in the cyclohexane hydrogenation product is the only apparent limit to the nanoclusters' TTOs [28]. This observation has suggested other, in-progress experiments which may allow the nanocluster to surpass the lifetime of even the 5%Rh(0) /Al_2O_3 heterogeneous catalyst.

We have also reported, and refer the interested reader to, the following additional nanocluster papers detailing: (i) the first new, and most detailed, mechanistic work in nearly 45 years on how modern transition metal nanoclusters grow, work that revealed a new mechanism consisting of (a) *slow, continuous nucleation*, followed by (b) *fast, autocatalytic surface-growth* [26]; (ii) the effects of H_2 mass-transfer limitations on nanocluster size dispersity (and the mechanistic insights into competing agglomeration and autocatalytic surface growth that this observation provides) [29]; (iii) the first designed, rational synthesis of a series of dispersions of nanoclusters centering about 4 adjacent magic number sizes [30]; (iv) the first mechanistic explanation for the common occurrence of "magic-number"-size nanoclusters [30] (it is a natural outcome of surface-autocatalytic growth) [26]; and (v) the first, mechanism-based scheme for the rational synthesis, at least in principle, of all possible geometric isomers of tri- or higher-multimetallic nanoclusters [30], materials of considerable interest as bi-, tri- and higher multi-metallic catalysts.

We have also shown that polyoxonion-stabilized nanocluster benzene hydrogenation catalysts can be made, and will report those results in due course. He benzene hydrogenation work is an outgrowth of our mechanistic work revealing that what was previously thought to be a putative $[(C_8H_{17})NCH_3]^+$ $[RhCl_4]$– "ion-pair" benzene hydrogenation catalyst is, in fact, a Cl^- and $[(C_8H_{17})NCH_3]^+$ stabilized $Rh(0)_n$ nanocluster benzene hydrogenation catalyst [31]. In unpublished work in progress, we have demonstrated that this result is more general in that other, putatively "mono-metallic, homogeneous" benzene hydrogenation catalysts are actually nanocluster catalysts as well.

In short, polyoxoanion-stabilized nanoclusters are presently the best examples of "soluble analogs of heterogeneous catalysts" [21a]. This long-sought goal is one that inorganic, organometallic and catalytic chemists have been pursuing for more than two decades, a goal popularized by, for example, Mutterties' writings on the cluster-catalysis hypothesis (i.e., that metal carbonyl clusters might fulfill this goal) nearly 20 years ago [32].[3] Hence, polyoxoanion-stabilized nanocluster "soluble heterogeneous catalysts" will continue to be the focus of our synthetic, characterization, catalysis (of oxidation, reduction as well as other reactions), and mechanistic studies for years to come.

4. Vanadium-Containing, Polyoxoanion-Based Record Catĕchol Dioxygenase Catalysts

4.1 INTRODUCTION AND BACKGROUND

A *dioxygenase* is, by definition, an oxygenation reaction in which both atoms of dioxygen, O_2, are inserted into the substrate without the addition of external reductants (e.g., without the $2e^-$ and $2H^+$, or their equivalent, that characterize *monooxygenases* and which, therefore, convert one atom of O_2 to H_2O). The economic, environmental, process and other advantages of a system that would use only substrate plus O_2 are perhaps obvious. The advantages of a true dioxygenase catalyst are why, for at least the >15 years since Groves' classic $Ru^{II/IV/VI}$ porphyrin system [33] and probably for >10 years before that (i.e., for ≥ 25 years), the Holy Grail of oxygenation catalysis has been the invention of highly active, long lived and eventually commercially useful dioxygenase catalysts. Reaffirmation of this Holy Grail of oxygenation catalysis has recently appeared elsewhere [34]. One key problem to be overcome, of course, is how to activate either O_2 or a given substrate without turning on the free-radical-chain autoxidation that ground-state triplet dioxygen diradical generally prefers.

---

3. The CO-poisoned nature of metal carbonyl clusters led, however, to limits to the activities, and type of catalytic reactions, observed for carbonyl clusters. Carbonyl clusters have, however, served as impressive, X-ray crystallographically characterizable models of the possible bonding modes of a wide range of ligand types to multimetallic complexes; see for example the work of R. Adams for lead references to an extensive series of papers [32].

Key Goals: Highly Selective and Catalytic Non-Enzymic Dioxygenases in an Oxidation-Resistant Ligand System

Dioxygenases: Biological Mimics[1]

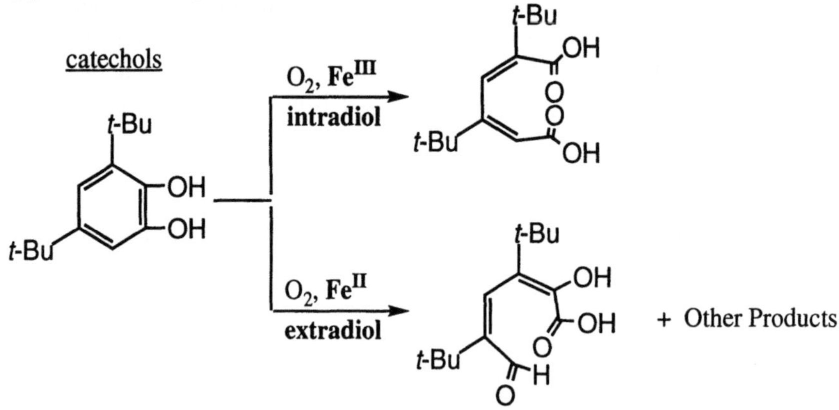

α-keto glutarate

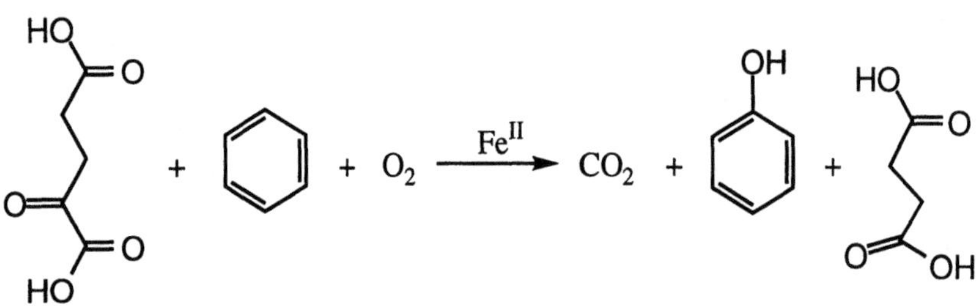

[1] See the ≥ 40 dioxygenase enzymes and reaction types in: M. Nozaki: *Topics Current Chem.* **78**, 145 (1979).

Dioxygenases: Abiological / Unfunctionalized Olefins[2]

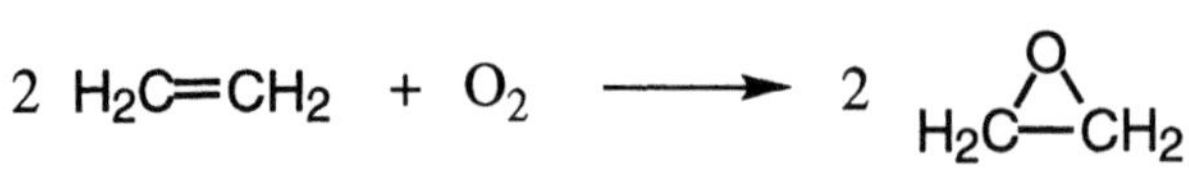

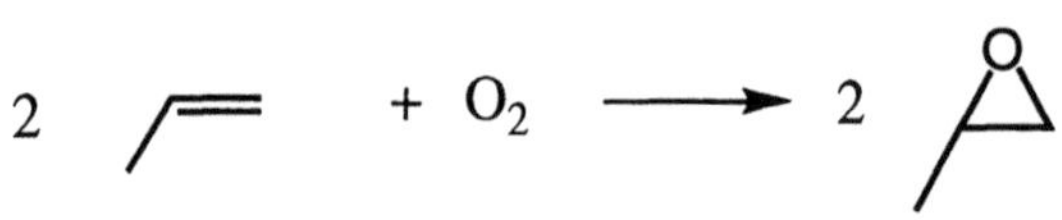

(and without allylic oxidation!)

[2] J. T. Groves, R. Quinn: *J. Am. Chem. Soc.* **103**, 5790 (1985); [Ru^{II}(tetramestiyl porphyrin)].

Fig. 8. Some key goals in abiological dioxygenase catalysis.

Nature has clearly solved the dioxygenase problem: there are more than 40 dioxgenase enzymes, many of which have been known for more than 20 years now [35,36,37]. Figure 8 summarizes three known dioxygenase reactions, Fe^{III}-based catechol intradiol-cleaveage dioxygenases (e.g., the enzymes catechol 1,2-dioxygenase or protocatechuate 3,4-dioxygenase), Fe^{II}-based extradiol dioxygenases (e.g., the enzyme metapyrocatechase), and the interesting case of α-ketoglutarate dioxygenase [36,37]. Note that in each case, and as required for a true dioxygenase, no external reductant is required and both oxygen atoms wind up in the substrate (in the case of the α-ketoglutarate dioxygenase, one O atom winds up in the benzene substrate and one in the oxidatively decarbonylated product, succinic acid). Of interest here is that the Fe^{II} extradiol cleaving dioxygenases are "more abundant than the intradiol enzymes, but they are less stable due to their sensitivity to oxidizing compounds" (see p. 46 elsewhere [37a]). The Fe^{II} extradiol cleaving enzymes are, therefore, less well studied mechanistically and, in general, less well understood than the intradiol-cleaving dioxygenase enzymes [36,37]. In addition, the α-keto acid-dependent dioxygenases "comprise one of the largest classes of non-heme iron proteins" (see p. 78 elsewhere [37a]), but are also relatively poorly studied and, therefore, not well understood either [36,37].

For some time now the PI's group has had ongoing synthetic, catalytic and mechanistic studies of polyoxonion-based oxidation catalysts with the sole long-term goal of inventing a long-lived, polyoxoanion- or other, all-inorganic- (and thus oxidation-inert) ligand-based dioxygenase. Not surprisingly, the slow, most time-consuming component of these studies has been the synthesis and detailed characterization of the resultant polyoxoanion precatalysts; however, that required phase of the work seems to be largely behind us at this point. Available elsewhere are our studies of polyoxoanion-supported olefin autoxidations [38], polyoxoanion-supported co-oxidative epoxidations (including the significant finding that the whole area of metal-catalyzed co-oxidative epoxidations appear to be inferior to simple, deliberately ROOH or AIBN-initiated co-oxidative epoxidations [39]), Nb-peroxo polyoxoanion-based allylic alcohol epoxidations using H_2O_2 [40], and the "inorganic porphyrin" oxidation catalysis mentioned earlier [11]. The mechanistic studies of autoxidation [38] are, and despite the 50+ year-old nature of autoxidation research, of interest since they provide a more simple, GLC-product-based way to unequivocally detect autoxidation than was previously available—including its more than 70 products in the case of cyclohexene autoxidation [38c]. Others entering the oxidation catalysis area will find that paper of interest since only when one *truly learns* the relevant free-radical- (e.g., Haber-Weiss chain) and related chemistry (which can easily be "disguised" to the uninitiated [41]), is one then in a position to avoid the otherwise substantial induction prior to doing new, interesting research in the mechanistically complex area of oxidation catalysis.

Returning to our dioxygenase catalysis, it occurred to us that no one had taken the logical tact of investigation of the key catechol, α-ketoglutarate and other prototype dioxygenase reactions (i) with Fe•polyoxoanions as catalysts or precatalysts, or (ii) even with the goal of demonstrating long-lived catalysis for these ostensibly simpler dioxygenase reactions (e.g., catechol dioxygenation being simplier than the more interesting, but also much more difficult—especially at our present level of knowledge—dioxygenation of unfunctionalized olefins or alkanes). We further reasoned that the surprisingly large gap in our knowledge of non-enzymic dioxgenase catalysis, and the underlying mechanisms, might be ideally filled in by studies using the oxdatively-resistant ligand systems provided by polyoxoanions—that is, studies in which the normally complicating factor of catalyst oxidative degradation would be minimized if not eliminated completely.

Dr. Heiko Weiner's search of the dioxygenase and related literature revealed three additional key insights: (iii) that Fe and V were the key ingredients of the prior, most active and interesting reports of, for example, the prototype 3,5-di-*tert*-butylcatechol (hereafter DTBC) oxygenation (see the tabulation of literature provided in the Supporting Materials elsewhere [42])—hence Fe and V became focal point of our initial studies; (iv) that the system to beat in terms of know lifetime (TTOs) was simple $VO(acac)_2$ and its report of 500 TTOs [43]; and (v) that even enzymic dioxygenases lack reports of demonstrated TTOs that are in even the thousands to tens-of-thousands (see footnote 27 elsewhere [42] for key references and a discussion of why this lack of TTO studies in dioxygenase enzymes is perhaps as expected).

The literature further indicates that (vi) sufficient, adjacent sites of coordinative unsaturation are crucial for creating at least bio-mimetic dioxygenase catalysis [36,37,44], so that this, too, became one of our guiding design principles. Note that in this regard Neumann's interesting, albeit slow, adamantane oxidation precatalyst [45], which yields an apparent dioxygenase, lacks adjacent sites of coordinative unsaturation and, therefore, is expected to be unreactive in the DTBC dioxygenase reactions investigated herein, and assuming that the Neumann precatalyst $[WZn_2Ru^{III}_2(OH)(OH_2)(ZnW_9O_{34})_2]^{11-}$ remains intact. In fact, control experiments have shown this to be the case: there is no DTBC dioxygenase reaction after 25 hours using the Neumann precatalyst [45].

In a seventh, potentially very important insight from studying the literature, the PI realized (vii) *that arguably <u>the</u> major mechanistic difference in dioxygenase vs. monooxygenases is that dioxygenases appear to avoid O–O bond-cleavage activation of O_2*, at least initially (i.e., early in the catalytic cycle)! This insight appears to not be widely appreciated (although strong hints are available in Funabiki's most recent review, however, where he notes that "activation of oxygen by a metal center is essential in the case of monooxygenases, but is not necessarily so in the case of dioxygenases"; see p. 107 elsewhere [37b]). Moreover, one mechanism commonly written for the Fe^{III}-intradiol cleaving enzymes, Scheme 1, due to Professor L. Que effectively contains this insight as well—note that *substrate activation*, but not O_2 activation, is key to the proposed mechanism [36a,46].

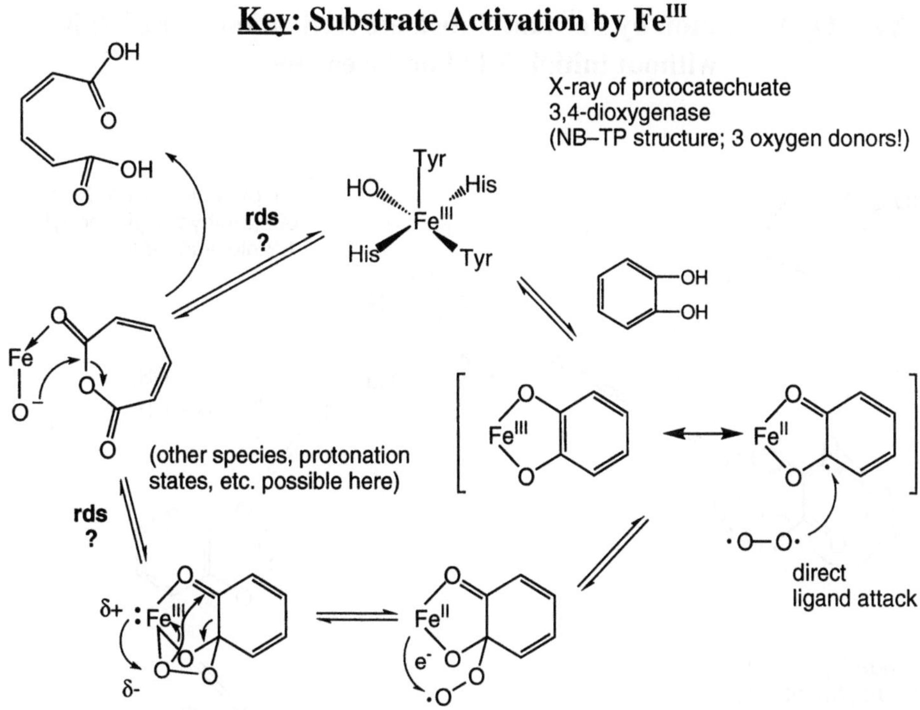

Scheme 1. Que et al.'s proposed mechanism for an Fe^{III} intradiol dioxygenase [46], adapted from Que and Ho's review [36a]. Note that somewhat different mechanisms are written by others, for example Funabiki [37a] who writes a mechanism involving only Fe^{III} and, to accomplish this, electron transfer to coordinated dioxygen and mono- (vs. the above indicated di-) coordination of catechol throughout the catalytic cycle.

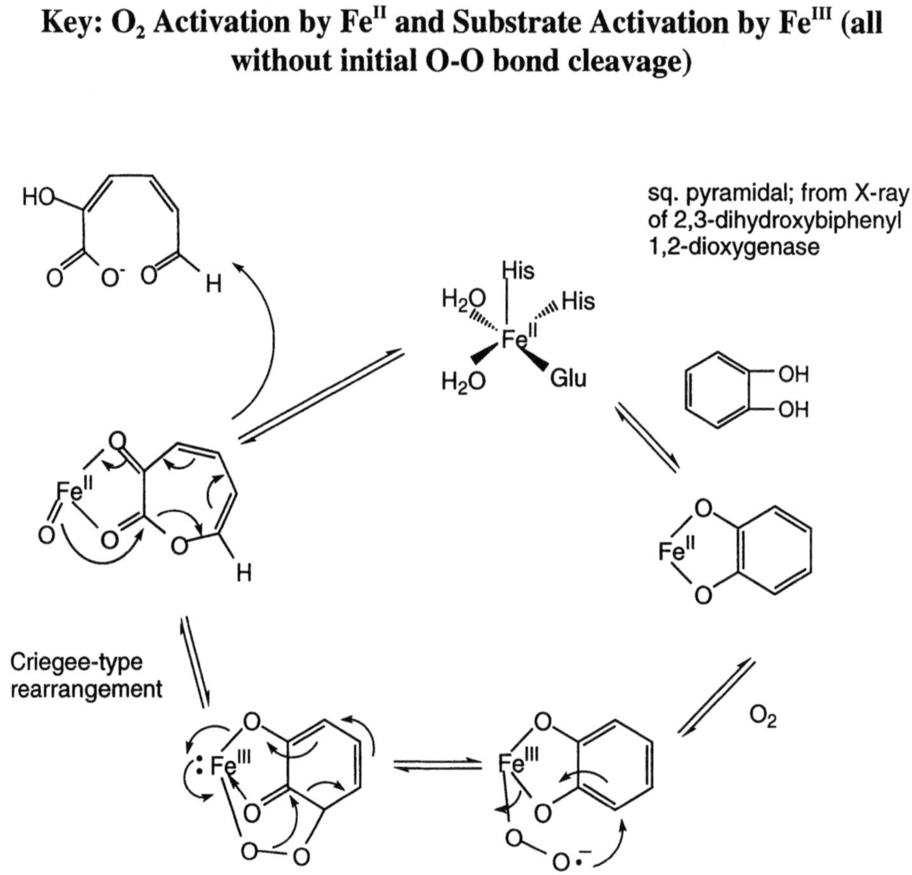

Scheme 2. The proposed, commonly written Fe^{II} extradiol dioxygenase mechanism derived from contributions by Arciero and Lipscomb [47], Bugg and co-workers [48] and as summarized in (and adapted from) Que and Ho's review [36a]. Note that a key feature is that while O_2 is activated in this case by electron transfer from $Fe^{II}{\bullet}O_2$ to form $Fe^{III}{\bullet}O_2^{-\bullet}$, there is no O-O bond cleavage (i.e., and no Fe=O intermediates) until *after* the substrate and the bound superoxide react, and until later in the catalytic cycle.

A final, key point then, and in our approach to developing polyoxoanion-based dioxygenases, is (viii) that no one appears to have utilized the mechanistic insights and predictions involved in the Fe^{III} and Fe^{II} dioxygenases literature mechanisms summarized in Scheme 1 and 2. Doing so became, therefore, another key component of our dioxygenase research plans.

4.2 A NOVEL POLYOXOANION-BASED IRON(II) CATECHOL EXTRADIOL DIOXYGENASE, $[(CH_3CN)_xFe^{II}{\bullet}P_2W_{15}Nb_3O_{62}]^{7-}$

Our studies of polyoxoanion-based dioxygenases began with our prototype polyoxoanion-support system, $P_2W_{15}Nb_3O_{62}^{9-}$ and Fe^{II}, namely

$[(CH_3CN)_xFe^{II}{\bullet}P_2W_{15}Nb_3O_{62}]^{7-}$. Again, the synthesis and characterization of this polyoxoanion-supported complex proved to be the slowest and most demanding step [49]. Elemental analysis and evidence from a useful O—>M charge transfer transition reported and used for the first time as part of the synthetic work [49] confirm the now very well precedented support of the transition metal atop the $P_2W_{15}Nb_3O_{62}^{9-}$ polyoxoanion [5; and references therein]. However, this 1 Fe^{II} / 1 $P_2W_{15}Nb_3O_{62}^{9-}$ complex could not be manipulated further since, when manipulated, it disproportionates", 2 {1Fe^{II} / 1 $P_2W_{15}Nb_3O_{62}^{9-}$} to 1 {$(Fe^{II})_2$ / 1 $P_2W_{15}Nb_3O_{62}^{9-}$} plus 1 {$P_2W_{15}Nb_3O_{62}^{9-}$} [49].

Despite its sensitivity, and despite proving to be a stoichiometric reagent and not a catalyst, this complex still provided results of interest [50]. The observed reaction stoichiometry is shown in Scheme 3.

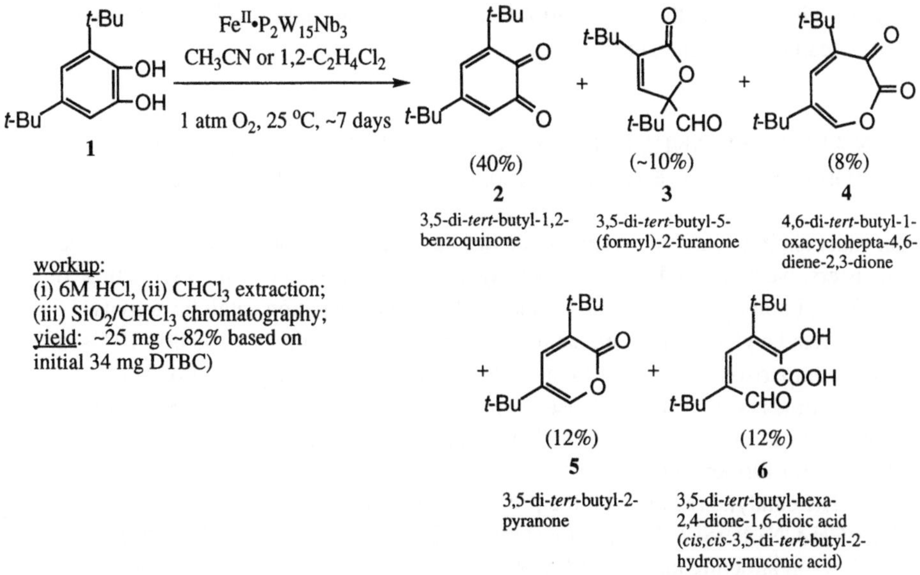

Scheme 3. Summary of the DTBC oxidation products that were identified: the DTBC autoxidation product **2**, and the extradiol products **3 - 6**. The compound numbers are the same as those used in the original publication for the convenience of the reader referring back to that paper [50].

Several items are of interest regarding the stoichiometry and other observations surrounding the reaction [50]: (i) the starting complex $[(CH_3CN)_xFe^{II}{\bullet}P_2W_{15}Nb_3O_{62}]^{7-}$ was found to bind exactly 1.0 equivalent of DTBC (see Figure 2 elsewhere [50]); (ii) after this, the $[DTBC{\bullet}Fe^{II}{\bullet}P_2W_{15}Nb_3O_{62}]^{7-}$ complex picks up 1.06 ± 0.10 equivalent of O_2, *the precise amount which defines a catechol dioxygenase*; and (iii) room temperature exposure of an unstirred solution of $[DTBC{\bullet}Fe^{II}{\bullet} P_2W_{15}Nb_3O_{62}]^{7-}$ in acetonitrile to O_2 produces three distinct color changes, corresponding to three distinct, independently observable reaction stages and intermediates, A $\rightarrow$ B $\rightarrow$ C $\rightarrow$ D. A novel feature of the reaction dynamics is that each reaction stage is slower than the proceeding stage, so that each reaction stage and intermediate could be studied, at least in principle.

Unprecedented features of this extradiol dioxygenase system, then, and in comparison to the extant literature [36,37], are: (a) the all-inorganic, polyoxoanion-based, and thus oxidation resistant ligand system; (b) the clean 1:1:1 DTBC : O_2 : Fe^{II} : $P_2W_{15}Nb_3O_{62}]^{7-}$ stoichiometry *involving a well-defined Fe^{II} precursor*, (c) the catechol *substrate-triggered* activation of dioxygen; (d) the proof that the system binds 1.0 equivalent of O_2 to (initially) Fe^{II}, and (e) the three distinct and resolvable reaction stages, each slower than the previous stage, a feature that makes the systems ideal for reaction intermediate and kinetic and mechanistic studies. The previous lack of clean, well defined systems, ones that would allow mechanistic studies of Fe^{II} catechol extradiol dioxygenase model reactions, is one main reason that extradiol dioxygenases have remained poorly understood[4] [36a]. The above polyoxoanion-based $[Fe^{II}•P_2W_{15}Nb_3O_{62}]^{7-}$ is, then, of additional significance mechanistically and given this system's features cited in (a)-(e) above.

4.3 AN ALL-INORGANIC, POLYOXOMETALATE-BASED CATECHOL DIOXYGENASE THAT EXHIBITS ≥100,000 CATALYTIC TURNOVERS

In our most exciting catalytic results since our development of polyoxoanion-stabilized nanoclusters, we combined the above 8 insights from the literature with a screening[5] of 24 Fe / V and other polyoxoanions for their DTBC dioxygenation catalytic activity [42]. The results yielded a record catalytic dioxygenase lifetime of >100,000 TTOs, a record in comparison to all either man-made or *demonstrated* enzymic dioxygenase catalyst lifetimes (see footnote 27 elsewhere for a discussion of this latter point[42]). The catalytic survey results also provided the best 4-5 polyoxoanion-based DTBC precatalysts for further studies [42].

The catalytic results with the prototype Fe/V polyoxoanion precatalyst, $(n\text{-}Bu_4N)_5[(CH_3CN)_xFe•SiW_9V_3O_{40}]$, are shown in Scheme 4. (This complex and its Wells-Dawson structure analog, $(n\text{-}Bu_4N)_5[(CH_3CN)_xFe•P_2W_{15}V_3O_{62}]$, were new complexes synthesized and characterized as part of this work and as detailed in our paper [42].) The "standard conditions" catalytic runs summarized in Table 1 of our

4. In their 1996 *Chem. Rev.*, Que and Ho noted, regarding the proposed Fe^{II} extradiol mechanism (reproduced as Scheme 2 herein), that "thus far, reaction conditions have not been identified that would allow some of the steps to be kinetically resolved" [36a].

5. (a) One thing that catalytic chemists understand, and have understood since long before combinatorial chemistry came onto the scene, is *the value and power of up-front catalyst screening*—the PI teaches this as the "lesson of Mittasch" in his catalysis class, Mittasch having discovered the apparently still best Fe-based ammonia synthesis catalyst from his screening in 1909 of 2500 catalysts in 6500 screening experiments (see the discussion and reference [50d] provided in footnote 17 elsewhere [21a]).
(b) Up-front screening by D. Edlund in the PI's labs, then at Oregon and more than 18 years ago, of a variety of non-basic to basic polyoxoanions, was also a crucial component leading to the discovery of polyoxoanion-stabilized nanoclusters and the presently preferred, basic, $P_2W_{15}Nb_3O_{62}^{9-}$ polyoxoanion [23, and references therein].

paper [42], and illustrated in Scheme 4, reveal ~3400 TTOs, a catalytic lifetime itself ~7 times larger than the previous record of 500 TTOs [43].

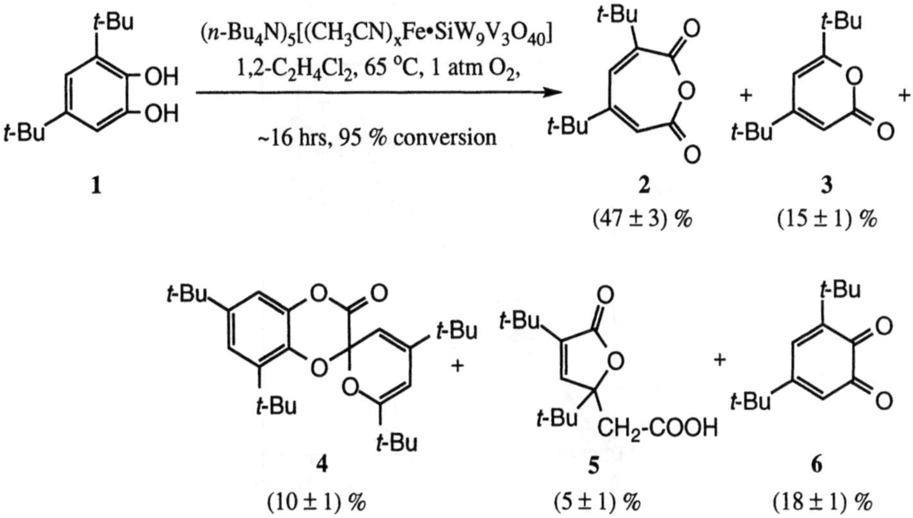

Scheme 4. Product formation in the catalytic oxygenation of 3,5-di-*tert*-butylcatechol, **1**, with the polyoxometalate-supported iron precatalyst, $[(CH_3CN)_xFe{\bullet}SiW_9V_3O_{40}]^{5-}$, and molecular oxygen. The compound numbers are the same as those used in the original publication for the convenience of the reader referring back to that paper [42] and, hence, the above compound numbers have no relationship to those in Scheme 3, for example.

A record >100,000 TTOs was demonstrated via the experiments summarized in Figure 9. Note that this figure contains duplicate runs, under identical conditions, of the previous best DTBC catalyst, $VO(acac)_2$, but that catalyst is clearly inferior (>30 fold slower). The TTOs demonstrated is about 3 orders of magnitude better than the typical 50-80 TTOs demonstrated in the literature for synthetic Fe complexes, and is 2.3 orders of magnitude better than the previous record of 500 TTOs [42].

The polyoxoanion-based DTBC catalytic studies revealed a number of other important results and insights as summarized elsewhere [42]; a few of the highlights are that: (i) the mass balance is very high, 95±5%, in comparison to earlier DTBC dioxygenase literature and even at a standard conditions DTBC : catalyst ratio of ca. 3400:1; (ii) the high mass balance and synthetically useful amounts of products (more than 10 g in some runs) allowed the isolation, purification and X-ray crystallographic and other full characterization of the main organic products, including the new, spiro product, **4** (and its isomer, **4'** [42]); and (iii) O_2 uptake stoichiometry studies were performed and prove that the net reaction has a ~1:1 O_2-to-DTBC stoichiometry and, therefore, behaves as a true dioxygenase. That the products are also those expected and seen in the literature for other dioxygenase systems fortifies the finding that the present reaction, Scheme 4, is a dioxygenase. The 18% conversion of DTBC to the benzoquinone product, **6**, is of course by definition an autoxidation, not a dioxygenase, product. Note, however, that a mechanistically very useful feature of selecting DTBC as a substrate is that *it is easily autoxidized and is, therefore, a sensitive probe of the always competing, undesired autoxidation pathway.* In this sense, then, the selectivity to dioxygenase products by the polyoxoanion-based precatalysts is high.

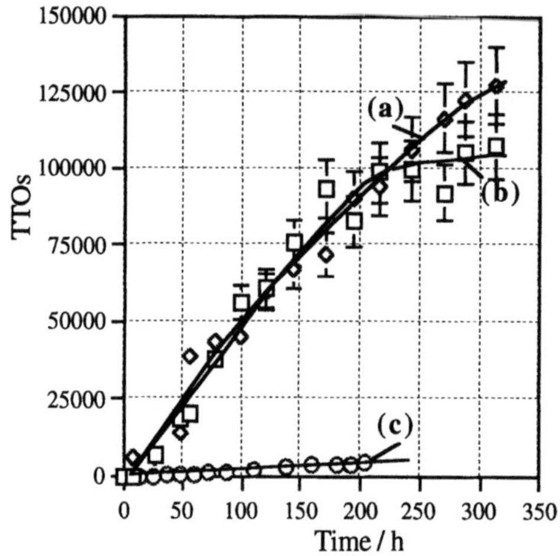

Fig. 9. Total turnovers (TTOs) vs. time plots obtained in a typical total turnover experiment using the Keggin-type polyoxometalate $(n\text{-}Bu_4N)_5[(CH_3CN)_xFe{\bullet}SiW_9V_3O_{40}]$. The experiment was carried out in 125 mL 1,2-$C_2H_4Cl_2$ with 14.0 g (62.97 mmol) DTBC and 1.83 mg (4.89 x 10^{-4} mmol) $(n\text{-}Bu_4N)_5[(CH_3CN)_xFe{\bullet}SiW_9V_3O_{40}]$ at 65°C and at 1 atm oxygen pressure. Approximately 95±5 % conversion of DTBC (curve a) was found after a total reaction time of 312 h, corresponding to ca. 127,000 total turnovers; ca. 25.5 mmol of products (**2**-**6**) were found at the end of the reaction corresponding to ca. 107,000 product-formation-based TTOs (curve b). For comparison, added to this Figure (curve c) is the best DTBC conversion, out of two independent experiments, using an equimolar concentration of the previously most highly catalytic catechol dioxygenase precatalyst described in the literature, $VO(acac)_2$, all under otherwise identical conditions.

Further findings of interest are that : (iv) the oxygen uptake studies, along with the high mass balance, allowed the net reaction to be dissected into five formally parallel reactions, (a)-(e), which yield the 5 major products, a dissection which allows one to understand, for the first time, how the ~1:1 O_2-to-DTBC stoichiometry actually arises; and (v) a variety of solvents were surveyed revealing that best solvents, such as 1,2-$C_2H_4Cl_2$, are those that are weakly coordinating, results again implicating that multiple adjacent sites of coordinative unsaturation are probably a crucial feature of this and other dioxygenase catalysts (recall the literature on this point discussed in the Introduction, section 4.1). The challenge of finding an environmentally more friendly, "greener", non-coordinating solvent than chlorinated hydrocarbon solvents [34b] remains, however.

We also reported some initial kinetic and mechanistic and catalyst isolation studies [42]; continuing those, and discovering the true identity of the active catalyst, are our two highest priorities currently in the polyoxonion-based dioxygenase area. The unusual, sigmoidal-shaped kinetic curves for the catechol dioxygenase catalyst beginning with $(n\text{-}Bu_4N)_7SiW_9V_3O_{40}$, **I**, (a), and $(n\text{-}Bu_4N)_5[(CH_3CN)_xFe{\bullet}SiW_9V_3O_{40}]$, **III**, (b), shown in Figure 10 are fairly well fit by a A —> B (rate constant k_1, induction period) step, then A + B —> 2 B (rate constant k_2) minimal ("Occam's Razor") kinetic mechanism. Note that the latter, A + B —> 2 B reaction is the elementary step which defines autocatalysis. The implication is that a product B is also a reactant which, therefore, then turns on the reaction after the induction periods seen in Figures 10a and 10b.

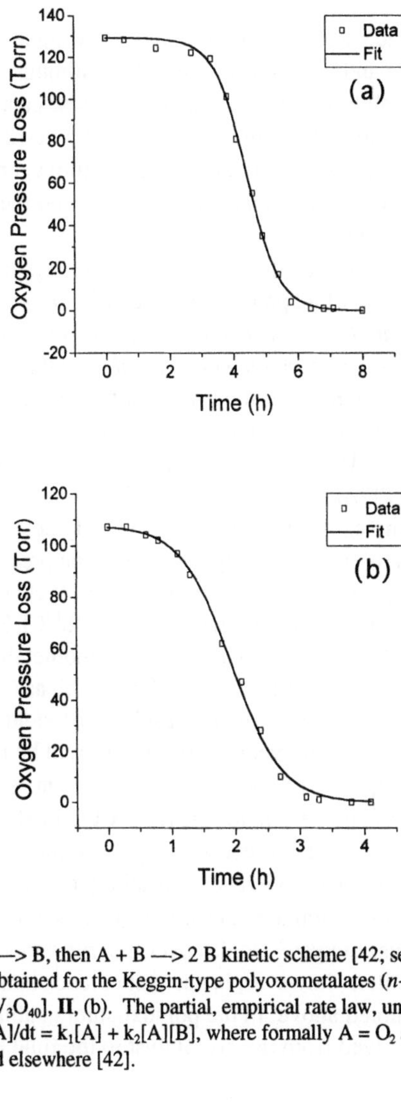

Fig. 10. Curve fits of an A —> B, then A + B —> 2 B kinetic scheme [42; see also reference 21 therein] to the oxygen uptake results obtained for the Keggin-type polyoxometalates $(n\text{-}Bu_4N)_7SiW_9V_3O_{40}$, **I**, (a), and $(n\text{-}Bu_4N)_5[(CH_3CN)_xFe\bullet SiW_9V_3O_{40}]$, **II**, (b). The partial, empirical rate law, under the experimental conditions employed, is therefore: $-d[A]/dt = k_1[A] + k_2[A][B]$, where formally $A = O_2$ and the possible identities of B are discussed in the text and elsewhere [42].

The partial, empirical rate law (and under the specific conditions of the kinetic measurements, detailed in the Experimental elsewhere [42]) is: $-d[A]/dt = k_1[A] + k_2[A][B]$, where A is formally O_2 (but where $\{-d[O_2]/dt\}/n = \{d[\mathbf{I}]/dt\}$ for precatalyst, **I**, for example and where n is a coefficient determined by the reaction stoichiometry [42]). The dependencies on DTBC, **I** or **III** and the other components of the complete rate law are under further investigation, as is the initial evidence [42] that B can be products like RCO_2H or (in our recent unpublished results) H_2O_2. Noteworthy here is that DTBC autoxidation to quinone probably produces H_2O_2 initially, that is, 1 DTBC + 1.0 O_2 —> 1.0 benzoquinone, **6**, + 1 H_2O_2, and even though the formal stoichiometry for DTBC oxidation with the minimal amount of O_2 is as written elsewhere (see Scheme 4, reaction (e), provided in reference [42]): 1 DTBC + 0.5 O_2 —> 1.0 benzoquinone, **6**, + 1 H_2O.

Even the initial kinetic studies are definitive in showing that the polyoxoanions are precatalysts and not the true catalysts—although they are highly effective precatalysts which lead to record lifetime dioxygenase catalytic activity. Fortunately, we have been able to isolate highly active, deep blue-black catalyst materials; the purification and full characterization of these dioxygenase catalysts, plus seeing if they too, or just their polyoxoanion precatalysts, exhibit the record lifetimes, are our highest priorities at present.

Note that another lesson of catalysis is repeated here, again one the PI teaches in his catalysis class, the principle of a "precatalyst reservoir": *often in catalysis it is sufficient, if not the best one can do, to prepare well-defined catalyst precursors that are ideally at most a few steps away from the active catalyst.* One then must perform kinetic and mechanistic studies to discover the true catalyst. (An alternative, often-stated form of this concept is: "what you can isolate or observe spectroscopically is only rarely the actual catalyst".) The classic example here is Wilkinson's hydrogenation catalyst, $RhCl(PPh_3)_3$, where Halpern's definitive mechanistic work showed that all 5 rhodium complexes that had been previously detected, and which were claimed to be "the catalyst", were actually just part of the precatalyst reservoir of 5 complexes that are 1 to 2 steps away from the true catalytic cycle (a cycle involving $Rh^ICl(PPh_3)_2$(solvent), its H_2 oxidative-addition product, $H_2Rh^{III}Cl(PPh_3)_2$(solvent), and so on; see Figure 10.1 on p. 532 of the Collman et al.'s organometallic textbook [51]).

In the case of the polyoxoanion-based complexes such as $(n\text{-}Bu_4N)_5[(CH_3CN)_xFe{\bullet}SiW_9V_3O_{40}]$, they are, then, "just" record-yielding catalyst precursors. Our in-progress kinetic and mechanistic studies will be required to identify the true catalyst, and then perhaps even improve upon the present record dioxygenase catalytic results. The same is of course true for the $[(1,5\text{-COD})M{\bullet}P_2W_{15}Nb_3O_{62}]^{8-}$ (M = Ir, Rh) precatalyst which yielded the unprecedented polyoxoanion-stabilized nanoclusters such as $Ir(0)_{\sim 300}$—it took a >5 year, painstaking mechanistic study, one which provided a more general solution to the classic mechanistic problem of "is it homogeneous or heterogeneous catalysis", to discover the true nanocluster catalysts [22]. The impressive finding there, however, is one that extends considerably the precatalyst reservoir concept noted above: in the polyoxoanion-stabilized nanocluster case, >>300 steps must be involved from the precatalyst to the final nanoclusters [26], an impressive example of catalyst self-assembly to catalysts that are only ±15% in both size dispersion and ≤±15% in reproducible catalytic activity [21a,22,23]!

5. Summary

The above account summarizes some highlights of efforts in the Finke research group to exploit the enormous versatility offered by polyoxometalates as catalysts or catalyst precursors. In a nutshell, our main accomplishments *in catalysis* are the development of 2 of the 8 newer classes of polyoxoanions in catalysis shown previously in Figure 1. A valuable side-product of this work has been the development of a number of other polyoxonion types and structures, for example the $[P_2W_{18}M_4O_{68}]^{10-}$ and $[P_4W_{30}M_4O_{112}]^{16-}$ series, polyoxoanions which have since been exploited by others in areas ranging from catalysis to polyoxoanion-based magnetic materials to polyoxoanion-based drug studies [2,3]. It is almost surely true that only a small fraction of the polyoxoanions that are possible have been discovered at this time, polyoxoanions being Nature's products as a result of adding high valent metals, M^{n+}, to the most abundant natural solvent, H_2O, and vs. pH.

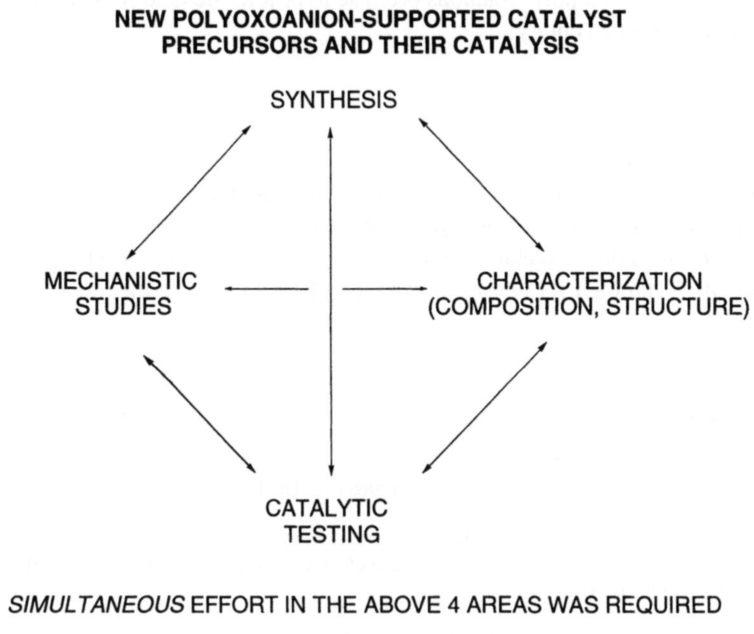

Fig. 11. Four key areas of effort required to develop the new polyoxoanion-based catalysts discussed in the present paper.

A very clear, take-home lesson emerges at this point about what is required to invent new polyoxoanion-based chemistry, Figure 11. It is the *combination, ideally nearly simultaneously,* of initial synthetic and purification work, the unequivocal characterization of new polyoxoanions—a lengthy and painstaking step especially in the cases where strongly diffracting, single crystals for X-ray crystallography are not forthcoming (e.g., see the "6 month problem areas" listed in footnote 14 elsewhere [5])—then catalysis survey studies, and finally the again often lengthy step of full kinetic and mechanistic studies. The willingness and hopefully ability to pursue each of these 4 areas is a perhaps distinguishing feature towards which the PI's polyoxoanion research group at least aspires. It is arguably the paradigm by which one invents all new catalysts these days, with the initial synthetic work being a distinguishing feature of programs around the world which are inventing new catalysts. Figure 11 is also a paradigm that works in other areas such as materials chemistry, but where one then replaces catalysts testing by appropriate testing of the desired physical properties of the new material. (The need for additional mechanistic studies in materials chemistry is another point that Figure 11 helps emphasize.)

Finally, Figure 11 is of course not "frozen in stone", but rather is part of a dynamic, continually developing paradigm. For example, one must add combinatorial methods under the synthetic vertex in Figure 11 to be state-of-the-art currently. And eventually, and even for materials the size of polyoxoanions and with the heavier elements such as tungsten, computational chemistry methods will become a 5th vertex added to, then, a 5 vertex figure derived from the 4 vertex one in Figure 11. Most important, however, is that polyoxoanions promise to be a fertile field for chemists to explore well into the 21st century!

Acknowledgments

The research discussed herein is due a talented group of graduate students and postdoctoral researchers whose names appear in the individual publications cited. Also greatly appreciated is the financial support from the NSF (grant CHE 953110), the DOE (grant DE-FG06-089ER13998) and from the PRF which made the research cited possible.

References and Notes

1. M. T. Pope: *Heteropoly and Isopoly Oxometalates*; Springer-Verlag: New York (1983).

2. *Polyoxometalates: From Platonic Solids to Anti-Retroviral Activity;* Proceedings of the July 15-17, 1992 Meeting at the Center for Interdisciplinary Research in Bielefeld, Germany, A. Müller, M. T. Pope (Eds.) Kluwer Publishers: Dordrecht, The Netherlands (1992).

3. C. L. Hill: Ed., Polyoxometalates *Chem. Rev.* **98**, 1-390 (1998).

4. Recent reviews of polyoxometalates in homogeneous and heterogeneous catalysis: (a) C. L. Hill, C. M Prosser-McCartha: *Coord. Chem. Rev.* **143**, 407 (1995). (b) A series of 34 recent papers in a volume devoted to polyoxoanions in catalysis: C. L. Hill: *J. Mol. Catal.* **114**, No. 1-3, 1-365 (1996). (c) N. Mizuno, M. Misono: *J. Mol. Catal.* **86**, 319 (1994). (d) T. Okuhara, N. Mizuno, M. Misono: *Adv. Catal.* **41**, 113 (1996). (e) I. V. Kozhevnikov *Catal. Rev.-Sci. Eng.* **37**(2), 311 (1995). (f) See also the papers on catalysis using polyoxoanions in a 1998 *Chem. Rev.* [3]. (f) R. Neumann: *Prog. Inorg. Chem.* **47**, 317 (1998).

5. R. G. Finke in reference [2], p. 267-280 therein.

6. D. E. Katsoulis in reference [3], p. 359-387. See the statements on p. 359 about the "nearly 600 refereed publications and over 120 patents", and that "the majority of the patent and applied literature is devoted to the applications of the Keggin-type heteropolyacids (HPAs) and their salts", a point that is also obvious when reading the literature. Note also the statement, on p. 361, that "about 80-85% of the patent and applided literature claims or investigates POMs (polyoxometalates) for their catalytic activity".

7. R. G. Finke, Keynote Address at the 215th ACS National Meeting in Dallas, Texas, March 29-April 2, 1998, titled "Polyoxoanions: Discrete, Soluble Metal Oxides with Applications in Catalysis", Abstract CATL #001.

8. H. Weiner, Y. Hayashi, R. G. Finke: *Inorg. Chem.* **38**, 2579 (1999).

9. (a) T. Okuhara, N. Mizuno, M. Misono *Adv. Catal.* **41**, 113-252 (1996). (b) N. Mizuno, M. Misono: "Heteropolyanions in Catalysis", *J. Mol. Catalysis*, **86**, 319 (1994), and references therein. (c) T. Okuhara, M. Misono: *Dynamic Processes on Solid Surfaces*, K. Tamaru, Ed.: Plenum Press, New York (1993), Chapter 10, p 259. (d) "Catalytic Chemistry of Solid Polyoxometalates and Their Industrial Applications", M. Misono: "*Polyoxometalates: From Platonic Solids to Anti-Retroviral Activity*", Proceedings of the July 15-17, 1992 Meeting at the Center for Interdisciplinary Research in Bielefeld, Germany, A. Müller, M. T. Pope (Eds.), Kluwer Publishers: Dordrecht, The Netherlands (1992), p. 255-265.

10. (a) Baker noted as early as 1973 that mono-metal-substituted polyoxometalates ligate the metal in a pseudo-porphyrin environment: Baker, L. C. W.: *Plenary Lecture, XV Int. Conf. on Coord. Chem., Proceedings,* Moscow (1973). (b) Two lead review to the extensive work of metalloporphyrin-catalyzed oxidations: (i) B. Meunier: *Catal. Met. Complexes* **17**, 1-47 (1994); (ii) B. Meunier: *Chem. Rev.* **92**, 1411-1456 (1992).

11. (a) For a lead reference to the work of Pope, Hill, Lyons, Neumann, and our own work see references 17-34 summarized in: D. K. Lyon, W. K. Miller, T. Novet, P. J. Domaille, E. Evitt, D. C. Johnson, R. G. Finke: *J. Am. Chem. Soc.* **113**, 7209 (1991). (b) D. Mansuy, J.-F. Bartoli, P. Battioni, D. K. Lyon, R. G. Finke: *J. Am. Chem. Soc.* **113**, 7220 (1991). (c) W. J. Randall, T. J. R. Weakley, R. G. Finke: *Inorg. Chem.*, **32**, 1068 (1993).

12. A lead reference, one which includes references to all of the earlier work (including that original work of Ishii and Venturello, and that of the French, British and Italian groups) is: D. C. Duncan, R. C. Chambers, E. Hecht, C. L. Hill: *J. Am. Chem. Soc.* **117**, 681 (1995).

13. A. R. Siedle, W. B. Gleason, R. A. Newmark, R. P. Skarjune, P. A. Lyon, C. G. Markell, K. O. Hodgson, A. L. Roe: *Inorg, Chem.* **29**, 1667 (1990), and the earlier papers in this series referenced therein.

14. (a) N. Mizuno, D. K. Lyon, R. G. Finke: U. S. Patent 5,250,739, Issued Oct. 5 (1993). (b) N. Mizuno, D. K. Lyon, R. G. Finke, *J. Catal.* **128**, 84-91 (1991). (c) N. Mizuno, H. Weiner, R. G. Finke: *J. Mol. Catal.* **114**, 15-28 (1996).

15. P. E. Elis Jr., J. E. Lyons: U. S. Patent 4,898,989, Feb. 6 (1990).

16. N. Mizuno, T. Hirose, M. Tateishi, M. Iwamoto: *J. Mol. Cat.* **88**, L125 (1994).

17. (a) C. L. Hill, A. M. Khenkin: *Mendeleev Comm.* **4**, 140 (1993). (b) X. Zhang, K. Sasaki, C. L. Hill: *J. Am. Chem. Soc.* **1996**, *118*, 4809. (c) X. Zhang, Q. Chen, D. C. Duncan, C. F. Campana, C. L. Hill: *Inorg. Chem.* **36**, 4208 (1997). (d) X. Zhang, Q. Chen, D. C. Duncan, R. J. Lachicotte, C. L. Hill: *Inorg. Chem.* **36**, 4381 (1997).

18. (a) R. G. Finke, M. W. Droege, P. J. Domaille: *Inorg. Chem.* **26**, 3886 (1987). (b) T. J. R. Weakley, R. G. Finke: *Inorg. Chem.* **29**, 1235 (1990). (c) W. J. Randall, M. W. Droege, N. Mizuno, K. Nomiya, T. J. R. Weakley, R. G. Finke: *Inorg. Syn.* **31**, 167 (1997).

19. A paper by R. Neumann et al. is notable both for its catalytic findings, the stability of the polyoxonion catalyst to the normally structure-disrupting H_2O_2, and also for its valuable mechanistic work: R. Neumann, M. Gara: *J. Am. Chem. Soc.* **117**, 5066 (1995).

20. This work also includes the Pd^{II}, Pt^{II} as well as the Mn^{II}, Ru^{II} and other metal-substituted analogs of this structural series: (a) R. Neumann, M. Gara: *J. Am. Chem. Soc.* **116**, 5509 (1994). (b) R. Neumann,. A. M. Khenkin, M. Dahan: *Angew. Chem. Int. Eng. Ed. 34* , 1587 (1995). (c) R. Neumann, M. Gara: *J. Am. Chem. Soc.* **117**, 5066 (1995). (d) R. Neumann, A. M. Khenkin: *Inorg Chem.* **34**, 5753 (1995); (e) See also reference [45] herein.

21. (a) J. D. Aiken III, Y. Lin, R. G. Finke: *J. Mol. Catal. A: Chemical* **114**, 29-51 (1996). ("A Pespective on Nanocluster Catalysis: Polyoxoanion and $(n\text{-}C_4H_9)_4N^+$ Stabilized $Ir(0)_{\sim 300}$ Nanocluster 'Soluble Heterogeneous Catalysts' "). (b) J. D. Aiken III, R. G. Finke: *J. Mol. Catal. A: Chemical* **145**, 1-44 (1999) ("A Review of Modern Transition-Metal Nanoclusters: Their Synthesis, Characterization, and Applications in Catalysis", including 197 references).

22. Y. Lin, R. G. Finke, *Inorg. Chem.* **33,** 4891 (1994).

23. Y. Lin, R. G. Finke, *J. Am. Chem. Soc.* **116,** 8335 (1994).

24. (a) T. Okuhara, N. Mizuno, M. Misono: *Adv. Catal.* **41**, 113-252 (1996). (b) N. Mizuno, M. Misono: "Heteropolyanions in Catalysis", *J. Mol. Catalysis*, **86,** 319 (1994), and references therein. (c) T. Okuhara, M. Misono: *Dynamic Processes on Solid Surfaces*, K. Tamaru, Ed: Plenum Press, New York (1993); Chapter 10, p 259. (d) "Catalytic Chemistry of Solid Polyoxometalates and Their Industrial Applications", M Misono: "*Polyoxometalates: From Platonic Solids to Anti-Retroviral Activity*", Proceedings of the July 15-17, 1992 Meeting at the Center for Interdisciplinary Research in Bielefeld, Germany, A. Müller, M. T. Pope, (Eds.) Kluwer Publishers: Dordrecht, The Netherlands (1992), p. 255-265.

25. (a) S. K. Tanielyan, R. L. Augustine Patent: WO 98/28074, July 2 (1998). (b) R. Augustine, S. Tanielyan, S. Anderson and H. Yang, *Chem. Comm.* 1257 (1999).

26. M. A. Watzky, R. G. Finke: *J. Am. Chem. Soc.* **119**, 10382 (1997).

27. J. D. Aiken III, R. G.Finke: *Chem. Mater.* **11**, 1035 (1999).

28. J. D. Aiken III, R. G. Finke: *J. Am. Chem. Soc.* **121**, 8803 (1999).

29. J. D. Aiken III, R. G. Finke: *J. Am. Chem. Soc.* **120**, 9545 (1998).

30. M. A. Watzky, R. G. Finke: *Chem. Mater.* **9**, 3083-3095 (1997).

31. K. S. Weddle, J. D. Aiken III, R. G. Finke, *J. Am. Chem. Soc.* **120**, 5653-5666 (1998).

32. Lead references: (a) E. L. Mutterties: *Bull. Soc. Chim. Belg.* **84**, 959 (1975). (b) E. L. Mutterties: *Science* **196**, 839 (1997). (c) E. L. Mutterties: *C&E News Special Report* Aug. 30, p. 28-41 (1982). (d) A. K. Smith, J. M. Basset: *J. Mol. Catal.* **2**, 229 (1977). (e) J. L. Vidal, W. E Walker: *Inorg. Chem.* **19**, 896, (1980), and all the later papers in the series from the Union Carbide group. (f) G. Süss-Fink, G. Meister: *Adv. Organomet. Chem.* **35**, 41 (1993). (g) R. D. Adams, F. A. Cotton, Eds, "Catalysis by Di- and Polynuclear Metal Complexes", Wiley, New York (1998).

33. Grove's classic Ru^{II}(tetramesitylporphyrin) catalyst which operates via a Ru^{II}, O=Ru^{IV}, and O=Ru^{VI}O intermediates but which is, however, too slow and too quickly deactivated and thus too short-lived to be commercial: (a) J. T. Groves, R. Quinn, *J. Am. Chem. Soc.* **107**, 3790 (1985); (b) J. T. Groves, A. Kwang-Hyun, R. Quinn, *J. Am. Chem. Soc.* **107**, 4217 (1988).

34. (a) An editorial: "On the trail of dioxygen activation", C. L. Hill, I. A. Weinstock: *Nature* **388**, 332 (1997). (b) See also the editorial on "Controlled Green Oxidation": C. L. Hill: *Nature* **401**, 436 (1999).

35. An older but still useful review of dioxygenases: M. Nozaki, *Topics in Current Chem.* **78**, 145 (1979).

36. Additional lead reviews of dioxygenases: (a) L. Que Jr, R. Y. N. Ho: *Chem. Rev.* **96**, 2607 (1996). (b) A. L. Feig, S. J. Lippard: *Chem. Rev.* **94**, 759 (1994). (c) *Microbial Degradation of Organic Molecules*, D. T. Gibson, Ed., Marcel Dekker, New York (1984). (d) L. Que Jr.: *Iron Carriers and Iron Proteins*, T. M. Loehr, Ed, VCH, New York (1989), pp 467-524; (e) J. D. Lipscomb, A.M. Orville: *Metal Ions Biol. Syst.*, **28**, 243 (1992).

37. (a) For a recent, comprehensive review on dioxgenases with 398 references, see: T. Funabiki: *Catalysis by Metal Complexes, Oxygenase and Model Systems*, T. Funabiki, Ed., Vol. 19, Kluwer Academic Press, Dortrecht, Holland (1997), Chapter 2, p. 19-104. (b) ibid, Chapter 3, p. 105-155. (c) A. Nishinaga: ibid, Chapter 4, p. 157-194

38.(a) N. Mizuno, D. K. Lyon, R. G. Finke: *J. Catalysis* **128**, 84-91 (1991). (b) N. Mizuno, D. K. Lyon, R. G., Finke: U. S. Patent 5,250,739, Issued Oct. 5 (1993). (c) H. Weiner, R. G. Finke, *J. Mol. Catal.* (2000), in press ("A simple, product-based, unequivocal test for the presence of olefin autoxidation: One of four key findings from product, kinetic, and mechanistic studies of polyoxoanion-based catalysis of cyclohexene oxidation").

39. N. Mizuno, H. Weiner, R. G. Finke: *J. Mol. Catal.* **114**, 15-28 (1996).

40. M. W. Droege, R. G. Finke: *J. Mol. Catal.* **69**, 323-338 (1991).

41. For recent examples of work demonstrating that what others believed was not free-radical chemistry is, however and in fact, the chemistry of freely diffusing radicals, see: (a) C. Walling, *Acc. Chem. Res.* **31**, 155 (1998). (b) P. A. MacFaul, D. D. M. Wayner, K. U. Ingold: *Acc. Chem. Res.* **31**, 159 (1998). (c) M. W. Grinstaff, M. G. Hill, J. A. Labinger, H. B. Gray: *Science* **264**, 1311 (1994). (d) P. A. MacFaul, K. U. Ingold, D. D. M. Wayner, L Que Jr: *J. Am. Chem. Soc.* **119**, 10594 (1997). (e) D. W. Snelgrove, P. A. MacFaul, K. U. Ingold, D. D. M. Wayner: *Tet. Let.* **37**, 823 (1996). (f) P. A. MacFaul, I. W. C. E. Arends, K. U. Ingold, D. D. M. Wayner: *J. C. S. Perkin Trans. 2* 135 (1997). (g) F. Minisci, F. Fontana, S. Araneo, F. Recupero, S. Banfi, S. Quici, *J. Am. Chem. Soc.* **117**, 226 (1995). (h) F. Minisci, F. Fontana, S. Araneo, F. Recupero, L. Zhao: *SYNLET*, **February**, 119 (1996). (i) M. Newcomb, P. A. Simakov, S.-Un Park: *Tet. Let.* **37**, 819 (1996).

42. H. Weiner, R. G. Finke *J. Am. Chem. Soc.* **121**, 983 (1999).

43. Y. Tatsuno, M. Tatsuda, S. Otsuka: *J. C. S., Chem. Comm.* 1100 (1982).

44. (a) The following work clearly shows that sites of coordinative unsaturation (i.e., presumably 3 total, 2 for catechol and 1 for dioxygen) are a key to achieving extradiol cleavage chemistry: M. Ito, L. Que Jr: *Angew. Chem. Int. Eng. Ed.* **36**, 1342 (1997).
(b) Note the system in the following paper does not have a site of for coordination of O_2 on the 6 coordinate Fe^{III}•catecholate complex; hence, the products are those resulting from intradiol cleavage. Of futher interest in the system in the following paper, one which exhibts 80 TTOs before deactivation to a Fe^{III}–O–Fe^{III} thermodynamic sink, is that it was cited at that time (i.e., prior to the >100,000 TTOs demonstrated catalysts

based on polyoxoanions [42]) as "highly reactive and catalytically active": M. Duda, M. Pascaly, B. Krebs *Chem. Commun.* 835 (1997).

45. (a) R. Neumann, M. Dahan: *Nature*, **388**, 353 (1997). (b) R. Neumann, M. Dahan: *J. Am. Chem. Soc.* **120**, 11969 (1998).

46. L. Que Jr., J. D. Lipscomp, E. Münck, J. M. Wood: *Biochem. Biophys. Acta* **485**, 60 (1997).

47. D. M. Arciero, J. D. Lipscomb: *J. Biol. Chem.* **261**, 2170 (1986).

48. (a) J. Sanvoisin, G. J. Langley, T. D. H. Bugg: *J. Am. Chem. Soc.* **117**, 7836 (1995). (b) T. D. H. Bugg, J. Sanvoisin, E. L. Spense: *Biochem. Soc Trans.* **25**(1), 81 (1997).

49. H. Weiner, Y. Hayashi, R. G. Finke: *Inorg. Chem.* **38** 2579 (1999).

50. H. Weiner, Y. Hayashi, R. G Finke: *Inorg. Chim. Acta* **291**, 426 (1999).

51. J. P. Collman, L. S. Hegedus, J. R. Norton, R. G. Finke, "Principles and Applications of Organotransition Metal Chemistry", University Science Books, Mill Valley, CA (1987).

Ribosomal Crystallography and Heteropolytungstates

DANIELA JANELL*, ANTE TOCILJ*, INGO KÖLLN*, FRANK SCHLÜNZEN*, MARCO GLÜHMANN*, HARLY A.S. HANSEN*, JÖRG HARMS*, ANAT BASHAN°, ILANA AGMON°, HEIKE BARTELS*°, MAGGIE KESSLER°, SHULAMITH WEINSTEIN°, FRANCOIS FRANCESHI^, ADA YONATH*°

** MPG for Ribosomal Struct., Notkestr. 85, 22603 Hamburg, Germany*

° Dept. Struct. Biol. Weizmann Inst., 76100 Rehovot, Israel

^ MPI for Mol. Genetics, Ihnestr. 73, 14195 Berlin, Germany

Send correspondence to

A. Yonath, Weizmann Institute, Rehovot 76100, Israel

Tel:972-8-9343028, Fax:972-8-9344154,

e-mail: yonath@mpgars.desy.de

Abstract. Heteropolytungstates play a dual role in ribosomal crystallography. Beside generating phases, one of them, $(NH_4)_6(P_2W_{18}O_{62})14H_2O$, was found to be extremely useful in inducing post crystallization rearrangements. These led to a significant increase in the internal order of crystals of the small ribosomal subunits from *Thermus thermophilus,* manifested in a dramatic extension of the resolution of their diffraction patterns, from the initial 7-9 Å to 3 Å. The current 3.3 Å electron density map of this particle, constructed using phases obtained from this W cluster together with other metal compounds, shows the recognizable overall morphology of the small ribosomal subunit. Over 96% of the nucleotides were traced and the fold of all proteins was determined fully or partially. Specific sites were determined independently by covalently bound heavy atom clusters, among them the surface of two proteins and a functional center, the gate for mRNA binding. All tungsten-cluster sites detected in this map are located in close proximity to the proteins of the particle, in positions that may have an influence on the stability and the rigidity of this rather flexible ribosomal subunit.

M.T. Pope and A. Müller (eds.), Polyoxometalate Chemistry, 391–415.

1. Introduction

Ribosomes are the universal intracellular molecular machines that are responsible for one of the most fundamental life processes, the translation of the genetic code into proteins. They are giant nucleoprotein organelles built of two independent subunits of unequal size which associate upon the initiation of protein biosynthesis. In bacteria their molecular weight is about 2.3 mega dalton and they are comprised of RNA and proteins at a 2:1 ratio. The small subunit (called 30S for eukaryotic ribosomes, according to its sedimentation coefficient), is of molecular weight of 0.85 mega dalton and consists of 20-21 proteins and one RNA chain of about 1500 nucleotides (called 16S RNA). It offers the site for the initiation and the progression of the biosynthetic process and facilitates the decoding of the genetic information. The large subunit (50S) catalyzes the formation of the peptide bond and provides the path for the progression of the nascent proteins. It is of molecular weight of about 1.45 mega dalton and contains two RNA chains (called 23S and 5S RNA) of a total of about 3000 nucleotides and 36-50 different proteins, depending on the source.

Crystals have been obtained from several ribosomal particles, diffracting best to around 3 Å, among them the small and the large ribosomal subunits from *T. thermophilus*, called T30S and T50S (Yonath et al., 1998; Bashan et al., 2000) and of the large ribosomal subunit from *Haloarcula marismortui*, H50S (von Böhlen et al., 1991). However, the bright synchrotron radiation that is essential for resolving the higher resolution terms, namely at the 3-5 Å shell, causes rapid radiation damage within a period sufficient for the collection of only small fractions of the data (about 1-3° of oscillation). Coupled with low level of isomorphism and/or space group variability (Makowski et al., 1987; Yonath et al., 1998; Ban et al., 1999), this extreme radiation sensitivity introduces substantial difficulties in the construction of complete data sets, mainly because the low level of isomorphism of the ribosomal crystals. To minimize the number of crystals needed for producing complete sets, they are irradiated by a beam with a cross-section smaller than their size. Once decay is observed, the crystals are being translated, and a new area is being irradiated. This approach was the key for crossing the 5 Å border in data collection, and for obtaining high quality data sets from the rod-like crystals of T30S, that may reach the dimensions of 500x60x60 microns.

The assignment of phases to the observed structure factor amplitude is the most crucial, albeit most complicated step in structure determination. Since the phases cannot be directly measured, their elucidation remains the least predictable task, even for average-size proteins. Multiple and single isomorphous replacement (MIR and SIR) and multiple anomalous dispersion (MAD), are the commonly used methods for phasing diffraction data from crystals of biological macromolecules. All require the preparation of heavy atom derivatives in which electron-dense atoms are inserted into the crystalline lattice at distinct locations. As the changes in the structure factor amplitudes resulting from the addition of the heavy atoms are being exploited, the derivatization reagents are chosen according to their potential ability to induce measurable signals. Whereas single metal atoms yield signal sufficient to phase data collected from crystals of average size proteins, owing to the large size of the ribosomal particles, dense heavy atom clusters seem to be advantageous (Thygesen et al., 1996). Heteropolytungstates were found to be extremely useful in this respect and a detailed account of these findings is given below. Additional examples are the smaller compounds, Ta_6Br_{14}, a tetra mercury compound, TAMM, and a tetra iridium cluster, TIR (Jahn, 1989). Among them, Ta_6Br_{14} yielded phase information for all ribosomal particles currently being studied, namely, T30S, T50S, T70S and H50S (Thygesen et al., 1996; Yonath et al., 1998; Clemons et al., 1999; Ban et al., 1999; Cate et al., 1999; Tocilj et al., 1999; Schlünzen et al, 2000).

The other two compounds, TIR and TAMM were designed to bind covalently to exposed sulfhydryls and led to the localization of two ribosomal proteins of the small subunit, S11 and S13 (Weinstein et al., 1999). In the studies of T30S, they also provided tools for targeting functional sites in which ribosomal RNA is involved (Weinstein et al., 1999; Auerbach et al., 2000; Bartels et al., 2000). Thus, they were bound to tailor made ligands, such as antibiotics or DNA oligomers complementary (cDNA) to exposed single strand rRNA regions, that were either co-crystallized with the T30S subunit or diffused into the already formed crystals. One of these is a 22-base DNA oligomer, complementary to the 3' end of the 16S RNA. This region, which is known to be rather flexible (Müller and Brimacombe, 1997), contains the anti Shine-Dalgarno sequence. Therefore the DNA oligomer complimentary to it was considered as the mRNA analog that participates in the formation of the initiation complex. The diffusion and hybridization of a heavily mercurated form of this oligomer into T30S led to a derivative diffracting very well to a high resolution. Therefore it is suggested that the hybridization of the 16S RNA with this oligomer limits the mobility of the flexible 3' arm of the 16S RNA, in a fashion that mimics the binding of mRNA.

2. Heteropolytungstates in Ribosomal Crystallography

The strategy proved, so far, most suitable for phasing ribosomal data is based on the determination of an initial phase set at low resolution. This may be extended later, by experimental or computational methods. Molecular replacement exploiting cryo EM reconstructions proved useful for H50S (Ban et al., 1999), T50S (Yonath and Franceschi, 1988) and for the whole ribosome from *T. thermophilus,* T70S (Cate et al., 1999; Harms et al., 1999). For T30S, however, despite extensive attempts, no solution was obtained, perhaps because of its multi-conformational nature and inherent flexibility (Gabashvili et al., 1999; Harms et al., 1999). Therefore phasing was performed by measuring data from derivatized crystals. Advantage was taken of dense compounds containing a large number of heavy-atoms, arranged in close proximity (Thygesen et al., 1996). These were used either for pre-crystallization covalent binding at preferred locations (Weinstein et al., 1999), or for traditional soaking experiments. Despite the large size of the clusters that may have hampered their penetration into the crystals, many of the soaking experiments were successful, presumably because the ribosomal crystals contain wide internal solvent channels. Furthermore, it seems that for crystals with wide solvent regions, the large size of these compounds is advantageous, as it limits their free movement and minimize multiple site binding.

Heteropolytungstates (e.g. Dawson, 1953; Pope and Papaconstantinou, 1967; D'Amour, H., 1976; Brown et al., 1977; Contant, R., 1990; Xin and Pope, 1994; Wei et al., 1997) yielded useful derivatives in crystallographic studies of several biological macromolecules, all by soaking experiments. These large anions are of exceptional stability over a wide range of pH and redox states. They posses a high degree of internal symmetry, and a correlation between it and their binding sites has been detected. In such cases they were found suitable for high resolution phasing. An example is the structure of riboflavin synthase (Ladenstein, et al., 1987) that possess an internal five fold symmetry, which coincided with that of $(NaP_5W_{30}O_{110})^{-14}$ (Alizadeh et al., 1985). However, in the absence of preferred orientation, the effective phasing resolution is limited to 4-5 Å, even when sophisticated spherical averaging techniques are being used (Fu et al., 1999). Nevertheless under favorable conditions, the W18 clusters did bind in a specific way, so that its individual W atoms could be resolved at resolution higher than 4.5 Å, and used for phasing.

In ribosomal crystallography the heteropolytungstates were found to be very suitable for phasing at low resolution, as well as for the validation of the results obtained by molecular replacement searches (Ban et al., 1999). So far, reports of materials that generated phases include: W12 ($Na_{16}[(O_3PCH_2PO_3)_4W_{12}O_{36}]40H_2O$) and W17 ($Cs_7(P_2W_{17}O_{61}Co(NC_5H_5))14H_2O$) (Yonath et al., 1998), as well as W9 $[(phSn)_4(AsW_9O_{33})_2]$ and W11Rh=$Cs_5(PW_{11}O_{39}[Rh_2CH_3COO_2]$ (Ban et al, 1999) were used in phasing the data of the halophilic large subunit. $H_4SiO_4[12WO]$ and $Li_{10}(P_2W_{17}O_{61})$ (Clemons et al., 1999) as well as W18 ($(NH_4)_6(P_2W_{18}O_{62})14H_2O$), and W4 ($(TMA)_2Na_2[Nb_2W_4O_{19}]18H_2O$) (Tocilj et al., 1999) were used for the determination of the medium resolution structure of the thermophilic small ribosomal subunit.

2.1 THE WONDERS OF W18

Neither of the ribosomal crystal types that diffract to molecular resolution was obtained solely from purified ribosomal particles. In all cases additives had to be used, each in a different fashion. Initially, small metal compounds were screened for their influence on the crystal's properties. Among them, minute amounts of Cd^{++} in the crystallizing droplet led to significant gain in the internal order of the H50S crystals, as expressed in resolution increase from 6-7 to 2.7-3 Å. A systematic search included larger and more complex materials, as well as various modes of their addition, led to the improvement of the crystals of T30S by post crystallization treatments by minute amounts of a heteropolytungstate cluster, W18.

As subjects for crystallization, the small ribosomal subunit is less suitable than the entire ribosome or its large subunits. Cryo electron microscopy (Stark et al., 1995; Frank et al., 1995), surface RNA probing (Alexander et al., 1994) and monitoring the ribosomal activity (Weller and Hill, 1992) showed that among the ribosomal particles, the small ribosomal subunit displays the highest conformational variability. This inherent flexibility may be the reason for the low resolution (about 10 Å) of the early crystals of T30S (Yonath et al., 1988; Trakhanov, et al., 1989). It also may account for the unsuitability of all the available cryo-EM reconstructions of the small ribosomal subunit for extracting initial phase sets, studies that, as mentioned above, were performed successfully for the large ribosomal subunit, H50S and T50S (Ban et al., 1999; Harms et al., 1999).

The dramatic improvement in crystal quality was not accompanied by changes in the unit cell dimensions or in the crystal symmetry (a=b=407 Å, c=176 Å, $P4_12_12$). However, data collected from the W18 treated crystals (called here Wative) could not be scaled to the data obtained from the original native crystals, indicating significant non-isomorphism and suggesting that a major conformational rearrangement occurred upon the W18 treatment. Among the many tungsten compounds tested by us (most of the material listed in TABLE I), so far only W18 was found suitable for the increase in resolution. Furthermore, soaking washed Wative crystals (called here back-soaked or BS crystals) in solutions containing $W9_2$, a compound closely related to W18, led to diffraction that could not be scaled with that obtained from Wative or BS crystals, most likely due to additional post-crystallization conformational rearrangements.

Interestingly, in studies performed independently on T30S crystals that were grown under the same conditions (Yonath et al., 1988), a related compound, $Li_{10}(P_2W_{17}O_{61})$, was used for phasing. This compound, however, was found to reduce, rather than to increase, the resolution (Clemons et al., 1999).

Conformational changes are not routinely induced within crystals due to the limitation of the motion imposed by the crystal network. However, realizing that the T30S crystals tolerate and even benefit from internal rearrangements, prompted us to induce reactivation of the T30S particles within the crystals. Controlled heating, the common procedure for functional activation of ribosomal particles (Zamir et al., 1971), was employed on entire crystals, enabling quantitative binding of compounds participating in protein biosynthesis or their analogs (Auerbach et al., 2000; Bashan et al., 2000).

It is conceivable that other metals could have led to a similar effect on ribosomal crystals. Thus, hints for the improvement of crystal order by Os hexamine chloride may be extracted from the facts that derivatization with this compound leads to non-isomorphous crystals that diffract to higher resolution (Clemons et al., 1999). This compound has already been used for improving RNA crystals (Cate and Doubna, 1996; Golden et al., 1998). It is known to interact with RNA chains in a specific fashion that may increase their rigidity. Hence, it is conceivable that the improvement of the internal order of the T30S subunits by this compound is linked to its binding property.

TABLE 1: TUNGSTEN COMPOUNDS USED IN THIS STUDY

Sodium tungstate

WAC = $W_3O_2(acetate)_6(H_2O)_3CF_3SO_3$

W4 = $(TMA)_2Na_2[Nb_2W_4O_{19}]18H_2O$ +

W11SiRh = $Cs_5H_xSiW_{11}O_{39}Rh^{III}CH_3COO(H)$ *

W12(Na) = $Na_{16}[(O_3PCH_2PO_3)_4W_{12}O_{36}]40H_2O$

W12(K) = $K_5H(PW_{12}O_{40})12H_2O$ +

PhSnW15 = $K_5H_4[(phSn)_3(P_2W_{15}O_{59})]16H_2O$

BuSnW15 = $K_5H_4((buSn)_3(P_2W_{15}O_{59})]15H_2O$

W17CsCo = $Cs_7(P_2W_{17}O_{61}Co(NC_5H_5))14H_2O$ +

W17LiCo = $Li_7(P_2W_{17}O_{61}Co(NC_5H_5))14H_2O$

BuSnW17 = $K_7[(buSn)(P_2W_{17}O_{61})]10H_2O$

W17-Th = $Cs_7(P_2W_{17}O_{61}Th(NC_5H_5))nH_2O$ *

W9$_2$ = $K_4H_3[(buSn)_3(a\text{-}SiW_9O_{31})_2]nH_2O$

W18 = $(NH_4)_6(P_2W_{18}O_{62})14H_2O$ +

W30 = $K_{14}(NaP_5W_{30}O_{110})31H_2O$

bu=butyl; ph=phenyl

TMA=tetra methyl ammonium

* Designed for covalent binding

+ Found useful for phasing or for map interpretation

2.2. DOUBLE-DERIVATIVE PHASING

2.2.1. Strategies in derivatization: A Flexible Definition of Native Crystals

The chemical basis for the increase in the order of the T30S crystals by W18 is still far from being fully understood. Nevertheless, careful studies led to several interesting observations. It was shown by inductively coupled plasma mass-spectrometry and atomic emission spectrometry, that although minute quantities of W18 are needed for successful treatment, relatively large amount of this cluster penetrate into the crystals and resides within them. These quantities are much higher than those detected by crystallographic methods, meaning that some W18 clusters are flowing in the solvent regions.

As the original T30S native crystals became obsolete by the W18 treatment, a new definition for native crystals had to be made. Initially, the Wative crystals were considered as native. These were further derivatized by soaking in solutions containing additional heavy atom derivatives (TABLE II). However, it was found that the flouting W18 clusters complicated the phasing process, since the non-bound W18 clusters led to high background, which, in turn, "masked" the contribution of the additional heavy atoms. Furthermore, these solubilized clusters generated measurable anomalous signals at low resolution which could not be separated from those originating from the specifically bound ones. Since the major phasing contribution of W18 is in the medium and the low resolution shells, a large fraction of the phasing information was lost, or led to confusion.

A more defined crystal-system was obtained by washing the Wative crystals. As the washed crystals diffract to resolution comparable to that obtained from the Watives, it was assumed that the main W18 sites that are contributing to the improvement of the crystal order are occupied even after the wash. This assumption was later verified by crystallographic anomalous measurements of the back-soaked crystals, as well as by the examination of the content of dissolved washed crystals (see below, in chapter 3.2). Indeed, it was found by the methods described above, that a large amount of W18 remains within the crystals even after applying an extensive washing procedure (twelve times during 40-50 hours).

Knowing that the Wative as well as the BS crystals contain significant amounts of W18, the choice, the combination and the design of the heavy atom derivatization was dictated not only by their potential ability to produce phases, but also according to the chemical properties of the combined systems. Consequently, the further derivatization steps were performed by soaking the Wative crystals in heavy atom solutions, in the presence or in the absence of W18, as dictated by the stability of the crystals.

Table II: Double Derivative Phasing Statistics.

Derivative	Unique Refl.	Resol. (Å)	Completeness (%)	R_{sym} (%)	R_{merge} (%)	Phasing power	R_{cullis}	Sites
	a) „Natives“							
BS (Native)	122788	3.8 (*)	86.4	12.7	-	-	-	-
Wative 1	206724	3.3	94.7	13.6	17.9	1.2	0.74	4
Wative 2	133199	3.5	80.5	10.6	16.6	1.5	0.77	3
Wative (ano)	42995	7.0	99.1	7.6	17.2	1.6/1.1 (†)	0.4/0.8 (†)	3
	b) Double soaks							
Ta_6Br_{14}	48132	5.0	56.1	14.3	26.6	1.4	0.84	3
$Hg_3C_6O_4H_8$ (‡)	17092	6.5	60.1	14.4	22.9	1.3	0.73	7
$C_2Hg_6N_2O_8$	55800	5.9	79.6	6.9	21.7	1.7	0.50	4
PIP (§)	62015	4.5	80.9	14.1	17.5	0.5	0.91	6
K_2O_4Os	49751	5.0	77.9	14.7	22.9	1.0	0.68	7
Pb_3-citrate (‡) (¶)	26982	6.4	92.4	11.1	14.6	1.1	0.79	8
$NaAuCl_4$	30078	5.5	81.0	13.6	15.5	1.0	0.78	1
W_4 (§)	26748	6.4	91.8	11.8	15.0	0.9	0.72	4
K_2PtCl_4 (‡)	23551	6.5	81.5)	11.1	20.9	0.8	0.90	7
AquaPt (§)	28439	5.0	49.3	15.9	26.1	0.6	0.85	5
CisPt (§)	19366	6.9	80.7	9.9	18.1	0.6	0.84	3
	c) Covalent binding							
OliT (‡) (¶) (§)	36083	4.8 (‖)	58.1	14.2	14.9	3.1	0.71	1
TAMM (§)	25797	7.0	60.3	10.7	16.4	0.6	0.94	2
TIR (‡) (§) (¶)	43424	4.5	57.3	8.5	21.4	0.6	0.88	2
M-HgAc (‡) (¶)	29719	6.0	80.6	13.7	16.2	1.2	0.66	6
U-Ac (‡) (¶)	30911	6.0	84.4	12.3	15.3	0.5	0.85	5

$R_{sym} = \Sigma \mid I - \langle I \rangle \mid / \Sigma \langle I \rangle$, $R_{merge} = \Sigma \mid F_{PH} - F_P \mid / \Sigma F_P$, R_{cullis}= <phase-integrated lack of closure>/ $\langle \mid F_{PH} - F_P \mid \rangle$, Phasing power = $\langle [\mid F_H$ (calc) $\mid$ / phase-integrated lack of closure] $\rangle$.
(*) Data were collected to 3Å, but not used in this study, (†) iso/ano, (‡) SIR, not incorporated yet, (§) See abbreviation, (¶) Wative was considered as native, (‖) Data were collected to 3.5Å, but not used in this study.

In a few cases the addition of the heavy atom cause crystal deterioration unless some of the internal W18 was removed before further soaking. The treatment was also applied to crystals containing analogs of compounds participating in protein biosynthesis as well as to those that were specifically modified by covalent binding.

2.2.2. Phasing at Narrow and Wide Resolution Ranges

Two approaches were taken for phasing the data obtained from the W18 treated derivatives T30S crystals. The "narrow range" strategy, employed only at the initial stages of these studies, was based on minimizing the contribution of the W18 to the diffraction patterns, regarding it as part of the crystal solvent. The borders of this range were determined according to two factors: the highest resolution of the derivatives and the range that was supposed to be effected by the presence of the W18, namely lower than 10-12 Å. Since at the initial stages of this study the heavy atom derivatized crystals diffracted to around 7 Å, the useable resolution shell for the narrow range phasing was between 7 to 12 Å.

The narrow range phasing procedure was found to be lengthy and demanding since in order to produce measurable signals from crystals of very large macromolecules which cannot be subdivided by non-crystallographic symmetry, multiple site derivatization is required. The identification and the refinement of these sites was found to be rather demanding, especially due to the requirement for careful cross-verifications (Schlünzen et al., 1999; Weinstein et al., 1999). Over fifty sites originating from four heavy atom derivatives were extracted from overcrowded and rather flat difference Patterson maps. These studies led to a 7.2 Å MIR map that could be partially interpreted, but suffered from considerable fragmentation (Schlünzen et al., 1999; Weinstein et al., 1999; Bashan et al., 2000).

The parallel approach, based on the incorporation of all available phase information was proved to be more suitable, despite the experimental and conceptual complications originating from the presence of unknown amount of W18 in the crystals. Owing to the non-isomorphism between the native and the Wative data, the sites of W18 in the Wative crystals could not be revealed by difference Patterson techniques. However, four sites with exceptionally high signals were identified in difference Patterson combined with cross Fourier maps, constructed from Wative and BS diffraction data (Figure 1).

W18, by itself, contributed significantly to the progress of the phasing process. As mentioned above, the resolution of the back-soaked crystals is as high as that of the Watives. Therefore it was assumed that the sites of W18 that cause the increase in resolution are occupied in the BS crystals despite the extensive washing steps.

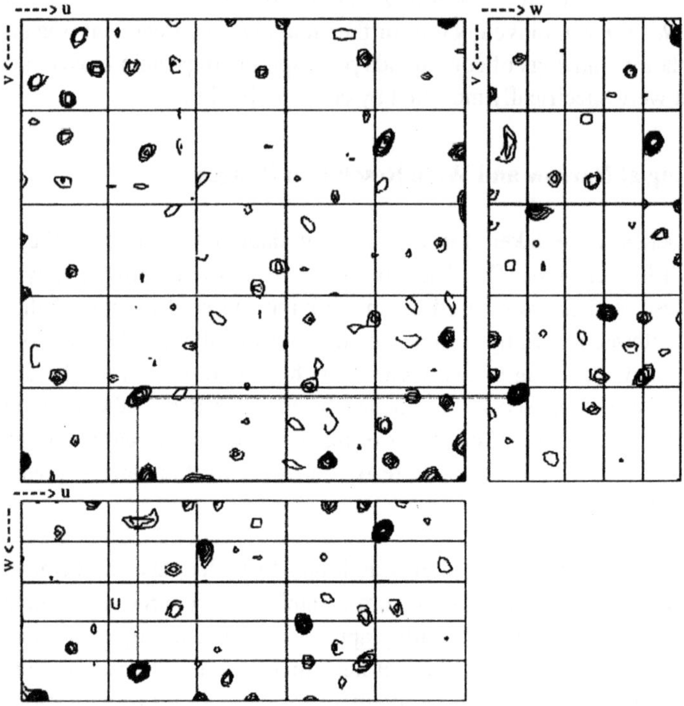

FIGURE 1: Three Harker sections of the difference Patterson map, using data collected from Wative and back soaked (BS) crystals, constructed at 9 Å resolution and showing the main W18 site.

Fluorescence spectra measured from the BS crystals indicated the presence of W18 in these crystals (Figure 2). Anomalous signals, originating from W18, were readily resolved by collecting data at two wavelengths, reconfirmed the above assumption. These data also cleared the space group ambiguities between the two enantiomers, $P4_12_12$ and $P4_32_12$ and their incorporation in a previous 7 Å map led to an increased level of detail (Figure 3).

The sites of the additional heavy atoms (TABLE II) were determined and verified by difference Patterson and cross Fourier procedures. The initial wide-range map was calculated at the 7-30 Å resolution shell, using phase information from W18 and Ta_6Br_{14}. The latter was instrumental in bridging between the lower resolution information, obtained from W18, and the higher resolution phases that were extracted from the data obtained from smaller heavy atom derivatives.

The exceptionally strong phasing power of the W18 cluster biased frequently the difference Patterson maps of the additional derivatives. Nevertheless, by careful interplay, most of the sites of the additional derivatives were revealed and reconfirmed. These enabled the computation of a 3.3 Å MIRAS map, phased exclusively by crystallographic methods, with no need for the incorporation of electron microscopy or otherwise constructed models. In these studies the clusters were first treated as group scatterers or a spherical averaged compounds. As such, they yielded phase information to about 7 Å. Later on we took advantage of those W18 clusters that were specifically bound and used them at close to atomic resolution.

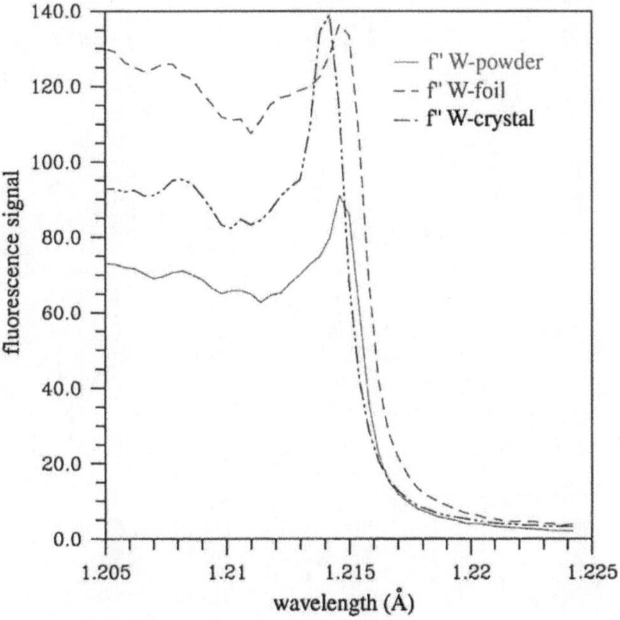

FIGURE 2: f" as a function of wavelength, as derived from fluorescence spectra measured from a W foil, W powder and a crystal soaked in a W18 solution, all measured around the W L_{III} edge.

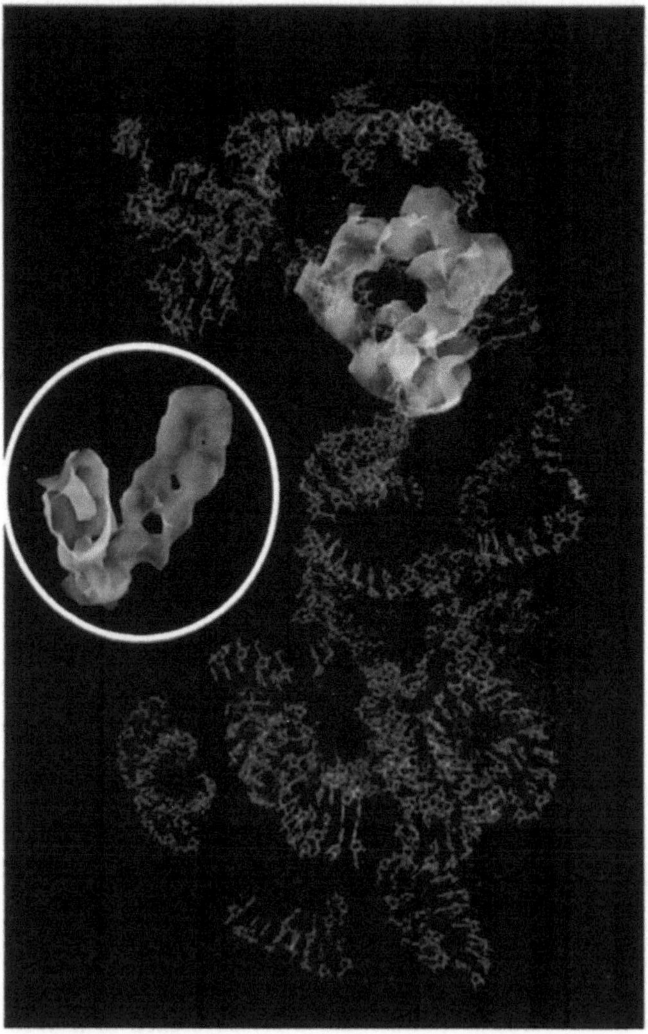

FIGURE 3: Part of the RNA chain together with one of the regions that became interpretable by the incorporation of the BS anomalous data at 7 Å (shown as rendered map in white). The circled insert focuses on a detail taken from a similar area (not shown here), exhibiting the typical features of helix-bulge-helix motif. Anomalous data were collected at the L_{III} edge of W (1.218 and 1.2734 Å). Experimental procedures are described in (Tocilj et al., 1999). All fitted models shown in this manuscript, were obtained interactively, with the program O (JONES et al., 1991).

3. The 3.3 Å electron density map

3.1. FEATURES SEEN IN THE SMALL RIBOSOMAL SUBUNIT AT 3.3 Å

The overall structure of the small ribosomal subunit, as seen in the 3.3 Å MIRAS map calculated with the wide-range phases (Schlünzen et al., 2000), is remarkably similar to most of the electron microscopical reconstructions of this particle at its functionally active conformation. It contains many of the recognizable features, including the traditional division into three main parts: a rather large head, a short neck and a bulky lower body (Stark et al., 1995; Frank et al., 1995; Gabashvili et al., 1999) (Figure 4 LEFT).

This map showed clearly the backbone of the RNA and in many regions bases were well separated and purines and pyrimidines could be assigned. Likewise, many of the proteins loop and side chains could be identified (Figure 4 RIGHT). The RNA chain was traced directly from the map and later compared with the available diagrams of the RNA secondary structure. Localization of the proteins was based on the large body of non-crystallographic information, as described in (Tocilj et al., 1999). Consequently this 3.3 Å map contains over 1450 (96%) of the nucleotides and the main fold of all 19 ribosomal proteins belonging to this subunit. It shows known as well as newly detected folding and packing motifs. It provides insight into the decoding mechanism and its universality, and highlights the role of selected components in maintaining the sophisticated architecture of the ribosome (Schlünzen et al., 2000).

Three long helices run parallel to the long axis of the subunit. Among them, two are located on the rather flat surface that faces the 50S subunit. These extended RNA helical elements transmit structural changes, correlating events at the particle's far end with the cycle of mRNA translocation at the decoding region, which is located about 150 Å away, at the connection between the body and the head. The three longitudinal helices are linked by transverse features, placed like ladder rungs between them. The head contains mainly short helices, in marked contrast to the long duplexes of the body. It has a bi-lobal architecture, with one helix serving as the bridge between the two hemispheres. The head joins the body through a single RNA helix which appears to act as a hinge.

The decoding center, which organizes mRNA and tRNA translocation and controls the fidelity in codon-anticodon interactions, is located at the upper part of the body and the lower part of the head. Its most prominent feature is the portion of the helix that forms most of the intersubunit contacts in the assembled ribosome H44 (also called the "penultimate stem"), which bends towards the neck. The channel through which the

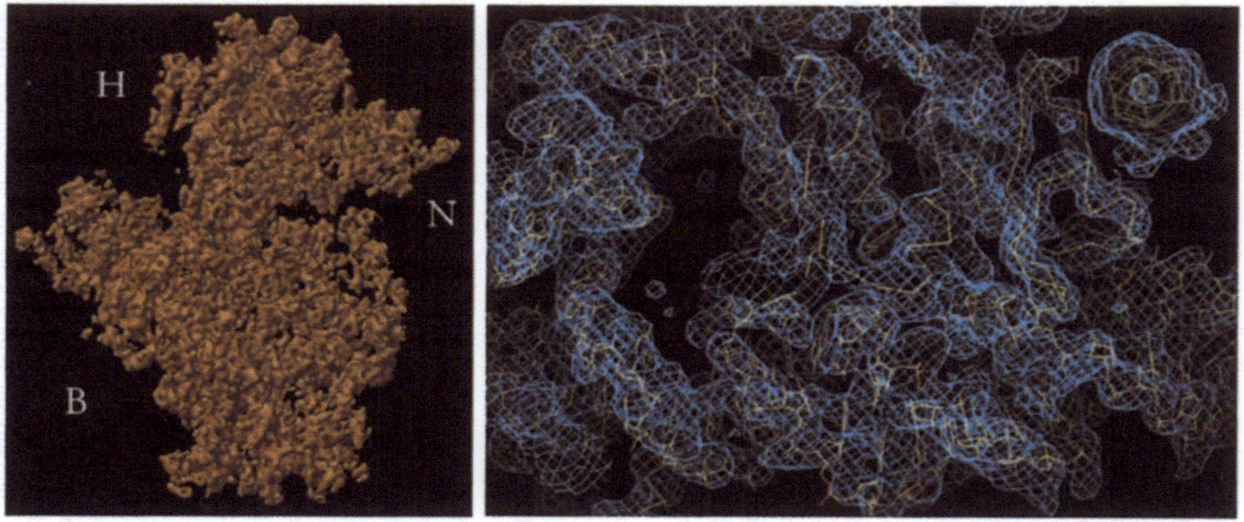

FIGURE 4: LEFT: The overall structure of the small ribosomal subunit, as extracted from the 3.3 Å map, with the traditional division into the three main parts: head (H), neck (N) and body (B). RIGHT: A region of the originally calculated 3.8 Å map showing RNA segments, the ribosomal Protein S8 and a tungsten cluster (green sphere).

messenger RNA (mRNA) progresses is located between the upper part of the body and the lower part of the head. The entrance to it can encircle the message when a latch-like contact, formed by the upper part of the body's shoulder and the lower part of the nose, closes.

All the major functional features of the subunit consist of RNA elements; the proteins appear to serve largely as struts, linkers and supports. Of interest are long extensions of proteins which penetrate into rRNA regions. Some of them reach distal proteins. Only one protein is located at the RNA-rich surface that interacts with the large subunit. Two additional proteins are located at the rims of the subunit interface region, and may be partially involved in tRNA binding or in inter subunit contacts, respectively. A few proteins may contribute to the fidelity and the directionality of the translocation. About half a dozen are peripheral, located on the particle's surface, at its solvent side. These were the hooks for W18 binding.

The smaller heavy atom markers led to independent positioning of two ribosomal proteins that possess exposed sulfhydryls, namely S11 and S13. It also revealed the location of the 3' end of the 16S RNA, highlighting the environment of the gate for mRNA binding, namely the Shine-Dalgarno sequence (Weinstein et al., 1999; Auerbach et al., 2000; Tocilj et al., 1999). The location of one of the two proteins, S13, found this way is in agreement with those suggested by neutron scattering (Moore et al., 1985), immunoelectron microscopy (Stöffler and Stöffler-Meilicke, 1986) and modelling based on crosslinking and enzymatic data (Müller and Brimacombe, 1997). For protein S11 the situation is somewhat different. Its position in the electron density map is in accord with those proposed by electron microscopy and by modeling, but differs from that obtained by neutron scattering, by a distance larger than the expected diameter of this protein.

3.2 DOES W18 STABILIZE FLEXIBLE REGIONS WITHIN THE SMALL RIBOSOMAL SUBUNIT?

The ability of W18 to induce controlled conformational rearrangements, in a fashion that increases dramatically the internal order within crystals of a cellular organelle as large as the small ribosomal subunit, is a remarkable property. Although still far from being fully understood, as shown in this study, the advantage gained by the exploitation of this property for the crystallographic studies of T30S is evident.

The strongest bound W18 molecules are those revealed in the BS crystals. It is conceivable that these are the sites that trigger the conformational changes that lead to the dramatic increase in resolution. All seven W18 sites that were detected, even after intensive back-soak, are located in close proximity to proteins, in positions which may significantly reduce the mobility of the entire T30S particles within the crystal network. Pairing of T30S particles around the crystallographic two fold axis is one of the main features of the map of T30S (Harms et al., 1999; Tocilj et al., 1999). The contacts holding the pairs together are extremely stable, so that they are maintained even after the rest of the crystal network is destroyed. Large proportions of butterfly-shaped pairs have been observed by electron microscopy in samples of thoroughly washed and dissolved T30S crystals (Figure 5). It was found that these pairing contacts are formed by the W18 clusters that are clearly observed at the interface between the particles in the electron density map (Figures 6 and 7).

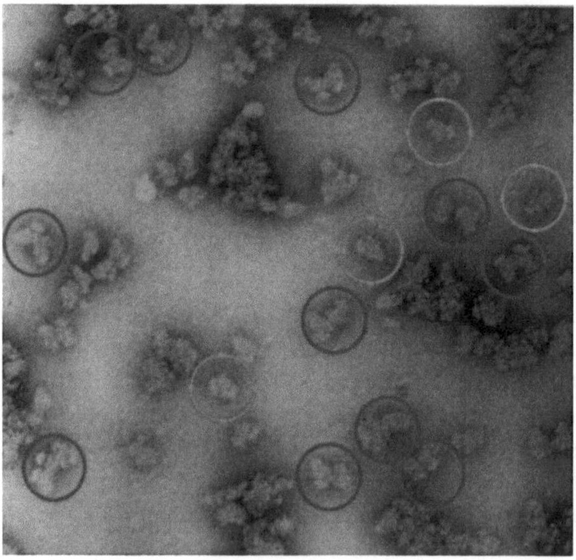

FIGURE 5: A negatively stained preparation of carefully dissolved Wative crystals as observed by transition electron microscopy, showing the butterfly-like T30S pairs (in blue circles) together with isolated particles (in cyan).

It should be noted that in the medium resolution (4.5 Å) and in the higher resolution (3.3 Å) map, all parts of the electron density of T30S seems to be equally resolved (Tocilj et al., 1999; Schlünzen et al., 2000). However, in the independently determined structure of T30S (Clemons et al., 1999) at 5.5 Å, a fair part of the "head" appears to be less well resolved. The higher clarity of our maps can not be simply connected to the higher resolution of our studies, since apart from the head, the overall shapes of our 4.5 Å and their 5.5 Å maps are in fairly good agreement. It is likely, however, that the gain in the internal order in this flexible feature is due to the W18 stabilization of the post-crystallization functionally activated particles.

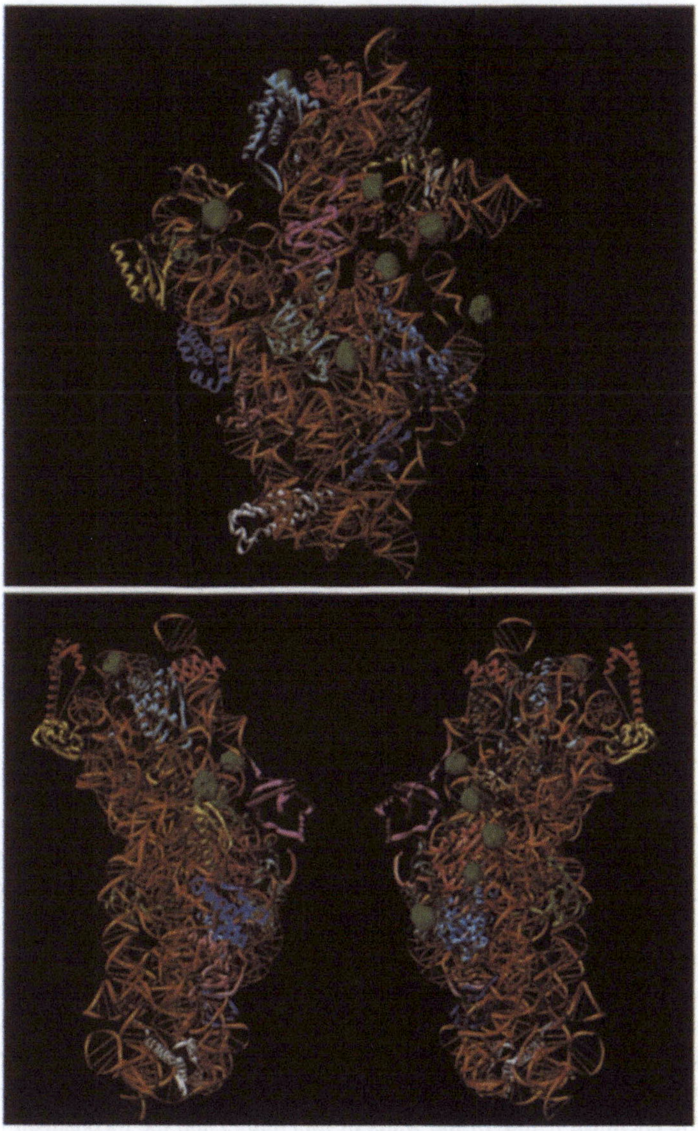

FIGURE 6: TOP: The complete model of T30S with the 16S RNA in orange and all fully or partially determined proteins in different colors. The positions of the W18 clusters are shown as green balls. BOTTOM: Two small ribosomal subunits placed in butterfly-like pairs. It is clearly seen that most of the positions of W18 clusters are on the contact side.

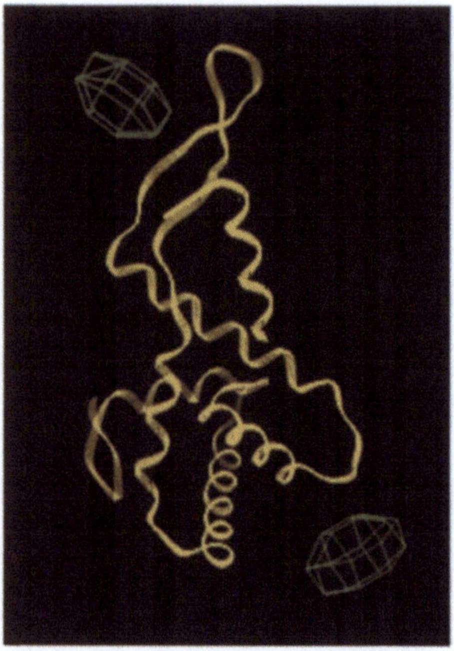

FIGURE 7: The position of two W18 clusters in close proximity to protein S7 (shown without residues 1-13).

The tendency of the heteropolytungstates to bind to relatively narrow solvent paths was detected not only at the tight interface area between the two particles consisting the T30S pairs, but also in internal cavities of other ribosomal crystals. One of these is the main tunnel of the large ribosomal subunit (Yonath et al., 1987). This tunnel is believed to provide the path used by nascent protein chains, once they are synthesized. Its approximate length, about 100 Å, and its diameter, up to 25 Å, were determined by three dimensional image reconstruction using diffraction data obtained from two dimensional sheets (Yonath et al., 1987). It took almost a decade until the existence of the this tunnel was reconfirmed by cryo electron microscopy (Stark et al., 1995; Frank et al., 1995) as well as by X-ray crystallographic studies (Yonath and Franceschi, 1998). However, the chemical nature of the walls of this tunnel and the mechanism that allows the movement of nascent proteins are still not known, although results of preliminary experiments suggested that part of these walls are built of RNA (Gewitz et al., 1988). Indications for the attachment of heteropolytungsten clusters (W30, W18 and W12) to the inner walls of this tunnel in H50S ribosomal subunits have been obtained at 7-8 Å resolution in our laboratory (data not shown), as well as at higher resolution. Thus, at 5 Å, four W11Rh molecules were detected within this tunnel (Ban et al., 1999).

4. Conclusions

We have shown that treatment of ribosomal crystals with a heteropolytungstate introduces significant internal order that lead to a striking increase in their resolution. Although the mechanism of this process has not been revealed yet, there are indications that the W18 cluster interacts with the ribosomal particle in a fashion that may reduces its internal mobility. Most of the interactions of W18 identified so far are made with ribosomal proteins. Despite the experimental complications generated by the W18 treatment, these heteropolytungstates were found to be suitable for crystallographic studies of very complex and sensitive biological macromolecules at high resolution.

Abbreviations: 70S, 50S, 30S: the whole ribosome and its two subunits from prokaryotes. A letter as a prefix to the ribosomal particles or ribosomal proteins represent the bacterial source (T=*Thermus thermophilus*; H=*Haloarcula marismortui*). tRNA and rRNA: transfer and ribosomal RNA. Small subunit proteins are named S and a running number, according to their sequence homology to *E. coli*; BS: back-soaked crystals; Wative: W18 treated crystals; SIR & MIR: single and multiple isomorphous replacement; SR: synchrotron radiation; TAMM: tetrakis(acetoxymercuri)-methane; TIR: a tetrairidium cluster; aquaPt: cis-di-aquacisplatin; cisPt: cis-[$Pt(NH_3)_4$)]; PIP: di-iododiplatinum (II) diethyleneamine; Hg_6: $C_2Hg_6N_2O_8$; Hg_3: $Hg_3C_6O_4H_8$; OliT: the TAMM modified DNA oligomer that compliments the 3'end of the 16S RNA

ACKNOWLEDGEMENTS

We thank M. Pope, W. Preetz and W. Jahn for their generous gifts of heavy atom compounds, M. Wilchek for indispensable advice, M. Safro and I. Levin for active participation, W. Traub for fruitful discussions, and R. Albrecht, T. Auerbach, H. Avila, W.S. Bennett, H. Burmeister, C. Glotz, Y. Halfon, K. Knaack, M. Laschever, S. Meier, J. Müssig, M. Pioletti, M. Peretz, C. Radzwill, M. Simitsopoulou and R. Zarivach for contributing to different stages of these studies, and to the synchrotron radiation facilities staff: EMBL & MPG beam lines at DESY; F1/CHESS; ID2 and ID14 at ESRF; ID19/APS at Argonne Nat. Lab. Support was provided by the Max-Planck Society, the US National Institute of Health (NIH GM 34360), the German Ministry for Science and Technology (BMBF 05-641EA) and the Kimmelman Center for Macromolecular Assembly at the Weizmann Institute. AY holds the Martin S. Kimmel Professorial Chair.

REFERENCES

Alexander, R. W., Muralikrishna, P. and Cooperman, B.S. (1994). Ribosomal components neighboring the conserved 518-533 loop of 16S rRNA in 30S subunits. *Biochemistry* **33**, 12109-12118.

Alizadeh, M.H., Harmalker, S.P., Jeannin, Y., Martin-Frere, J. and Pope, M.T. (1985). A heteropolyanion with fivefold molecular symmetry that contains a nonlabile encapsulated sodium ion. The structure and the chemistry of $(NaP_5W_{30}O_{110})^{-14}$. *J. Am. Chem. Soc.* **107**, 2662-2669.

Auerbach, T., Pioletti, M., Avila, H., Anagnostopoulos, K., Weinstein S., Franceschi, F. and Yonath, A. (2000). Genetic and biochemical manipulation of the small ribosomal subunit from *T. thermophilus* HB8. *Biomolecular Structure and Dynamics* **17**, 617-628.

Ban, N., Nissen, P., Capel, M., Moore P. and Steitz, T. (1999). Placement of protein and RNA structures into a 5 Å resolution map of the 50S ribosomal subunit, *Nature* **400**, 841-847.

Bartels, H., Glühmann, M., Janell, D., Schlünzen, F., Tocilj, A., Bashan, A., Levin, I., Hansen, H.A.S., Harms, J., Kessler, M., Pioletti, M., Auerbach, T., Agmon, I., Avila, H., Simitsopoulou, M, Weinstein, S., Peretz, M., Bennett, W.S., Franceschi, F. and Yonath, A. (2000). Targeting exposed RNA regions in crystals of the small ribosomal subunits at medium resolution. *Cellular and Molecular Biology,* in press.

Bashan, A., Pioletti, M., Bartels, H., Janell, D., Schlünzen, F., Glühmann, M., Levin, I., Harms, J., Hansen, H.A.S., Tocilj, A., Auerbach, T., Avila, H., Simitsopoulou, M., Peretz, M., Bennett, W.S., Agmon, I., Kessler, M., Weinstein, S., Franceschi, F. and Yonath, A. (2000) The identification of selected ribosomal components in crystallographic maps of prokaryotic ribosomal subunits at medium resolution. In The Ribosome: Structure, Function, Antibiotics and Cellular Interactions, R. Garrett, S. Douthwaite, A. Liljas, A. Matheson, P. Moore and H. Noller, eds. (Washington D.C., *ASM Press*), pp. 21-33.

von Böhlen, K., Makowski, I., Hansen, H.A.S., Bartels, H., Berkovitch-Yellin, Z., Zaytzev-Bashan, A., Meyer, S., Paulke, C., Franceschi, F. and Yonath, A. (1991). Characterization and Preliminary Attempts for Derivatization of Crystals of Large Ribosomal Subunits from Haloarcula marismortui, Diffracting to 3 Å Resolution. *J. Mol. Biol.* **222**, 1-15.

Brown, G.M., Noe-Spirlet, M.R., Busing W.R. and Levy, H.A. (1977). Dodecatungstophosphoric acid hexahydrate $(H_5O_2^+)_3(PW_{12}O_{40}^{-3})$. The true structure of Keggin's 'pentahydrate' from single-crystal X-ray and neutron diffraction data. *Acta Cryst.* **B33**, 1038-1046.

Cate, J.H. and Doubna, J.A. (1996). Metal binding sites in the major groove of a large ribozyme domain. *Structure* **4**, 1221-1229.

Cate, J.H., Yusupov, M.M., Yusupova, G.Z., Earnest, T.N. and Noller, H.F. (1999). X-ray crystal structure of 70S ribosome functional complexes, *Science* **285** 2095-2104.

Clemons, W. M., May, J.L.C., Wimberly, B.T., McCutcheon, J.P., Capel, M.S. and Ramakrishnan, V. (1999). Structure of a bacterial 30S ribosomal subunit at 5.5 Å resolution, *Nature* **400**, 833-840.

Contant, R., (1990). Potassium Octadecatungstodiphosphates(V) and Related Lacunary Compounds. *Inorg. Synth.* **27**, 104-111.

D'Amour, H., (1976). Comparison of the Heteropolyanions [PMo9O31(H2O)3]3-, [P2Mo18O62]6- and [P2W18O62]6-, *Acta Cryst.* **B32**, 729-740.

Dawson, B. (1953). The structure of the 9(18)-heteropoly anion in potassium 9(18)-tungstophosphate, $K_6(P_2W_{18})O_{62})14H_2O$. *Acta Cryst.* **6**, 113-126.

Gabashivili, I.S., Agrawal, R.K., Grassucci, R. & Frank, J. (1999). Structure and structural variations of the E. coli 30S ribosomal subunit as revealed by three-dimensional cryo-electron microscopy. *J. Mol. Biol.* **286**,1285-1291.

Gewitz, H.S., Glotz, C., Piefke, J., Yonath, A. and Wittmann, H.G. (1988). Two-dimensional crystalline sheets of large ribosomal subunits containing nascent protein chain, *Biochimie* **70**, 645-651.

Golden, B.L., Gooding, A.R., Podell, E.R. and Cech, T.R. (1998). A preorgenized active site in the crystal structure of the tetrahymena ribozyme, *Science* **282**, 259-264.

Frank, J., Zhu, J., Penczek, P., Li, Y., Srivastava, S., Verschoor, A., Radermacher, M., Grassucci, R., Lata, A.R and Agrawal, R.K. (1995). A model of protein synthesis based on cryo electron microscopy of the *E. coli* ribosome. *Nature* **376**, 441-444.

Fu, J., Gnatt, A.L., Bushnell, D.A., Jansen, G.J., Thompson, N.E., Burges, R.R., David, P.R. and Korenberg, R.D. (1999). Yeast RNA Polymerase II at 5 Å resolution. *Cell* **98**, 799-810.

Harms, J., Tocilj, A., Levin, I., Agmon, I., Kölln, I., Stark, H., van Heel, M., Cuff, M., Schlünzen, F., Bashan, A., Franceschi, F. and Yonath, A. (1999). Elucidating the medium resolution structure of ribosomal particles: an interplay between electron-cryo-microscopy and X-ray crystallography. *Structure* 7, 931-941.

Jahn, W. (1989). Synthesis of water soluble tetrairidium cluster for specific labelling of proteins. *Z. Naturforsch.* **44b**, 79-82.

Jones, T.A., Zou, J.-Y., Cowan, S.W. & Kjeldgaard, M. (1991). Improved methods for building protein models in electron density maps and the location of errors in these models. *Acta Cryst.* **A47**, 110-119.

Ladenstein, R., Bacher, A. and Huber, R. (1987). Some observations of a correlation between the symmetry of large heavy atom complexes and their binding sites on proteins. *J. Mol. Biol.* **195**, 751-3.

Liljas, A. and Al-Karadaghi, S. (1997). Structural aspects of protein synthesis. *Nature Structural Biology* **4**, 767-771.

Makowski, I., Frolow, F., Saper, M.A., Shoham, M., Wittmann, H.G. and Yonath, A. (1987). Single crystals of large ribosomal particles from Halobacterium marismortui diffract to 6 Å. *J. Mol. Biol.* **193**, 819-821.

Moore, P.B., Capel, M.S., Kjeldgaard, M. and Engelman, D.M. (1985). A 19 protein map of the 30S ribosomal subunit of E. coli. In *"Structure, Function & Genetics of Ribosomes"* (B. Hardesty & G. Kramer Eds.) Springer Verlag, Heidelberg & NY pp. 87-100.

Müller, F. and Brimacombe, R. (1997). A new model of the three-dimensional folding of *E. coli* 16S ribosomal RNA. II The RNA-protein interaction data. *J. Mol. Biol.* **271**, 524-544.

Nyborg, J., Nissen, P., Kjeldgaard. M., Thirup. S., Polekhina. G., Clark B.F.C. and Reshetnikova L. (1996). Structure of the ternary complex of EF-TU - macromolecular mimicry in translation. *Trends in Biochemical Sciences* **21**, 81-82.

Pope, M.T. and Papaconstantinou, E., (1967). Heteropoly Blues. II. Reduction of 2:18-Tungstates, *Inorg. Chem.* **6**, 1147-1152.

Ramakrishnan, V. and White, S.W (1992). The structure of ribosomal protein S5 reveals sites of interaction with 16S RNA. *Nature* **358**, 768-771.

Ramakrishnan, V. & White, S.W (1998). Ribosomal protein structures: insights into the architecture, machinery and evolution of the ribosome, *Trends in Biochemical Sciences* **3**, 208-212.

Schlünzen, F., Kölln, I., Janell, D., Glühmann, M., Levin, I., Bashan, A., Harms, J., Bartels, H., Auerbach, T., Pioletti, T., Avila, H., Anagnostopoulos, K., Hansen, H.A.S., Bennett, W.S., Agmon, I., Kessler, M., Tocilj, A., Peretz, M., Weinstein, S., Franceschi, F. and Yonath, A. (1999). The identification of selected components in electron density maps of prokaryotic ribosomes at 7 Å resolution. *J. Syn. Radiation* **6**, 928-941.

Schlünzen, F., Tocilj, A., Zarivach, R., Harms, J., Glühmann, M., Janell, D., Bashan, A., Bartels, H., Agmon, I., Franceschi, F. and Yonath, A. (2000). Structure of a functionally activated small ribosomal subunit at 3.3 Å resolution. *Cell*, in press.

Stark, H., Müller, F., Orlova, E.V., Schatz, M., Dube, P., Erdemir, T., Zemlin, F., Brimacombe, R. and van Heel, M. (1995). The 70S *E. coli* ribosome at 23 Å resolution: fitting the ribosomal RNA. *Structure* **3**, 815-821.

Stöffler, G. and Stöffler-Meilicke, M. (1986). Immuno electron microscopy on E. coli ribosomes. In "*Structure, Function and Genetics of Ribosomes*" (B.Hardesty and G.Kramer Eds.) Springer Verlag, Heidelberg and NY. pp.28-46.

Tocilj, A., Schlünzen, F., Janell, D., Glühmann, M., Hansen, H.A.S., Harms, J., Bashan, A., Bartels, H., Agmon, I., Franceschi F. and Yonath, A. (1999). The small ribosomal subunit from *Thermus thermophilus* at 4.5 Å resolution: pattern fittings and the identification of a functional site. *PNAS* **96**, 14252-14257.

Trakhanov, S.D., Yusupove, M.M., Shirokov, V.A., Garber, M.B., Mitscher, A. Ruff, M., Tierry, J.-C. and Moras, D.(1989). Preliminary X-ray investigation on 70S ribosome crystals. *J. Mol. Biol.* **209**, 327-334.

Thygesen, J., Weinstein, S., Franceschi, F. and Yonath, A. (1996). On the suitability of multi metal clusters for phasing in crystallography of large macromolecular assemblies. *Structure* **4**, 513-518.

Wei, X., Dickman, M.H. and Pope, M.T. (1997). New Routes for Multiple Derivatization of Polyoxometalates: Diacetato-dirhodium-11-tungstophosphate *Inorg. Chem.* **36,** 130-131.

Weinstein, S., Jahn, W., Glotz, C., Schlünzen, F., Levin, I., Janell, D., Harms, J., Kölln, I., Hansen, H.A.S., Glühmann, M., Bennett, W.S., Bartels, H., Bashan, A., Agmon, I., Kessler, M., Pioletti, M., Avila, H., Anagnostopoulos, K., Peretz, M., Auerbach, T., Franceschi, F. and Yonath, A. (1999). Metal compounds as tools for the construction and the interpretation of medium resolution maps of ribosomal particles. *J. Struct. Biol.* **127,** 141-151.

Weller, J. W., and Hill, W.E. (1992). Probing dynamic changes in rRNA conformation in the 30S subunit of the E. coli Ribosome. *Biochemistry* **31**, 2748-2757.

Xin, F. and Pope, M.T. (1994). Polyoxometalate derivatives with multiple organic groups. 1. Synthesis and structure of tris(organotin) beta-keggin and alpha-Dawson tungstophosphates. *Organometallic.* **13**, 4881-4886.

Yonath, A. Glotz, C. Gewitz, H.S. Bartels, K. von Boehlen, K. Makowski, I. & Wittmann, H.G. (1988). Characterization of Crystals of Small Ribosomal Subunits, *J. Mol. Biol.* **203**, 831-833.

Yonath, A. Leonard K.R. and Wittmann, H.G. (1987). A tunnel in the large ribosomal subunit revealed by three dimensional image reconstruction *Science* **236,** 813-817.

Yonath, A. and Franceschi, F., (1998). Functional Universality and Evolutionary Diversity: Insights from the Structure of the Ribosome, *Structure* **6,** 678-684.

Yonath, A., Harms, J., Hansen, H.A.S., Bashan, A., Peretz, M., Bartels, H., Schlünzen, F., Kölln, I., Bennett, W.S., Levin, I., Krumbholz, S., Tocilj, A., Weinstein, S., Agmon, I., Piolleti, M., Janell, D., Auerbach, T. and Franceschi, F. (1998). The quest for the molecular structure of a large macromolecular assembly exhibiting severe non-isomorphism, extreme beam sensitivity and no internal symmetry. *Acta Cryst.* **54A,** 945-955.

Zamir, A., Miskin, R. and Elson, D. (1971). Inactivation and reactivation of ribosomal subunits: amino acyl-transfer RNA binding activity of the 30S subunit of *E. coli. J. Mol. Biol.* **60(2),** 347-64.

Photocatalytic Decontamination by Polyoxometalates

A. HISKIA, E. ANDROULAKI, A. MYLONAS, A. TROUPIS AND E. PAPACONSTANTINOU
Institute of Physical Chemistry, NCSR Demokritos, 15310, Greece

Abstract. Polyoxometalates (POM), at least $PW_{12}O_{40}^{3-}$, $SiW_{12}O_{40}^{4-}$ and $W_{10}O_{32}^{4-}$, are effective photocatalysts for the mineralization of diversified organic pollutants, such as, lindane, cresol, phenol, chlorophenols and polychlorinated phenols, and chloroacetic acid. Key reactions are the formation of OH radicals, the high affinity of organic pollutants for POM, and the regeneration of catalyst by dioxygen. The mineralization (i.e. formation of CO_2, H_2O and inorganic anions) proceeds via several intermediates resulting from H-atom abstraction, hydroxylation and to a lesser extent dehalogenation. The breaking of the aromatic ring is followed by the formation of several saturated and unsaturated organic acids. In all cases so far, ethanoic acid has been detected. The formation of $-CH_2$ and $-CH_3$ groups from aromatic carbons (i.e. -CH groups) suggests that a reductive pathway accompanies the oxidation process. The overall photobehavior of POM resembles the highly publicized photodecomposition of organic pollutants by TiO_2.

Key words. Polyoxometalates, Photocatalysis, Decontamination, Organic pollutants.

Introduction

Contamination of water resources by a variety of organic compounds has become a serious problem in industrialized areas, as well as in areas with intense agricultural activities. In order to address this significant problem, various methods have been applied. The most common of them involve treatment with oxidizing reagents, mainly chlorine or ozone, to degrade the organic contaminant. The disadvantages that appear in this case, are summarized in attaining only incomplete purification, giving rise to the production of toxic metabolites and by the need of large quantities of the oxidizing reagent consumed in the process.

Alternatively, other methods more effective and at the same time friendly to the environment are under development. They are generally referred to as Advanced Oxidation Processes (AOP) and involve mainly UV light, UV light in the presence of hydrogen peroxide or ozone and UV and near-visible light in the presence of titanium dioxide (TiO_2) [1-3]. These methods lead more or less to mineralization of the pollutants and are based mainly on the formation of the highly oxidizing OH radicals.

Recently, a new Advanced Oxidation Process has been demonstrated using UV and near-visible light in the presence of polyoxometallates (POM) [4-6]. POM, mainly of molybdenum and tungsten, become powerful oxidizing reagents [7, 8] upon excitation with near visible and UV light, capable of destroying a great variety of toxic compounds in the aquatic environment [4-6, 9]. The main oxidant seems to be OH radicals generated by reaction of POM with H_2O [4-6, 10]. POMs are at least as effective as the widely publicized TiO_2. Dioxygen is important in the reoxidation (regeneration) of the catalyst

M.T. Pope and A. Müller (eds.), Polyoxometalate Chemistry, 417–424.

and through reductive activation may or may not participate further in the process, depending on the substrate.

This paper gives an overall account on the mineralization of a variety of diversified organic pollutants with UV and near visible light, in presence of representative POM, $PW_{12}O_{40}^{3-}$, $SiW_{12}O_{40}^{4-}$, $W_{10}O_{32}^{4-}$. It also reports on the intermediates involved, as identified by HPLC-DAD, GC-MS and suggests the possible mechanistic route of the photocatalytic degradation process.

Experimental

Aqueous solutions of various organic substrates (chloroacetic acid, lindane, cresol, phenol, chlorophenols and polychlorinated phenols), in presence of the catalyst ($PW_{12}O_{40}^{3-}$, 7 x 10^{-4} M), were made by dissolving certain quantities of substrate in $HClO_4$, 0.1M. Samples of 4.0 ml of the above solutions were added to a spectrophotometer cell (1-cm path), with total volume of about 8 ml. After 20 minutes of deaeration or oxygenation the cell was covered with an airtight serum cap. Photolysis experiments were performed with an Oriel 1000 W Xe arc lamp, at 20° C, equipped with a cool water circulating filter to absorb the near IR radiation and 320, 345 nm filters to avoid possible direct photolysis of the organic substrates. The incident radiation was reduced to about 40% with a slit diaphragm, in order to obtain reasonable photolysis times. The solution was magnetically stirred throughout the experiment. For experiments at higher pH, $SiW_{12}O_{40}^{4-}$ was used. Limited work was done with $W_{10}O_{32}^{4-}$ which is unstable in aqueous solutions.

A Waters apparatus equipped with a PDA detector carried out HPLC analysis, for the determination of intermediates produced during photolysis. GC analysis, for the determination of lindane, was carried out by a Varian Model 3400 gas chromatograph equipped with ECD, split/splitless injection port and a DB-1 fused silica capillary column. GC analysis, for the determination of CO_2, was carried using TCD and a 2 m Porapack Q column. Carbon dioxide content was calculated using a calibration curve, made of known quantities of CO_2 and processed under the same experimental conditions.

Identification of intermediates was performed using a Micromass Platform II quadrupole mass spectrometer, equipped with a DB-5 fused silica capillary column.
The formation of maleic, oxalic, ethanoic and formic acids were analysed by suppressed ion chromatography, performed with a Dionex apparatus. In order to have adequate quantities for the identification of intermediates by GC-MS, samples of 30 ml of the 2,4,6-TCP, 10^{-3} M, were photolysed under the same conditions. The photolysed solution was extracted with dichloromethane (3 x 20 ml). The organic layers were combined, dried through sodium sulfate and evaporated to 100 μl.

Chloride ions were analysed spectrophotometrically [11].
The degree of reduction of POMs in photolyzed deaerated solutions was calculated from the known extinction coefficients of the blue products ($PW_{12}O_{40}^{3-}$: ϵ_{751nm} = 2 x 10^3 $M^{-1}cm^{-1}$, ϵ_{653nm} = 4.4 x 10^3 $M^{-1}cm^{-1}$ and $SiW_{12}O_{40}^{4-}$: ϵ_{730nm} = 2.1 x 10^3 $M^{-1}cm^{-1}$, ϵ_{625nm} = 3.8 x 10^3 $M^{-1}cm^{-1}$ for the one- and two-electron reduction products, respectively.)[12]

Results and Discussion

The excited state of POM arising from absorption of light at the O → M CT band (near visible and UV light) is a powerful oxidizing reagent. The oxidizing ability is manifested, mainly, through formation of OH radicals arising from the reaction of the excited POM with adsorbed water. The existing prevailing mechanism of H-abstraction as the initial reaction of excited POM with organic substrates (mainly alcohols) has been modified by addition of one more step that involves the formation of OH radicals [10]. This radical, as is well known, reacts with organic substrates, mainly alcohols, by H-abstraction. The formation of OH radicals, and the high oxidizing ability of the excited POM is responsible for the mineralization, i.e. formation of CO_2, H_2O and inorganic anions of a great variety of organic compounds and for that matter organic pollutants.

Several experimental approaches have suggested the formation of OH radicals: a) The detection of hydroxylation intermediates [5,6,16], b) ESR data [13], c) The fact that the excited state potentials of, practically, all POM are more positive than the reaction

$$OH + H^+ + e^- \rightarrow H_2O \quad E = 2.8 \text{ V vs NHE} \qquad [10]$$

and d) results concerning photodegradation of *p*-nitrosodimethylaniline, a known trapping reagent for OH radicals [14,15].

Following the first results on the subject [4], several other publications provided data on the mineralization of a great variety of organic pollutants with POM, such as phenols [6], chlorinated [5] and polychlorinated phenols [16], cresols [6], chloracetic acids [17], organochlorine incecticides (lindane, aldrin, endosulfan, DDT) [18,19], heavily chlorinated compounds (HCB) [16] and herbicides(atrazine) [20].

A typical experiment showing the effect of POM in the photodegradation of organic pollutants is presented in Fig. 1.

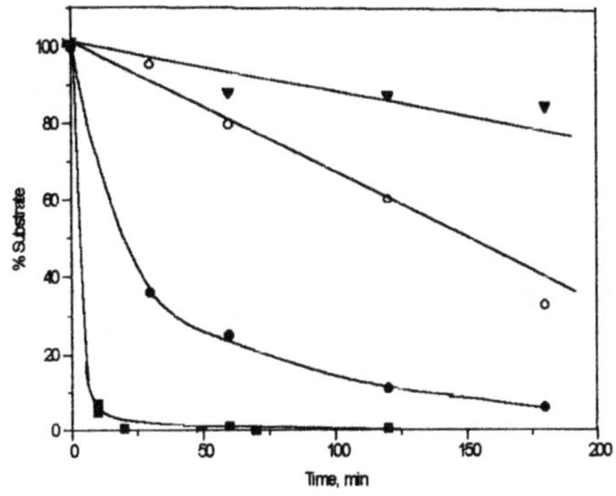

Figure 1. Photodecomposition of aqueous solutions of 2,4DCP 4 x 10^{-4} M: (▼) oxygenated solution 345 nm cut off filter, no catalyst; (○) oxygenated solution, 320 nm cut off filter, no catalyst. In presence of catalyst, 7 x 10^{-4} M, with 320 nm cut off filter: (●) Deoxygenated solution, (■) oxygenated solution. T ≅ 20 ^{0}C.

As model compound 2,4 Dichlorophenol (2,4DCP) has been chosen. It can be seen that the presence of catalyst can seriously accelerate the photodegradation process. In the presence of dioxygen, more effective photocatalytic degradation of substrate takes place.

Dioxygen's main function is the regeneration of the catalyst [21], participating in some cases in the initial decomposition of the substrates [6]. No reaction takes place between POM and 2,4 DCP in the dark.

Fig. 2 indicates the decay of 2,4DCP and the gradual formation of CO_2 and Cl^-. The mineralization proceeds via formation and decay of intermediates resulting from H-abstraction, hydroxylation and dehalogenation. The formation and decay several of them are presented in Fig. 3.

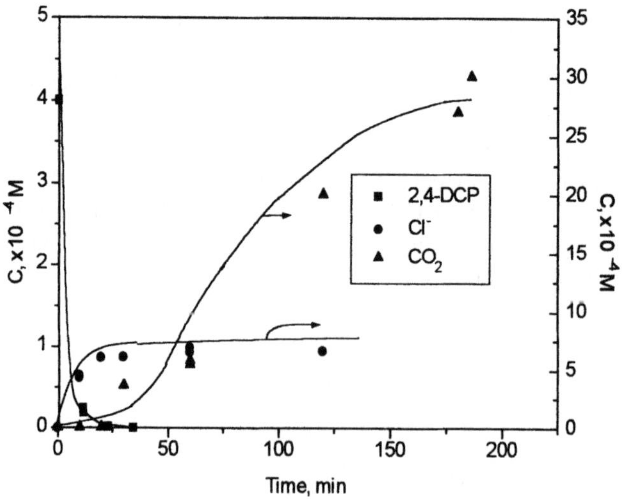

Figure 2. Formation of CO_2, Cl^- and decay of substrate upon photolysis of aqueous oxygenated solution in presence of $PW_{12}O_{40}^{3-}$ 7 x 10^{-4} M, 2,4 DCP 4 x 10^{-4} M, pH 1 ($HClO_4$); T 20 ^{0}C; $\lambda > 320$ nm [16].

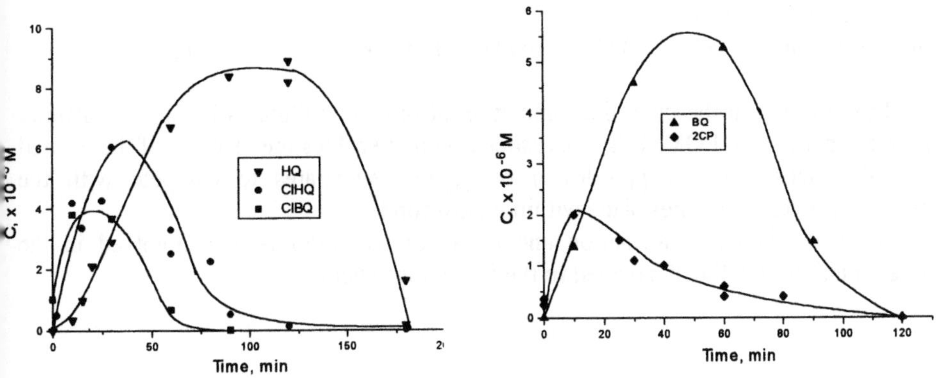

Figure 3. Formation and decay of several intermediates upon photolysis of aqueous oxygenated solution of 2,4 DCP in presence of $PW_{12}O_{40}^{3-}$. 2,4DCP 2 x 10^{-3} M, catalyst 7 x 10^{-4} M, pH 1 ($HClO_4$), T 20 ^{0}C, λ > 320 nm [16].

The breaking of the aromatic ring is followed by the formation of various saturated and unsaturated aliphatic acids. In all studies, so far, ethanoic acid has been detected.

The overall reactions involved are:

$$POM + h\nu \rightarrow POM^* \quad (1)$$

$$POM^* \rightarrow POM \quad (2)$$

Direct reaction of POM* with substrate

$$POM^* + S \rightarrow POM(e^-) + S(+) \text{ (i.e. various oxidation products)} \quad (3)$$

Indirect reaction of POM with substrate, through OH radicals

$$POM^* + H_2O \rightarrow POM(e-) + OH + H+ \quad (4)$$

OH + S → hydrogen abstraction, hydroxylation, dehalogenation, breaking of the aromatic ring, aliphatic acids; (final products: CO_2, H_2O and inorganic anions). (5)

Regeneration of catalyst

POM(e-) + oxidant → POM + oxidant products (6)

The key step in the photodecomposition of organic pollutants is the high affinity of organic substrates to POM. It is this association that enables the reaction of the excited POM with traces (ppm or ppb level) of organic substrates to compete with the deactivation process that takes place within nanoseconds.

Detailed studies of formation and decay of the intermediates involved in the photodegradation of 2,4DCP have led to the following scheme:

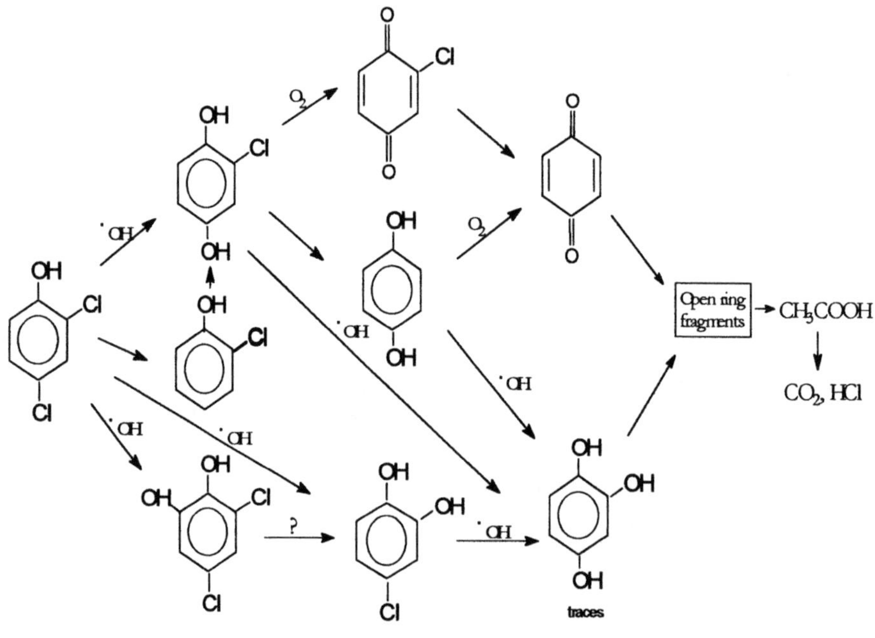

Scheme 1: Detected intermediates during photocatalytic degradation of 2,4DCP with POM [16].

Other pollutants, whose solubility is limited to ppb values, undergo effective mineralization in presence of POM, as has been stated earlier. Two typical cases are lindane and HCB whose photodecay is depicted in Figures 4 and 5 together with the formation of CO_2 and Cl^-.

The overall behavior of POM, as far as photodecomposition of organic pollutants is concerned, is reminiscent of the widely publicized TiO_2. For instance:

(a) The final products in both cases are CO_2, H_2O and inorganic anions.

(b)Both methods proceed to mineralization via similar intermediates arising from H-abstraction, hydroxylation and to a lesser extent dehalogenation. However, the detail pathways to mineralization may present differences.
(c) The initial photodecomposition of organic pollutants by POM and TiO_2, follows first order kinetics.

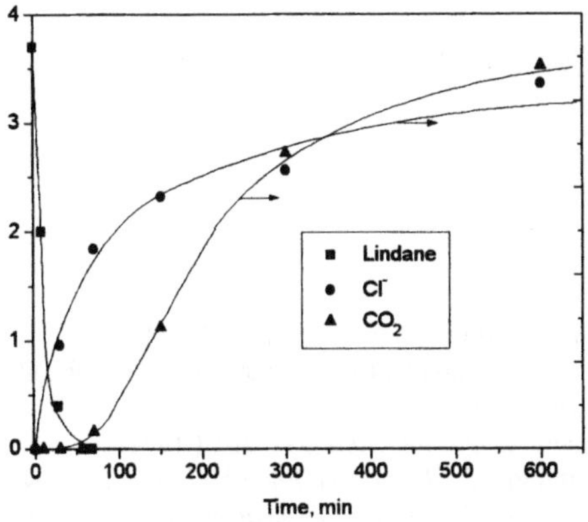

Figure 4. Formation of CO_2, Cl^- and decay of substrate upon photolysis of aqueous oxygenated solution in presence of $PW_{12}O_{40}^{3}$ 7 x 10^{-4} M, Lindane, 3.7x10^{-5} M, pH 1 ($HClO_4$); T 20 ^{0}C; $\lambda > 320$ nm [16].

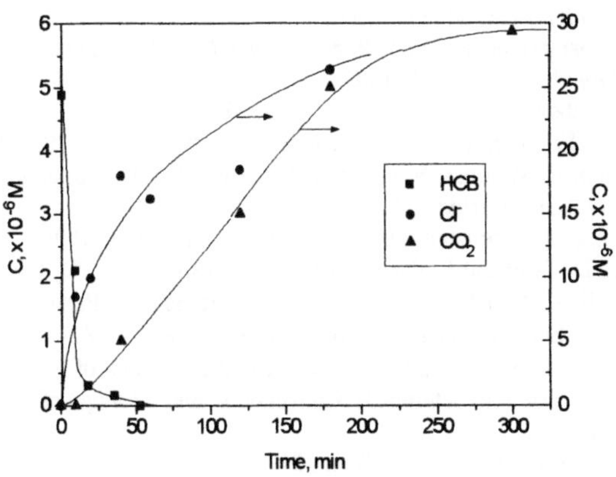

Figure 5. Formation of CO_2, Cl^- and decay of substrate upon photolysis of aqueous oxygenated solution in presence of $PW_{12}O_{40}^{3}$ 7 x 10^{-4} M, HCB, partially dissolved, 7 x 10^{-6} M, pH 1 ($HClO_4$); T 20 ^{0}C; $\lambda > 320$ nm [16].

(d) The formation of $-CH_2$ and $-CH_3$ groups from aromatic carbons (i.e. -CH groups) implies that in photocatalytic processes by POM, like in the case of TiO_2, there is a reductive (hydrogenation) pathway, going along the oxidation processes.
(e) Another important aspect as far as similarity of these two processes is concerned, is the decomposition of nitrogen containing aromatic rings. In particular, atrazine photodecomposes within a few minutes by both methods, to cyanuric acid. Cyanuric acid, on the other hand, resists photodegradation for hours either in presence of TiO_2 or POM.

Acknowledgements
We thank the Ministry for Development, General Secretariat of Research and Technology for financing part of this work.

References

[1] O. Legrini, E. Oliveros, and A.M. Brown, *Chem. Rev.* **93,** 671 (1993),.
[2] M.R. Hoffmann, S.T. Martin, W. Choi and D. Bahnemann, *Chem. Rev.*,**95**, 69 (1995).
[3] P.V. Kamat, *Chem. Rev.*, **93**, 267 (1993).
[4] A. Mylonas and E. Papaconstantinou, *J. Mol. Catal.*, **92,** 261.(1994)
[5] A. Mylonas and E. Papaconstantinou, J. *Photochem. Photobiol. A: Chem.*, **94**, 77 (1996).
[6] A. Mylonas E. Papaconstantinou and V. Roussis, *Polyhedron*, **15**, 3211 (1996).
[7] E. Papaconstantinou, A. Argitis, D. Dimotikali, A. Hiskia and A. Ioannidis, in *Homogeneous and Heterogeneous Photocatalysis,* (Eds. E. Pelizzetti, N. Serpone), D. Reidel Publ. Co., Dordrecht 1986, p. 415.
[8] E. Papaconstantinou, *Chem. Soc. Rev.*, **18,** 1 (1989).
[9] H. Einaga and M. Misono, *Bull. Chem. Soc. Jpn.*, **69**, 3435 (1996).
[10] A. Mylonas, A. Hiskia, E. Androulaki, D. Dimotikali, E. Papaconstantinou, *Phys. Chem. Chem. Phys.*, **1**, 437 (1999).
[11] T. M. Florence, *Anal. Chim. Acta*, **54**, 373 (1971)
[12] G.M. Varga, E. Papaconstantinou and M.T. Pope, *Inorg. Chem.*, **9**, 662 (1970).
[13] T. Yamase, *Inorg. Chim. Acta*, , **76**, L25 (1983)
[14] I. Kraljik and C.N. Trumbore, *J.Am.Chem.Soc.*, , **87**, 2547.(1965)
[15] M. Hatada, I. Kraljik, A. El Samahy and C.N. Trumbore, *J. Phys. Chem.*, **78**, 888 (1974).
[16] A. Hiskia, E. Androulaki, A. Mylonas, S. Boyatzis, D. Dimoticali, C. Minero, E. Pelizzetti and E. Papaconstantinou, *Res Chem. Intermed.*, 1999, in press.
[17] A. Mylonas, A. Hiskia, E. Papaconstantinou, *J. Mol. Catal.*, **114,** 191 (1996).
[18] A. Hiskia, A. Mylonas D. Tsipi and E. Papaconstantinou, *Pestic. Sci.*, , **50**, 171 (1997).
[19] A. Hiskia, A. Mylonas E. Androulaki, D. Dimotikali, D. Tsipi and E. Papaconstantinou, 5th Symp. On Environmental Science and Technology, Lesbos, 1997, Vol.A, 309.
[20] A. Hiskia, A. Kokorakis, H. Hennig and E. Papaconstantinou, Book of Abstracts, 12th International Conference on Photochemical Conversion and Storage of Solar Energy, Berlin, Germany **3,** W77.(1998)
[21] A. Hiskia and E. Papaconstantinou, *Inorg. Chem.*, **31,** 163. (1992)

Index

MIX
Papier aus verantwortungsvollen Quellen
Paper from responsible sources
FSC® C105338

If you have any concerns about our products,
you can contact us on
ProductSafety@springernature.com

In case Publisher is established outside the EU,
the EU authorized representative is:
Springer Nature Customer Service Center GmbH
Europaplatz 3, 69115 Heidelberg, Germany

Printed by Libri Plureos GmbH
in Hamburg, Germany